Lecture Notes in Engineering

Edited by C. A. Brebbia and S. A. Orszag

46

M. Gad-el-Hak (Editor)

Frontiers in Experimental Fluid Mechanics

Springer-Verlag
Berlin Heidelberg New York
London Paris Tokyo Hong Kong

Editor
Prof. Mohamed Gad-el-Hak
Dept. of Aerospace & Mechanical Engineering
University of Notre Dame
Notre Dame, IN 46556
USA

ISBN-13:978-3-540-51296-7   e-ISBN-13:978-3-642-83831-6
DOI: 10.1007/978-3-642-83831-6

Lecture Notes in Engineering

The Springer-Verlag Lecture Notes provide rapid (approximately six months), refereed publication of topical items, longer than ordinary journal articles but shorter and less formal than most monographs and textbooks. They are published in an attractive yet economical format; authors or editors provide manuscripts typed to specifications, ready for photo-reproduction.

PREFACE

One cannot overemphasize the importance of studying fluids in motion or at rest for a variety of scientific and engineering endeavors. Fluid mechanics as an art reaches back into antiquity, but its rational formulation is a relatively recent undertaking. Newton's (1687) *Principia* provides the cornerstone of modern science, and about 150 years after its publication the complete equations describing fluid motion were derived from first principles. Another 150 years have elapsed with only about 70 particular solutions of these nonlinear, partial differential equations found. Increasingly, therefore, we must rely on experimental and numerical simulations for answers to our most pressing problems.

The speed and capacity of today's supercomputers have improved by roughly a factor of 10^3 in the last 20 years. Nevertheless, direct integration of the equations of fluid motion at realistic Reynolds numbers is still a remote prospect, let alone prohibitively expensive. For high Re turbulent flows, the equations are averaged and the additional unknowns resulting from the nonlinearity of the governing equations are heuristically modeled in terms of the original dependent variables. Experimental input is required for both developing and validating the different turbulence models used to close the problem.

Much of the physics of a particular flow situation can be understood by conducting appropriate experiments. Of course measuring a flow quantity can itself be an end product, providing useful predictive or postdictive information such as flow rate, aerodynamic forces, etc. In an article appearing in <u>Fluid Mechanics Measurements</u>*, Roger Arndt provides a useful summary of what we measure and why. Wind and water tunnels are among the most useful tools available to the experimentalist. In a turbulent flow, where flow quantities are in general random in space and time, one must consider the spatial and temporal resolutions of the probes used as well as the acquisition and analysis of required colossal amounts of data. The problem is further complicated if the measurements are to be carried out in a separated flow, a two-phase fluid, or in flow around a minute, and live, body. Dynamical systems theory and flow control are two research areas of great current interest. These and other special situations are among the topics covered in this volume of Springer's <u>Lecture Notes in Engineering</u>. The book is, in the words of Peter Bradshaw, a collection of ten noninteracting review articles. Each emphasizes the use of experiments to achieve better physical understanding of a particular class of flow problems. A companion volume entitled "<u>Advances in Fluid Mechanics Measurements</u>" contains twelve articles with more emphasis on the development of particular measuring techniques.

Chapter 1 outlines some developments in nonlinear dynamical systems theory which have recently had a strong influence on the design of experiments and the processing of data. This is followed in Chapters 2 and 3 by descriptions of two of the most important experimental tools available to fluid dynamicists, the wind tunnel and the water tunnel. The physics of turbulent

---

* R.J. Goldstein, Editor, pp. 1-42, Hemisphere, Washington, 1983.

boundary layers is discussed in Chapter 4. Chapter 5 follows with a broad overview of flow control techniques available for bounded shear flows to achieve a variety of desired goals. Particular flow control methods are reviewed in the next two chapters: microbubbles drag reduction (Chapter 6); and control of wakes (Chapter 7). Vortex dynamics of delta wings are discussed in Chapter 8, followed by the marvels of insect flight in Chapter 9. The subject of trailing edge flow and their associated aeroacoustics is reviewed in the final chapter.

The topics covered were chosen because of their importance to the field, recent appeal, and potential for future development. The articles are comprehensive and coverage is pedagogical with a bias towards recent developments. Whenever possible, the manuscripts were each reviewed by two independent referees. The reviewers' constructive comments were incorporated in the final version.

The editor wishes to express his deep appreciation to all the authors and the reviewers who unselfishly gave much of their time and effort. Their reward is in helping the reader interested in a particular topic through the ever expanding maze of available literature.

Mohamed Gad-el-Hak
Notre Dame, Indiana
October 1988

LIST OF CONTRIBUTORS

Mr. C.W. Amato
Fluid Dynamics Research Center
IIT
Chicago, IL 60616

Dr. S.M. Batill
Department of Aerospace
  and Mechanical Engineering
University of Notre Dame
Notre Dame, IN 46556

Dr. M.L. Billet
Applied Research Laboratory
Pennsylvania State University
State College, PA 16804

Dr. W.K. Blake
Ship Acoustics Department
Naval Ship Research and
  Development Center
Bethesda, MD 20084

Dr. S. Deutsch
Applied Research Laboratory
Pennsylvania State University
State College, PA 16804

Professor M. Gad-el-Hak
Department of Aerospace
  and Mechanical Engineering
University of Notre Dame
Notre Dame, IN 46556

Mr. J.L. Gershfeld
Ship Acoustics Department
Naval Ship Research and
  Development Center
Bethesda, MD 20084

Professor C.-M. Ho
Department of Aerospace Engineering
University of Southern California
Los Angeles, CA 90089

Dr. G.C. Lauchle
Applied Research Laboratory
Pennsylvania State University
State College, PA 16804

Dr. M. Lee
Advanced Technology Division
Digital Equipment Corporation
Cupertino, CA 95014

Professor M.W. Luttges
Department of Aerospace
  Engineering Sciences
University of Colorado
Boulder, CO 80309

Professor C.L. Merkle
Department of Mechanical Engineering
Pennsylvania State University
University Park, PA 16802

Professor R.C. Nelson
Department of Aerospace
  and Mechanical Engineering
University of Notre Dame
Notre Dame, IN 46556

Professor M. Sen
Department of Aerospace and
  Mechanical Engineering
University of Notre Dame
Notre Dame, IN 46556

Professor K.R. Sreenivasan
Mason Laboratory
Yale University
New Haven, CT 06520

Professor D.R. Williams
Fluid Dynamics Research Center
IIT
Chicago, IL 60616

TABLE OF CONTENTS

The Influence of Developments in Dynamical Systems Theory on Experimental Fluid Mechanics

Mihir Sen

Department of Aerospace and Mechanical Engineering
University of Notre Dame
Notre Dame, IN 46556

Abstract

An experimenter can use the tools of his trade in much the same way a theoretician can use analysis. An enriched background permitting new directions in experimental fluid mechanics has been provided by significant advances in the theory of differential equations in the last few decades. In this review we outline some of the concepts which are of special importance, and include a discussion of the possible kinds of behavior to be expected in actual measurements. Examples in which the support of theoretical developments has contributed to the experimental discovery of new results are used as illustrations.

1. Introduction

The experimental and theoretical parts of any mature science are strongly related in that they are but different paths to the same truth. Thus, travelers on one road can learn much from those on the other about their glimpses of the common goal. This is the message of this work, in particular about how experimentalists working in the area of fluid mechanics can benefit from some recent theoretical progress in the area of dynamical systems.

The historical evolution of fluid mechanics has taken its unique trajectory within the history of science. It has long served as a focal point for the development of many widely used mathematical methods and techniques. One may refer to the relation between singular perturbation analysis and Prandtl's boundary layer theory as a fairly recent example. Much of the basic mathematical development went hand in hand with fluid mechanics, though there are, of course, many other applications of the theory. It would not be far from the truth to say that some of the theory was developed to fill a specific need in fluid mechanics. The opposite is also often true; fluid mechanics, like any other science, benefits from progress in mathematics which is driven either by other applications or for its own sake. Here we will try to show that these benefits influence not only the theoretical part of the science but also its experimental aspects. Thus, the hypothesis here is that theoretical developments in mathematics, or in related fields, can be used to enrich the course of experimental investigations in fluid mechanics, not simply to predict results to be checked by experiment, but to help find new directions and phenomena. We will look upon experiments as one way of obtaining results of basic equations without solving them or even knowing what they are.

Although the governing equations for a Newtonian fluid have long been known, exact solutions can be obtained only for the simplest of situations. Approximate solutions using simplifications like the boundary layer theory are usual, while numerical methods of all sorts are becoming commonplace. However, fundamental barriers still remain in many areas like turbulence and multi-phase flows for which it is fruitless to search for analytical answers. Since engineering needs require information like drag forces and pressure drops which are beyond our capacity to predict, experimental data are empirically correlated and used as far as possible in practical situations. Though there is nothing fundamentally wrong with that, it is not intellectually satisfying; it is also extremely difficult to come up with meaningful and universally valid results in practice without the support of some theory or rationalization, however approximate.

A close working relation between theory and experiment is perhaps the hallmark of a true science. Taylor (1974) reviewed the interaction between experiment and theory, restricting himself only to a discussion of developments in the field of fluid mechanics. Since that time, however, many theoretical developments have become widely known in areas not specifically related to fluid mechanics, but which could be of considerable benefit to the experimentalist in the field. Though there is no question that an experimentalist with wide-ranging and profound knowledge of mathematical theory is at an advantage, it is also true that most have to be content with only a working acquaintance or even restrict their study to areas with the most promise of application to their work. For fluid mechanics, a significant part of useful recent progress lies in the *qualitative* or *geometrical* theory of differential equations. This permits us to make general statements about solutions without actually obtaining them and about the kinds of solutions that one could conceivably get. It tells an experimenter what the possibilities are, so that he can not only look out for them in actual measurements but also actively seek them. These methods are not of recent origin; they have been studied by Poincaré in the last century and a large number of subsequent workers have contributed to the significant growth and maturity of this branch of mathematics. The following section gives a brief outline of its principal features.

2. Background on Dynamical Systems

First we consider a *one-dimensional map* of the type

$$x_{n+1} = F(x_n) \tag{1}$$

which has many interesting properties when F is nonlinear. A *fixed point* is defined as one which maps onto itself. For $F(x) = 4\lambda x(1-x)$, where λ is a control parameter, the resulting equation (1), called the *logistics map*, is archetypical of one-dimensional quadratic maps. It has a single stable fixed point for $0 < \lambda < \frac{3}{4}$ to which all initial conditions in $(0,1)$ tend after a large number of iterations. On increasing λ beyond $\frac{3}{4}$, there appears first an oscillating pattern, followed by one whose period is double, then quadruple, and so on. Eventually this period doubling sequence leads to aperiodicity at $\lambda = \lambda_\infty$. If λ_n is the parameter value for the n'th period doubling, it is found that δ_n defined by

$$\delta_n = \frac{\lambda_{n+1} - \lambda_n}{\lambda_{n+2} - \lambda_{n+1}} \tag{2}$$

quickly approaches the constant $\delta = 4.669\ldots$ as $n \to \infty$. This is a universal property of all quadratic maps (Feigenbaum, 1983). In this kind of aperiodicity the value of x_n obtained after a given number of iterations is very sensitive to the initial conditions. Any uncertainty in the initial conditions or errors in the computations would lead to a large uncertainty in the value of x_n. This unpredictability (Lighthill, 1986) is often referred to as *deterministic chaos*, emphasizing the fact that it is obtained from a deterministic equation such as equation (1). This is a phrase that has now entered the scientific vocabulary and is the subject of extensive research and reviews. A number of books are available, some for the serious layman (Gleick, 1987), others with an engineering flavor (Thompson & Stewart, 1986), and still others for more mathematically oriented readers (Guckenheimer & Holmes, 1983; Lichtenberg & Lieberman, 1983).

It is more common in fluid mechanics to study differential equations. Suppose that a dynamical system can be represented by a set of N ordinary differential equations

$$\frac{dx_i}{dt} = f_i \ (i = 1 \text{ to } N) \tag{3}$$

The system is *autonomous* if f_i does not explicitly contain time t. It is called *conservative* if $\Sigma \partial f_i/\partial x_i = 0$, and *dissipative* if it is < 0. The $x_i(t)$ can be thought of as physical quantities such as fluid velocity or temperature, measured at N given locations within the flow. These variables can be represented as a moving point in an N-dimensional *phase space*. In dissipative systems the volume of a set of initial conditions in phase space eventually goes to zero. Strictly speaking an infinite number of equations are required due to the continuum nature of a fluid. Furthermore, using basic conservation principles we know what the equations are. However, we assume here that we can put our sensors only at a finite number of points and that the governing equations are either unknown or unsolvable. Thus, from the experimental point of view, direct measurement of $x_i(t)$ is a substitute for knowledge which would otherwise be obtained from the equations. It is important then to know the kind of solutions one can conceivably expect to have and how to interpret them.

Let us examine the possibilities by looking at solutions of equation (3) for different f_i.

We will assume first of all that f_i depends on a set of M parameters λ_i (i = 1 to M). These might include the flow Reynolds number, Rayleigh number of the imposed heating, Prandtl number of the fluid or geometrical parameters such as aspect ratio or tilt angles. The parametric space is then M-dimensional. It is important to remember that for many flow situations it can be shown that the solution is unique as one parameter becomes vanishingly small. For example, this is true for an incompressible fluid within a finite boundary as the reciprocal of the viscosity tends to zero. However, as this parameter is increased, other solutions appear. The term bifurcation was originally introduced by Poincaré (1885) to describe a similar splitting of equilibrium solutions in the problem of a rotating mass of fluid. In the century since then, use of the word has multiplied within the context of nonlinear differential equations as a wide variety of such behavior has become known. We use it now to denote the appearance of multiplicity of steady or unsteady solutions near a specific value of a suitable bifurcation parameter. Some of these multiple states may be stable, others unstable. If there is more than one stable state, initial conditions will determine the state to which the system will finally evolve. Thus each stable solution has its own *basin of attraction*, which is defined by the set of the initial conditions which lead to it. New solutions created by the bifurcation may be relatively close to or far from the original solution. Thus experimentally, one may have either a gradual change or a jump into another state as the bifurcation parameter is varied.

2.1 Steady States

The possible steady states are first obtained by taking $dx_i/dt = 0$ and solving for the zeros or singularities of f_i. This is the subject of singularity theory (Golubitsky & Schaeffer, 1985), the fluid mechanical aspects of which have been discussed by Benjamin (1978a,1978b). For a nonlinear f_i, there may be multiple solutions with transitions from one to the other at specific bifurcation points. One simple example is obtained by taking $f(x) = -x^3 + (\lambda-\lambda_0)x$, for which we obtain the pitchfork bifurcation shown in Fig. 1. Up to the bifurcation point λ_0, x is unique but after that there are three different possibilities. Of these three, a linearized stability analysis will show that the trivial one is unstable, so that any general initial condition will be *attracted* by either the positive or the negative solution. The darker and lighter lines show the stable and unstable solutions respectively. Since the x = 0 solution cannot be crossed, the sign of the initial value will determine the sign of the final steady value, defining thus the two basins of attraction.

We have obtained all this information without actually solving the differential equation (1). In fact in this particular case we can get the exact integral, which is

$$x^2 = \frac{\lambda}{1 - C \, \exp\{-2(\lambda-\lambda_0)t\}} \tag{4}$$

from which we can verify our previous conclusions. Of course, (4) also provides detailed information about the instantaneous value of x for any t.

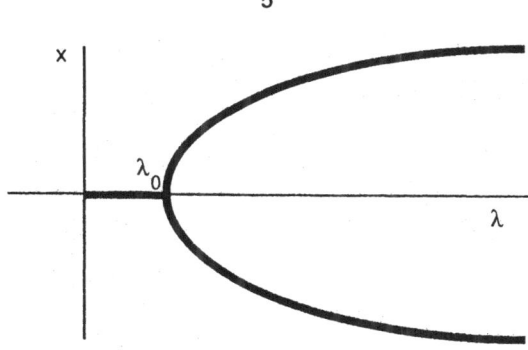

Fig. 1: Pitchfork bifurcation

Although by assumption the exact solution corresponding to (4) is not generally available, information regarding the diagram in Fig. 1 may be obtained from an experiment for which the differential equation is a model. In a series of runs, not only must the parameter λ be varied, but also the initial condition. Only stable states will be observed; the x = 0 solution for $\lambda > \lambda_0$ can only be deduced as the boundary between the basins of attraction. And, unless it is somehow known that f is cubic in x, it is not possible to guarantee experimentally that all possible stable steady states have been found.

Measurements of infinite precision in λ are necessary near $\lambda = \lambda_0$ to define the exact shape of the diagram near this point, since a small change in λ leads to a large change in x. Such measurements are not possible. They may not even be necessary since a slight change in the form of $f(x)$ can result in a drastic qualitative change in the nature of the solutions in this region. These changes are referred to as *unfoldings* of the bifurcation diagram. Since $f(x)$ is a model of a physical system, relaxation of any of the assumptions made in its derivation may lead to such unfolding. If we consider $f(x) = -x^3 + (\lambda-\lambda_0)x - \alpha_1$, the unfolded diagram is as shown in Fig. 2 which is often referred to as an imperfect bifurcation since the parameter α_1 can be thought of as an imperfection in the model.

There are parameter values for which three real solutions exist and values for which there is only one. In terms of the parameters λ and α_1, the cusped curve

$$4(\lambda - \lambda_0)^3 = 27\,\alpha_1^2 \tag{5}$$

in parametric space separates the two regions.

If $f(x) = -x^3 + (\lambda-\lambda_0)x - \alpha_1 - \alpha_2\,x^2$, there are different unfoldings as indicated in Fig. 3 by the diagrams for different combinations of α_1 and α_2. These unfoldings are called *universal* since in a sense they cover all possibilities.

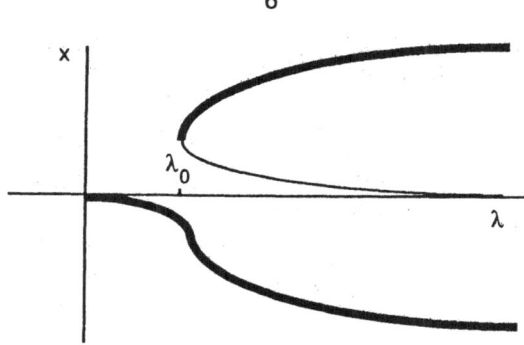

Fig. 2: Imperfect bifurcation

The *primary* branch in any diagram is defined as that which connects to the unique small λ solution. If, in an experiment, λ has been slowly changed from a zero value, it is very likely that the solution obtained will be on the primary branch. For the diagrams (3) and (4) in Fig. 3, one can observe that on continuously increasing λ along the primary branch there are discontinuous changes in x. These jumps exhibit hysteresis in that they are at different places depending on whether λ is increasing or decreasing. *Secondary* modes, which are those not connected to the primary branch, can only develop beyond a certain critical value of λ.

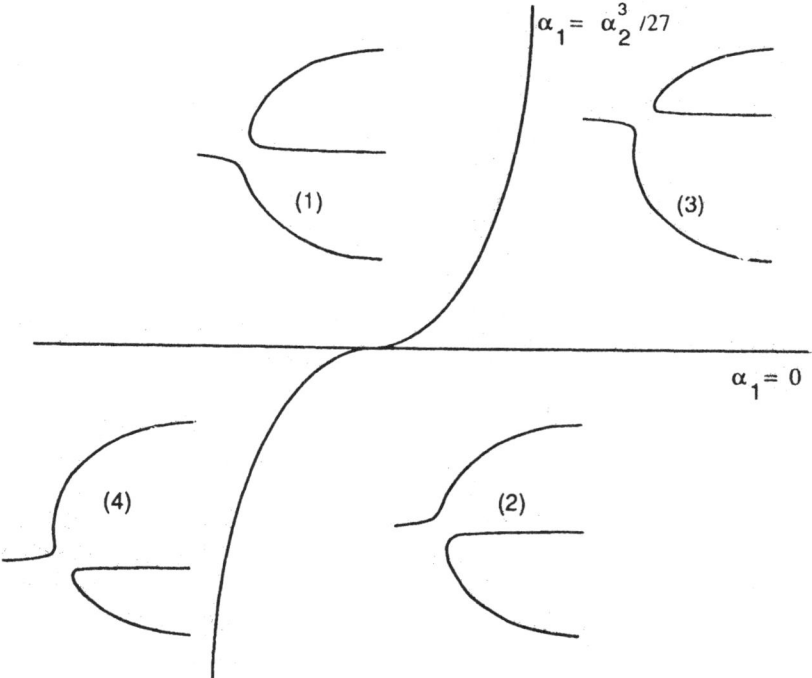

Fig. 3: Universal unfoldings of the pitchfork
(from Golubitsky & Schaeffer, 1985)

For the special case of $\alpha_1 = 0$ and $\alpha_2 > 0$ the diagram looks like Fig. 4. This is an asymmetrical or transcritical bifurcation. In comparison with Fig. 1, the nose of the curve has shifted from $(\lambda_0, 0)$ to $(\lambda_1, -\alpha_2/2)$ where $\lambda_1 = \lambda_0 - \alpha_2^2/4$. It also illustrates the fact that, even though the $x = 0$ solution loses linear stability at $\lambda = \lambda_0$, it can become *subcritically unstable* in the region $\lambda_1 < \lambda < \lambda_0$ for a large enough perturbation.

On comparing Fig. 1 with Figs. 2, 3 and 4, one can observe the large qualitative change in the bifurcation diagram with the addition of small terms in the governing equations. This is equivalent to saying that the mathematical model which produced Fig. 1 is not *structurally stable;* the behavior is not typical and may not be observed in an experiment. Of course, if α_1 and α_2 are small, it will be very difficult to experimentally distinguish between Fig. 1 and any one of its perturbations. On the other hand, the experiment under consideration might correspond to large values of α_1 or α_2, in which case hysteresis and subcritical instability, which were absent in Fig. 1, may be observed.

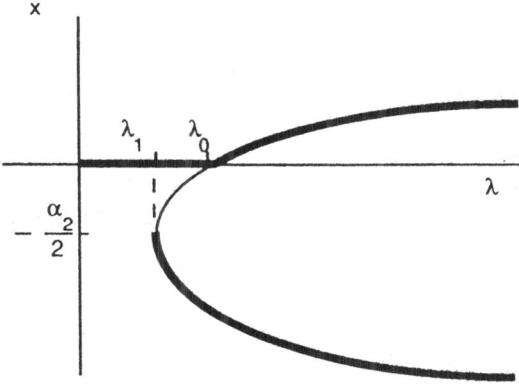

Fig. 4: Bifurcation diagram with $\alpha_1 = 0$ and $\alpha_2 > 0$
(from Golubitsky & Schaeffer, 1985)

2.2 Time-Dependent Motion

The range of possible system behavior is increased if we include time dependent solutions. These can be represented in the same kind of bifurcation diagrams as Figs. 1 to 4 if a suitable distinguishing characteristic such as amplitude is plotted. One way to analyze unsteady phenomena is to construct a phase portrait by plotting the motion of the point x_i ($i = 1$ to N) in an N-dimensional space. This is often difficult to do experimentally since not all of the dependent variables can be measured continuously. An approximate phase space portrait can be reconstructed from a single measured signal $x(t)$. It has been shown that on introducing an arbitrary time delay T, the phase portrait obtained from the vectors {x(t_k), x(t_k+T), ..., x(t_k+nT)}, k = 1,2, ..., ∞ will have many of the properties as one constructed from the measurement of N independent variables, if n ≥ 2N (Packard *et al.*, 1980; Eckmann & Ruelle, 1985). In practice, n may be increased by one until additional structure fails to appear when an extra dimension is added.

One simple kind of bifurcation occurs when a steady state loses stability to a finite amplitude periodic oscillation (*Hopf bifurcation*). Consider the two-dimensional van der Pol equation where $f_1 = x_2$ and $f_2 = -(x_1^2 - \lambda) x_2 + x_1$. For $\lambda < 0$, all transients die down to $x_1 = x_2 = 0$. The point thus obtained in phase space is called an *attractor* because it draws neighboring initial conditions to it. For $\lambda > 0$, a new *limit cycle* attractor appears, representing oscillations the amplitude of which do not depend on initial conditions. Figure 5 shows an oscillation of period $\tau = 6.67$ in a phase plot of $x_1(t)$ vs. $x_2(t)$ for $\lambda = 1$ that would be obtained after initial transients have died down.

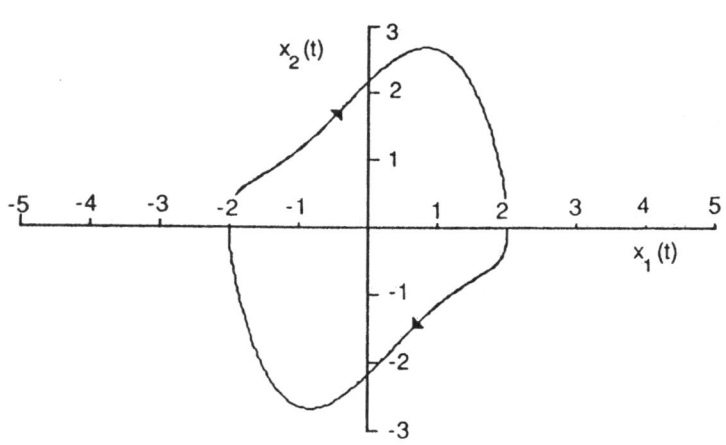

Fig. 5: Phase plot $x_1(t)$ vs. $x_2(t)$ for $\lambda = 1$.

If, however, only one variable, $x_1(t)$ say, can be measured, it is still possible to get a phase portrait by plotting $x_1(t)$ vs. $x_1(t-T)$ as shown in Fig. 6 for two different values of arbitrarily chosen T. The value of T should not alter the topological nature of the attractor, i.e. the curves can be changed from one to the other by suitable stretching and turning. It is evident from Fig. 6 that the solution is periodic and that the period can be easily determined.

Even though the nonlinear oscillations shown in Figs. 5 and 6 have a single period, a large number of frequencies are present. Frequency content of the time series $x_1(t)$ is shown in Fig. 7 by its power spectral density, E(f), in arbitrary logarithmic units. Discrete frequencies are detected from the presence of spikes and periodicity is inferred from the presence of harmonics of the fundamental frequency, in this case only the odd ones being present.

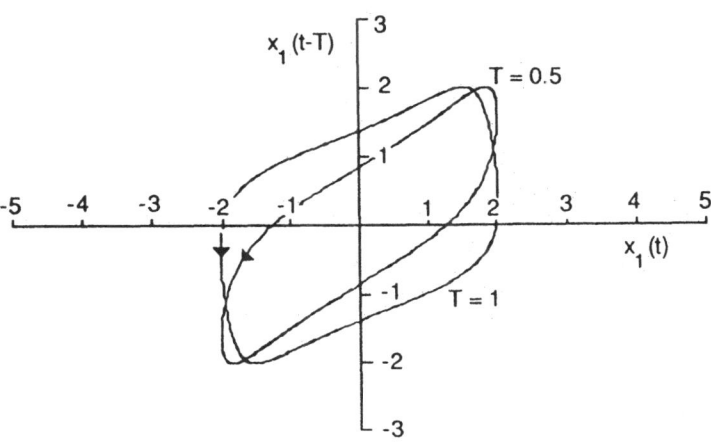

Fig. 6: Reconstructed phase plot from single variable $x_1(t)$

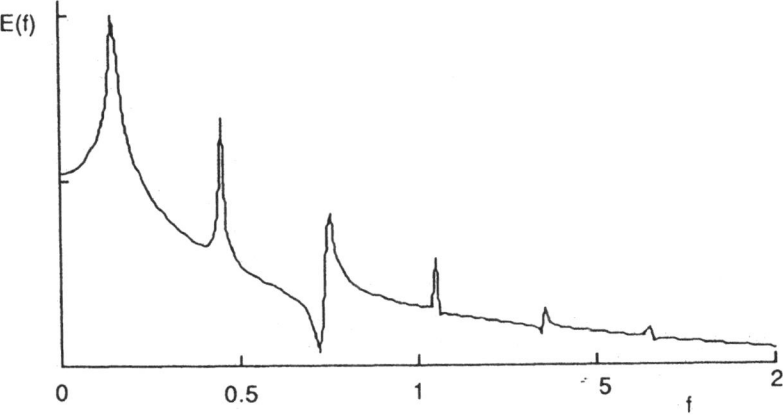

Fig. 7: Power spectral density of $x_1(t)$

Only steady states and limit cycles can occur in autonomous two-dimensional systems since the phase space motion is on a plane and the path cannot intersect itself. In three dimensions or more, the trajectory in phase space can become much more complicated. Consider, as an example, the three dimensional dissipative system: $f_1 = x_2 - x_1$, $f_2 = \lambda^{-1} - x_1 x_3$, $f_3 = 1 - x_1 x_3$ which appears in an analysis of a toroidal natural circulation loop (Sen et al., 1985). There are two steady solutions, only one of which is stable for $\lambda < 1$. Above $\lambda = 1$ a periodic limit cycle orbit appears. A reconstructed phase plot using a time delay of $T = 1$ is shown in Fig. 8(a) for $\lambda = 1.65$. A limit cycle of period $\tau = 6.04$ is observed. On increasing λ beyond $\lambda_1 = 1.68$, this bifurcates into a period-2 cycle as shown in Fig. 8(b), which in turn transforms into a period-4 cycle illustrated by Fig. 8(c) for $\lambda > \lambda_2 = 1.87$.

In some dynamical systems it is possible to have two fundamental frequencies which are incommensurably related to each other. This *quasiperiodicity* would mean that the motion passes through every point on the surface of a torus. The section of the torus would show up on a *Poincaré return map* which is obtained by marking the intersection of the phase space trajectory with a fixed plane.

In the present example, the sequence of period doublings eventually gives way to a chaotic behavior just as it happened for the one-dimensional logistics map. In fact a value of $\lambda_\infty = 1.92$ at which chaos will appear can roughly be estimated from λ_1 and λ_2, using the Feigenbaum universal property for period doublings, equation (2). Figure 8(d) shows the reconstructed phase plot for $\lambda = 10$. This kind of figure is often referred to as a *chaotic* or *strange attractor*. The strangeness refers to the *fractal* structure (Mandelbrot, 1983) of the attractor, a characteristic which does not always accompany chaos.

Chaotic behavior has a number of features, one of which is that the power spectral density is broad-band on which a spiky structure could also be found superposed. Figure 9 shows the power spectral density $E(f)$ of the time series $x_1(t)$ of Fig. 8(d) obtained by using a FFT routine with 4096 data points.

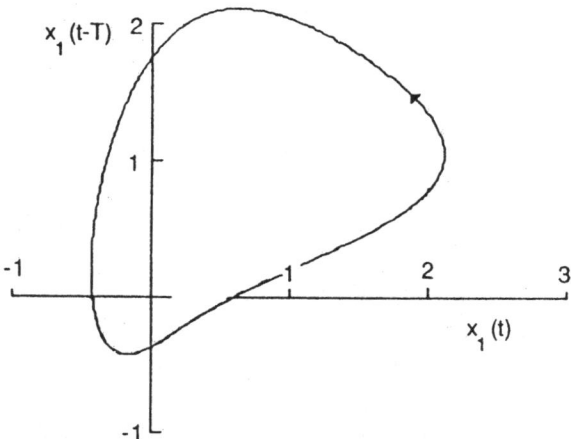

Fig. 8(a): Reconstructed phase plot for $T = 1$ and $\lambda = 1.6$.

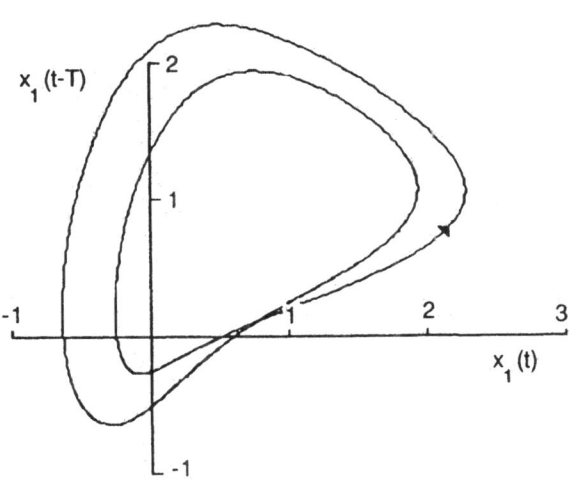

Fig. 8(b): Reconstructed phase plot for T = 1 and λ = 1.85

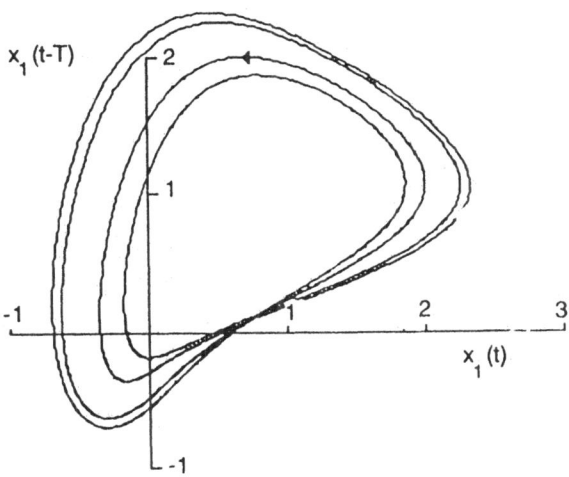

Fig. 8(c): Reconstructed phase plot for T = 1 and λ = 1.90

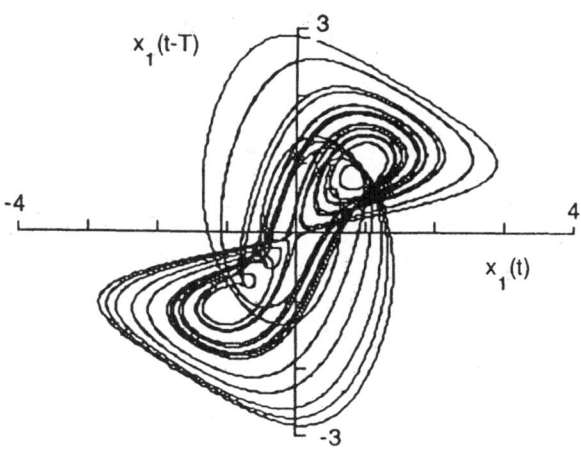

$x_1(t-T)$

Fig. 8(d): Reconstructed phase plot for T = 1 and λ = 10

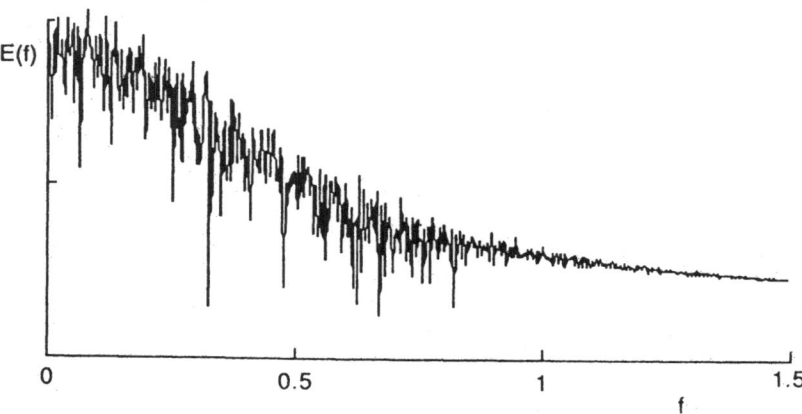

Fig. 9: Power spectral density of $x_1(t)$

There are other characteristics of deterministic chaos that can be used to identify it from experimental data. One common method is to determine *Lyapunov exponents* which essentially represent the average rates at which neighboring points move away from each other in phase space in each one of the coordinate directions (Eckmann & Ruelle, 1985; Wolf, 1986). For chaos, the largest exponent should be positive. Another useful quantity is the *dimension* of the attractor in phase space. A dimension of zero, one, two and three represent a point, line, surface and volume respectively. Thus, the limit cycle shown in Fig. 5 would have a dimension of one. There are many different definitions of dimensions which can be used to carry out computations from discrete time series information (Farmer *et al.*, 1983). In the method suggested by Grassberger & Procaccia (1983), a correlation C(r) of a set of points x_i is determined, where

$$C(r) = \lim_{N \to \infty} \frac{1}{N^2} \left\{ \text{number of pairs of points } (x_i, x_j) \text{ with } |x_i - x_j| < r \right\} \qquad (6)$$

The dimension is found from $C(r) \sim r^\nu$, where ν is the dimension. Each definition of dimension gives integer values for the points, lines, etc. and gives a noninteger value for the strange attractor since it is a fractal.

Although the periodic doubling route to chaos of Feigenbaum has been mentioned, it is not the only scenario that has been found in dynamical systems. Table 1 shows some of the others that have been postulated. The arrows indicate bifurcations or transitions in the qualitative nature of the flow.

Table 1: Models for the transition to chaos and their mechanisms
(adapted from Lichtenberg & Lieberman, 1983)

Model	Mechanisms				
Landau	Stationary point	→ Singly periodic orbit	→ Doubly periodic orbit	→ Triply periodic orbit	→...→ Turbulent motion
Ruelle-Takens-Newhouse	Stationary point	→ Singly periodic orbit	→ Doubly periodic orbit	→ Strange attractor	
Feigenbaum	Stationary point	→ Singly periodic orbit (period T)	→ Singly periodic orbit (period 2T)	→ Singly periodic orbit (period 4T)	→...→ Strange attractor
Pomeau-Manneville	Stationary point	→ Singly periodic orbit	→ Intermittent chaotic motion		

Thus, experimental investigation of a fluid mechanical phenomenon leading to chaos can focus on a determination of the Lyapunov exponents and dimension of the attractor as well as on the route to chaos that actually takes place. Turbulence is a prime though not the only example of disordered behavior of fluid systems. Even though it is chaotic, one must not use the terms turbulence and chaos synonymously, but distinguish carefully between a spatio-temporal hydrodynamic phenomenon on the one hand and the temporal complexity obtained from low dimensional dynamical systems on the other. What connection there is between the two is still not entirely clear. To add to the confusion, the vocabulary itself is in a state of flux and one can read about the strange attractor theory of turbulence, when what is really meant is chaos (Lanford, 1982).

3. Experimental Study of Bifurcation

In principle, any of the factors which affect or influence the outcome of an experimental result could be a bifurcation parameter λ. However, analysis around the bifurcation point or points is experimentally possible only if the parameter can be varied continuously to a reasonable precision within a small neighborhood.

Flow velocity is one of the simplest parameters to control experimentally, since it often involves the adjustment of an electric motor or the gradual opening or closing of a valve. A large number of studies have been made with this as a continuously varied quantity. Examples are boundary layer flows wherein the free-stream velocity is changed, and a rotating Taylor-Couette flow with changing rotational speeds of the cylinders. In any case, at extremely low speeds the solution of the governing equations is unique and so is the flow. At higher speeds the flow becomes unstable and bifurcates to either another steady state or a time-dependent flow. Eventually a series of bifurcations leads to turbulence. Events leading up to the onset of turbulence are of considerable interest in the application of the theory of dynamical systems, though it seems at present that fully developed turbulence itself is not easily tackled (Aubry *et al.*, 1988).

Other quantities may be less easy to vary but could nevertheless be used as bifurcation parameters. Inventiveness in finding these, and experimental ingenuity in carrying out experiments with them, could shed new light on previously well-studied flows. As an example, the geometry of a flow domain is not normally considered continuously variable, but which in some circumstances can be made to change in a controlled fashion. The change must necessarily be made with the fluid in motion so that the appropriate initial conditions are obtained. Since there is a possibility of other solutions in the vicinity, the changes should be slow compared to the time of response of the flow, so that the flow is not jerked out of the basin of attraction of the solution being sought.

Though the seminal work of Lorenz (1963) was on a low-dimensional model of convection, it took more than a decade for experimentalists in fluid mechanics to become aware of the implications for fluid mechanics of the results from dynamical systems theory. In recent years, however, they have become increasingly conscious of its importance. More than providing new theoretical results for comparison, there are now new ways of looking at data. Investigators have become more interested in the processes leading up to chaos. They are assessing the significance of discrete frequencies appearing in the power spectral density and have begun to plot their measurements in terms of phase diagrams and return maps.

4. Fluid Mechanical Examples

There are many examples in the fluid mechanics literature in which experimenters have obtained significant results along the lines of dynamical systems theory, many of them without the advantage of detailed theoretical predictions. A few selected illustrations are discussed below. The list is by no means complete nor representative of all the possibilities. Significant omissions include external or open flows such as the wake of a circular cylinder (Sreenivasan, 1985; Van Atta & Gharib, 1987) and multifractal studies of the dissipation field in turbulent flows (Prasad *et al.*, 1988).

4.1 Dripping Faucet

The problem of a dripping faucet is one for which the relevant physics and governing equations are known, but are mathematically intractable. It has been shown by Shaw (1984) that even then experimental information can be processed and analyzed in a meaningful manner. In this experiment, water from a large tank was allowed to pass through a motor controlled valve and then allowed to drip through a 1 mm orifice. The flow rate was the bifurcation parameter which could be varied at will by adjusting the valve. The falling drops were timed as they passed through a light beam which produced pulses in a photocell signal. The time interval between two drops T_n was measured. T_n was plotted against T_{n+1} to obtain a one-dimensional map of the kind indicated by equation (1). The results show a wide range of behavior for increasing flow rate. At low flow rates, the phenomenon is periodic and the measurements appear as a point in a (T_n, T_{n+1}) plot. At larger flow rates, a period doubling bifurcation occurs and two drops fall before the system repeats itself. The data are now clustered around two separate points. Other interesting regimes such as chaos can also be found with plots similar to those obtained in a quadratic one-dimensional map. The fluid system also exhibits other nonlinear phenomena such as hysteresis and multiple basins of attraction, so that the behavior often depends on initial conditions. It is shown by Shaw (1984) that even though the complete system is complex, a low-dimensional approximation will predict the results for small flow rates.

4.2 Natural Convection

Buoyancy driven convection in closed enclosures is a problem of considerable interest to experimentalists. A complete discussion of the structures of transitions between steady states in this and other flows is given by Shirer & Wells (1983). The controlling parameters for the system are the Rayleigh number, Ra, which measures the strength of the thermal driving force, the Prandtl number of the fluid and the geometry of the enclosure. One simple geometry is offered by a closed natural circulation loop of any shape, heated and cooled over different parts of its length. The tilt of the loop with respect to the gravity vector can be used as a geometrical parameter that is easy to change continuously in an experiment. The flow direction, which can be easily observed, enables two different solutions to be distinguished. Under these conditions, multiple steady states have been theoretically predicted (Damerell & Schoenhals, 1979) and experimentally observed (Acosta *et al.*, 1987; Sen & Torrance, 1988). From an experimental point of view it is important to note that the secondary states do not appear if one starts increasing Ra at fixed inclination. They can be observed only if the inclination is changed once steady state has been reached, using in effect nonzero initial conditions. In searching for the secondary states, the change in the inclination should be slow and gradual since any sudden movement may send the system into its primary state.

Under simplifying assumptions of one-dimensional flow and prescribed wall temperature (Yorke & Yorke, 1981) or heat flux (Sen *et al.*, 1985) the governing equations of toroidal natural convection loops reduce to three ordinary differential equations for which chaotic solutions can be shown to exist. Although experiments corresponding to these heating modes have not yet been made, other heating and cooling arrangements for which the governing differential equations remain partial, but which are easier to set up experimentally, have yielded experimental evidence of chaos (Gorman *et al.*, 1986).

For enclosures of rectangular geometry, the aspect ratio is defined as the ratio of the horizontal to vertical dimensions. Small aspect ratio layers have been found to behave like low-dimensional systems near the onset of chaos (Gollub, 1983). On increasing Ra, the flow first becomes time-dependent when a periodic limit cycle with fundamental frequency f_1 appears. On increasing Ra it becomes quasiperiodic with the appearance of another incommensurate frequency f_2. There is phase locking at higher Ra so that f_2/f_1 becomes the ratio of two small integers. Ultimately, chaos with a broad band frequency spectrum appears. Extensive measurements have been made of the route to chaos in Rayleigh-Bénard experiments using liquid helium (Maurer & Libchaber, 1979; Ahlers, 1980), mercury (Stavans *et al.*, 1985; Jensen *et al.*, 1985), silicone oil (Dubois & Bergé, 1986) and other fluids. A number of investigators have determined the dimension of the attractor in the chaotic regime. Malraison *et al.* (1983) found a value of slightly less than three for certain conditions using the definition of Grassberger & Procaccia (1983). The measured dimension was not essentially changed as the dimension of the phase space used to calculate it was increased beyond four. Giglio *et al.* (1984) have also used the same method to measure values of the dimension and found it to be low but nonintegral. The value increases with increasing Ra, being slightly less than five even for Ra eighty times that for the onset of convection.

4.3 Taylor-Couette Flow

The rotational Taylor-Couette flow is one that has been long studied. Represented by the ratio of the length of the cylinders to the radial gap between them, the aspect ratio Γ in most analyses was taken to be infinite while in all experiments is finite. Non-uniqueness of the flows has been shown by Coles (1965) who found that, depending on the path followed in parameter space, up to 26 distinct stable flow states could be obtained for the same Reynolds number R. Though considered a relevant factor, the aspect ratio was not taken to be a bifurcation parameter since it is not as easy to change continuously as much as the rotational speeds of the cylinders are. Benjamin and coworkers (Benjamin, 1978c; Benjamin & Mullin, 1982; Mullin, 1982) were able to do just that, and were in a position to study the effect of the height of the liquid annulus by varying it continuously in a special apparatus where the outer cylinder and end walls were stationary and the inner cylinder was rotating. The primary mode consisting of either two- or four-cell motion develops for low R while similar secondary modes are obtained above a critical value of R. Two other three- and four-cell anomalous secondary modes spiralling in directions opposite to those of the normal four-cell mode were also discovered. Hysteresis between two-cell and four-cell forms of the primary mode is obtained by varying the length of the liquid annulus. In the (R,Γ) plane, the change between these steady states can be represented as shown schematically in Fig. 10. The curve forms a bifurcation set in that every point on it is a bifurcation point representing a qualitative change between two-cell and four-cell motion. A cusp similar to that represented by equation (5) can be identified.

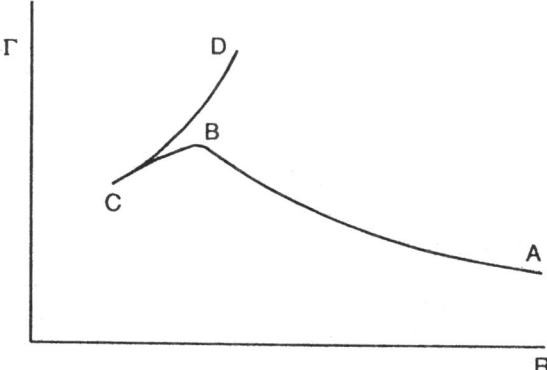

Fig. 10: Experimentally determined bifurcation set for the Taylor-Couette flow
(adapted from Benjamin, 1978c)

On taking Γ = constant sections, as shown in Fig. 11 (Zeeman, 1982; Mullin, 1984), one can observe the parallelism with the bifurcation diagrams of Fig. 3. A suitable variable, such as the inward radial component of the velocity halfway up the cylinders, v, distinguishes between the modes. The cell numbers of each mode are shown and the arrows indicate jumps in the velocity on changing R. The stable and unstable modes are in darker and lighter lines respectively. In each case the secondary mode is isolated from the primary, and the only way to reach it by slow changes is to change the aspect ratio Γ. For example, in order to reach the stable secondary four-cell mode in Fig. 11(a), one must start from conditions corresponding to the primary four-cell mode in Fig. 11(d), and then change Γ slowly so as to get the conditions of Fig. 11(a).

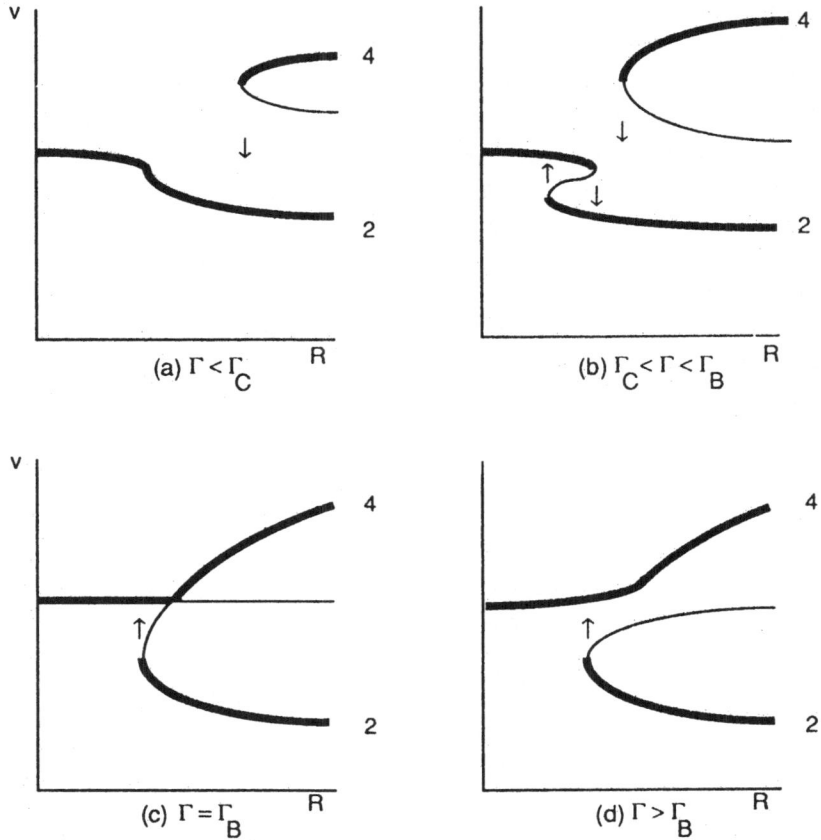

Fig. 11: Schematic of bifurcations in Taylor-Couette flow for different Γ
(adapted from Zeeman, 1982)

The remarkable fact is that some of these experimental results were qualitatively conjectured by Benjamin (1978b) from a study of a cubic algebraic equation as indicated in §2.1. Since then, detailed numerical analyses of the complete partial differential equations have confirmed and extended the results (Cliffe, 1983; Hughes *et al.*, 1985).

For both cylinders rotating independently, Andereck *et al.* (1986) have obtained a fairly complex diagram of steady and time-dependent flow regimes that are observed. Some of these time-dependent flows are chaotic. Brandstäter *et al.* (1983) used three methods to determine the dimension of the chaotic flow. Even for R thirty percent above that for the onset of chaos, the dimensions were less than five, indicating that the phenomenon up to that stage is low-dimensional. They have also used their experimental data to determine the largest Lyapunov coefficient and to verify that it is positive in the chaotic regime (Brandstäter *et al.,* 1984).

4.4 Stirring

Mixing of miscible fluids is normally associated with turbulence. It is, however, possible to mix highly viscous fluids by low-speed unsteady stirring which would not be considered as turbulence. Recent progress in the mathematical theory of chaos in conservative dynamical systems (see, for example, Lichtenberg & Lieberman, 1983), a topic we have not discussed here, has significant potential applications to this process. In a two-dimensional incompressible flow the position of a particle of fluid (x,y) is given by

$$\frac{dx}{dt} = -\frac{\partial \psi}{\partial y}$$

and (7)

$$\frac{dy}{dt} = \frac{\partial \psi}{\partial x}$$

where ψ is a stream function. This is a Hamiltonian system which is conservative since the divergence of the RHS is identically zero. Volumes do not shrink in phase space which, in this case, is the physical space itself. If $\psi = \psi(x,y)$ the system is integrable, while if $\psi = \psi(x,y,t)$ the flow can be nonintegrable, a condition necessary for chaos. Laminar mixing of this kind, which can occur even in Stokes flow, was termed *chaotic advection* by Aref (1984) and *Lagrangian turbulence* by Chaiken *et al.* (1986). Some theoretical support for the idea has been provided by Aref (1984) who considered the specific example of two point vortices which are alternately on and off. The motion of the particle after each cycle is governed by a two-dimensional map of the kind $x_{n+1} = F(x_n, y_n)$ and $y_{n+1} = G(x_n, y_n)$, which is also area preserving because of mass conservation. Chaotic particle paths can be produced under some circumstances. In general then it is possible to choose time-dependent boundary conditions which can result in efficient mixing due to spatial chaos in a two-dimensional flow field. Chaotic mixing in Stokes flow conflicts with the idea of reversibility which was demonstrated by Taylor (1960) for steady flow between concentric cylinders. That flow was integrable, but for a nonintegrable chaotic system the exponential divergence of neighboring trajectories causes extreme sensitivity to perturbations from external sources, and reversibility is not practically possible.

Experiments in a rectangular cavity with alternate periodic motion of the opposite walls were carried out by Chien *et al.* (1986). Flow visualization was employed with a tracer of low diffusivity and approximately the same density and viscosity as the main fluid. They looked at the deformation of material lines and blobs. The efficiency of mixing was found to depend strongly on the nondimensional frequency of oscillation of the walls. Chaiken *et al.* (1986) carried out experiments of flow between eccentric cylinders which were driven by computer controlled stepper motors. The cylinders rotated alternately through a small angle. Considerable mixing of a tracer dye was obtained, except for small elliptic islands which were untouched.

5. Concluding Remarks

Many different theoretical considerations are involved in the design and operation of an experiment. First there is the aspect of the experimental technique itself using sophisticated instrumentation along with advanced data processing methods, both of which involve considerable physical insight and mathematical technique. There are also the theoretical ideas which are a motivation and guide for the design of the experiment, the variables that it will measure and the parameters that will be varied. We have reviewed some of the kind of behaviors that can be expected from actual flows and the way in which they can be interpreted. Since the subject matter is much larger than can be handled here, only the principal ideas were discussed and the interested reader is directed to the references for further study. The approach has been that of an investigator who obtains solutions to the equations governing a physical system, not through analytical or numerical but experimental means. With the experimentally obtained solution, he can go through many of the steps that would have to be followed in a similar theoretical study. In fact, since experimentation is the art of finding solutions without actually solving equations, it is irrelevant in principle whether the governing equations are known or not.

Two different classes of results are pointed out as being useful, one for the steady state and the other time-dependent. Actual complex flows are seen to share many of the characteristics of simple mathematical models, indicating perhaps that low-dimensional dynamics often govern the flow. Though we have mostly discussed examples of flows with finite domains, attempts are being made to seek applications to other types of flows. As time goes on, many more applications will inevitably be found for these ideas.

Acknowledgment

The author would like to thank Prof. K.R. Sreenivasan of Yale University for his careful reading of the manuscript and for constructive suggestions.

References

Acosta, R, Sen, M. & Ramos, E. (1987) Single-phase natural circulation in a tilted square loop, *Wärme- und Stoffübertragung* **21**, 269-275.

Ahlers, G. (1980) Onset of convection and turbulence in a cylindrical container, in *Systems Far from Equilibrium*, (Ed.) L. Garrido, Lect. Notes in Phys., Vol. 132, Springer-Verlag, Berlin, pp. 143-161.

Andereck, C.D., Liu, S.S. & Swinney, H.L. (1986) Flow regimes in a circular Couette system with independently rotating cylinders, *J. Fluid Mech.* **164**, 155-183.

Aref, H. (1984) Stirring by chaotic advection, *J. Fluid Mech.* **143**, 1-21.

Aubry, N., Holmes, P., Lumley, J.L. & Stone, E. (1988) The dynamics of coherent structures in the wall region of a turbulent boundary layer, *J. Fluid Mech.* **192**, 115-173.

Benjamin, T.B. & Mullin, T. (1982) Notes on the multiplicity of flows in the Taylor experiment, *J. Fluid Mech.* **121**, 219-230.

Benjamin, T.B. (1978a) Applications of generic bifurcation theory in fluid mechanics, in *Contemporary Developments in Continuum Mechanics and Partial Differential Equations*, (Eds.) G.M. de la Penha, L.A. Madeiros, North-Holland, Amsterdam, pp. 45-73.

Benjamin, T.B. (1978b) Bifurcation phenomena in steady flow of a viscous fluid, Part 1, Theory, *Proc. R. Soc. Lond.* A **359**, 1-26.

Benjamin, T.B. (1978c) Bifurcation phenomena in steady flow of a viscous fluid, Part 2, Experiments, *Proc. R. Soc. Lond.* A 359, 27-43.

Brandstäter, A., Swift, J., Swinney, H.L. & Wolf, A. (1984) A strange attractor in a Taylor-Couette experiment, in *Turbulence and Chaotic Phenomena in Fluids*, (Ed.) T. Tatsumi, North-Holland, Amsterdam, pp. 179-184.

Brandstäter, A., Swift, J., Swinney, H.L., Wolf, A. Farmer, J.D., Jen, E. & Crutchfield, P.J. (1983) Low-dimensional chaos in a hydrodynamical system, *Phys. Rev. Lett.* **51**, 1442-1445.

Chaiken, J. Chevray, R., Tabor, M. & Tan, Q.M. (1986) Experimental study of Lagrangian turbulence in Stokes flow, *Proc. R. Soc. Lond.* A **408**, 165-174.

Chien, W.-L., Rising, H. & Ottino, J.M. (1986) Laminar mixing and chaotic mixing in several cavity flows, *J. Fluid Mech.* **170**, 355-377.

Cliffe, K.A. (1983) Numerical calculations of two-cell and single-cell Taylor flows, *J. Fluid Mech.* **135**, 219-233.

Coles, D. (1965) Transition in circular Couette flow, *J. Fluid Mech.* **21**, 385-425.

Damerell, P.S. & Schoenhals, R.J. (1979) Flow in a toroidal thermosyphon with angular displacement of heated and cooled sections, *ASME J. Heat Transfer* **101**, 672-676.

Dubois, M. & Bergé, P. (1986) Rotation number dependence at the onset of chaos in free Rayleigh-Bénard convection, *Phys. Scr.* **33**, 159-162.

Eckmann, J.-P. & Ruelle, D. (1985) Ergodic theory of chaos and strange attractors, *Rev. Mod. Phys.* **57**, 617-656.

Farmer, J.D., Ott, E. & Yorke, J.A. (1983) The dimension of chaotic attractors, *Physica* **7D**, 153-180.

Feigenbaum, M.J. (1983) Universal behaviour in nonlinear systems, in *Nonlinear Dynamics and Turbulence*, (Eds.) G.I. Barenblatt, G. Iooss & D.D. Joseph, Pitman, Boston, pp. 101-138.

Giglio, M., Musazzi, S. & Perini, U. (1984) Low-dimensionality turbulent convection, *Phys. Rev. Lett.* **53**, 2402-2404.

Gleick, J. (1987) *Chaos: Making a New Science*, Viking, New York.

Gollub, J.P. (1983) Recent experiments on the transition to turbulent convection, in *Nonlinear Dynamics and Turbulence*, (Eds.) G.I. Barenblatt, G. Iooss & D.D. Joseph, Pitman, Boston, pp. 156-171.

Golubitsky, M. & Schaeffer, D.G. (1985) *Singularities and Groups in Bifurcations Theory*, Vol. 1, Springer-Verlag, New York.

Gorman, M, Widmann, P.J. & Robbins, K.A. (1986) Nonlinear dynamics of a convection loop: a quantitative comparison of experiment with theory, *Physica* **19D**, 255-267.

Grassberger, P. & Procaccia, I. (1983) Characterization of strange attractors, *Phys. Rev. Lett.* **50**, 346-349.

Guckenheimer, J. & Holmes, P. (1983) *Nonlinear Oscillations, Dynamical Systems and Bifurcations of Vector Fields*, Appl. Math. Sci., Vol. 42, Springer-Verlag, New York.

Hughes, C.T., Leonardi, E., de Vahl Davis, G. & Reizes, J.A. (1985) A numerical study of the multiplicity of flows in the Taylor experiment, *Physico Chemical Hydrodynamics* **6**, 637-645.

Jensen, M.H., Kadanoff, L.P. & Libchaber, A. (1985) Global instability of the onset of chaos: results of a forced Rayleigh-Bénard experiment, *Phys. Rev. Lett.* **55**, 2798-2801.

Lanford III, O.E. (1982) The strange attractor theory of turbulence, *Ann. Rev. Fluid Mech.* **14**, 347-364.

Lichtenberg, A.J. & Lieberman, M.A. (1983) *Regular and Stochastic Motion*, Appl. Math. Sci., Vol. 38, Springer-Verlag, New York.

Lighthill, J. (1986) The recognized failure of predictability in Newtonian dynamics, *Proc. R. Soc. Lond.* A **407**, 35-50.

Lorenz, E.N. (1963) Deterministic nonperiodic flow, *J. Atmos. Sci.* **20**, 130-141.

Malraison, B., Atten, P., Bergé, P. & Dubois, M. (1983) Dimension d'attracteurs étranges: une détermination expérimentale en régime chaotique de deux systèmes convectifs, *C.R. Acad. Sc. Paris* **C297**, 209-214.

Mandelbrot, B.B. (1983) *The Fractal Geometry of Nature*, W.H. Freeman, New York.

Maurer, J. & Libchaber, A., (1979) Rayleigh-Bénard experiment in liquid helium; frequency locking and the onset of turbulence, *J. Phys. Lett.* **40**, L419-L423.

Mullin, T. (1982) Mutations of steady cellular flows in the Taylor experiment, *J. Fluid Mech.* **121**, 207-218.

Mullin, T. (1984) Cell number selection in Taylor-Couette flow, in *Cellular Structures in Instabilities*, (Eds.) J.E. Wesfreid & S. Zaleski, Springer-Verlag, New York, pp. 75-83.

Packard, N.H., Crutchfield, J.P., Farmer, J.D. & Shaw, R.S. (1980) Geometry from a time series, *Phys. Rev. Lett.* **45**, 712-716.

Poincaré, H. (1885) Sur l'équilibre d'une masse fluide animée d'un mouvement de rotation, *Acta Math.* **7**, 259-380.

Prasad, R.R., Meneveau, C. & Sreenivasan, K.R. (1988) The multifractal nature of the 'dissipation' field of passive scalars in fully developed flows, *Phys. Rev. Lett.* **61**, 74-77.

Sen, M. & Torrance, K.E. (1988) Analytical and experimental study of steady-state convection in a double loop thermosyphon, *Int. J. Heat Mass Transfer* **31**, 709-722.

Sen, M., Ramos, E. & Treviño, C. (1985) The toroidal thermosyphon with known heat flux, *Int. J. Heat Mass Transfer* **28**, 219-233.

Shaw, R. (1984) *The Dripping Faucet as a Model Chaotic System*, Aerial Press, Santa Cruz, CA.

Shirer, H.N. & Wells, R. (1983) *Mathematical Structure of the Singularities at the Transitions Between Steady States in Hydrodynamic Systems,* Lect. Notes in Phys., Vol. 185, Springer-Verlag, Berlin.

Sreenivasan, K.R. (1985) Transition and turbulence in fluid flows and low-dimensional chaos, in *Frontiers in Fluid Mechanics*, (Eds.) S.H. Davis & J.L. Lumley, Springer-Verlag, New York, pp. 41-67.

Stavans, J., Heslot, F. & Libchaber, A. (1985) Fixed winding number and the quasiperiodic route to chaos in a convective field, *Phys. Rev. Lett.* **55**, 596-599.

Taylor, G.I. (1960) Low Reynolds number flow (16 mm film), Educational Services Inc., Newton, MA.

Taylor, G.I. (1974) The interaction between experiment and theory in fluid mechanics, *Ann. Rev. Fluid Mech.* **6**, 1-16.

Van Atta, C.W. & Gharib, M. (1987) Ordered and chaotic vortex streets behind circular cylinders at low Reynolds numbers, *J. Fluid Mech.* **174**, 113-133.

Wolf, A. (1986) Quantifying chaos with Lyapunov exponents, in *Chaos*, (Ed.) A.V. Holden, Princeton University Press, Princeton, pp. 273-290.

Yorke, J.A. & Yorke, E.D. (1981) Chaotic behavior and fluid dynamics, in *Hydrodynamic Instabilities and the Transition to Turbulence*, (Eds.) H.L. Swinney & J.P. Gollub, Springer-Verlag, New York, pp. 77-95.

Zeeman, E.C. (1982) Bifurcation, catastrophe, and turbulence, in *New Directions in Applied Mathematics*, (Ed.) P.J. Hilton & G.S. Young, Springer-Verlag, New York, pp. 109-153.

LOW SPEED, INDRAFT WIND TUNNELS

by

Stephen M. Batill
Robert C. Nelson
Department of Aerospace and Mechanical Engineering
University of Notre Dame
Notre Dame, Indiana USA 46556

1. Historical Perspective

Since the beginning of time mankind has marvelled and envied the flight of birds. The literature and artwork of many early civilizations depicted mankind's keen interest and desire to fly. The ancient Greek tale of Icarus and Daedalus is typical of the folklore dealing with men flying like birds. Through the centuries the invention of kites and hot air balloons kept the interest in flight alive. It would take until the 15th century A.D. before serious scientific study of flight began to develop. One of the first men to study scientifically the flight of birds was Leonardo da Vinci. He drew over 500 sketches and wrote down hundreds of ideas on the concept of manned flight. Much of his work dealt with ornithopters, man powered machines utilizing flapping wings. Many men tried to use da Vinci's ideas, however, ornithopters proved to be impractical.

The successful flight of a man carrying, heavier than air, powered aircraft by the Wright brothers on December 17, 1903, was made possible by the development of ground based experimental facilities. The first facility developed for measuring aerodynamic reactions in a controlled environment was the whirling arm. The whirling arm was developed by Benjamin Robins in 1746 and is shown by way of a sketch in Figure 1. It consists of a horizontal arm that rotates about the vertical axis due to the tension in the string that is attached to a falling weight. Upon release of the weight, the arm would begin to rotate and would reach a steady state within several revolutions. Robins measured the air resistance of the test object by observing the change in the speed of rotation of the arm. He used the whirling arm apparatus to measure the drag force on various blunt body shapes. His results showed that different shaped bodies having the same frontal area would not necessarily have the same drag. This observation was contrary to existing theoretical thought. Later, Sir George Cayley used a whirling arm to measure the lift and drag characteristics of airfoil shapes. From his measurements Cayley was able to design and build the first successful glider. Another contribution of his work was the recognition that propulsion was needed only to overcome the air resistance and hence powered flight would be possible without a propulsive force in the vertical direction. This opened up the possibility of other types of designs besides the ornithopter.

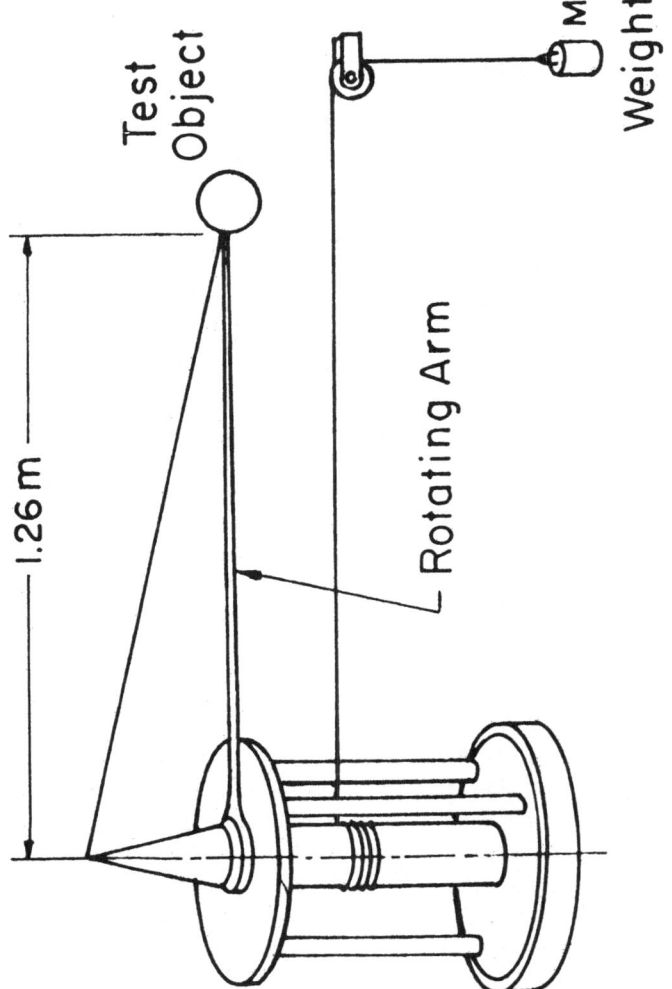

Figure 1 Whirling Arm Test Apparatus

The whirling arm apparatus proved to be the first technique for obtaining systematic aerodynamic data on bodies and airfoils. The technique was not without its limitations. The biggest limitation was that the models flew in their own wakes. Experimenters had difficulty determining the relative velocity of the model and the aerodynamic forces. Because of these restrictions, a new testing device was sought. The wind tunnel provided a facility free of the problems associated with the whirling arm. Credit for the design of the first wind tunnel is usually given to Frank H. Wenham of the United Kingdom. Wenham built a wind tunnel consisting of a fan-blower driven by a steam engine followed by a 305 mm (12 in) square channel that was 3.66 m (12 ft) long. The fan directed air down the channel to the model. With this first wind tunnel, Wenham and his associates discovered the effect of aspect ratio on wing performance. In addition, their data showed that long narrow wings provided more lift than short stubby wings of equal area. Their experiments revealed that higher lift forces were possible than those predicted by Newton's theory of lift. This revelation encouraged many early pioneers that powered aircraft were indeed possible.

The Wright brothers' success in designing the first powered man carrying aircraft was significantly enhanced by their use of a wind tunnel. The Wright brothers' initial glider designs were based upon aerodynamic data available during that time period. They were disappointed in the performance of their designs and began to doubt the accuracy of the existing aerodynamic data base. Because of their lack of confidence with published aerodynamic data, the Wright brothers began a series of experiments. Their first experiments were conducted, appropriately enough, with a modified bicycle. They used a bicycle with a third wheel mounted horizontally on the forward part of the frame. To the horizontal wheel they attached two different airfoils one hundred and eighty degrees apart. The relative performance of the two airfoils was determined by pedaling the bicycle up to approximately 15 miles per hour and observing the rotation of the horizontal wheel. Although the data from this experiment was rather crude, it confirmed the Wright brothers' suspicion of the existing aerodynamic data. The Wright brothers then designed several wind tunnels and a balance system. Their indraft tunnel consisted of a fan located upstream of the test section, followed by screens and a honeycomb to minimize the turbulence entering the 406 mm (16 in) square test section. Using their tunnel, they systematically evaluated the performance of many airfoil shapes. With the data from their experiments, they designed their 1902 glider. This design proved to be a very successful glider. The following year they added an engine and twin counterrotating propellers to the 1902 glider design to achieve their historic flight at Kitty Hawk, North Carolina on December 17, 1903.

In the United States, the Wright brothers' airplane was considered a novelty of little practical value. However, the European nations were quick to recognize that the airplane had great potential for military reconnaissance. The enthusiasm and interest of Europe for developing aircraft was due in part to the Wright brothers' barnstorming tour through Europe in 1908. The Wright brothers demonstrated their airplane to many of the government leaders and monarchs of Europe. The enthusiasm for aviation in Europe also led to the development of many wind tunnel laboratories.

Several of the major wind tunnel laboratories that would provide significant aeronautical leadership during this early period, as well as in the following decades, include the laboratories started by Thomas Stanton in 1903 at the National Physical Laboratory, England, Ludwig Prandtl in 1908 at Gottingen, Germany, and Gustave Eiffel in 1909 at Champs de Mars, France. On the other hand, major wind tunnel facilities in the United States were almost nonexistent prior to World War I. The notable exception were the wind tunnels developed by Albert Zahm. Dr. Zahm started his aeronautical studies while he was a professor of Physics at the University of Notre Dame. Through his experiments with gliders and the whirling arm apparatus he was able to develop the theoretical basis of aircraft static stability. His first wind tunnel was built at Catholic University in 1901 and a second tunnel was built at the United States Naval Yard in Washington D.C. in 1908. Later in his position as director of the Smithsonian, Zahm would play an important role in convincing congress to create the National Advisory Committee of Aeronautics. With the creation of the NACA in 1917, the United States started the development of major wind tunnel laboratories. The NACA Laboratories, as well as the industrial and university wind tunnels, helped America regain leadership in the development of aircraft.

The following sections serve to describe the primary topic of the paper, the low speed indraft wind tunnel, and to introduce a number of topics associated with the design and operation of this type of experimental facility. The authors assume a basic familiarity with the fundamental nomenclature for wind tunnels as presented in some of the excellent and exhaustive references devoted to this subject such as Rae and Pope [1983] or Bradshaw and Pankhurst [1964]. Certain terms and interpretations which will be particular to this paper will be presented and should provide adequate background for a general introduction to indraft wind tunnels.

In order to set a framework for this discussion, it is important to establish the primary purpose of the wind tunnel. Whether the tunnel is designed to conduct basic fluid mechanics studies in a research laboratory or to perform production testing in an engineering company, the purpose of the tunnel is to provide a controllable flow of air in a test or working section. This is usually done in order to simulate conditions which occur outside of the wind tunnel as a body moves through a fluid. This requires that the tunnel and associated control and support equipment must be capable of providing:

(1) specified spatial velocity distribution (usually uniform although in some situations a specific, nonuniform velocity profile is desired),

(2) specified temporal variation of velocity (also usually constant but a time varying velocity may also be required),

(3) specified level of disturbances (velocity, pressure, acoustic, etc.),

(4) specified density, temperature and pressure, and

(5) an environment which will not be adversely affected due to the introduction of the test object or model.

The specific application may help establish allowable levels of variations in these parameters or flow quality requirements. A set of conventional flow quality standards as indicated by Mort et.al. [1972] of $\Delta\alpha = \pm0.1°$ and mean velocity variation of $\Delta u/V = \pm0.5\%$ may be suitable for aircraft configurational testing but may be unacceptable for another type of test. The design of the facility will be significantly influenced by the selection of the design criteria, particularly flow quality, so the designer must be very careful to establish realistic values for a specific application. As the discussion of types and components of wind tunnels proceeds, further discussion of performance criteria will be given as well as the influence of the wind tunnel design on achieving those criteria.

2. Indraft Wind Tunnels

The open circuit wind tunnel is perhaps the simplest of all wind tunnel types. An open circuit wind tunnel is a channel which pulls air from the atmosphere, the air then enters a test chamber in which, hopefully, controllable test conditions are achieved and then the air is returned to the atmosphere. This was the first type of wind tunnel developed and is often referred to as an "Eiffel" or "NPL" tunnel (after the tunnel designed by Gustave Eiffel in France and tunnel at the National Physical Laboratory in Teddington, England, UK). The distinction between these two names is associated with the wall boundaries in the test section, as will be discussed later.

The air is typically moved by a fan or blower which is part of the test channel. The fan can be located upstream of the test section in a manner in which the air is "pushed" into the test section. This type of open circuit tunnel is referred to as a blow-down or blower tunnel. The fan or power source can also be located downstream of the test section in such a manner that the air is "pulled" into the test section. The use of the fan as a suction device is the more widely used approach in wind tunnel design. For purposes of this paper, this latter configuration of an open circuit tunnel, with the power source downstream of the test section, will be referred to as an indraft tunnel and will be the primary focus of the discussion.

One of the principal concerns associated with the operation of an open circuit wind tunnel is the "condition" of the air in the reservoir from which the air is taken and to which the air is exhausted after passing through the tunnel. Depending upon the size of the wind tunnel, there are a number of possible configurations which influence the characteristics of the upstream and downstream reservoir.

(1) The entire tunnel is located within a single, closed room.

(2) The entrance to the tunnel is in a single, closed room which is somehow connected to another closed room into which the tunnel exhausts.

(3) The entrance to the tunnel is within a "porous" building (i.e. somewhat controlled environment) and the exhaust is to the atmosphere outside of the building and therefore subjected to uncontrollable disturbances.

(4) The entrance to the tunnel is outside of the building and the exit is inside a porous building.

(5) Both the entrance and exhaust portions of the wind tunnel are open to the atmosphere.

Each of these possible cases presents certain problems with regard to the design and operation of an open circuit wind tunnel. Strictly speaking, the first two types indicated above could be considered as closed loop, or closed circuit, wind tunnels with the room, or rooms, which contain the inlet and exhaust acting as the "return leg". But since these rooms would not be specifically designed to serve only as a part of the wind tunnel circuit, for purpose of discussion, it will be considered an open circuit tunnel.

Upon initial inspection, one might consider an indraft wind tunnel as being inefficient in comparison to a closed or continuous circuit tunnel. In a closed circuit tunnel, the air is continuously recirculated through the test section and thus should already have "momentum". Therefore the closed circuit tunnel should require less "power" to operate. The fact that the air is recirculated presents both advantages and disadvantages for the closed circuit tunnel. These will be discussed below, but "power" considerations do not necessarily dominate. This is because closed circuit tunnels must decelerate the air after leaving the test section in order to pass through the "closed loop" without significant flow losses and in order for the fan to operate in an efficient manner. Only a small amount of energy, on the order of 10-15 percent, is actually conserved, as noted by Pope and Harper[1966].

One of the primary features of the closed circuit tunnel is to provide a reservoir of air in which conditions can be controlled. The working fluid is contained in a closed region so that it can be pressurized, heated, cooled, filtered, dried, etc. depending upon the particular application. High Reynolds number can be achieved by testing in a pressurized tunnel or in a wind tunnel using a gas other than air, or a combination of the two. Pressurizing the tunnel above atmospheric conditions causes a decrease in the kinematic viscosity and an increase in the Reynolds number per unit length. The same effect can be accomplished by using a different working gas for the tunnel. For example, if freon is used at standard atmospheric conditions, the Reynolds number per unit length is increased by a factor of three. A combination of these two techniques can yield a significant increase in the Reynolds number/length. Another approach to achieving dynamic similarity in configurational testing involves cooling the working fluid in the tunnel, typcially nitrogen, to cryogenic conditions in order to reduce the viscosity and thus raise the Reynolds number. Both of these approaches require sophisticated and expensive wind tunnel facilities and introduce additional complications with respect to wind tunnel model design and data collection instrumentation.

In order to evaluate the characteristics of the indraft tunnel, it is beneficial to review some of the advantages and disadvantages of its counterpart, the closed circuit wind tunnel. A schematic of a "typical" closed circuit wind tunnel is shown in Figure 2. After the air passes through the test section in a closed circuit wind tunnel, it must be decelerated and passed through a fan or some other "energy" source, to increase its pressure. It is then turned through 360°, and reintroduced to

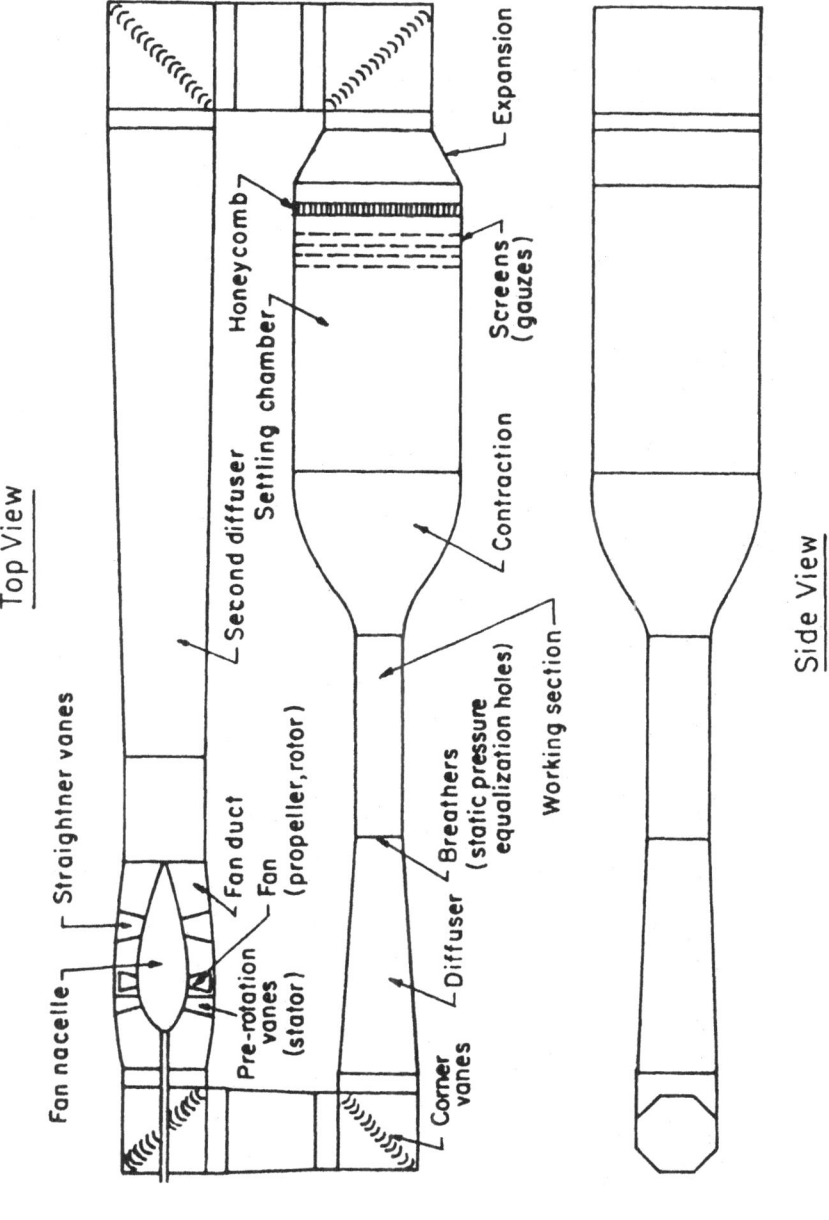

Figure 2 Typical Closed Circuit Wind Tunnel

the test section. In passing around the circuit, momentum is lost, disturbances over a large range of scales are introduced, and , as energy is added, the temperature may be increased. The flow is typically turned through a series of relatively sharp corners of approximately 90°. In most cases the corners are modified with turning vanes which serve to direct the flow around the corner and may also act as cooling vanes. One of the segments of the closed circuit will also contain the fan used to drive the tunnel. The fan usually has both fixed and rotating vanes and along with increasing the pressure in the moving stream, it introduces significant disturbances into the flow in the form of swirl and blade wakes. The air is eventually returned to a plenum or reservoir which then introduces the air to the inlet and the test section. Though there are many sources of disturbances, it is often felt that because they arise from known sources, i.e. fan, corners, etc., they can be controlled and thus the required flow quality can be achieved. It is the active and passive control of these disturbances which challenges the designer of a closed circuit tunnel.

Usually the most significant argument in favor of an indraft wind tunnel is associated with initial development and construction cost. Pope and Harper [1966] indicate that a closed circuit tunnel can cost from 60% to 100% more than a indraft tunnel of comparable size. Mort et.al. [1972] state that an open circuit tunnel can be constructed for 20 to 30% less than a closed circuit tunnel. Though the development and construction cost savings must be weighed in comparison with performance and operational costs, these statistics indicate why indraft tunnels of significant size have been developed.

2.1 Indraft Tunnel Types

There are two primary types of indraft wind tunnels and these are differentiated by the type of test section. The open jet or Eiffel tunnel, shown schematically in Figure 3, effectively develops a free jet in which the testing is performed. The test section is surrounded by an enclosure and the jet is formed as the flow leaves the inlet or contraction section. The jet is bounded by a free shear layer and thus distortion of the jet due to the presence of test model is not subjected to solid boundary interference in the test section. The second type of indraft tunnel commonly referred to as the NPL tunnel design incorporates a closed test section.

There are also advantages and disadvantages associated with the indraft wind tunnel. The main advantages of the indraft tunnel are its low construction cost and its suitability for flow visualization provided that the tunnel is exhausted to the atmosphere. The disadvantages of this type of tunnel are that it requires more energy to operate than a comparable closed circuit tunnel, and it may be susceptible to atmospheric influences which can cause disturbances in the tunnel.

2.2 Basic Indraft Wind Tunnel Components

The two basic types of indraft wind tunnels ane shown in Figures 3 and 4. Though the wind tunnel is a complete system, it is often useful to identify and examine the individual components in order to see how each contributes to the performance of the overall system. The following sections

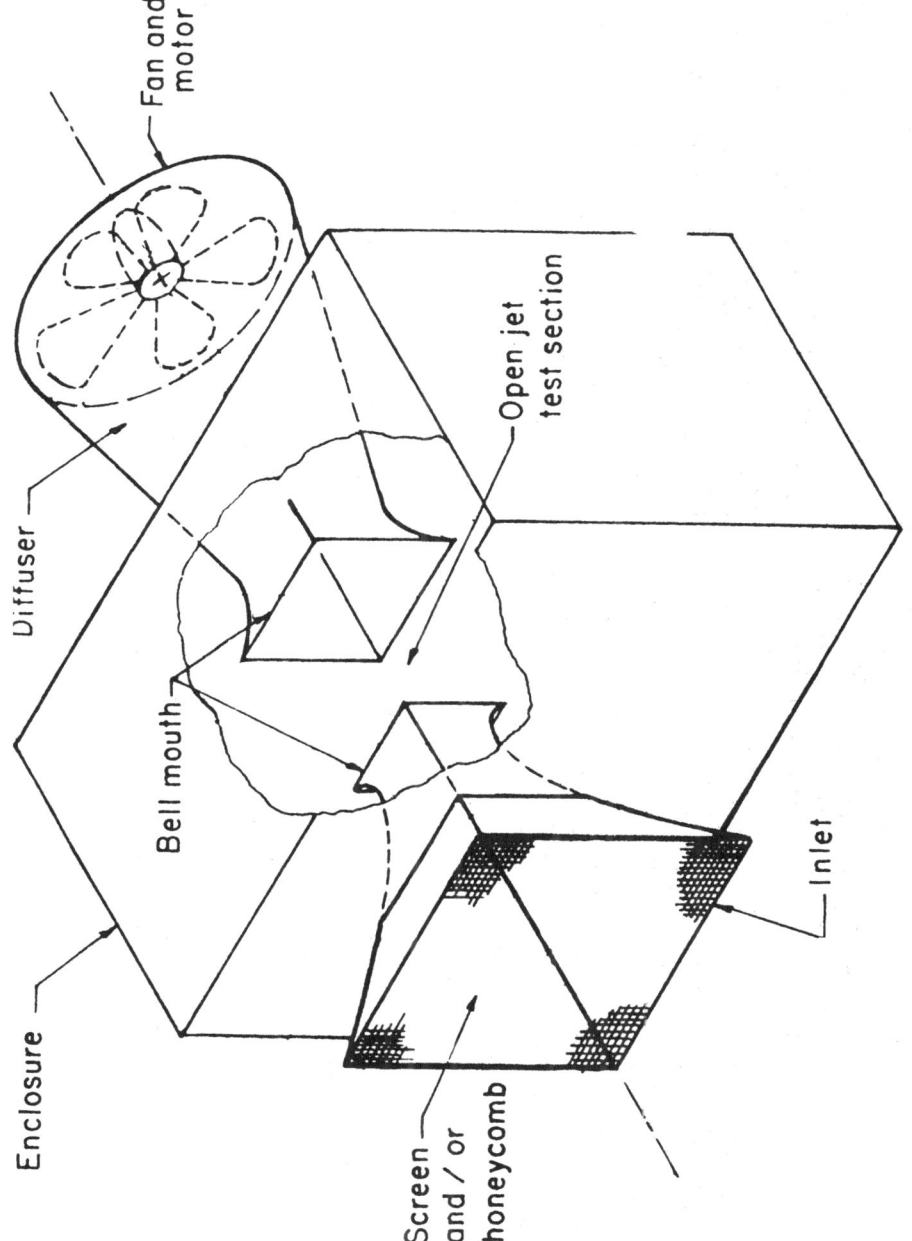

Fan and motor

Diffuser

Enclosure

Open jet test section

Bell mouth

Inlet

Screen and / or honeycomb

Figure 3 Open Jet Tunnel (Eiffel Design)

34

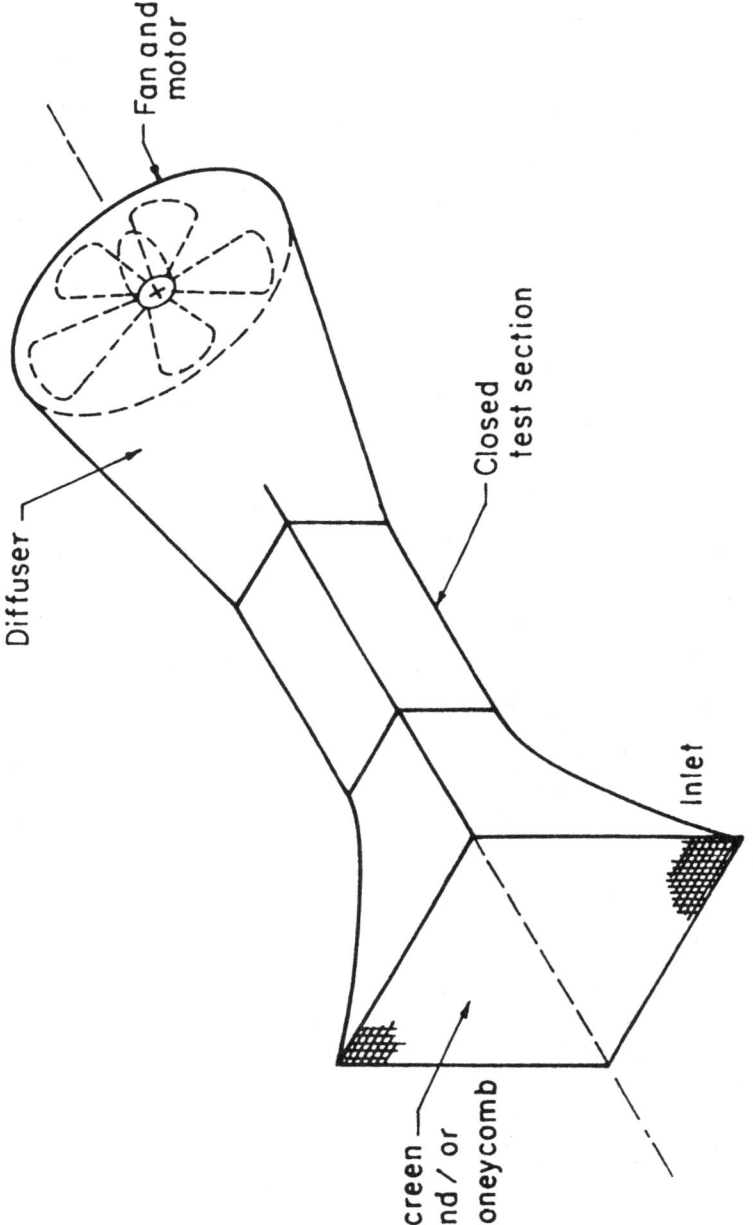

Fan and motor

Diffuser

Closed test section

Screen and / or honeycomb

Inlet

Figure 4 Closed Jet Tunnel (NPL Design)

briefly overview the role each component plays in the system performance. More detailed discussion related to the design of each component is included later in the paper.

2.2.1 *Screens and Turbulence Management*

The uniformity of the flow and the turbulence level in the test section can be influenced by placing screens, honeycomb material, or a combination of the two, upstream of the inlet. Both screens and honeycomb can be used to reduce the axial and lateral velocity fluctuations. Screens create a large pressure loss in the flow direction and are effective in producing uniform axial velocity profiles. Because of the large pressure losses associated with screens, they should be used in regions of low velocity. Multiple screens are often used and the spacing between the screens is determined by the decay of the grid turbulence produced by the screen. Typically, the space between screens should be approximately 500 wire diameters to ensure that the screen wake has decayed. Although screens are very effective in managing turbulence in the flow direction, they are not suitable for controlling swirl or lateral velocity fluctuations. Honeycomb can be used for controlling lateral velocity fluctuations with a relatively small pressure loss as compared to screens. The length of the honeycomb is usually of the order of seven to eight cell diameters. The effect of cell size and cross section on honeycomb performance has been studied extensively by Loehrke and Nagib [1972]. Because screens will collect dust and debris, they need to be cleaned periodically. Therefore, access for removing screens for cleaning must be incorporated into the design of the tunnel.

2.2.2 *Contraction Sections*

After the air passes through the section upstream of the contraction section, the flow is accelerated by the contraction section. The contraction increases the velocity in a continuous manner from the entrance to the test section. The proper aerodynamic design of the contraction section is essential to achieving quality flow in the test section. Geometry of the contraction section is important and will influence a number of the flow characterisitics in the test section. The shape of the contraction must be selected so that flow separation of the boundary layer along the walls is prevented and that secondary flows in the corners are minimized. In addition, the flow must be accelerated so that it is uniform across the test section.

2.2.3 *Test Sections*

The test section is the "heart" of the wind tunnel. It is in this region that the controlled flow must be achieved and the experiments performed. As mentioned above, there are two basic types of test sections: an open or free jet and a closed or solid wall. Additional factors affecting test section performance are tunnel access, leakage and viewing area. In order to facilitate model installation and model configurational changes, the test section is usually equipped with access doors. These doors must be designed to seal tightly to prevent flow leakage. Another important feature that may often be incorporated into the test section design is a viewing area or window for flow visualization and laser doppler anemometry measurements.

2.2.4 *Diffusers*

The role of the diffuser is to decelerate the air and recover static pressure after the air leaves the test section. There is usually a transition in cross sectional area from that of the test section to that of the fan. Factors affecting the performance of the diffuser include the area ratio and the expansion angle. To avoid steady state or intermittent separation along the diffuser walls, the expansion must be very gradual and abrupt changes in cross section are not possible. A typical "rule of thumb" limits the total included angle between opposite walls to less than 7 degrees. The length of the diffuser is set by the area ratio and building constraints.

2.2.5 *Model and Fan Protection*

In most wind tunnels some form of model protection system is incorporated into the tunnel design. For indraft tunnels the inlet screens act to protect the model from foreign object damage. In addition, a course screen is usually placed downstream of the test section to protect the fan and motor in the event that there is a model structural failure.

2.2.6 *Fan, Motor and Controller*

This system provides the power to the wind tunnel. The fan can be either axial or centrifugal. The axial fan has been the most popular, primarily because of its lower noise levels. The fan section should be connected to the diffuser with a flexible coupling to limit fan or motor vibrations from being transmitted through the diffuser walls to the test section.

2.2.7 *Exit Section*

Often this section may be eliminated, but it can play an important role in the tunnel performance. Open return wind tunnels that exhaust to the atmosphere are sensitive to external winds. The influence of the wind gusts can be minimized by proper treatment of the exhaust section, see Eckert et.al.[1976]. A vertical exit system has been found to be much less sensitive to wind gusts than a horizontal exit.

2.3 *Special Applications*

The majority of wind tunnels used by industry and government laboratories are closed circuit. Indraft tunnels are usually considered for special applications, not generalized production testing. It is therefore useful to consider some of these special applications. Each of the applications indicated below could be discussed at length but are simply introduced here in order to appreciate the breadth of applications which can be considered for indraft wind tunnels.

2.3.1 *Flow Visualization*

From the earliest studies in fluid mechanics, flow visualization has proved to be a valuable tool for the experimentalist. In low speed wind tunnels smoke is often used to mark the flow. Though a number of individuals have been responsible for the development of facilities and techniques for smoke flow visualization, many contributions were pioneered by F.N.M. Brown at the University of Notre Dame, as can be found in Mueller [1978]. The work of Brown, and the success and failures of others who have developed smoke visualization tunnels, has demonstrated the importance of the turbulence management devices and inlet design for this special class of subsonic wind tunnels. The most successful type of smoke visualization in use today utilizes the indraft subsonic wind tunnel with a high contraction ratio ($\approx$20) inlet and numerous turbulence management devices. The smoke is generated outside of the tunnel and introduced upstream of the inlet in the region of very low velocity flow. Other techniques in other types of tunnels, such as smoke-wires or smoke wands, are much more limited with respect to their applicability due to limitations on speed, interference effects and smoke quantity. In the past few years, there has been a renewed interest in the use of smoke visualization for many different aerodynamics applications. Examples can be found in Asanuma[1977], Merzkirch[1980] or Veret [1986]. This interest has brought about the design and fabrication of a number of new facilities. Most current flow visualization tunnels are relatively small in scale with test sections on the order of 610 x 610 mm (2 ft x 2 ft). There is interest in extending these techniques to significantly larger facilities as well as to much higher velocities. For these larger facilities, the designer must be much more sensitive to the tunnel design. Larger contraction ratios result in larger overall facilities and associated installation and operation costs. More importantly, the larger the tunnel, the longer the residence time of the smoke filaments in the inlets. These smoke filaments are then allowed to diffuse or be disturbed for a longer period of time and this increases the difficulty associated with introducing coherent smoke streaklines into the desired locations within the test section. The design criteria related to the inlet for a smoke visualization tunnel are, therefore, different and often more demanding than those for other types of facilities. The effects of the inlet and the turbulence management devices on an indraft tunnel are not easily separated. Although much of the discussion in this paper will focus on the inlet design, this fact must be recognized.

Due to the interest in using indraft wind tunnels for smoke flow visualization additional emphasis is placed on that application in this paper. Sections are included to describe special requirements on the test section as well as photographic equipment and procedures. The wind tunnel is a critical part of this complete experimental system.

2.3.2 *Unsteady Flows*

If a particular test requires an unsteady flow, an indraft tunnel often provides the best combination of flow quality and tunnel speed control. There are a number of ways to generate unsteady flow in an otherwise steady uniform flow tunnel. Oscillating vanes, turbulence grids,

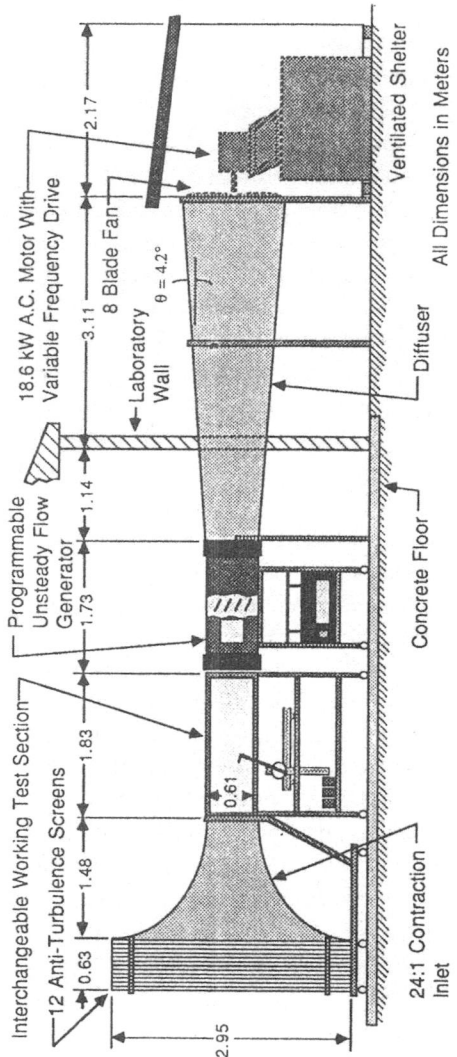

Figure 5a Unsteady Flow Tunnel
(Courtesy of T. J. Mueller and M. Brendel,
University of Notre Dame)

39

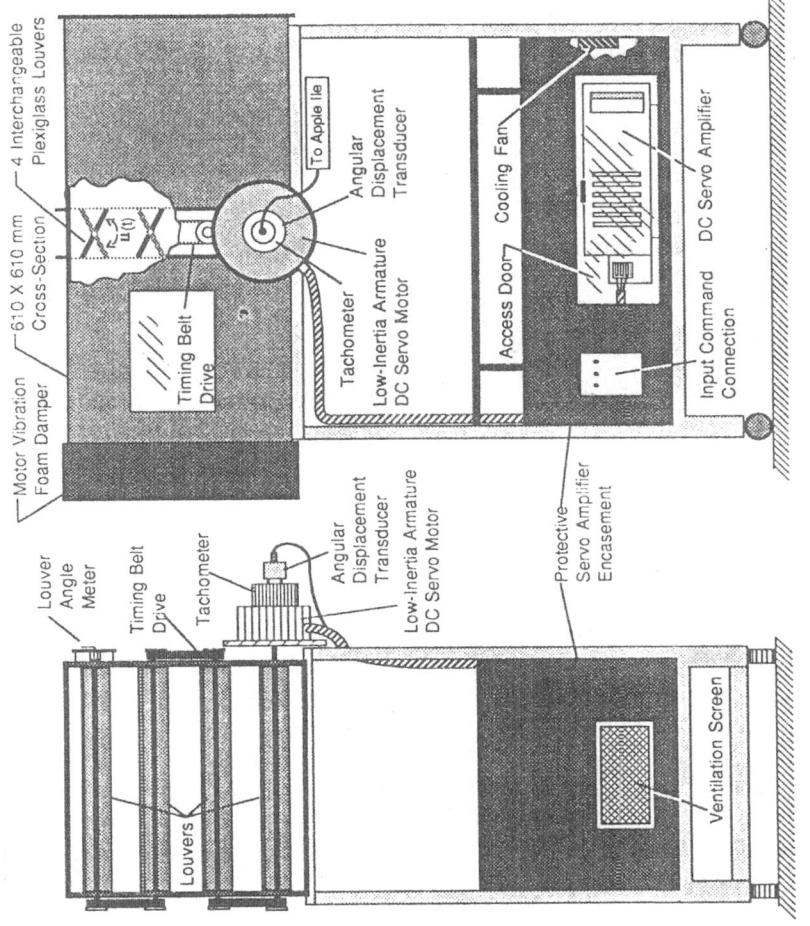

Figure 5b Details of Unsteady Flow Generator
Courtesy of T. J. Mueller and M. Brendel,
University of Notre Dame

spires, jets and surface roughness located upstream of the test section are just a few of the techniques used. Figure 5 is a sketch of an unsteady flow generator used to produce streamwise variations in the tunnel velocity from the work of Brendel and Mueller [1988}. The unsteady flow is created by controlling the pressure drop to the fan. The blockage is varied with four louvers mounted downstream of the test section. The shafts of the louvers are externally connected by a series of timing belts to a low inertia printed-circuit -armature DC servo motor. Computer control of the system permits a wide variety of unsteady flow profiles.

2.3.3 *Propulsion System Testing*

Similar to the situation which exists in flow visualization, testing of propulsion systems may require injection into the flow of the products of combustion of the propulsion system. These products of combustion cannot be reintroduced into the engine as would result with a closed circuit tunnel. Hence, the requirement of an indraft tunnel for certain types of propulsion testing is obvious. See Baals and Corliss,[1981].

2.3.4 *VSTOL*

As mentioned in the previous sections, if, during the course of a particular test, undesirable contaminates are introduced into the flow in the test section, an indraft tunnel may provide the best answer for eliminating the contaminates. One of the most undesirable contaminates is large aerodynamic disturbances which would be difficult to eliminate. This is the case for various types of VSTOL or rotorcraft types of tests. The indraft tunnel can be designed to provide an environment where the air entrained into the inlet is not influenced by the disturbances generated in the test section and passed through the diffuser and fan section. See Baals and Corliss,[1981].

2.4 *Examples of Indraft Wind Tunnels*

Rae and Pope [1984] include an extensive listing of wind tunnel facilities throughout the world and a review of this list indicates the extensive use of indraft wind tunnels. The purpose of this section is to illustrate the wide range of applications of the indraft wind tunnel concept ranging from a major government research facility to small university research wind tunnels.

2.4.1   *NASA 80'x120'- NASA Ames Research Laboratory*

The largest wind tunnel in the world is part of NASA's National Full-Scale Aerodynamic Complex. Figure 6 is a schematic of the wind tunnel complex. The tunnel can be operated as either a closed circuit tunnel having a 40' x 80' test section with a maximum speed of 345 mph or as an indraft tunnel having an 80' x 120' test section with a maximum speed of 115 mph. The indraft tunnel is operated by opening the indraft wall closure and exhaust louvers and closing the return circuit wall louvers. The 80' x 120' tunnel was designed for testing full scale VSTOL aircraft.

41

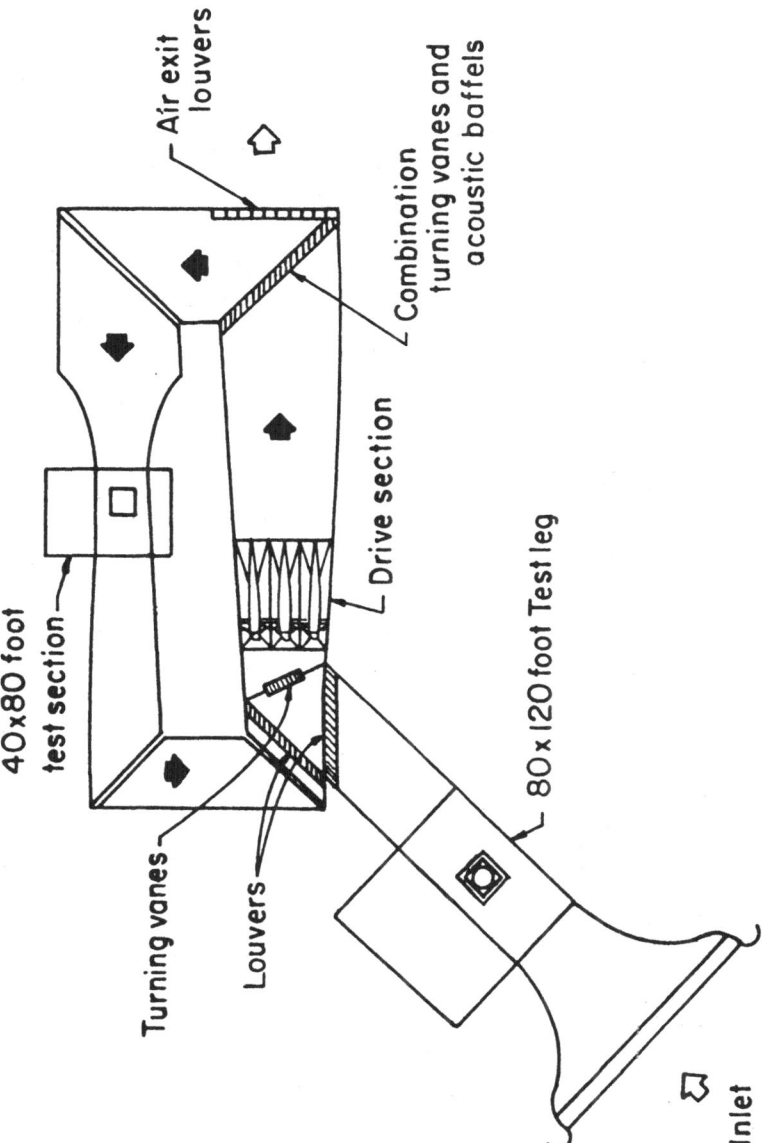

Figure 6 NASA National Full-Scale Aerodynamic Complex
NASA Ames Research Center

Extensive studies were conducted during the design of this facility in order to integrate the indraft leg of the 80' x 120' into the existing closed circuit tunnel. The orientation and inlet design were considered critical design areas.

2.4.2   Subsonic Aerodynamic Research Laboratory (SARL)- Wright Aeronautical Laboratories

The Air Force Flight Dynamics Laboratory completed the design and fabrication of a large indraft subsonic wind tunnel in 1987. The Subsonic Aerodynamics Research Laboratory (SARL) wind tunnel is shown in Figure 7. This tunnel is the largest wind tunnel to be designed with flow visualization as one of its major design objectives. The Air Force Flight Dynamics Laboratory built this tunnel to provide a large subsonic experimental facility for obtaining detailed high quality aerodynamic data for computation fluid dynamics code validation.

The tunnel consists of an entrance of honeycomb and screens, a 30:1 contraction, a 7' x 10' octogonal test section, diffuser, fan, and 20,000 hp motor. The tunnel is capable of test velocities corresponding to a Mach number of 0.6. The test section incorporates 28 windows and as a result 56% of the test section surface area is available for flow visualization viewing and laser doppler anemometry measurements.

2.4.3   Low Velocity Airflow Calibration and Research Facility - National Bureau of Standards

This wind tunnel, shown schematically in Figure 8, was designed and built in the 1970's with the primary purpose of providing accurate airflow measurements at very low velocities (Purtell and Klebanoff [1979]). A lattice of furnace filters is located in front of the entrance cone. The area ratio of the entrance cone is 2.14 to 1. The flow then enters the settling chamber which houses a honeycomb section and ten 24 mesh-per-inch stainless steel screens having a wire diameter of 0.0075 inches. Next the flow enters a 5.36 to 1 contraction followed by a twenty foot long test section. The side walls diverge from 36 inches at the beginning to 39 inches at the end of the test section, while the top and bottom walls remain parallel. A 36 inch diameter, variable pitch, twelve-bladed axial fan is directly coupled to a 30 hp DC motor. The pitch of the fan and voltage to the motor are accurately controlled by the control system.

2.4.4   Smoke Flow Visualization Tunnels - University of Notre Dame

During the past fifty years, a series of low speed indraft wind tunnels have been designed at the Aerospace Laboratory of the University of Notre Dame. The evolution of these designs resulted in the development of a very low turbulence smoke flow visualization tunnel. Figure 9 is a schematic of the Notre Dame smoke tunnel. The inlet of the tunnel consists of a set of twelve screens at the entrance of a 24:1 contraction section. The combination of screens and the large contraction ratio yields a uniform velocity profile in the test section with a turbulence intensity of less

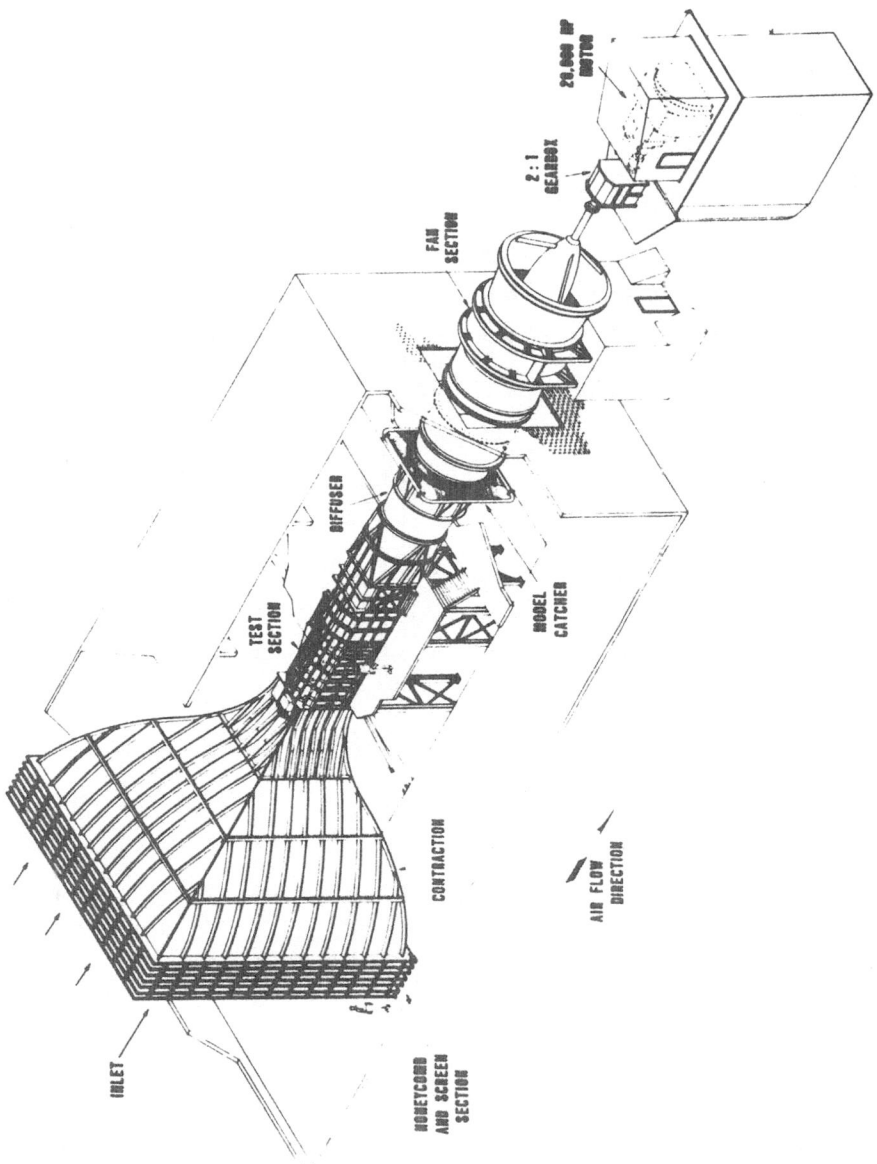

Figure 7 Wright Aeronautical Laboratories Subsonic
Aerodynamic Reserach Laboratory (SARL)

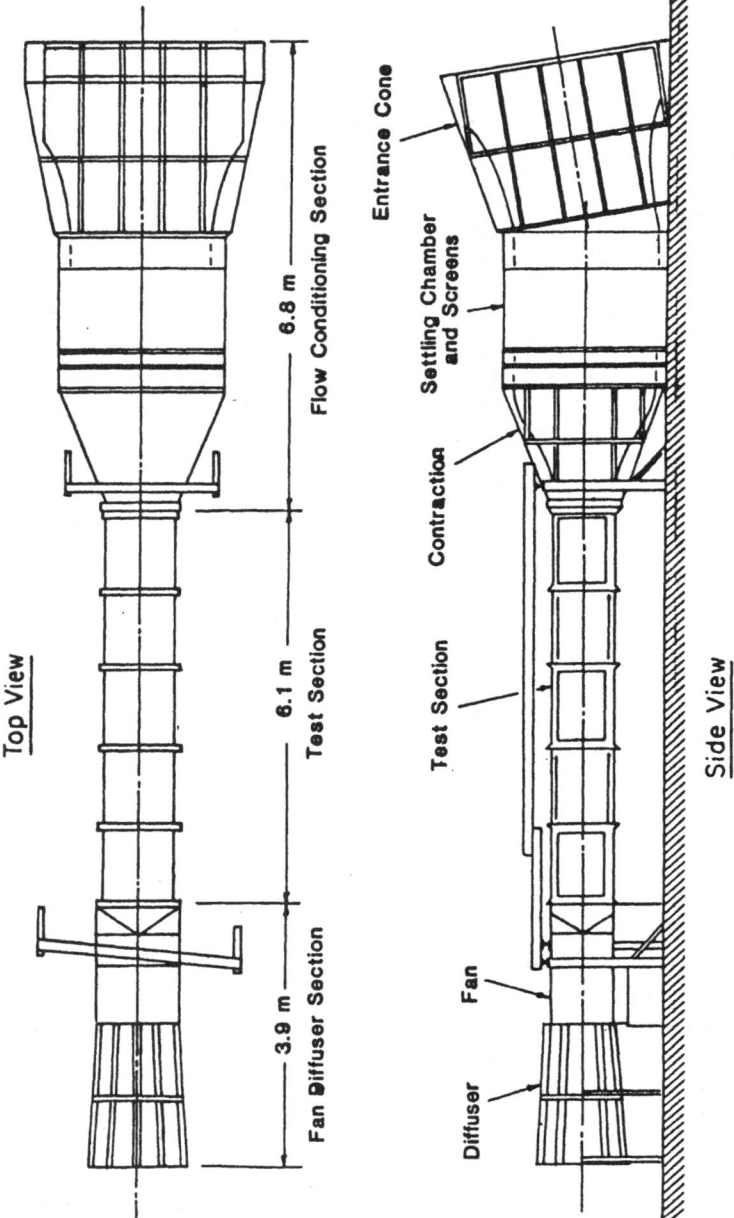

Figure 8 National Bureau of Standard Low Velocity
Airflow Calibration and Research Facility

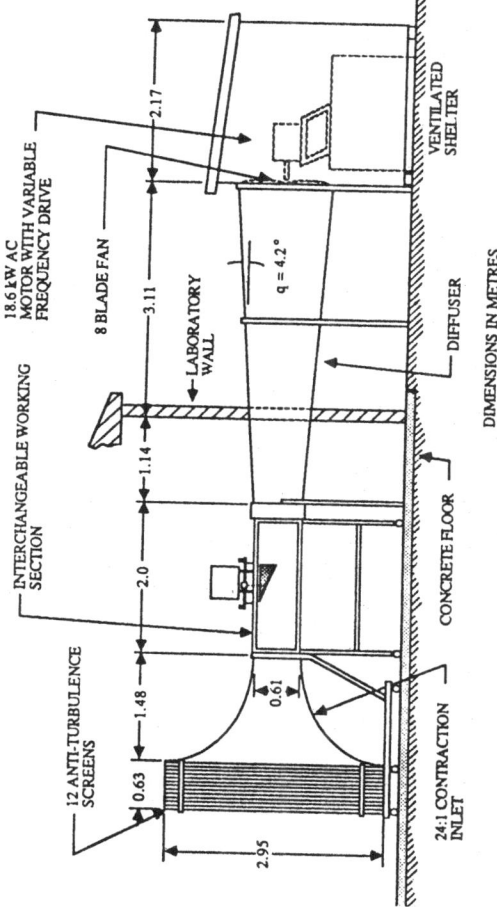

Figure 9 University of Notre Dame Low Turbulence Smoke Tunnel

than 0.1%. Another feature of these tunnels is the extremely short distance between the entrance to screens, the point where the smoke filaments are normally injected, and the test section. This reduces the smoke residence time and minimizes the smoke diffusion allowing for more coherent smoke streaklines. The tunnel test sections are interchangeable. The typical test section has dimensions of 1.83 m (6 ft) in length with a 0.61 m x 0.61 m (2 ft x 2 ft)cross section and include plate glass windows for viewing the smoke flow visualization. Downstream of the test section the flow is expanded in a 4.2 m (13.78 ft) diffuser with a 4.2 degree divergence angle. The tunnel is powered by an eight-bladed 1.2 m (3.94 ft) diameter fan coupled directly to a 18.6 kW AC induction motor located at the end of the diffuser section. The motor is controlled by an adjustable frequency AC drive. This control system allows for very accurate tunnel speed control.

3. Indraft Wind Tunnel Design

The purpose of this paper is to introduce, in some detail, a number of topics associated with the design and performance of low speed, indraft wind tunnels. As indicated above, this type of wind tunnel has played an important role in the history of aviation and aeronautical engineering and it will continue to do so. There are a number of excellent references which present detailed information for a much wider class of wind tunnels, Rae and Pope [1984] or Bradshaw and Pankhurst [1964], but it is our goal to focus on those topics particularly associated with the indraft tunnel. The discussion is intended to provide some background for anyone interested in the design of this type of wind tunnel. It is not intended to present an exhaustive treatment of design methods, nor to propose the "ideal" indraft wind tunnel. Since each wind tunnel is unique and presents different challenges, we hope only to present some of the "questions" which a potential facility designer must address and to provide some insight into problems or opportunities presented by the indraft wind tunnel.

This section of the paper is intended to introduce a number of design considerations for each major component of the indraft wind tunnel. Of particular concern will be the inlet or contraction section of the tunnel. The results of a number of recent numerical and experimental studies are included. The emphasis on the inlet section is due to the authors' experiences in this area and are not intended to indicate the lack of importance of the other components of the tunnel. Each component must be properly designed and integrated into the complete facility in order to develop a complete and effective wind tunnel system.

3.1. *General Design Considerations*

There has been surprisingly little effort expended by the aeronautical engineering community towards the aerodynamic design and component integration of one of their most valuable tools, the wind tunnel. To quote from Bradshaw and Pankhurst [1964] , "Wind tunnel design lies somewhere between an art and a science, with occasional excursions into propitiatory magic". This is not to say that most current wind tunnel facilities are not well designed, or operate in a less than satisfactory manner, but on occasion the development of these facilities has been a painful trial and error procedure.

There are two good reasons for this current state of affairs. First, the fluid dynamic phenomena occurring within the wind tunnel are quite complex. There have been numerous studies directed at inlet, fan and diffuser flow field analysis and design as well as techniques developed for turning and conditioning the flow in a wind tunnel. Each of these areas address some of the most challenging of current problems in fluid mechanics such as three-dimensional flows, unsteady and separating boundary layers and the structure and control of turbulence. These are all very sophisticated problems which will present challenges to aerodynamicists for years to come. A second, more subtle reason, is that the facility designer is often somewhat inexperienced with the many phases of tunnel design since most organizations do not regularly design and fabricate wind tunnels. The designer, therefore, must frequently rely on the experience of others and use proven methods. For these reasons, there has been little revolution, and only a minor amount of evolution, in subsonic wind tunnel design in recent years.

Much of this paper is directed toward aspects of the aerodynamic design of indraft subsonic, wind tunnels. It is not intended to present a step-by-step procedure for design but to simply introduce those features of each component which must be considered during the design process, identify important design parameters and design criteria and suggest sources for particular design methods.

It is important to realize that the wind tunnel is a complete system which is composed of individual elements which operate in a "series", and, like a chain, will only be as effective as its "weakest link". The wind tunnel must be designed as a system and the interaction between components must be considered. In this respect, there is a distinct advantage in the indraft wind tunnel. Unlike the closed circuit tunnel, where a flow separation in the diffuser will influence the fan performance, modify the flow in the return leg and eventually be "felt" in the test section, feedback through the complete flow circuit is not a direct factor in the indraft tunnel. This also presents an additional challenge. Unlike the closed circuit tunnel, where "fixes" in the return circuit can be employed to improve tunnel flow quality, Shindo. et al. [1978], the individual components of the indraft tunnel must all perform to design specification.

In the following sections, individual tunnel components are identified and design criteria and variables discussed. In most cases the discussion is somewhat general except for the wind tunnel inlet. Due to the important role the inlet plays in the indraft tunnel, particular emphasis is given to this component. Purtell and Klebanoff [1979] indicate that the "contraction is perhaps the most critical of the various subsections from the design point of view" and for this reason some of the results of a recent numerical and experimental study, Batill, Caylor and Hoffman [1983] , are presented in detail.

In order to begin the design process, the designer needs to be able to establish a set of criteria to guide the design decisions as well as a method for quantifying those criteria. Terms such as adequate, good, reasonable, etc., are typically useless in the design process. The following are general criteria that can influence the design of the entire wind tunnel system.

(1) Performance Characteristics
    a) speed range
    b) Reynolds number/units length
    c) ease of operation

(2) Size
    a) height
    b) width
    c) length

(3) Quality of flow
    a) steadiness
    b) uniformity
    c) angularity
    d) turbulence level and content
    e) susceptibility to separation
    f) boundary layer growth
    g) maintain coherent smoke streaklines

(4) Construction and Operation
    a) materials
    b) cost
    c) maintenance and repair
    d) accessibility for model changes

The geometric criteria (height, length, width) are easily quantified and upper bounds on allowable values are often set by the buildings which will house the facility. The question of "how small?" then becomes critical and will have an influence on cost and manufacturing.

The criteria associated with performance and flow quality are more complex. They are related to the size and, in some cases, can be quantified in the form of allowable velocity profiles, turbulence intensities and spectra and pressure distributions. In the following sections, design criteria for each component will be discussed as well as the design variables which can be selected in order to satisfy the required criteria.

3.2. *Site or location*

When designing a new wind tunnel, the building in which the tunnel is housed will play a major roll in defining geometric constraints that will affect the construction costs and the performance of the wind tunnel. If a new building is to be constructed as part of the project, then many issues such as environmental impact, community noise and the influence of tunnel exhaust on surrounding buildings must be addressed in determining an appropriate site for the building and tunnel. On the other hand, incorporating a new wind tunnel into an existing building has all of the above problems plus the additional constraints associated with space limitations, obstructions, existing facilities, etc. posed by the current facility. Critical factors that will influence the design are listed below.

    (1)   inlet interference
    (2)   exit interference
    (3)   overall tunnel size
    (4)   noise (generated by the tunnel and other equipment)
    (5)   ventilation (tunnel influence on heating and air conditioning systems)
    (6)   lighting (particular to flow visualization and LDV experiments)

The tunnel must be located so that obstacles such as building support columns do not interfere with the flow entering the inlet or leaving the exit. In addition, the room dimensions may place constraints on some or all of the wind tunnel components. For an indraft tunnel that exhausts to the atmosphere, the building must be modified to permit adequate ventilation.

3.3   *Turbulence management devices*

Almost all practical wind tunnels have damping screens or honeycombs upstream of the contraction to straighten the flow and reduce the scale of the turbulence. The use of screens is particularly critical in obtaining low turbulence levels in the test section. Numerous studies have been conducted to determine the effect of damping screens in reducing wind tunnel turbulence. They include a variety of experiments to examine the effect of screen wire size, mesh solidity, spacing, and positioning in the settling chamber. The early work of Dryden and Schubauer [1947] and Schubauer, Spangenberg and Klebanoff [1950] and the more recent effort of Loehrke and Nagib [1972] and Nagib, Marion and Tan-atichat [1984] have been significant in the development of our understanding of the role of screens and other devices in the management of wind tunnel turbulence and the integration of these devices into the complete wind tunnel system.

The turbulence level in the test section can have a large affect on many aerodynamic or fluid dynamic measurements. Prior to the 1940's, most wind tunnels had turbulence levels of 0.5% or greater. However, the use of screens at the National Bureau of Standards in 1934 and again in 1938 showed how effective screens could be in reducing test section turbulence levels. The National Bureau of Standards in cooperation with the National Advisory Committee for Aeronautics performed systematic investigations of screens on turbulence levels. With the use of screens, wind tunnels have been designed in which the turbulence intensity is of the order of 0.02 to 0.05 percent.

The freestream turbulence and mean velocity profile in the test section can be controlled by the use of turbulence manipulators such as screens, perforated plates, porous foam, and honeycomb. The turbulence manipulators are located in the settling chamber or in front of the tunnel inlet. The purpose of these manipulator devices is to reduce the incoming turbulence to acceptable levels within the test section. This is accomplished by transforming large scale incoming turbulence to smaller turbulent scales characteristic of the manipulator device. For example, screens smooth the incoming air by decreasing the turbulence with scales larger than the mesh size of the screen while introducing smaller scale turbulent motions. The screens are in effect transforming the scale of the turbulence. Because small scale turbulence decays more rapidly than the large scale turbulence, the turbulence level downstream of the screens is reduced. The reduction of turbulence levels by turbulence manipulators is not without its drawbacks. There is a pressure drop across the manipulator device that must be made up for by the tunnel drive system. The pressure loss across the screen is a function of the open area ratio (screen solidity) and the wire Reynolds number. In general, any of the manipulator devices should be placed in regions of the tunnel circuit where the velocity is low to avoid large pressure losses. The settling chamber of a closed circuit tunnel or in front of the inlet of an open circuit tunnel are the appropriate locations for the turbulence manipulators.

A complete discussion of the role of these devices is beyond the scope of this text. The references cited above provide both basic discussions as well as detailed results and additional references. The design and integration of turbulence management devices should be a major concern of any tunnel designer early in the design process and should not be considered as an "afterthought" or as a means of fixing inadequate tunnel performance.

3.4   Inlet or Contraction Section

The purpose of the wind tunnel inlet is to align and accelerate the fluid into the test section so that proper and controllable test conditions can be generated. The contraction section must be considered in conjunction with any flow conditioning devices such as screens or honeycomb. Both the conditioning devices and the area reduction associated with the contraction have significant influence on the flow within the wind tunnel. The management of spatial and temporal irregularities in the flow field, both small scale (turbulence) and large scale (gust, wind or swirl) and the elimination of such irregularities are influenced, in part, by the inlet design (Comte-Bellot and Corrsin[1966]).

The function of the inlet in each type of tunnel is somewhat different depending upon the tunnel type. In the indraft tunnel , the air is entrained from the region around the inlet and therefore, can be influenced by wind, objects, motion, etc. in the vicinity of the entrance to the inlet. Some of these factors can be controlled and some cannot. The inlet serves the purpose of aligning and accelerating the air and it must be able to cope with various types of upstream conditions. This is often accomplished through the use of various flow management devices but the aerodynamics of the inlet, the surrounding fluid and the flow management devices are closely coupled. The contraction ratios on this type of tunnel are usually rather high, between 10 and 30, with some recent designs using contractions as high as 150 (Batill, Nelson and Mueller [1981]).

The inlet on the blowdown tunnel (i.e., an open circuit tunnel with the fan upstream of the inlet and test section) and on the closed circuit tunnel encounters different inflow conditions. The disturbances created by the fan, diffuser, turning vanes and all other hardware that is upstream of the inlet are present at the entrance to the inlet. The inlet is not intended to eliminate these disturbances and, in certain cases, can actually amplify their influence. There are numerous studies associated with the influence of area contractions on the development or decay of grid or isotropic turbulence (Klien and Ramjee [1973]), but in most cases these have dealt with situations in which the scale of the turbulence was much smaller than the scale of the region of the flow. The effect of the inlet on the large scale disturbances is still a matter for future research.

Inlet design on all subsonic facilities must be approached from the point of view that a well designed inlet will not compensate for problems associated with other sections of the tunnel, but an inadequately designed inlet can create significant problems and degrade the tunnel performance. The influence of the inlet design on overall tunnel performance seems to vary with both the type of facility and with personal experience. Opinions range from those who believe that almost any inlet contour which is "reasonable" will perform in an adequate fashion, (and obviously the term reasonable is based on experience) to those cited earlier which indicate the inlet may be the most important single component.

Prior to a more detailed discussion of some aspects of the physics of the flow in wind tunnel contraction sections, it is useful to consider how the inlet design influences or is influenced by the overall facility design. The tunnel design will most likely start with decisions relative to the type of testing and the size of the test section which will be required for certain types of models. Since subsonic tunnels often deal with scale model flight vehicles, the matching of important scaling parameters, such as Reynolds number, will influence the size and maximum speeds required in the test section. The selection of tunnel type, either open or closed circuit, is often dictated by the type of testing to be conducted, power requirements and, quite often, by the overall size of the facility. It is not unlikely for the total length and width of a wind tunnel to be an order of magnitude greater than the respective length, width and height of the test section. It is at this point that the inlet begins to influence the basic design of a tunnel. The selection of the inlet contraction ratio, where bigger is often considered better, will significantly influence the overall size of the tunnel as well as set performance parameters, such as allowable velocities in the plenum or stilling chambers, and subsequent power requirements. The remaining inlet parameters which must then be determined are length, cross-sectional geometry and finally, wall contour. How does the selection of these parameters influence the tunnel performance? Before this can be determined, one needs to establish what inlet parameters influence the wind tunnel performance characteristics and which of the inlet parameters are most important.

The primary function of the inlet is to modify the flow velocity within the tunnel through a reduction in cross-sectional area. In a subsonic wind tunnel inlet, the influence of the contraction is felt both upstream and downstream of the inlet. Consider the case of a tunnel with an extended plenum of constant cross-sectional area upstream of the inlet. If the plenum is very long (neglecting the development of the wall boundary layers), upstream of the inlet there would exist a uniform, low velocity flow. It is in this region where screens and honeycomb are located so that low velocities are required to maintain small pressure drops across the screens and thus minimize power requirements. Since the flow is subsonic, the fluid in the center of the section "senses" the upcoming area reduction associated with the inlet and begins to accelerate. In order to satisfy conservation of mass, the flow near the wall decelerates in the constant or even mildly decreasing area region near the entrance to the inlet. The resulting decrease in velocity brings about an adverse pressure gradient along the walls and particularly in the corners of the inlet. The boundary layer developing along the walls is then subjected to an adverse pressure gradient and, depending upon the velocity distribution in the boundary layer and the strength of the adverse pressure gradient, separation may occur. The separation can result in unsteady free shear layers and recirculating regions which can effectively alter the geometry of the inlet. This type of inlet separation can be intermittent and thus bring about significant unsteadiness in the test section.

As the flow continues to accelerate into the inlet, the fluid near the wall encounters large convex curvature and, near the exit plane of the inlet (i.e. the entrance to the test section), the velocity along the wall can become greater than the mean exit velocity. The flow near the wall must then be decelerated as it enters the test section which brings about another possible region of adverse pressure gradient and provides another opportunity for boundary layer separation. These problems are increased in three-dimensional inlets where the streamwise curvature in the corner region is greater than the curvature along either of the walls themselves. The magnitude of the adverse pressure gradients can be reduced by lengthening the contraction but this leads to increased boundary layer growth, reduced boundary layer stability, and increased tunnel size and cost.

The nonuniformity of the velocity near the entrance plane of the inlet also creates difficulty with respect to the design and function of the screens and honeycomb upstream of the inlet. Nonuniformity at the exit plane influences the length of the test section, and possibly, the allowable model locations. The inlet is often characterized by a length to height (L/H) or a length to effective diameter (L/D) ratio. This ratio has shown to have an influence on the turbulence reduction associated with both the screens and the inlet. Some of the other problems which must be considered when designing the inlet are:

(1) influence on the acoustic environment within the tunnel;

(2) the development of longitudinal (Goertler) vortices within the regions of concave curvature (see Lin [1955]);

(3) the influence of changing cross-sectional geometry (i.e. circular to rectangular, square to hexagonal, etc.) on the wall boundary layer.

Each of these will be influenced by the inlet design. Currently there is little information available which can be directly applied by the facility designer, whereby the influence of these effects on the performance of a particular facility can be predicted. Generally, the typical situation is one in which the designer "hopes" such effects will not be significant and, if problems arise subsequent to the construction of the facility, he must then resort to a number of traditional fixes.

Many different techniques have been used to design the inlets for the subsonic tunnels in use today (Varner et.al.[1982]). These techniques range from "sketching" a smooth curve to detailed three-dimensional flow field calculations and extensive model and prototype testing. There appears to be no consensus as to the "best" methods available. A brief overview of some of the inlet analytic techniques will be discussed later, but it appears that most often the results of a particular analysis procedure are tempered by the experience of the wind tunnel designer.

The following are the list of possible areas for which design criteria for an inlet can be established:

(1) Size
    a) height
    b) width
    c) length
    d) wall geometry
    e) cross section shape

(2) Performance
    a) steadiness
    b) uniformity
    c) angularity
    d) turbulence level and content
    e) susceptibility to separation
    f) boundary layer growth
    g) ability to maintain coherent smoke streaklines

The geometric criteria (height, length, width) are easily quantified and are influence by the overall wind tunnel facility size. The question of size again becomes critical since it will have an influence on cost and manufacturing. Some of the other design criteria related to the geometry of the inlet may be based on an existing wind tunnel for which an new inlet would be required.

The criteria associated with flow quality are more difficult to deal with and quantify. They are related to the tunnel performance and, in some cases, can be quantified in the form of velocity profiles, turbulence intensities and spectra and pressure distributions. There are actually only a few design parameters which the designer can control in order to meet these design criteria. They are:

(1) contraction ratio;
(2) cross-sectional geometry;
(3) length;
(4) wall contour;
(5) surface type (i.e., a polished/smooth/rough, etc.)

With so few design parameters to select, it appears as if it should be a rather straightforward problem. Unfortunately there may not exist a direct correspondence between each of the design parameters and the design criteria.

Two design methods which have been used in the past involve a direct and an indirect approach. The indirect approach centers on the definition of certain flow field parameters, most commonly the axial velocity distribution, and an attempt is made to determine the wall contour required to yield that velocity field. Depending upon the way in which the velocity field is defined there may or may not be a contour which provides satisfactory results. The direct approach involves the selection of a contour and then the subsequent determination of the flow field parameters. Such a method could be repeated until an acceptable design is achieved with respect to the selected criteria. A problem occurs in that there is an infinite number of possible cross-sectional shapes and wall contours which could describe the geometry of the walls between the entrance and exit of the inlet. The obvious solution is to select a particular family of wall contours, for which there are a reasonable number of parameters which describe the contour and then limit the inlet design to this family. This is the procedure described in this paper. It is therefore limited by the suitability of the selected family of contours as well as by the ability to include and quantify all the appropriate design criteria.

The following sections present some additional detailed information related to inlet design. A discussion of the development of inlet design methods is included along with some of the results of an experimental study on a high contraction ratio indraft tunnel inlet. A series of design charts are presented which have been developed from the numerical solution to the steady, inviscid flow in a three dimensional inlet. Only a sample of the more complete set of charts provided in Batill, Caylor and Hoffman [1983] are included in this paper, but they illustrate a procedure for inlet design.

3.4.1. *Background - Inlet Design*

The aerodynamic design of wind tunnel contraction sections has evolved over the years as somewhat of an inexact science. Around 1940, the first serious attempts were made at developing methods for the design of wind tunnel inlets. Much of the work at that time focused on solving the equation for incompressible, inviscid, and irrotational flow (i.e., the two-dimensional or axisymmetric form of the Laplace equation). The general solution method was to specify a centerline velocity distribution in terms of axial distance along the inlet and then find a solution for the stream function in series form. Tsien [1943] published work based on this design philosophy. He proposed the design of an axisymmetric contraction with a monotonically increasing axial velocity. Szczeniowski [1943] solved the Stokes-Beltrami equation using a different axial velocity distribution. The resulting streamlines were expressed in terms of Bessel functions. In a similar method, Batchelor and Shaw [1944] examined the theoretical flow through an axisymmetric contraction by using a relaxation method for solving the governing differential equation. They were particularly interested in minimizing the adverse pressure gradients which occur in the inlet and exit regions of the inlet.

At about this same time Smith and Wang [1944] proposed an interesting technique for the design of a two-dimensional or axisymmetric contraction cone with a high degree of exit flow uniformity. They made use of the analogy between the magnetic field that is created by two coaxial and parallel coils carrying electric current and the velocity field that is created by two corresponding ring vortices. Several of the inlets that are in current use at the Notre Dame Aerospace Laboratory were designed using Smith and Wang's technique. These large, three-dimensional inlets were constructed by joining together walls with the same two-dimensional wall contour. These inlets have proven very successful in providing the test section flow quality necessary for smoke flow visualization, although there are significant problems associated with using smoke near walls or in corner regions of these tunnels.

Very few papers on wind tunnel contraction design were written in the ensuing years. It was not until the mid-70's that interest in contraction design was renewed. Recently, several independent pieces of work have been published on the problem of inlet design with the aid of large scale computing machinery. These more recent works have attempted to overcome some of the deficiencies of the earlier work by adding more practical constraints to the inlet flow solutions. Many of the earlier design methods assumed potential flow in contractions of infinite length and gave little attention to real flow phenomena such as boundary layer growth along the inlet walls.

In one of the more recent works, Chmielewski [1974] specified a streamwise acceleration distribution. He used two parameters to choose the shortest contraction that avoided boundary layer separation. His two-dimensional study carried contraction design methodology beyond that of previous investigations by including a quantitative consideration of boundary layer behavior. Borger [1976] used a polynomial of the fifth-degree to describe the contraction contour. The coefficients of the polynomial were chosen to produce the minimum length contraction that avoids boundary layer separation at both the inlet and exit and provides uniform flow at the exit plane. Borger's design calls for a slight expansion at the exit to improve exit flow uniformity. The work of Mikhail and Rainbird [1978] was based on the hypothesis that it is possible to control the wall pressure development and the flow nonuniformity by controlling the wall curvature distribution. By optimizing this distribution, a short contraction can be used that keeps both the adverse pressure gradients and flow nonuniformity within tolerable ranges. They considered the optimum contraction to be the shortest one that satisfied specified flow quality requirements in the test section.

The papers by Chmielewski [1974], Borger [1976], and Mikhail and Rainbird [1978] have presented more reasonable requirements on exit flow uniformity and the influence of the boundary layer on the inlet flow field. Two of the more significant of the later works are those of Morel [1975,1976]. Morel also viewed design as a search for the optimum wall shape leading to the minimum inlet length required for a given purpose. He formulated a set of design criteria to judge flow qualities. Morel felt, like most other designers, that obtaining exit flow uniformity and avoiding separation were the two primary goals of the contraction design. In addition, he specified minimum contraction length and minimum exit boundary layer thickness as secondary goals to be satisfied. This set of criteria was used by Morel to develop his practical design technique.

Morel's two-dimensional and axisymmetric design methods are based on the results of numerical solutions which were incorporated into a set of design charts that could be used in actual design studies. The contraction contours are formed by two cubic arcs joined at a common point called the match point. This match point is the single parameter used in defining Morel's family of wall shapes. The design charts are used to determine the required contraction length and the position of the match point in terms of allowable pressure coefficients used to indicate the susceptibility of the inlet to boundary layer separation. These pressure coefficients are determined by boundary layer and flow uniformity requirements. The match point, the contraction ratio, and the contraction inlet height are the three values needed to completely define the geometry of one of the matched cubic inlets.

All of the methods that have been discussed have dealt with the design of two-dimensional or axisymmetric contraction cones. In reality, however, most practical inlets are three-dimensional with square or rectangular cross-sections. Expanding a two-dimensional design technique to account for three-dimensional flow effects in wind tunnel contractions is not a trivial task, even with today's advanced computing capabilities. The most significant problems in three-dimensional inlets with non-circular cross-sections occur in the corner regions. In contractions of polygonal cross-sections, severe secondary flows can exist in the boundary layer near the corners. Even if such crossflows can be avoided, as in a "two-dimensional" contraction, the boundary layers in corner regions will be more susceptible to separation than the boundary layers near planes of symmetry. A additional complexity is the definition of the upstream and downstream boundary conditions.

To compromise between the difficulties of the construction of axisymmetric inlets and the undesirable boundary layer effects in the corners of rectangular contractions, some contractions have been built with octagonal cross-sections (Cohen and Ritchie [1962]). For design purposes, it was assumed that both the potential flow and the boundary layer behave as in a contraction of circular cross-section. This design philosophy, however, is only a compromise. What is needed is a methodology for the design of three-dimensional contractions, regardless of cross-sectional shape.

In order to develop an aerodynamic design of an inlet for a specific wind tunnel application, one would like to be able to predict the performance of the inlet and be able to evaluate that performance against a set of defined criteria. There are a number of current aerodynamic prediction methods which appear suitable for fully three-dimensional, subsonic wind tunnel inlets. Two such methods were evaluated and are discussed in detail in Batill, Caylor and Hoffman[1983]. In this study, computer programs were developed using both a source panel, distributed singularity approach and a finite difference potential flow solution of Laplace's equation for inlets of arbitrary rectangular cross-sections. Both methods were considered in some detail since each presented certain promise. The finite difference procedure was eventually selected for use in the development of the design data which will be discussed later. It should be noted that these methods are not new. Their applicability to this type of flow field has been demonstrated in the past, but they have not been used to systematically study the parameters which influence the inlet flow fields.

As a parallel effort with the analytic program development in Batill, Caylor and Hoffman[1983], an experimental program was pursued which was intended to provide both improved understanding for the physics of the inlet flow field and a benchmark for evaluation of the analysis procedures. Unlike in the analysis, where the inlet can be effectively "isolated" from the performance of the remainder of the tunnel components, the interaction between the inlet and turbulence screens became important in the experimental study.

Surprisingly, little experimental work has been done to evaluate existing inlet designs. A few papers have been published which discuss experimental work carried out to examine individual aspects of contraction design. Most of these have focused on the effect of the inlet on the turbulence level in the test section. Uberoi [1956] examined the effect of contraction ratio on isotropic turbulence by conducting experiments in three square contractions of different contraction ratios. Klien and Ramjee [1973] studied the effect of contraction geometry on non-isotropic turbulence by using eight circular nozzles of different geometries all with the same contraction ratio. Ribner and Tucker [1953] employed a spectrum concept to study the selective effect of the contraction on the components of turbulent velocity fluctuations.

The work done to explore various effects on wind tunnel turbulence has provided valuable results. The turbulence in an inlet, however, is just one aspect to be considered and researched. Very little experimentation has been done to tie together all the important aspects of a contraction design, i.e., pressure gradients, exit flow quality, boundary layer behavior, turbulence levels, etc. Also, few experiments have been performed to correlate real inlet flows with theoretical predictions.

Many design techniques have been proposed in an effort to obtain a high quality flow in the test section. A good inlet design should consider all of the following performance and physical specifications:

(1)   a high degree of exit flow uniformity,
(2)   no flow separation or unsteadiness,
(3)   minimal boundary layer growth,
(4)   reduction in turbulence intensity, and
(5)   shortest possible contraction length.

In reality, an "ideal" inlet which meets all of the above specifications is difficult, if not impossible, to achieve. Often trade-offs and compromises must be made in order to design a satisfactory contraction for a given application.

Many early design concepts were concerned primarily with obtaining an axially aligned, uniform flow at the contraction exit plane. This was, and still is, the most important function of the contraction section. Aerodynamic measurements in the test section could be adversely affected if the oncoming airstream is misaligned or possesses a non-uniform velocity distribution.

Later design work was also concerned with the presence of adverse pressure gradients and the possibility of boundary layer separation. Minimizing the boundary layer growth along the inlet walls is advantageous for two reasons. First, the smaller the boundary layer thickness the less likely are the chances of separation in regions of adverse pressure gradients. Second, the boundary layer in the contraction continues to grow as it enters the tunnel working section. The presence of a relatively large boundary layer can effectively modify the geometry of the inlet and alter the exit plane velocity distribution. Also, the boundary layer can affect measurements in the test section by directly interfering with the model, pitot tube, hot-wire probe, etc.

Early NACA work in various wind tunnels demonstrated that turbulence could also affect measurements on airfoil models (Dryden and Abbot [1949]). Recently, research has been conducted by Mueller and Pohlen [1983] to study the effects of turbulence on low Reynolds number airfoil performance in wind tunnels. These sets of experiments have indicated the need to reduce the amount of turbulent fluctuations in the airstream. Ideally, the flow in the test section should be practically turbulence free in order to simulate the conditions of atmospheric flight.

The wind tunnel contraction section serves to modify the turbulence levels in the flow in the test section. The degree of turbulence in the air flow is specified by the turbulence intensity - the ratio of the fluctuating velocity component to the mean velocity. The turbulence intensity is reduced as it passes through the length of the contraction. The length of the inlet serves to decrease the turbulence levels in the test section by providing a finite distance over which the turbulence is allowed to decay. The area reduction of a contraction tends to stretch the vortex filaments associated with turbulence. Several experimental studies have shown that contractions exhibit a selective effect on the three components of turbulent fluctuations. Prandtl was the first to point out that a sharp decrease in the cross-sectional area of a pipe with a consequent increase in the mean speed of the flow tends to smooth out flow irregularities. Thus, the contraction reduces the turbulence intensity level by increasing the mean flow speed. The scale of the initial turbulence can be reduced through the use of damping screens upstream of the inlet. Smaller scale fluctuations tend to decay more quickly than larger ones, Ramjee and Hussain [1976].

3.4.2   *Some Experimental Results*

A series of experiments were conducted by Batill, Caylor and Hoffman [1983] in order to evaluate the performance characteristics of a number of indraft wind tunnel inlets designed using Morel's two dimensional technique. Morel's technique was chosen because two inlets designed earlier by his method performed satisfactorily. The two inlets, a 75:1 contraction and a 150:1 contraction, were built for use with a transonic smoke flow visualization wind tunnel (Batill, Nelson and Mueller [1981]). Such radically high contraction ratios were necessary to allow for smoke flow visualization in the test section and to avoid excessive pressure losses through the screens. Morel's method was chosen because his design charts made the method relatively simple to use in practical design applications. Morel's results had to be extrapolated to accommodate the higher contraction ratios, though. The design concept was expanded to the three-dimensional case by matching four contraction walls of the same contour. The resulting inlets each had square cross-sections. Using a similar approach, a contraction ratio of 30:1 inlet was built specifically for this experimental investigation. This large contraction ratio is similar to that currently in use in the SARL facility at Wright-Patterson Air Force Base, Ohio.

Several aspects of contraction performance were examined and are briefly reviewed in the following sections. In particular, the 30:1 contraction ratio inlet design was evaluated in terms of:

    (1)   exit flow uniformity,

    (2)   adverse pressure gradients and wall separation,

    (3)   effect of the contraction on turbulence,

    (4)   boundary layer behavior and

    (5)   the influence of damping screens on inlet flow quality.

The experiments were conducted using a single inlet designed using Morel's two-dimensional approach of having a matched cubic wall geometry. A schematic of a general three dimensional wind tunnel inlet is shown in Figure 10 along with some of the notation used. The inlet was designed using conservative values for the inlet and exit wall pressure coefficients. The contraction was 56.6" long and the inlet and exit heights were 47.0" and 8.58" respectively. The match point was located at X = 0.71 (i.e. 40.2 inches from the entrance plane). The wall contours for the matched cubic wall geometry are given by:

$$H = H_e + (H_i - H_e) [1 - x^3 /(X^2 L^3)] \qquad \text{for } x/L \le X$$

$$H = H_e + (H_i - H_e) (1 - x/L)^3 / (1 - X)^2 \qquad \text{for } x/L \ge X$$

where $\qquad\qquad X = x_m / L$

x is measured along the centerline of the inlet and x_m is the location of the "match point". The top and bottom of the contraction were constructed from sheets of 1/8" masonite and the sides were formed from clear sheets of Lexan. The Lexan was flexible enough to form the wall contour and also possessed the optical quality needed for flow visualization work. The structure was supported externally so as not to interfere with the flow in the interior of the inlet. Particular care was taken to allow viewing of the corner regions. A special test section and diffuser unit was built to make the inlet compatible with existing wind tunnel facilities. The wall contour of the 30:1 inlet is shown in Figure 11 and includes the location of various measurement points. The damping screens were made of two types of mesh. Five screens were made of a multistrand nylon mesh grid with a 0.002" diameter strand in a 26 per inch grid. The six screens attached upstream of the nylon screens were made of an aluminum mesh with a 0.009" diameter aluminum wire in an 18 per inch grid. The screens were mounted in two different ways as shown schematically in Figure 12. Much of the

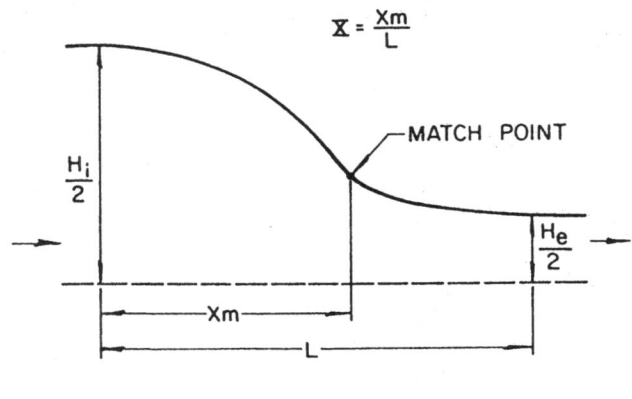

$$\overline{X} = \frac{Xm}{L}$$

a.

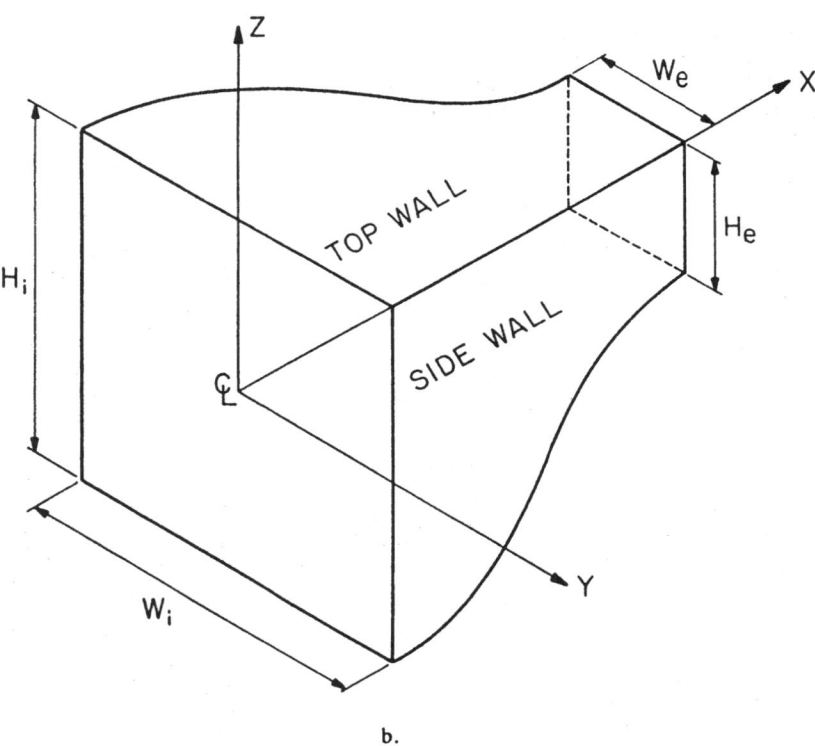

b.

Figure 10. Schematic of Three-Dimensional Wind Tunnel Geometry

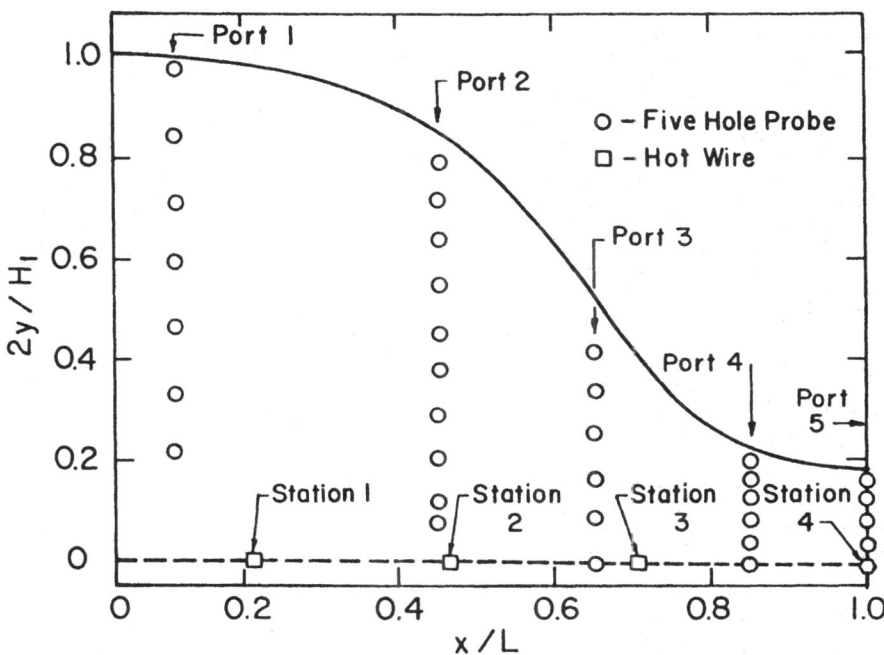

Figure 11. Wall Contour of 30:1 Contraction Ratio Inlet and Measurement Locations

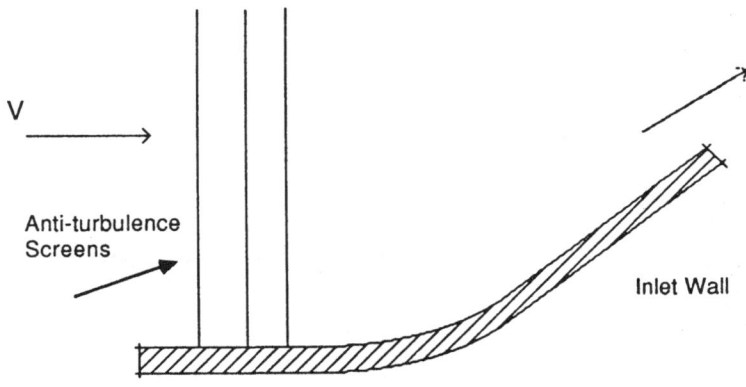

V

Anti-turbulence
Screens

Inlet Wall

a. Without Wall Discontinuity at Entrance Plane

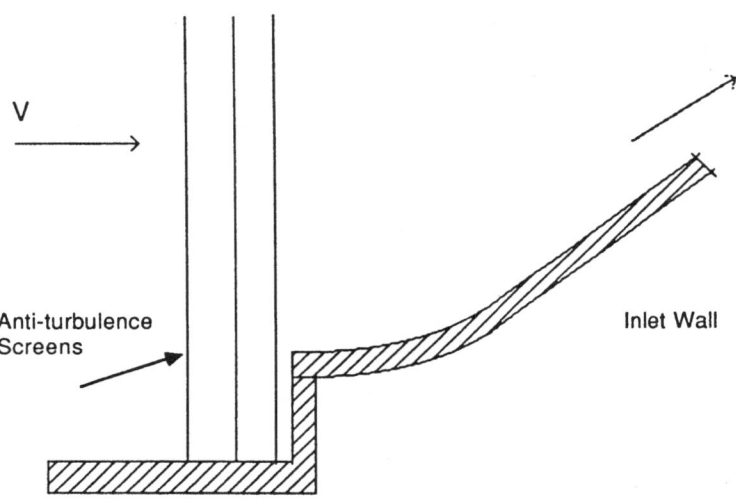

V

Anti-turbulence
Screens

Inlet Wall

b. With Wall Discontinuity at Entrance Plane

Figure 12. Wall Details Near Inlet Entrance Plane

testing was conducted in order to compare the characteristics of the inlet with no screens present to the case when all eleven screens were present and the influence of the screen placement.

Measurements of steady surface pressure distrubutions, inlet flow field velocities and free stream turbulence were conducted. The mean flow field velocities were measured using a five-hole pressure probe. The probe was calibrated to a range of ±30° included angle between the probe axis and the flow velocity and proved to be accurate to within ±0.5°. The turbulence measurements were conducted along the inlet centerline using a single wire probe.

The five-hole probe data at the exit plane, Figure 13, indicated that the contraction exit flow was uniform and aligned for the 11 screen case. The pitch and yaw angles at all points in the profile were within ± 1 degree of 0 degrees. The precision of these results corresponds well with the accuracy level of the five-hole probe system. In Figure 13a the velocity is shown to decrease for the last data point near the wall. A velocity profile taken with the hot-wire, however, did not indicate this decrease, even at points closer to the wall. The discrepancy in the results was probably caused by boundary interference on the measured probe pressures. If the probe is placed within 5-probe head diameters of a wall, the calculated velocities could be as much as 4% low. The data point in question in Figure 13a was located 0.3 inches from the wall; this is just 2.4 probe diameters. Surface pressure measurements at the wall indicated a non-uniform lateral velocity distribution at $x/L = 0.95$. At this axial location the wall velocity varied by 2.5% from the centerline to the port one-inch from the corner. The lateral velocity profile suggests that a non-uniform wall velocity distribution may exist at the contraction exit plane.

In a two- or three-dimensional contraction, regions of adverse pressure gradient will occur along the wall at both the inlet and exit of the contraction. The wall pressure measurements indicate that the damping screen configuration directly affects the wall pressure gradient in the upstream portion of the contraction. For the optimum screen configuration (all 11 screens flush with the inlet lip) no adverse gradient was detectable. The sharpest negative velocity gradient occurred for the case where there was the "forward facing step" discontinuity between the most downstream screen and the entrance to the inlet. For this case the wall velocity slowed by almost 70% over a distance about equal to 0.23L. The longest adverse gradient occurred when no damping screens were used; in this case a negative velocity gradient existed over a distance of about 0.3L.

Pressure measurements in the inlet showed no evidence of an adverse pressure gradient along the contraction walls near the exit plane. Since the constant area exit region of the inlet is very short, an adverse pressure gradient would also take place over a very short region. The inability to detect the adverse gradients in the inlet was due to an insufficient number of static ports near the exit. The external support structure of the inlet made it difficult to instrument the inlet with extra pressure taps. Separation along the walls of a contraction could occur if a positive pressure gradient is severe enough. Smoke flow visualization was used help identify regions of separated flow in the inlet. With the screens flush, no significant separation was observed. When the configuration with the wall discontinuity at the entrance to the inlet was used, however, a separation

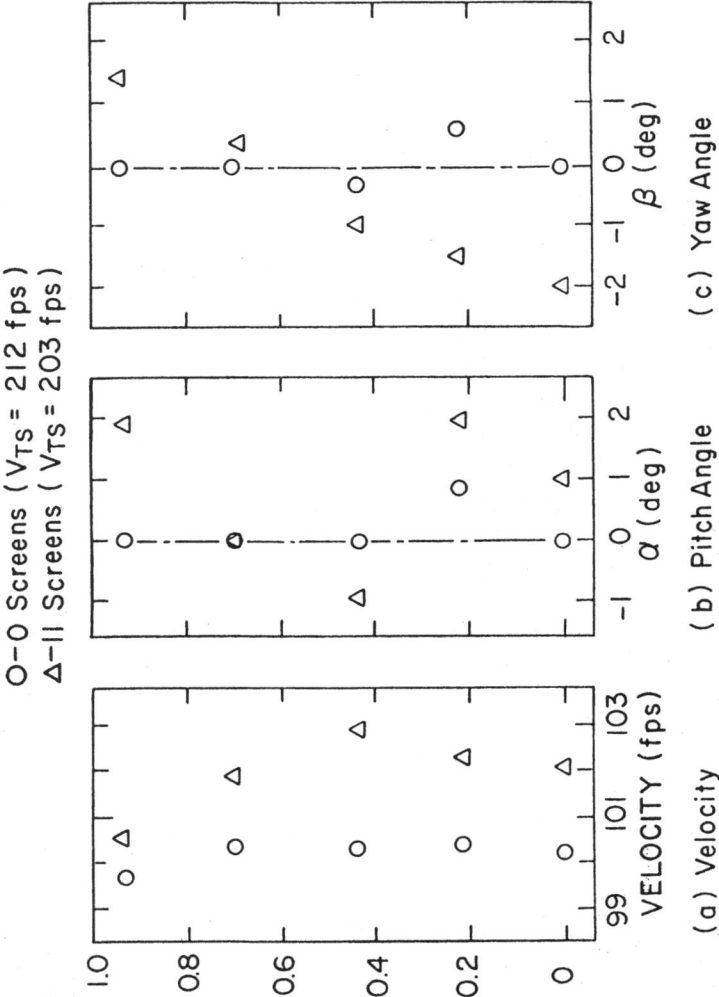

Figure 13. Five-Hole Pressure Probe Measurements at Inlet Exit Plane

bubble near the inlet lip was clearly evident. This phenomena is pictured in Figure 14. A positive pressure gradient was formed when the inlet flow encountered an effective area increase on the downstream side of the separation bubble causing the flow velocity near the wall to decrease. This same type situation probably exists for the no screen case as well, except the separated region is larger and not as well defined).

The effect of the contraction on turbulence levels was also studied. The local turbulence intensities in the inlet were shown to decrease toward the contraction exit. The magnitude of the fluctuating axial velocities, u', on the other hand, increase through the contraction, especially in the region of greatest area reduction. The influence of the contraction depends upon the magnitude or content of the initial turbulence. The experiments show that for higher initial turbulence intensities the percent reduction in the intensities was greater. When one screen is used, the intensity at the contraction exit is less than 1/3 its initial value. For the 11 screen case, however, the reduction in turbulence intensity is only about 40%. Using turbulence intensities as a means of evaluating the effectiveness of the contraction in reducing turbulence is misleading because the intensities are referenced to the local mean velocity. Another way of assessing the effect of the contraction on turbulence is to look at the behavior of the rms velocities for the different screen combinations. When only one screen was in place the magnitude of u' increased about six times through the length of the contraction. When all 11 screens were used, u' at the exit plane was found to be almost twenty times its upstream value. This jump is attributed more to acoustic excitation than to the effect of the contraction. Nevertheless, the magnitude of the fluctuating component of velocity continuously increases through the inlet for all combinations of screens.

Comparing these experimental results with those of Uberoi [1956] and Klien and Ramjee[1973] shows that the effect of the contraction on turbulence depends on the contraction ratio and the nature of the turbulent fluctuations. Uberoi examined the effect of contraction ratio on isotropic turbulence. He used three square inlets with contraction ratios of 4:1, 9:1, and 16:1. For the two smaller inlets Uberoi's results agreed with predictions from linear theory for isotropic turbulence, i.e., the magnitude of u' decreases through the contraction. For the 16:1 contraction, however, u' first decreased and then increased to a final value which was 1.2 times the initial value. Since the inlets currently being studied have such high contraction ratios, the increase in u' through the inlets may be due to high contraction ratio effects.

Klien and Ramjee[1973] also studied the effects of contraction geometry on non-isotropic turbulence. They found that the contraction ratio was the governing parameter and not the wall geometry. Their results showed that, for a contraction ratio of 10, the magnitude of u' increased 6-fold through the contraction while the turbulence intensity continuously decreased to about 1/5 of its initial value. These results compare favorably with those obtained in the 30:1 inlet. In most practical applications the turbulence in front of the contraction is non-isotropic since damping screens have also been shown to produce non-isotropic turbulence, (Ribner and Tucker [1953]). The degree of

a. Entire Inlet

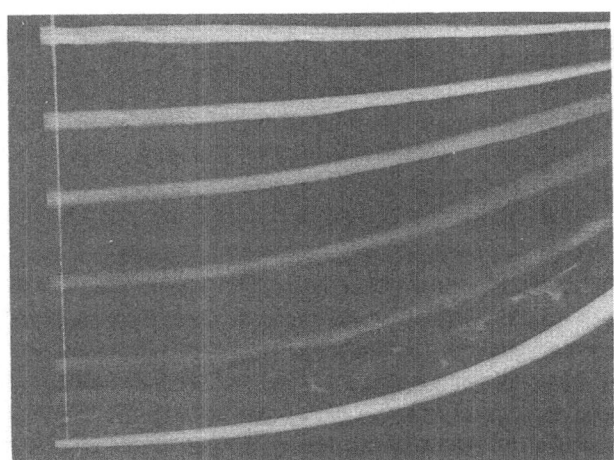

b. Entrance Plane Detail

Figure 14. Flow Visualization for 30:1 Inlet with
Entrance Plane Wall Discontinuity

anisotropy in the inlet was not measured because only a single wire hot-wire probe was used in the experiments.

Smoke flow visualization also proved useful in examining the boundary layer. Figure 15, shows an interesting flow phenomena along the walls of the inlet. In these two photographs, smoke streaklines are introduced as close as possible to the vertical, transparent wall of the inlet with and without the "step" wall discontinuity caused by the screen placement. In the contracting region a cross-flow entrained in the boundary layer can be seen. The lateral motion of the entrained flow is away from the corners towards the wall centerline.

The use of damping screens upstream of the contraction cone is known to improve the flow quality in the test section. This was evaluated with particular emphasis on flow uniformity and steadiness. Five-hole probe measurements were used to compare the flow uniformity and angularity for the cases of 0 and 11 screens. There was a decrease in the flow speed due to the associated total pressure losses through the screens for a fixed fan speed. The effect of the screens in modifying the flow angularity is very evident as shown in Figure 16. In the upstream section of the inlet, the screens decrease the pitch angle of the flow by more than 10 degrees. Fluctuations in the measured pitch angles in the 0 screen case ranged up to 3 degrees from the mean value, whereas in the 11 screen case the deviation was approximately 2 degrees. The screens had similar effects on the yaw angularity of the flow, but the effects were not as pronounced due to the symmetric location of the data points (in the vertical centerplane). The screens improve the flow quality at the contraction exit plane in terms of both velocity uniformity and flow angularity.

Some earlier flow visualization studies were conducted at Notre Dame to show the effect of the damping screens, see Batill, Nelson and Mueller[1981]. All of the previous work, though, was concerned with the coherence of the streaklines in the test section. The advantage of the transparent inlet is that it allows viewing of the streamtubes where they are the thickest, just downstream of the screens. Figure 17 shows the streamtubes in the inlet when 2, 4, and 11 screens were used. For the 2 screen case, the smoke began to break up within the first 20% of the inlet and had diffused by the time it reached the test section. When 4 screens were used, the smoke tubes remained intact but still exhibited some signs of flow unsteadiness. The use of all 11 screens produced well defined, coherent streamtubes throughout the length of the inlet and into the test section.

The experience gained in the experiments discussed above may have generated more questions than it answered. The performance of the inlet is strongly influenced by the character of the incoming flow. The upstream conditions, both in analysis and experiments, influence the inlet flow field. In the analysis, the upstream conditions can be "controlled" in a different manner than in the experiment. The experiments demonstrated that this inlet design, when coupled with adequate screening, would perform well. The screens are obviously vital, but the role they play in affecting both velocity fields and turbulence levels is not fully understood. Both the inlet and the turbulence managing devices must be adequately designed for a tunnel to function properly.

a. Without Wall Discontinuity

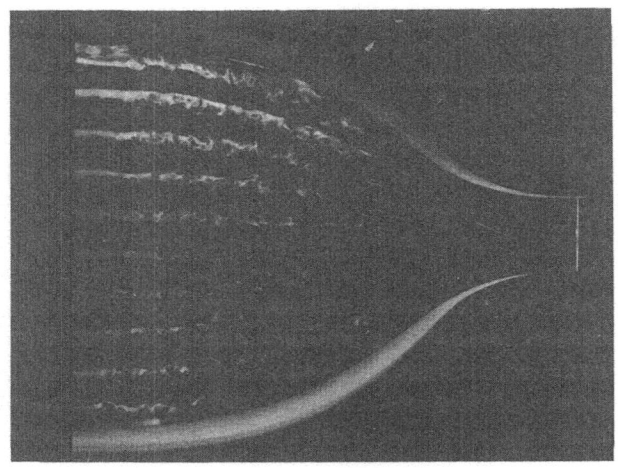

b. With Wall Discontinuity

Figure 15. Flow·Visualization of Inlet Wall Boundary Layer

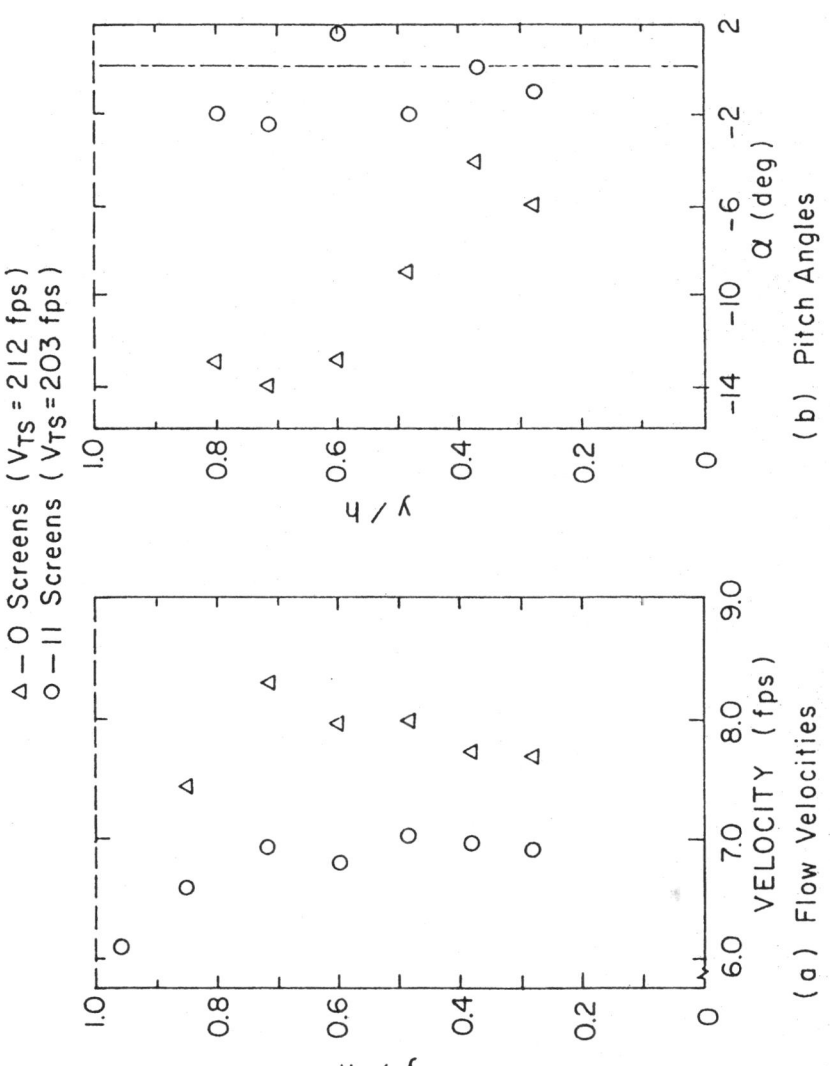

Figure 16. Influence of Screens on Entrance Plane Flow Field

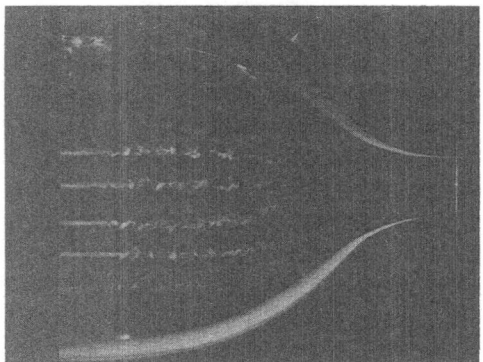

a. 2 screens

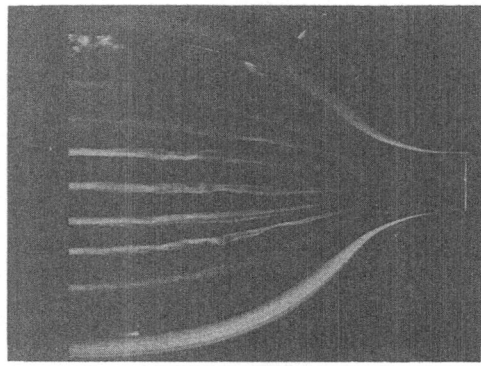

b. 4 screens

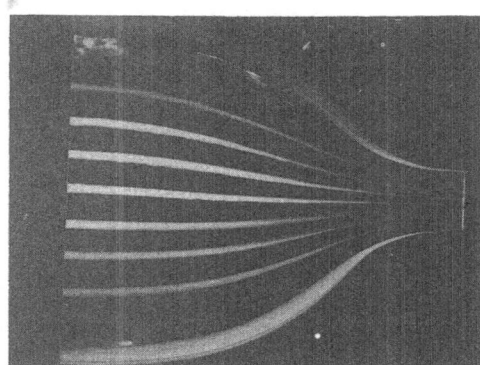

c. 11 screens

Figure 17. Flow Visualization in Inlet Flowfield, Screen Influence

3.4.3 *Inlet Design - Inlet Geometry Definition*

The preceding work was prompted by the desire to develop a set of practical guidelines for wind tunnel inlet design. This would become a formidable task unless one would limit the type and application of the tunnel facility for which the inlet is to be designed. The approach taken to develop the design data was to first characterize the inlet geometry with a relatively simple set of parameters and then to use the numerical prediction techniques to predict performance for a range of these parameters. The inlet performance was "measured" with a set of parameters related to flow field uniformity and adverse wall pressure gradients. The designer would then be able to select a set of performance goals, and use the results of the numerical simulations which have been presented in chart form to select the required inlet size and shape parameters.

The application for the design data is the relatively high contraction ratio tunnel. The results may prove suitable for either indraft or closed circuit design, though indraft tunnels were the primary application. The numerical procedure has been demonstrated for cases in which the upstream boundary conditions were more characteristic of the closed circuit tunnel but the experimental studies demonstrated reasonable agreement for an indraft tunnel design. The range of contraction ratios considered (10-40) for both square and rectangular cross-section inlets is suitable for subsonic tunnels which could operate over a relatively wide speed range. The low end of the range may be more suitable for closed circuit tunnels where the higher contraction ratios are characteristic of values used in some recent indraft tunnel designs, particularly those designed for smoke flow visualization.

The development of the set of rational preliminary design guidelines or procedures required the selection of a set of parameters which characterize the inlet geometry, and the selection of a set of parameters which define the inlet performance. The suitability of the subsequent design procedure depends on these two sets of parameters. Those selected were the result of the experience gained in the analytic work and experimental studies, but in no way represent the "best" set.

The inlet geometries considered are square or rectangular cross-sections as shown in Figure 10. The basic geometric parameter is the area contraction ratio, $CR = (H_iW_i) / (H_eW_e)$. The ratio of height to width (H/W) sets an "aspect ratio" for the cross-section. The length of the inlet can be scaled by the height (L/H) so that the basic size parameters are all non-dimensional. The H/W ratio could be a function of axial distance as would be the case in an inlet which would be used to transition from a square plenum to a rectangular cross-section. This type of inlet was not considered in the work presented here, but it is consistent with the prediction methods and, therefore, could be considered in the future.

The selection of the wall geometry is a much more difficult task. There are obviously a large number of candidate wall geometries from arbitrary "smooth curves" to sophisticated high order polynomials. The matched cubic geometry detailed in the previous section was selected for two reasons. First, it allows for a single parameter, X, the nondimensional position of the match point, to be used to completely define the wall geometry. Second, it had proven successful for a number of recent tunnel designs. From a practical consideration, it is realized that the curve formed by the matched cubic is more complex than some possible geometries such as those formed by circular arc and straight line segments. This does complicate fabrication but the fact that is satisfies two important "boundary" conditions and can be defined by a single parameter may outweigh the added complications. These boundary conditions are the zero wall slopes (i.e., alignment with the test section longitudinal axis) at the upstream and downstream end in the inlet. The need for zero wall slope at the downstream end is obvious, and recent experience has shown benefits of zero wall slope at the inlet entrance, particularly with screens or other turbulence management devices present. Reasonable values for X could range from 0.2 to 0.8 which effectively spans the range of very rapid contractions with mild curvature in the downstream section to inlets with most of the area reduction in the downstream section.

3.4.4. Inlet Design Parameters

The following design parameters were selected in order to provide a simple set of parameters which would be related as closely as possible to measures of "good" inlet performance. These parameters were then related to flow quality and susceptibility to flow separation.

Flow uniformity can be measured in terms of a maximum variation of velocity at a given cross-section. Two parameters were defined.

$$u_i = \frac{(V_{cl\,i} - V_{cor\,i})}{U_{m\,i}} \times 100 \ (\%)$$

$$u_e = \frac{(V_{cor\,e} - V_{cl\,e})}{U_{m\,e}} \times 100 \ (\%)$$

These represent the percent of maximum variation in velocity at the entrance (subscript i) and exit (subscript e) plane cross-sections. At the entrance plane the greatest velocity occurs at the centerline, $V_{cl\,i}$, and the smallest velocity at the corner, $V_{cor\,i}$. The opposite situation occurs at the exit plane, with the greatest velocity occurring at the corner, $V_{cor\,e}$. The two parameters are expressed in terms of a percent variation from the mean velocity at the given cross-section. The mean velocity was determined by determining the volume flux and dividing by the cross-sectional area.

The exit plane uniformity is important due to its influence on the test section flow quality. Although the nonuniformity decays as the flow continues in the constant area test section as a result of streamline curvature, the degree of nonuniformity will influence the length of the test section and allowable model locations. The entrance plane uniformity, though not apparently as critical a parameter, can effect the selection of the turbulence management devices. Since u_i can be greater than 100%, the variation in speed across the entrance plane can be quite large. This can affect the size and placement of screens or honeycomb or possibly set an upper limit on tunnel operating speed.

The second set of parameters selected to describe the inlet performance are related to the development of the wall boundary layers. It would be ideal to use the results of the flowfield predictions to perform detailed calculations of the development of the boundary layers in order to predict separation or displacement thickness. This type of calculation may be beyond the scope of a preliminary design study described here, but there are simpler techniques for predicting boundary layer behavior. It is important to eliminate the possibility of separation of the wall layer at any point in the inlet due to the influence separated regions have on the test section flow quality. The behavior of the boundary layer is dependent upon the character of the layer as the flow enters the inlet. This is dependent upon tunnel type and upon disturbances created upstream of the inlet. The development of the layers within the inlet will be influenced by the pressure distribution imposed on the layer and by the wall roughness. Experiments and numerical studies have shown the presence of adverse pressure gradients near both the entrance and exit planes of the inlets whose walls are matched cubics. Figure 18 illustrates an example of the adverse pressure gradient near the inlet entrance for a specific inlet design. The pressure coefficient Cp_{ui} is based on the mean entrance plane velocity.

$$Cp_{ui} = 1 - (V / U_{m\,i})^2$$

Figure 18 shows the pressure coefficient at the inlet centerline, the wall centerline and along the corner region for a CR = 30 inlet with a match point at X = 0.71 and an L/H = 1.204. This is similar to the inlet used in the experiments described above. A series of calculations for a wide range of inlet geometries were performed and are described below. These numerical results indicated that for the match point forward in the inlet X $\approx$ 0.2, there is a very strong adverse pressure region which begins well upstream of the inlet entrance plane and reaches a peak at x/L $\approx$ 0.1. For the match point well downstream in the inlet, X $\approx$ 0.80, the magnitude of the adverse region is much less but it extends over a greater length of the inlet. These numerical solutions were developed for a range of contraction ratios from 10 to 40 and the general form of the axial pressure distributions was consistent across this range. The magnitudes and locations of the peaks are quite comparable indicating a very weak dependence on CR for this range of values.

The type of data shown in Figure 18 indicate the environment seen by the boundary layer but it alone does not indicate whether or not separation occurs. There are a few simple methods available to help predict separation. One method which has been used in similar studies in the past is Stratford's[1959] turbulent boundary layer separation criteria. The Stratford criteria is straightforward and easy to apply. It shows that separation depends upon the strength of the

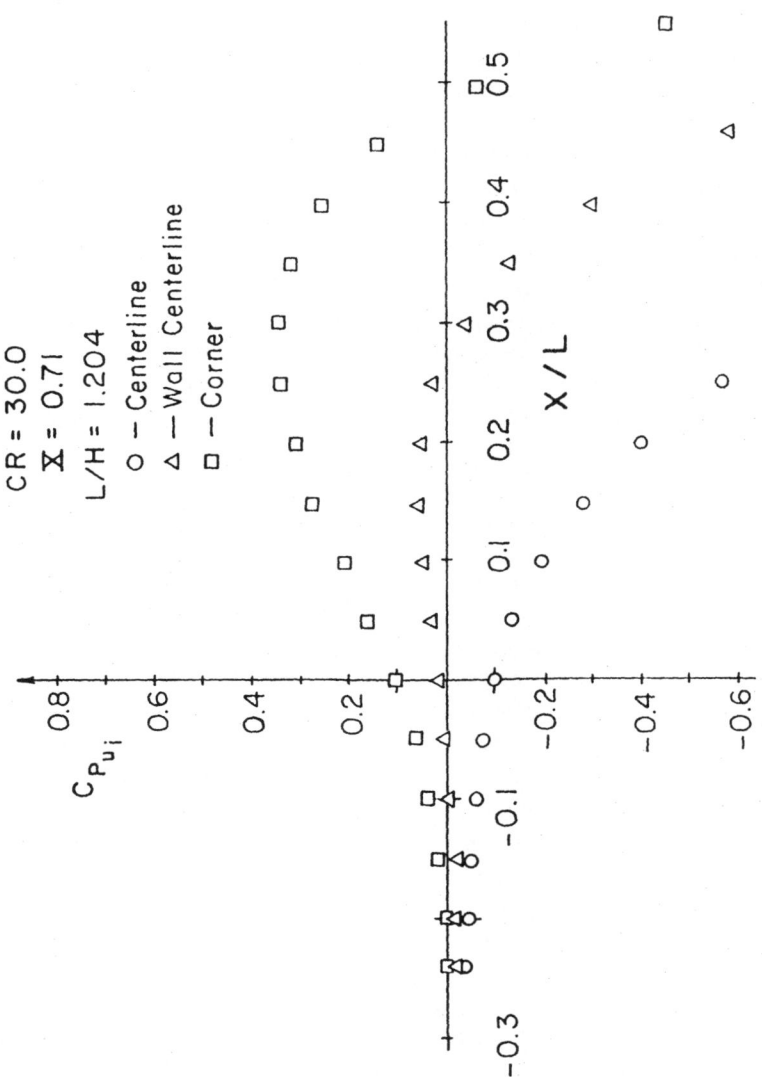

Figure 18. Pressure Distribution Near Inlet Entrance Plane

adverse pressure gradient and length of the developing turbulent boundary layer. Separation is oredicted to occur at a point where

$$C_p (x \, dC_p / dx)^{1/2} = 0.39 (10^{-6} Re_x)^{1/10}$$

The empirical constant (0.39) is changed to (0.35) for cases where for ($d^2p/dx^2 < 0$). The distance x is measured from the "origin" of the turbulent boundary layer and the relation is valid for Reynold's numbers based on x, Re_x, on the order of 10^6. This criteria was applied to a group of inlet designs as will be discussed in the next section. The location of the origin of the boundary layer is usually not known in the inlet design problem, and the region of most severe adverse pressure is the corner, where the flow is highly three-dimensional. Although this criteria is useful in developing a "feel" for boundary layer behavior, a more detailed approach would be required for accurate prediction of separation.

Morel [1975] did show that the magnitude of the adverse pressure region could be correlated to two simple parameters. These two parameters are used to study the sensitivity of the pressure distribution on the inlet geometry. There is a simple parameter for the entrance region, Cp_i ,and one for the exit region, Cp_e.

$$Cp_i = 1 - (V_{min} / U_{m\,i})^2$$

$$Cp_e = 1 - (U_{m\,e} / V_{max})$$

The values $U_{m\,i}$ and $U_{m\,e}$ are the mean values at the entrance and exit planes respectively. The values V_{max} and V_{min} are the maximum and minimum flow velocities occuring anywhere along the inlet wall. These extreme velocities will occur along the corner and are characteristic of the magnitude of the adverse pressure gradient occurring near the inlet entrance and exit. The four parameters u_i, u_e, Cp_i, Cp_e can be used to describe the aerodynamic performance of the inlet. A designer can select allowable values for each design criteria and then determine the required tunnel geometry to achieve each.

3.4.5. *Inlet Design Charts*
As indicated in the earlier discussion, prediction of the three dimensional flow field in a wind tunnel inlet is a complex problem. Analytic techniques are often limited by the inlet geometry and the description of the upstream and downstream boundary conditions. Numerous numerical techniques exist which can be effectively applied to this internal flow problem. These include panelling techniques for inviscid flows Hess and Smith[1967], Renken[1976], Hess et.al.[1979] and Lee[1981] and finite difference and finite element methods which could be applied to either viscous or inviscid flow fields (Van Den Broek[1973]). Each of these approaches could be used to evaluate

the flow field details in a given wind tunnel inlet, and for a detailed design study each would have advantages and disadvantages. Batill, Caylor and Hoffman[1983] considered a number of these methods and were particularly concerned with selecting an efficient approach which would be suitable for application to a large number of inlet geometries in order to develop the information necessary for the preliminary design of relatively high contraction ratio inlets. Adapting a method presented by Sanderse and van der Voreen [1977], they developed a program to solve the three dimensional, velocity potential in an internal, inviscid flow. Laplace's equation in three dimensions,

$$\Phi_{xx} + \Phi_{yy} + \Phi_{zz} = 0$$

was solved using finite diffence techniques. The potiential equation, inlet geometry and associted boundary conditions were tranformed into a rectangular region and a fully conservative field form of the equations were solved. The uniform flow upstream and downstream boundary conditions were applied in the constant cross sectional area regions "far" upstream and downstream of the contration section. The goal of the solutions was not to achieve "exact" solutions to the inlet flowfield for specific inlets but to allow for calculation of the inlet flowfield for a wide range of inlet geometries. This allowed for an evaluation of the sensitivity of the inlet flowfield to the basic design variable and and assessment of the influence the design variables had on the design parameters as defined above.

Using the performance parameters discussed in the previous paragraphs, the results of a series of calculations were summarized into a set of design charts. The following shows the range of the parameters used in developing the design charts given in Batill, Caylor and Hoffman[1983].

CR	L/H	H/W
10	0.6 → 1.6	1.0
25	0.6 → 1.6	1.0 → 1.67
40	0.6 → 1.6	1.0

These charts are shown for the set of parameters discussed earlier in Figure 19, for a contraction ratio 10 inlet with a square cross section for an L/H range of 0.6 to 1.6. A complete set of design charts for the complete range of parameters indicated above is included in Batill, Caylor, Hoffman [1983]. The figure has four parts representing the variation of each parameter, u_i, u_e, Cp_i, Cp_e with match point and length to height ratio for a given contraction ratio. These curves can be used to select the preliminary inlet geometry for a given wind tunnel application.

As would be expected, the degree of entrance or exit plane uniformity are competing parameters. Entrance plane uniformity can be improved by moving the match point rearward, but this increases the exit plane nonuniformity. Both parameters are improved by increasing the inlet length. Entrance and exit plane pressure distributions, as reflected in the parameters Cp_e and Cp_i,

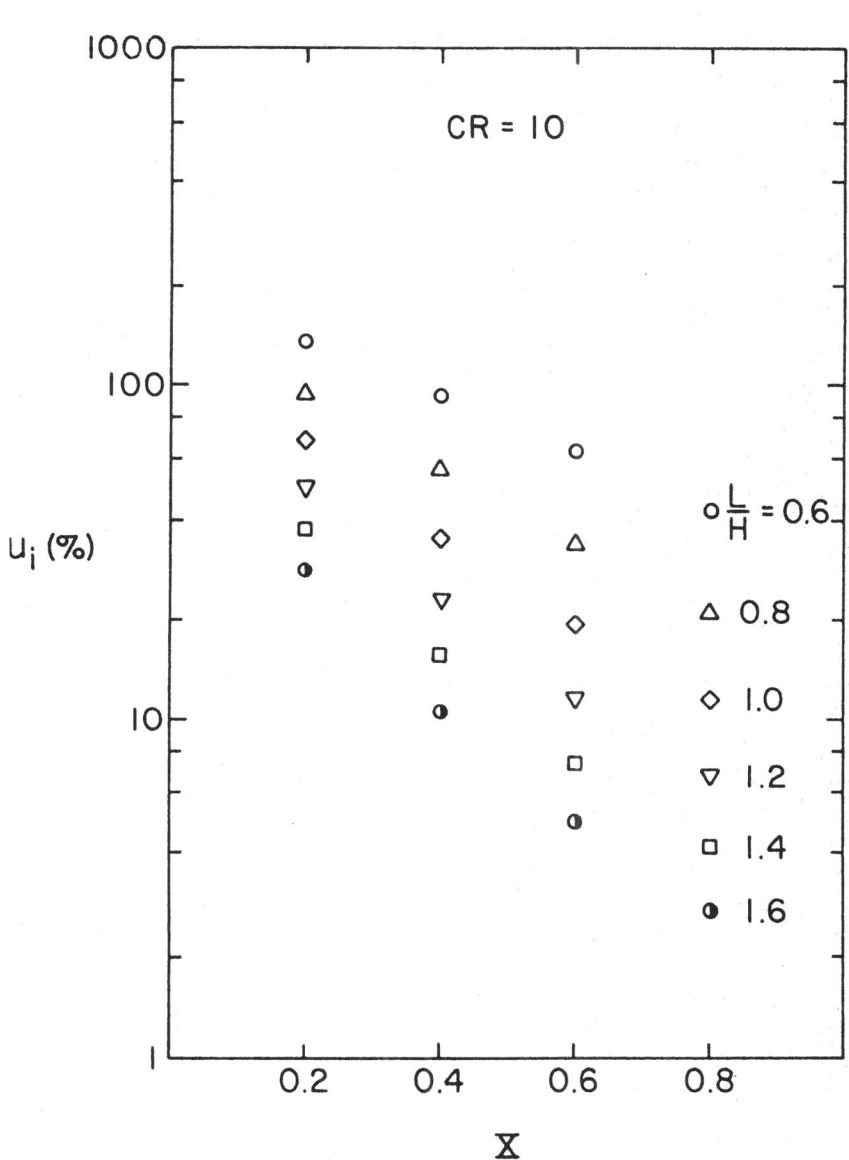

a. u_i - Entrance Plane Velocity Uniformity

Figure 19a. Inlet Design Data

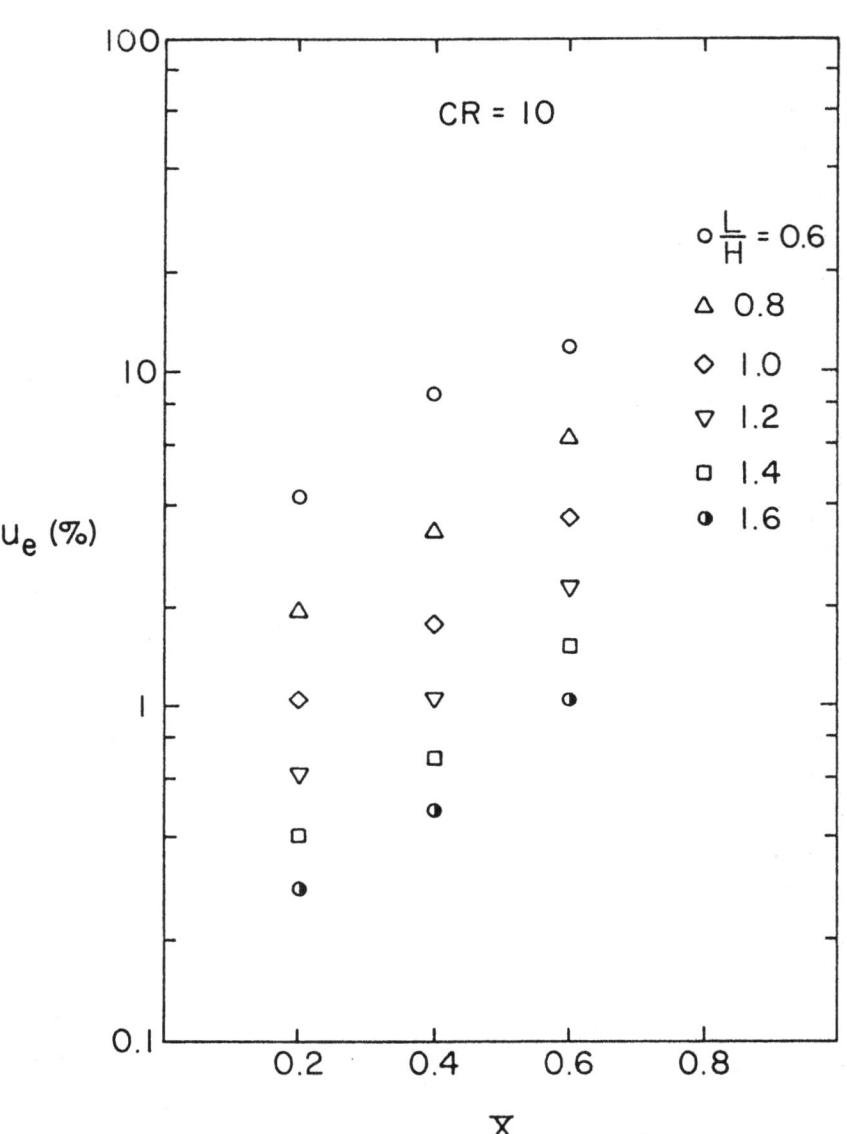

b. u_e - Exit Plane Velocity Uniformity

Figure 19b. Inlet Design Data

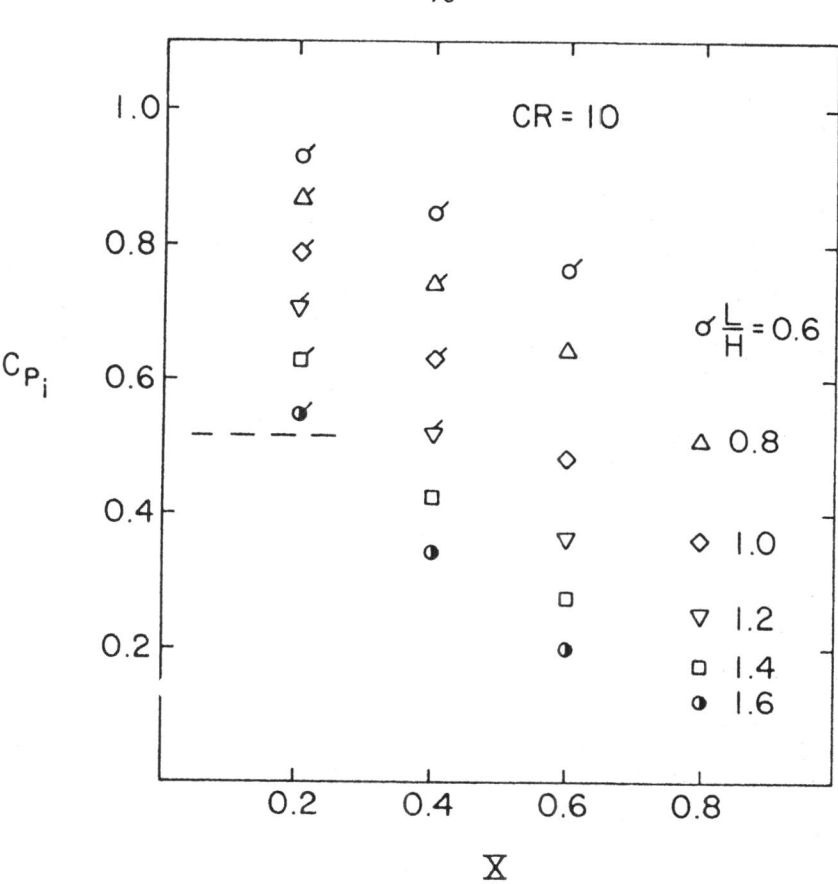

c. C_{p_i} - Entrance Region Pressure Coefficient

Figure 19c. Inlet Design Data

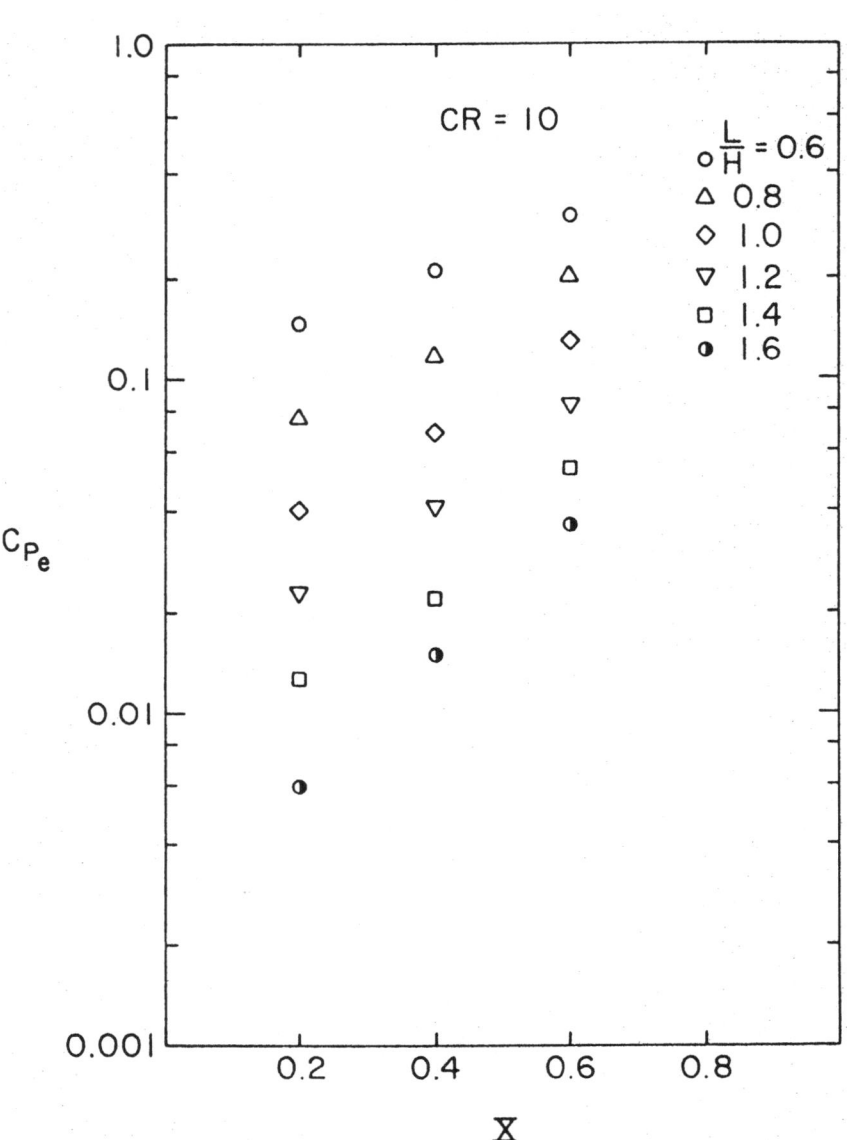

d. C_{p_e} - Exit Region Pressure Coefficient

Figure 19d. Inlet Design Data

are also competing parameters. Forward positions of the match point result in low values for Cp_e (small adverse pressure gradients in the downstream end of the inlet) but this also results in large values for Cp_i. Again both parameters are reduced by increasing the inlet length.

Additional calculations were performed applying Stratford's criteria to the pressure distribution along the corner region and the results are included with Cp_i (part c of Figure 19). Assuming an effective origin of the turbulent boundary layer equal to 50% of the inlet height upstream of the entrance plane, the flagged symbols represent designs for which turbulent boundary layer separation is predicted. This seems to correlate very well with the parameter Cp_i and on Figure 19c the upper bound for Cp_i is shown. This upper bound does depend upon the selection of position of the virtual origin of the boundary layer and should therefore be used as only a rough estimate of a limit on Cp_i.

Similar calculations were not performed for the boundary layer near the exit plane. Since the "history" of this layer is so complex it was felt that it would be inappropriate to attempt to apply Stratford's criteria in this region. Allowable values for Cp_e will have to be developed from experience or from more detailed boundary layer calculations.

The dependence of each parameter on contraction ratio can be determined by interpolating within the contraction ratio range which was evaluated. The parameter u_i is not a strong function of contraction ratio, but the other three parameters show a stronger dependence. The only calculations for rectangular (non-square) cross section inlets were performed for a contraction ratio of 25. The values for u_i and u_e decrease for increased H/W ratios. It appeared that as H/W increases the inlet becomes more "two-dimensional" for a given contraction ratio, even though the ratio of H/W remains constant along the inlet. The influence of the corners is not as strong. The parameter Cp_i showed a decrease with increases in H/W as would be expected if the influence of the corners is reduced. This implies that a shorter rectangular inlet could be used rather than a square cross-section for a similar contraction ratio.

The inlet design is highly dependent upon the selection of the target parameters for the design criteria. The results summarized above have indicated that a relatively simple set of design criteria can be established and that these simple criteria can be quantified through a combination of analytic, numerical and experimental studies. The inlet design data included above will not be suited for all inlet designs. Improved numerical techniques may be employed to provide more detailed flowfield data for a specific design. This type of data may prove useful in establishing a preliminary inlet design so that the number of expensive detailed calculations can be reduced. Considering the cost associated with the development of most wind tunnel facilites, detailed flowfield calculations for the final inlet design should be considered prior to fabrication or during developmental testing (Kaul et.al.[1985]). Though a number of inlets have been designed using the preliminary design methods outlined above, it will only be through experience and additional analysis that more confidence in acceptable values of these design criteria can be achieved.

3.5 *Test Section*

The test section size and type, i.e., open or closed jet, is determined by the kind of experiments planned for the tunnel. Other factors affecting the design of the test section include tunnel speed and Reynolds number requirements. A list of some of the design variables are given below.

    (1) Wall boundary conditions (Open, closed , porous)
    (2) Cross sectional geometry
    (3) Length
    (4) Accessability (Doors or removable walls)
    (5) Model or instrumenation support requirements
    (6) Windows for viewing, lighting or optical instrumentation

Many different cross sectional shapes such as round, elliptical, square, rectangular, octagonal, etc., have been successfully used for wind tunnel test sections. The actual cross sectional geometry depends upon special testing requirements. However, for large wind tunnels, flat floors or floor inserts are required for work crews to install and modify the test models.

The major factor affecting the power required to move the flow through the test section is the surface area exposed to the flow. The power losses in the test section are directly related to the skin friction drag. The difference in skin friction coefficient between the different cross sectional shapes is quite small and therefore the power loss is directly related to the wetted area.

Another factor affecting the performance of the test section is the growth of the boundary layer along the walls of the section. The increase in boundary layer thickness with increasing length causes the potential core velocity to increase which then lowers the static pressure. The static pressure variation produces an additional component of drag on a model in the test section. This additional drag is called the horizontal buoyancy and the drag measurements must be corrected for this interference. The buoyancy effect can be minimized by tapering the test section walls to account for the boundary layer growth. Obtaining a zero static pressure gradient over a wide range of tunnel velocities is difficult to accomplish without using some form of adaptive wall geometry. A common practice is to let the walls diverge about 1/2 degree .

Since the test section is where the model will be installed, a support system will be needed. Care must be taken in the design of the support to insure that the support system does not interfere with the flow around the model. The support system should present a minimum blockage to the flow but must also be strong and rigid. These are, of course, conflicting requirements and present the designer with a strong challenge.

Flow visualization requires test sections with ample lighting and viewing area. For flow visualization as well as laser Doppler anemometry experiments, the test section should incorporate windows that are as large as possible. Flush mounted lighting incorporated into the tunnel walls can also be considered.

83

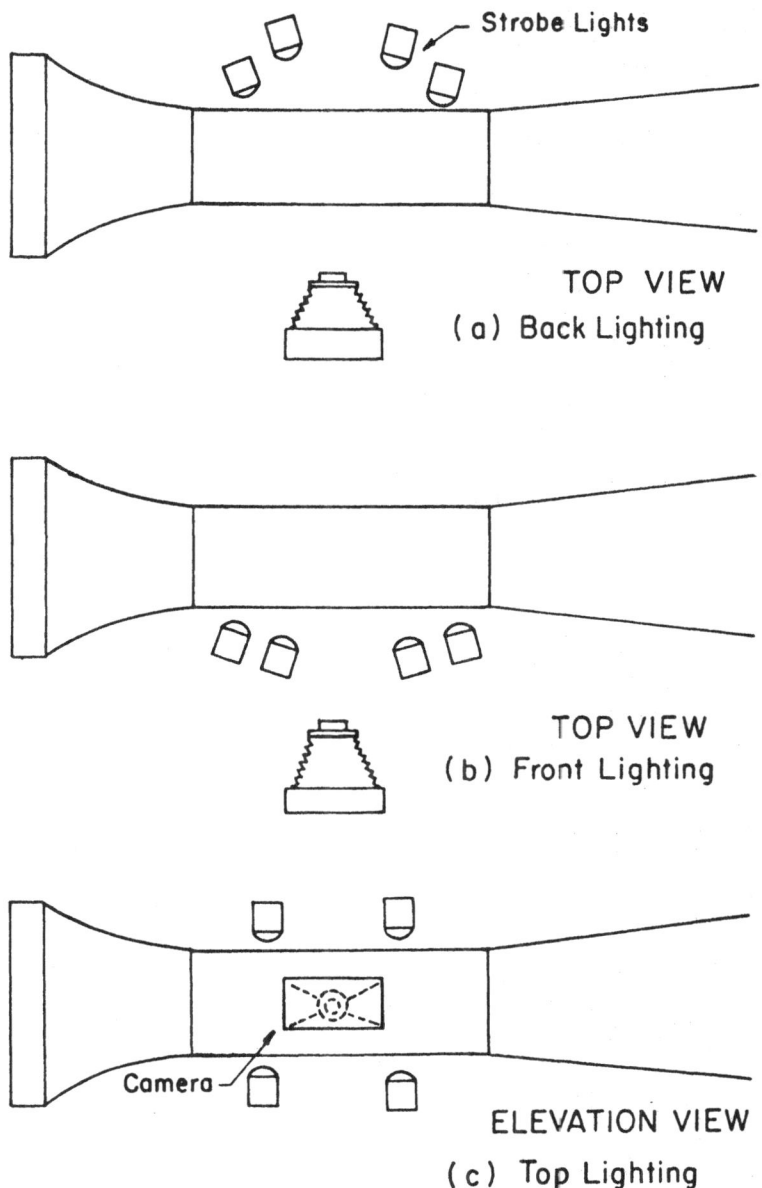

Figure 20 Lighting Arrangements

3.5.1. *Smoke Flow Photography*

With recent emphasis on flow visualization technques and the usefullness of indraft tunnels for smoke flow visualization, the following sections have been included to provide some perspective on the design of indraft tunnels for this special application. It is important to realize that the use of an indraft tunnel for flow visualization requires an understanding of the wind tunnel and data collection requirements.

One of the most critical aspects of flow visualization photography is in obtaining sufficient light on the subject of interest. In wind tunnel applications this problem is often the most difficult to overcome because of space limitations. The main problem is getting sufficient light intensity for proper illumination of the smoke streaklines, as well as in maintaining high contrast between the smoke and the background. As most large wind tunnels have very limited viewing areas, one is generally forced to use less light than desirable. Space limitations often restrict the placement of lights and cameras.

Lights that are flush mounted into the walls, ceiling and floor of the tunnel should be considered when designing a wind tunnel having a large test section. For smaller tunnels, lights are usually mounted outside of the tunnel. Figure 20 illustrates the three most commonly used lighting arrangements. When the lights are positioned on the opposite side of the tunnel and out of the direct field of view of the camera, the arrangement is called "back lighting". Glare from the model can be minimized using the back lighting arrangement. If lights are positioned on the same side of the tunnel as the camera, the arrangement is called "front lighting". Major difficulties with front lighting are in the reflections and glare from the model and tunnel windows. These problems can be reduced by proper positioning of the lights and camera. However, when there is a limited viewing area and space is restricted, it may not be possible to position the lights properly. The final lighting arrangement is called "top" or "bottom lighting". This arrangement can be used in any test section where the flow can be illuminated in a plane perpendicular to the normal viewing direction. The test section can be illuminated through this transparent slit. An advantage of top and bottom lighting is that the light source is normal to the field of view and interference such as reflection and glare from the model is minimized.

The type of light source required for smoke visualization tests depends upon the particular test being conducted. The two most widely used light sources are stroboscopic and high intensity incandescent lamps. Short duration flash lamps such as the General Radio 1540 strobolumes or their equivalent have been found to produce satisfactory results. These lamps operate from an electrical source of 120 volts and 60 cycles. These lighting systems can provide either single flashes or repeated flashes at rates to 25000 flashes per minute. The beam intensity of an single lamp is rated at 0.5×10^6 to 16×10^6 candela depending upon the flash rate. The rated duration of the flash is less than 15 micro-seconds. For still photographs, the strobolumes are synchronized with the camera shutter and operate similarly to any camera using an electronic flash. With movies, the flash rate can be synchronized to the framing speed. The number of lamps required depends on the nature of the test being conducted. For continuous lighting, high intensity spot lamps are required. A summary type of the photographic equipment needed for smoke flow photography is included in Table I.

Another light source that has become quite popular due to unique applications is the use of pulsed or continuous lasers. High powered lasers can be used to create light sheets for illuminating planes within the tunnel. By passing the laser beam through a cylindrical lens, one can produce a thin light sheet. Figure 21 shows several photographs of the leading edge vortices on a sharp edge delta wing. In the upper photograph the laser sheet is normal to the wing and illuminates the smoke entrained in the vortices in the cross flow plane. For the lower photograph the laser sheet has been rotated so that the light sheet lies in the plane of the vortices. The light sheet can also be created by an oscillating mirror. With this technique the laser beam is rotated back and forth throughout the region of interest at very high frequencies. Pulsed laser can be used to obtain high intensity light for short durations.

---

1. Cameras/Video Equipment
   a) A 4" x 5" view or 35mm camera with an f/4.5, 6" lens in a shutter provided with a synchronizing switch suited to a short interval flash lamp.
   b) A high speed motion camera capable or high speed video equipment with associated synchronizing equipment

2. Lights
   a) High intensity stroboscopic lights.
   b) High intensity flood lights for continuous operation.
   c) High powered laser with optical scanner.

3. Darkroom
   a) Darkroom should be located near the wind tunnel laboratory.
   b) Tank processing for 4" x 5", 35mm and 16mm black and white film must be available.
   c) A high quality enlarger and equpment for processing black and white glossy prints is required.
   d) It is essential that the technician in charge of the darkroom be under the direction of the smoke tunnel engineer.

Table I. Recommended Equipment for Smoke Flow Photography

3.5.2. *Photographic Procedures*

Over the years, many universities and research laboratories (both government and private) have attempted to obtain flow visualization data at low speeds. Many of these efforts have ended in failure for a variety of reasons. Even in tunnels where quality smoke is available, only poor smoke flow photographs were obtained. One can only speculate on the reasons for such failure. However, one of the difficulties is probably due to the fact that the experimentalist could see the flow pattern

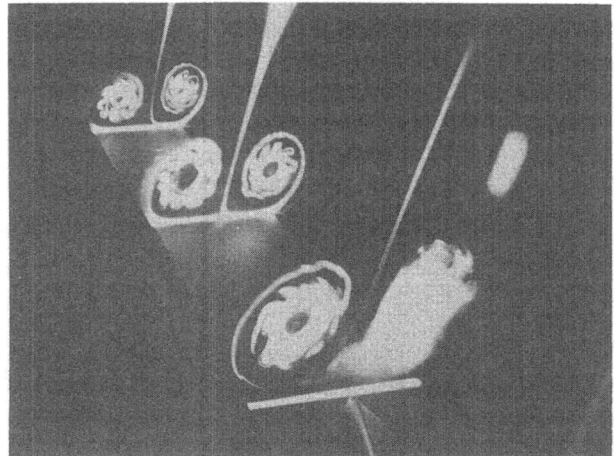

a) Laser sheet normal to model surface

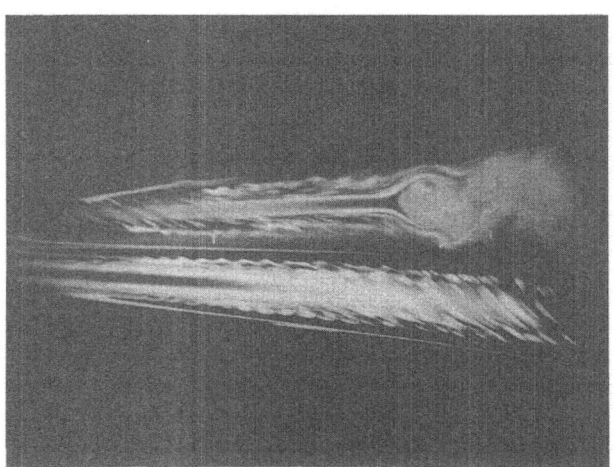

b) Laser sheet in the plane of the vortices

Figure 21 Laser Light Sheet Illumination of Leading Edge Vortices

with his eye but was unable to obtain good photographic records. This is usually because, in many installations, the photographic work is left to a photographic technician. To achieve quality smoke photographs, the experimental aerodynamicist should also be the photographer. An experimental aerodynamicist is familiar with the phenomena and their influences on the smoke pattern in the flow field. This is not to say that a photographic technician cannot be used; however, the aerodynamicist must be willing to spend enough time with the photographer so the technician knows what is required. Some of the difficulties in obtaining quality smoke flow photographs are included in Table II.

1.	Poor Smoke Quality	
	a)	High turbulence level in the tunnel.
	b)	Inadequate smoke generation.
	c)	Smoke rake located inside the tunnel circuit.
2.	Poor Photographic Procedures	
	a)	Heavy reliance on photographic technicians.
	b)	Heavy reliance on commercial photographic processing.

Table II. Sources of Problems in Flow Visualization Experiments

Processing the photographic data presents another challenge. Due to the difficulties in lighting, the photographic plates are usually underexposed and the exposure is generally uneven. Therefore, to produce a quality print, an uneven exposure of the negative is required. This procedure is referred to as "photographic dodging". It enables one to enhance the photographic image, i.e., to capture all the data on the print. Again, the person processing the film must know what he or she is looking for in the print, otherwise much, if not all, useful data will be lost.

One final point: the use of flow visualization techniques in the study of complicated flow patterns is as much an art as it is a science. Obtaining good flow visualization data requires a great deal of patience and time. Each new application will present the experimental aerodynamicist with new photographic difficulties. However, with patience and some effort, meaningful flow visualization data can be acquired.

3.6. *Diffuser*

The purpose of the diffuser is to convert the kinetic energy of the air leaving the test section as efficiently as possible into pressure energy. This is accomplished using an expanding duct that reduces the flow velocity which results in a static pressure recovery. The performance of the diffuser is very sensitive to design errors. A poor diffuser design may cause flow separation to occur either in a continuous or intermittent manner. In either case the diffuser will not be satisfactory.

The factors affecting the diffuser performance include:

(1) Velocity distribution at the diffuser entrance

(2)   Equivalent cone angle

(3)   Length

(4)   Area or expansion ratio

(5)   Shape transition (square to circular, etc.)

Although the flow entering the test section will be uniform with a low turbulence intensity for a properly designed inlet, the flow leaving the test section will in general be non uniform at higher turbulence levels due to the wake from the model and support system in the test section. The major design factors over which the tunnel designer has control are the expansion or equivalent cone angle and the area ratio. The equivalent cone angle of the diffuser is determined by replacing the diffuser with an equivalent conical section having the same length and area ratio as the diffuser. A large cone angle or expansion will lead to flow separation in the diffuser. A commonly used rule of thumb is to keep the equivalent cone angle to be less than 7 degrees. The area ratio establishes the static pressure recovery that is possible for a given diffuser. Obviously, there is a trade-off between the expansion angle and the area ratio. If one selects a large area ratio and keeps the expansion angle to within recommended limits, the length of the diffuser may become quite long. A long diffuser will be susceptible to boundary layer flow separation, due to the adverse pressure gradient within the diffuser. Furthermore, a long diffuser will probably be impractical due to tunnel length constraints. For the case where a large diffusion must occur in a short distance the designer must incorporate flow control devices within the diffuser to miimize the regions and extent of flow separation. This can be accomplished by using vortex generators along the diffuser walls.

3.7. *Power System - Fan and Motor*

The power required to maintain a given flow speed through the test section is equal to the total power losses throughout the circuit. The kinetic energy is dissipated by viscous losses within the tunnel circuit. The loss in kinetic energy appears as a decrease in the total pressure. The total pressure loss in the circuit must be compensated for by a pressure rise across the fan. The following outlines some of the considerations associated with tunnel power requirements.

The losses through the tunnel circuit can be evaluated by using a combination of theoretical and empirical techniques to estimate the loss of each of the major wind tunnel components. The loss in energy can be expressed in terms of a static pressure loss, Δp, or as a loss coefficient defined as, $k = \Delta p/q$ where q is the local dynamic pressure. The power delivered by the motor and fan must balance losses in the circuit.

$$\eta P = \Sigma \text{ Circuit Losses}$$

where η is the efficiency of the fan and P is the shaft power of the motor. The flow energy through the test section can be expressed as

$$E_o = \frac{1}{2} \rho_o A_o V_o^{\;3}$$

The energy loss in each section can be written as,

$$\Delta E_i = \frac{1}{2} \rho_i A_i V_i^3$$

but,

$$k_i = \frac{\Delta P}{q_o}$$

therefore,

$$\Delta E_i = k_i \frac{\frac{1}{2} \rho_o V_o^2}{\frac{1}{2} \rho_i V_i^2} (\frac{1}{2} \rho_i A_i V_i^3)$$

or

$$\Delta E_i = k_i (\frac{1}{2} \rho_o V_o^2) A_i V_i$$

From the continuity equation we have,

$$\rho_i V_i A_i = \rho_o V_o A_o$$

Upon substitution into the energy loss equation yields,

$$\Delta E_i = k_i (\rho_o/\rho_i) (\frac{1}{2} \rho_o V_o^3 A_o)$$

For low speed wind tunnels M < 0.3, the flow can be assumed to be incompressible, i.e., $\rho_o/\rho_i = 1$. The power required can then be expressed as

$$\eta P = \frac{1}{2} \rho_o V_o^3 A_o \sum_{i=1}^{n} k_i$$

Detailed methods for estimating the loss coefficients for the various tunnel components are given by Eckert, Mort and Jope [1976] and Rae and Pope[1983].

Having estimated the power requirements, the next step is to design a motor and fan system that can deliver the required power. Methods for designing the fan can be found in Rae and Pope [1983] and performance information on commercially available fans can be readily obtained from the manufacturer.

Once the motor and fan have been selected, the next consideration is the drive system control. The control system should incorporate safety features to protect the motor and permit safe operation of the tunnel. In addition the control system should incorporate a feedback system so that the test section velocity can be accurately controlled and maintained.

3.8. *System integration, support facilities and Safety*

The final topic to be discussed is that of system integration. The design variables and criteria for each component of the tunnel have been discussed in the previous sections. Because the wind tunnel is a system of these components, trade-offs are required between the components in order to satisfy the design objectives and constraints of the tunnel. System integration is the most important aspect of the design process. Without attention to system integration one can design a wind tunnel that is composed of components that are optimum in their own right but as a system there performance may be unsatisfactory. For example, if an optimum diffuser length forces the inlet design to be very short due to a tunnel length constraint, then the performance of the tunnel may be unsatisfactory due to flow separation in the inlet.

In the area of system integration one must also be concerned with additional factors such as environmental control, instrumentation, lighting, and space and structural limitations of the building intended to house the wind tunnel. Environmental control for low speed tunnels includes heating and cooling considerations. If the tunnel flow is exhausted outside the building then large temperature changes may occur during the operation of the tunnel, both in the laboratory room as well as the tunnel test section. With proper planning both the laboratory and test section temperature can be controlled.

Instrumentation of the tunnel must include as a minimum instruments that yield information on the test section speed, and performance of the fan. With the motor speed controllers that are available today it is possible to accurately control the test section velocity through a velocity feedback control system.

If flow visualization testing is a requirement of the tunnel design, then the room housing the tunnel should incorporate features for making the room light tight. This will also require electrical switches so that all the room lights can be readily controlled by the experimenter at the test section.

4. Summary

The successful design of an indraft subsonic wind tunnel requires the proper selection of many design parameters. Although each new wind tunnel design will in general have its own set of unique performance objectives and design constraints, the tunnel design requirements outlined in this paper should provide useful insight into the critical issues associated with tunnel design and sources for information related to the design of specific components.

The design variables for each of the major wind tunnel components were identified and their influence on tunnel performance was assessed. In addition, the importance of system integration was addressed throughout the article. The authors hope that this article will provide a potential wind tunnel designer with a good overview of the important design variables that must be considered for a successful wind tunnel design.

5. Acknowledgements

The authors wish to acknowledge the Department of Aerospace and Mechanical Engineering at the University of Notre Dame and the United States Air Force Flight Dynamics Laboratory, Wright Patterson Air Force Base, Ohio for their contribution in support of the reserach related to wind tunnel inlets and reported in this article. They also wish to thank the Air Force Flight Dynamics Laboratory for drawings and information related to the SARL wind tunnel facility.

The authors dedicate this brief overview to those wind tunnel designers and engineers who have devoted their professional careers to the development and operation of wind tunnels. Their past accomplishments and future work are vital to the developments in aeronautics.

6. References

Asanuma, T., (Editor), (1977) "Flow Visualization", Proceedings of the First International Symposium of Flow Visualization", Tokyo, Japan, Hemisphere Publishing Corp.

Baals, D.D. and Corliss, (1981) Wind Tunnels of NASA , NASA SP-440.

Batchelor, G. K. and Shaw, F. S., (1944) "A Consideration of the Design of Wind Tunnel Contractions", Australian Council for Aeronautics, Report ACA-4.

Batill, S.M., Caylor, M.J., Hoffman,J.J., (1983) "An Experimental and Analytic Study of the Flow in Subsonic Wind Tunnel Inlets", Air Force Wright Aeronautical Laboratory, Report No. AFWAL-TR-83-3109, Wright Patterson Air Force Base, OH.

Batill, S., Nelson, R., and Mueller, T., (1981) "High Speed Smoke Flow Visualization", Air Force Wright Aeronautical Laboratory, Report No. AFWAL-TR-81-3002, Wright Patterson Air Force Base, OH.

Borger, G. G., (1976) "The Optimization of Wind Tunnel Contractions for the Subsonic Range", NASA TRF-16899.

Bradshaw, P., and Pankhurst, B. C., (1981) "The Design of Low-Speed Wind Tunnels", Prog. in Aero. Sci., Vol 5, pp. 1-67.

Brendel, M. and Mueller, T.J. (1988) "Boundary Layer Measurements on an Airfoil at a Low Reynolds Number in and Oscillation Free Stream" AIAA Journal, Vol. 26, No.3, pp.257-263.

Chmielewski, G. E., (1974) "Boundary Layer Considerations in the Design of Aerodynamic Contractions", J. of Aircraft, II, No. 8.

Cohen, M. J. and Ritchie, N.J.B., (1962) "Low Speed Three-Dimensional Contraction Design", J.Royal Aero. Soc., 66, pp. 231-236.

Comte-Bellot, G. and Corrsin, S., (1966) "The Use of a Contraction to Improve the Isotropy of Grid Generated Turbulence", J. Fluid Mech., 25.

Dryden, H. L. and Schubauer, G. B., (1947) "The Use of Damping Screens for the Reduction of Wind-Tunnel Turbulence", paper presented at the Aerodynamics Session, Fifteenth Annual Meeting, I.A.S., New York.

Dryden, H. L. and Abbot, I. H., (1949) "The Design of Low-Turbulence Wind Tunnels", NACA Tech. Rep. 940.

Eckert, W.T., Mort, K.W., Piazza, J.E., (1976) "An Experimental Investigation of End Treatment for Nonreturn Wind Tunnels", NASA TM X-3402.

Eckert, W.T., Mort, K.W. and Jope, J. (1976) "Aerodynamic Design Guidelines and Computer Program for Estimation of Subsonic Wind Tunnel Performance", NASA TN D-8243.

Hess, J. L. and Smith, A.M.O., (1967) "Calculation of Potential Flow About Arbitrary Bodies", Prog. Aero. Sci., 8, pp. 1-138.

Hess, J. L., Mack, D. P. and Stockman, N. O., (1979) "An Efficient User-Oriented Method for Calculating Compressible Flow In and About Three-Dimensional Inlets - Panel Methods", NASA CR-159578.

Kaul,U.K., Ross,J.C., Jacocks,J.L., (1985) "A Numerical Simulation of the NFAC (National Full-Scale Aerodynamics Complex) Open Return Wind Tunnel Inlet Flow", AIAA Paper 85-0437, 23rd Aerospace Sciences Meeting, Reno, Nevada.

Klien, A. and Ramjee, V., (1973) "Effects of Contraction Geometry on Non-Isotropic Free-stream Turbulence", Aeronautical Quarterly, 3, pp. 34-38.

Lee, K. D., (1981) "Numerical Simulation of the Wind Tunnel Environment by a Panel Method", AIAA Jour., Vol.19, No. 4, pp. 470-475.

Lin, C.C.,(1955), *Hydrodynamic Stability*, Cambridge University Press, Cambridge, England.

Loehrke, R. I. and Nagib, H. M., (1972) "Experiments on the Management of Free Stream Turbulence", AGARD Report No. 598.

Merzkirch, Wolfgang (Editor), (1980) "Flow Visualization I", Proceedings of the Second International Symposium on Flow Visualization, Bochum, West Germany, Hemisphere Publishing Corp.

Mikhail, M. N. and Rainbird, W. J., (1978) "Optimum Design of Wind Tunnel Contractions", AIAA Paper No. 78-819.

Morel, T., (1975) "Comprehensive Design of Axisymmetric Wind Tunnel Contractions", J. Fluids Engr., ASME Transactions, pp. 225-233.

Morel, T., (1976) "Design of Two-Dimensional Wind Tunnel Contractions", ASME Paper No. 76-WA/FE-4, pp. 107.

Mort, K.W., Eckert, W.T., Kelly, M.W., (1972), "The Steady-State Flow Quality of an Open Return Wind Tunnel", Canadian Aeronautics and Space Journal, Vol. **XX**, No. **XX**, pp. 285-289.

Mueller, T. J., (1978) "Smoke Visualization of Subsonic and Supersonic Flows (The Legacy of F.N.M. Brown),", Final Report UNDAS TN-34121-1, AFOSR-TR-78-1262.

Mueller, T. J., (1980) "On the Historical Development of Apparatus and Techniques for Smoke Visualization of Subsonic and Supersonic Flows", AIAA Paper No. 80-0420, 11th Aerodynamic Testing Conference.

Mueller, T., Pohlen, L., et. al., (1983) "The Influence of Free-stream Disturbances on Low Reynolds Number Airfoil Experiments", Exp. in Fluids, 1, No. 1, pp. 3-14.

Nagib,H.M., Marion,A., Tan-atichat,J.,(1984), "On the Design of Contractions and Settling Chambers For Optimal Turbulence Manipulation In Wind Tunnels", AIAA Paper 84-0536, 22nd Aerospace Sciences Meeting, Reno, Nevada.

Pope, A., Harper, J.J., (1966) *Low-Speed Wind Tunnel Testing* , John Wiley and Sons, New York.

Purtell, L. P. and Klebanoff, P. S., (1979) "A Low-Velocity Airflow Calibration and Research Facility", Natl. Bur. of Standards Tech. Note 989.

Rae, W.H., Pope, A.,(1983) *Low Speed Wind Tunnel Testing*, John Wiley and Sons, New York.

Ramjee, V. and Hussain, A., (1976) "Influence of the Axisymmetric Contraction Ratio of Free Stream Turbulence", J. Fluids Engr., 98 .

Renken, J., (1976) "Investigation of the Three-Dimensional Flow in a Duct With Quadrangular, Variable Cross-Section By Means of the Panel Method", Deutsche Forschungs -und Versuchsanstalt fur Luft -und Raumfahrt, Institut fur Luftstrahlantriebe, Cologne, West Germany, Report No. DLR-FB 75-46.

Ribner, H. S. and Tucker, M., (1953) "Spectrum of Turbulence in a Contracting Stream", NACA Tech. Rep. 1123.

Sanderse, A. and van der Voreen, J., (1977) "Finite Difference Calculation of Incompressible Flows Through a Straight Channel of Varying Rectangular Cross-Section, With Application to Low Speed Wind Tunnels", NLR TR 77109 U.

Schubauer, G. B., Spangenberg, W. G., and Klebanoff, P. S., (1950) "Aerodynamic Characteristics of Damping Screens", NACA TN 2001, Natl. Bur. of Standards, Washington, D.C.

Shindo,S., Rae, W.H., Aoki, Y.,and Hill, E.G.,(1979) "Improvement of Flow Quality at the University of Washington Subsonic Wind Tunnel", AIAA Jour. of Aircraft, Vol.16, No.7, pp. 419-420.

Smith, R. H. and Wang, C., (1944) "Contracting Cones Giving Uniform Throat Speeds", J. Aero. Sci., pp. 356-360.

Stratford, B. S., (1959) "The Prediction of Separation of the Turbulent Boundary Layer", J. Fluid Mech., 5 .

Szczeniowski, B., (1943) "Contraction Cone for a Wind Tunnel", J. Aero.Sci., pp. 311-312.

Tighe,T., (1978) "The Philip P. Antonatos Subsonic Aerodynamic Research Laboratory (SARL)", AIAA Student Journal, Winter 1987/1988, pp.31-33.

Tsien, H., (1943) "On the Design of the Contraction Cone for a Wind Tunnel", J. Aero. Sci., pp. 68-70.

Uberoi, M. S.,(1956) "Effect of Wind-Tunnel Contraction on Free-Stream Turbulence", J. Aero. Sci., 23, pp. 754-764.

Van Den Broek,G.J. (1973) "The Calculation of a New Contraction Section For The CSIR Trisonic Wind Tunnel", Council for Scientific and Industrial Research, CSIR Report ME 1219, Pretoria, South Africa.

Varner,M.O., Summers,W.E., Davis,M.W., (1982) "A Review of Two-Dimensional Nozzle Design Techniques", AIAA Paper-82-0609, 12th Aerodynamic Testing Conference, Williamsburg, Virginia.

Veret, C. (Editor), (1980) "Flow Visualization IV", Proceedings of the Fourth International Symposium on Flow Visualization, Paris, France , Hemisphere Publishing Corp.

HIGH-REYNOLDS NUMBER LIQUID
FLOW MEASUREMENTS

Gerald C. Lauchle
Michael L. Billet
Steven Deutsch
Applied Research Laboratory, The Pennsylvania
State University, P.O. Box 30, State College, PA 16804

1.0 INTRODUCTION

Liquid flow facilities are usually categorized as towing tanks, seakeeping and maneuvering basins, circulating water channels, blowdown facilities, and water tunnels. A very complete listing of such facilities has been compiled by the 16th International Towing Tank Conference (ITTC) (1985) and it includes most of the major water facilities in the world. A description of new and planned cavitation facilities is given in Holl and Billet (1987). The intent of this chapter is to discuss the various experimental techniques used in hydrodynamic testing at high Reynolds numbers. This usually implies that water tunnels or tow tanks are used. Because Gad-el-Hak (1987) recently reviewed tow tank testing, our emphasis here is on water (or other liquid) tunnel measurements.

Although, in principle, a water tunnel is just a wind tunnel filled with water, there are many substantive differences in their operation that result from the differences in the properties of water and of air. Water, for example, is about 800 times as dense as air. Since the stresses associated with inertia are a function of fluid density, stresses in water (and, in particular, turbulent stresses) and any vibration they cause, will be larger by a factor of about eight hundred. Thus, to measure the statistical characteristics of turbulent fluctuating pressure fields, for example, a water flow will result in a significantly better signal level from the transducer than would be the case in air.

At the same time, the kinematic viscosity of water is lower than that of air by a factor of about 15; hence, a water tunnel is an ideal facility in which to carry out high-Reynolds-number experiments. From the point of view of fluctuating-stress measurements, even at the same Reynolds numbers, the stresses will be some 3-1/2 times larger in water than in air.

Because the speed of sound is about four and a half times greater in water than in air, the Mach number at a given velocity is 0.22 times that in air; hence, compressibility effects are much smaller. At the same Reynolds number, the Mach number will be 68 times smaller in water than in air; hence, radiation of sound to the farfield from turbulence-produced noise is substantially reduced (since this radiation depends on a high power of the Mach number).

Regarding structural excitation, the impedance match between water and most structural materials (being dependent upon the product of density and the speed of sound) is far better than that between air and most structural materials. At either the same Reynolds number or at the same stress levels, the velocities of the flows in air and water are far different; therefore, the modes excited in the structure will be quite different. Thus, the entire problem of hydroelastic interaction is quite different in water-tunnel operations.

In water the Prandtl number is nearly an order of magnitude higher than it is in air. This means that in water, the transport of heat by molecular motion is about an order of magnitude less effective than the transport of momentum; in air they are comparable. This makes some types of flow visualization, such as Schlieren and holographic photography, and dye injection much more effective in water tunnel testing than in wind tunnel testing.

1.1 Typical Water Tunnels

The Applied Research Laboratory at Penn State University (ARL Penn State) supports and maintains the operation of four (4) water tunnels and a glycerine tunnel. These tunnels are used for both basic and applied research. The Garfield Thomas, 1.22 m diameter Water Tunnel is the largest at ARL Penn State and is also recognized as the largest high-speed water tunnel in the free world. Its basic design and operation are discussed by Lehman (1959). The test section of the 1.22 m tunnel is 4.27 m long and supports water velocities of up to 18.3 m/s. Figure 1 summarizes other physical features of this water tunnel. The next largest water tunnel at ARL Penn State is schematically described in Fig. 2. Its circular test section is 0.30 m diam. X 0.76 m long. Also in use is a rectangular section 0.51 m X 0.11 m X 0.76 m long. Speeds of 24.4 m/s are achieved in the 0.3 m water tunnel. Figure 3 shows a schematic of the 0.15 m water tunnel. Its test section is 0.61 m long and water speeds of 21.3 m/s are achieved. The fourth water tunnel in use at ARL Penn State is an ultra-high speed cavitation tunnel, Fig. 4. Speeds of 83.8 m/s are achieved in the 3.8 cm diameter test section. The facility is particulary well suited to cavitation damage research (Stinebring, et.al., 1980) although some work on cavitation in liquid freon has also been performed, e.g., Holl, et.al. (1975). Used exclusively for fundamental turbulence research, the Boundary Layer Research Facility (BLRF) supports fully-developed turbulent pipe flow of glycerine. Bakewell and Lumley (1967) designed and fabricated this highly unique glycerine tunnel (Fig. 5) which has a test section 0.30 m in diameter X 7.6 m long with a mean flow velocity of 6.1 m/s. The velocity remains constant and the Reynolds number is varied by altering the viscosity of the glycerine through temperature control.

The use of water or liquid tunnels, such as those described above, in high-Reynolds number hydrodynamics and hydroacoustics research is the subject of this chapter. Five general areas of measurement are described: Body Force Measurements, Boundary Layer Measurements, Multiphase Flow Measurements, Hydroacoustic Measurements and Internal Flow Measurements. In any given typical test program, there may be much overlap among these various measurements, but each has its own unique set of procedures. Those techniques, illustrated by examples, are presented here. The chapter represents an extension of two previously published articles on water tunnel

1.22 m WATER TUNNEL (Garfield Thomas)

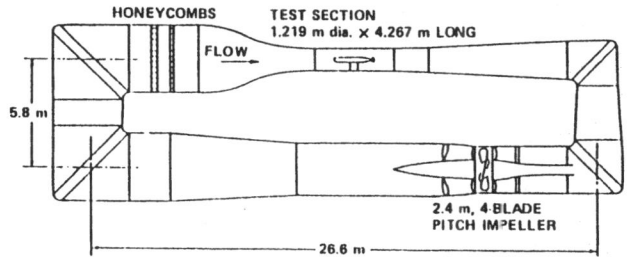

DESCRIPTION OF FACILITY: Closed Circuit, Closed Jet

TYPE OF DRIVE SYSTEM: 4- Blade Adjustable Pitch Impeller

TOTAL MOTOR POWER: 2000 HP Variable Speed (1491 kw)

WORKING SECTION MAX. VELOCITY: 18.29 m/s

MAX. AND MIN. ABS. PRESSURES: 413.7 to 20.7 kPa

Figure 1. The 1.22 m Diameter Garfield Thomas Water Tunnel

30.5 cm WATER TUNNEL

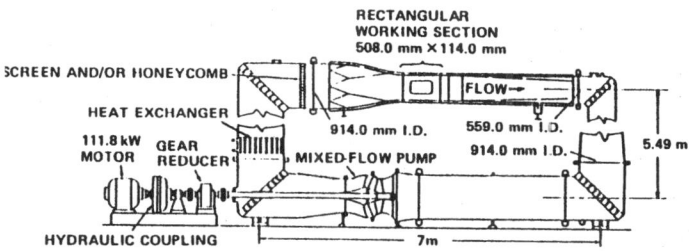

DESCRIPTION OF FACILITY: Closed Circuit, Closed Jet

TEST SECTIONS: 1) Circular: 304.8 mm dia. × 762.0 mm long

    2) Rectangular: 508.0 mm × 114.3 mm
      × 762.0 mm

TYPE OF DRIVE SYSTEM: Mixed Flow Peerless Pump

TOTAL MOTOR POWER: 150 HP (111.8 kw)

WORKING SECTION MAX. VELOCITY: 24.38 m/s

MAX. AND MIN. ABS. PRESSURES: 413.7 to 20.7 kPa

Figure 2. The 30.5 cm ARL Penn State Water Tunnel

15.2 cm WATER TUNNEL

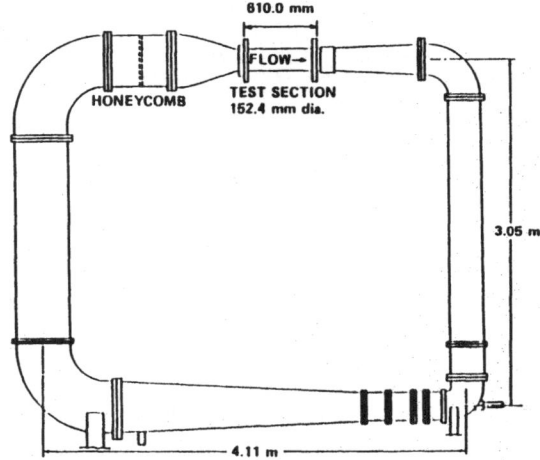

DESCRIPTION OF FACILITY:	Closed Circuit, Closed Jet
TYPE OF DRIVE SYSTEM:	Axial-Flow Pump
TOTAL MOTOR POWER:	25 HP (18.64 kw)
WORKING SECTION MAX. VELOCITY:	21.34 m/s
MAX. AND MIN. ABS. PRESSURES:	861.9 to 20.7 kPa

Figure 3.   The 15.2 cm ARL Penn State Water Tunnel

ULTRA-HIGH SPEED WATER TUNNEL

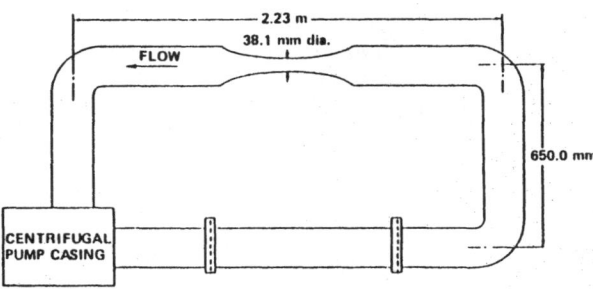

DESCRIPTION OF FACILITY:	Closed Circuit, Closed Jet
TYPE OF DRIVE SYSTEM:	Centrifugal Variable Speed Drive
TOTAL MOTOR POWER:	75 HP (55.9 kw)
WORKING SECTION MAX. VELOCITY:	83.8 m/s
MAX. AND MIN. ABS. PRESSURES:	8274.0 to 41.4 kPa

Figure 4.   The 3.8 cm ARL Penn State Ultra-High Speed Water Tunnel

BOUNDARY LAYER RESEARCH TUNNEL (GLYCERINE)

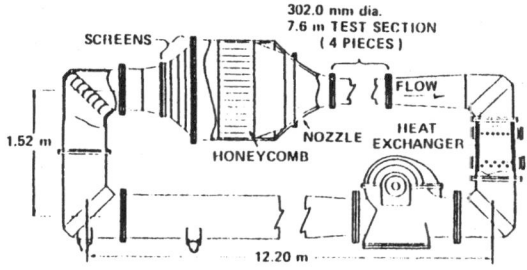

DESCRIPTION OF FACILITY:	Constant Dia. Test Section, Closed Circuit
TYPE OF DRIVE SYSTEM:	Gould Centrifugal Pump
TOTAL MOTOR POWER:	100 HP (74.6 kw)
WORKING SECTION MAX. VELOCITY:	7.62 m/s
MAX. AND MIN. ABS. PRESSURES:	586.0 to 101.4 kPa
TEMPERATURE RANGE:	35^0C to 46^0C
REYNOLDS NUMBER RANGE:	10,000 - 26,500

Figure 5.   The 30 cm ARL Penn State Glycerine Tunnel

testing by Henderson and Parkin (1982) and Parkin (1983). As a preface to specific types of measurement, a brief description of typical water tunnel operation is presented.

2.0 WATER TUNNEL OPERATION

The closed-circuit water tunnel is usually composed of a series of interconnected cylindrical shells of varying cross section. The test section, or working section, is usually a true cylinder. The settling section and nozzle are upstream of the test section and is where turbulence management apparatus is located. Downstream of the test section the flow velocity is slowly decreased by a diffuser section. In order to complete the closed circuit loop, turns are required. These turns must be designed with turning vanes to minimize losses. The pump is located in the lowest leg of the tunnel where the hydrostatic pressures are highest to help suppress pump cavitation. Also the cross sectional areas are largest in this leg so that the flow velocity is minimized.

The two fundamental variables in the test section are velocity and pressure. A well-designed tunnel provides for accurate control of these variables over a wide range. Speed control in the 1.22 m diameter water tunnel at ARL Penn State is achieved by varying the pump impeller rotational speed and the pitch of the impeller blades. This pump is driven by a 2000 HP variable-speed (0-180 rpm) induction motor.

The pitch can be varied over 28 deg. by means of a hydraulic servomechanism which is operated remotely from the tunnel operating console.

The static pressure in the test section must be maintained at a pre-determined value regardless of velocity. Using as an example the 1.22 m water tunnel, this pressure control is achieved by varying the air pressure on the top of a large pressure-regulating tank connected to the bottom leg of the tunnel. This tank (approximately 1 m in diameter X 4 m high) has in it a redwood float that prevents the air from coming in contact with a large surface area of water; a necessary feature when concerned with gas content effects in some hydrodynamics work.

A schematic of the pressure regulating control system is shown in Fig. 6. The air pressure in the pressure-regulating tank, and thus the tunnel pressure, is controlled by a differential pressure transmitter (23), which is supplied with two pressures, one from the working section and the other from the pressure-regulating valve on the console (22). The output from the transmitter is an air pressure whose value indicates whether the working-section pressure is equal to or deviates from the desired pressure. The output is delivered to a controller (25), which will adjust the working-section pressure if a difference exists. The adjustment is achieved by supplying an operating pressure to control valves (15) and (16), which connect the low-pressure-regulating tank with the low- or intermediate-pressure receivers. If, for example, the working-section pressure is low, the valve (15) to the low-pressure receiver, which in normal operation is slightly open, will close and so stop the flow

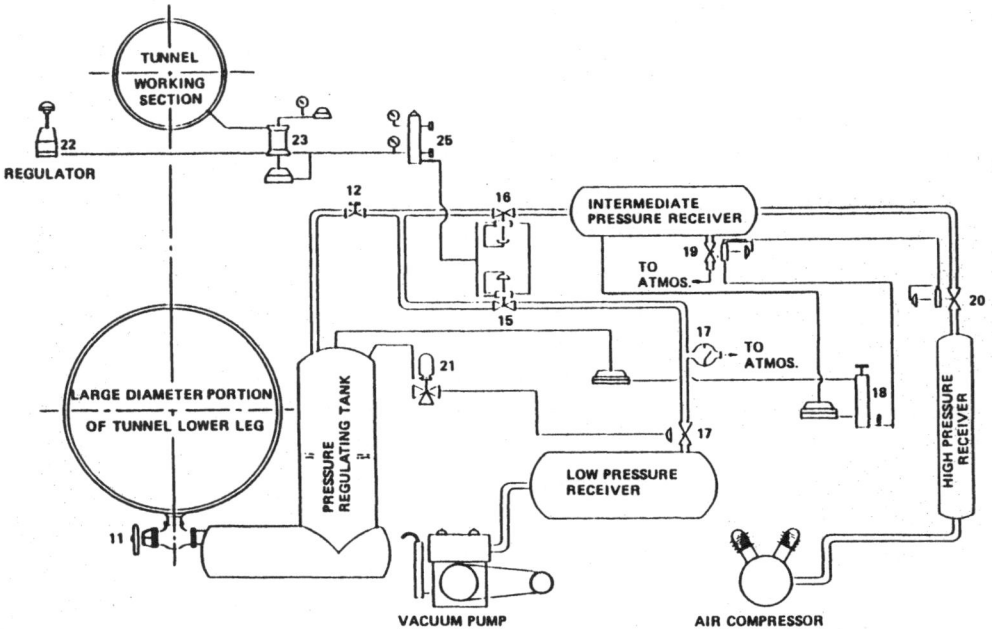

Figure 6.   Control System for Tunnel Pressure Regulation
(from Lehman, 1959)

from the regulating tank to the low-pressure receiver. At the same time, the valve (16) - which also normally operates while slightly open - will be opened wider so as to permit a greater flow from the intermediate-pressure receivers to the pressure-regulating tank. This situation will continue until the transmitter senses that the tunnel has reached the desired pressure and causes the controller to restore the control valves to their neutral positions. In addition to the control of pressure and velocity, there are other variables that must be considered in specific types of testing. The more important of these are now discussed.

2.1 Water-Conditioning

A water-conditioning system, or bypass system, is shown for the 1.22 m Water Tunnel in Fig. 7. It is used to filter the water, degas the water and if required change the temperature of the water. Because the flow through such a system may be significant (typically 65 l/s), the automatic pressure control system is tied into the bypass circuit (11). With the water level in the pressure-regulating tank equal to the set value, the throttle valve (39) will be adjusted to maintain the flow from the tunnel exactly equal to the flow to the tunnel, that is, from the degasser through the adjustable speed pump to the water tunnel. As the water level deviates from the set value, the throttle valve will open or close as required, increasing or decreasing the flow from the tunnel to the degasser so as to restore normal conditions.

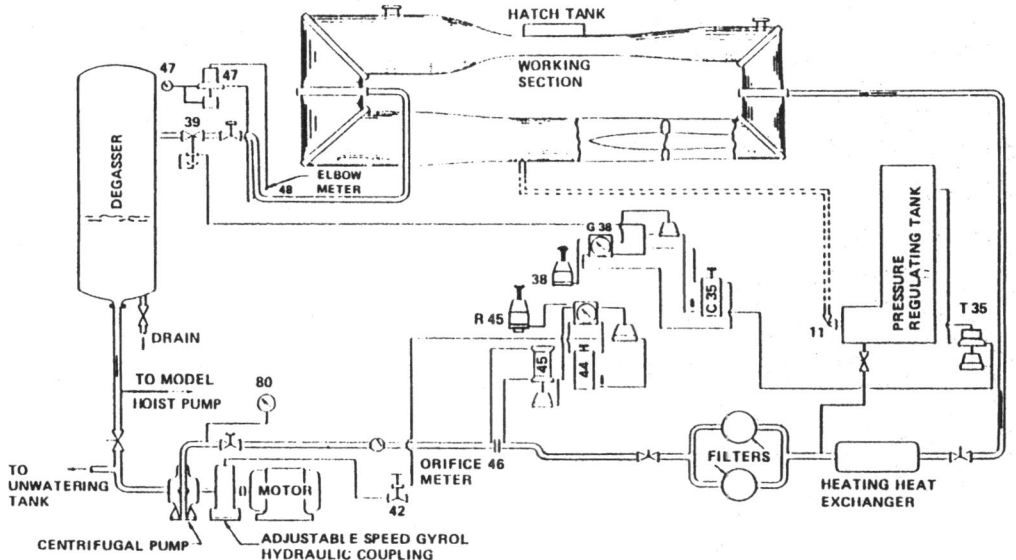

Figure 7. Tunnel Bypass System (from Lehman, 1959)

All water that enters the tunnel from the main lines or well must enter through the bypass system, in which there are two filters of 10 μm porosity. With the bypass system operating, tunnel water is continually filtered. Of particular importance in cavitation research and in some types of drag reduction research is the air content level of the water. Gas removal from the water is accomplished with a Cochrane Cold-Water Degasifier located in the bypass circuit. The degasser consists of an upright cylindrical tank approximately 2 m in diameter and 6 m high. The upper portion of this tank is filled with plastic saddles which are approximately 3.8 cm high and 7.6 cm in outside diameter. Water taken from the tunnel through the bypass system is sprayed into the top of the tank, the interior of the tank being maintained at a high vacuum of approximately 71 cm of mercury. The water spray falling over the saddles distributes the water in thin films, and exposure of these films to the vacuum permits the rapid removal of dissolved air. The water collects in the bottom of the degasser and returns to the bypass system.

Operation of the degasser permits the lowering of the gas content in the tunnel from about 15 parts per million by molecular weight to around two or three parts per million in approximately three hours. On occasions when an increase in the gas content is desired, one of two methods are utilized. Small increases in gas content are achieved by using the Cochrane Cold-Water Degasifier located in the bypass circuit with the tank pressurized. When levels near saturation are required, fresh water is added to the tunnel circuit. Air is not normally added to the tunnel to increase the air content level because often times large bubbles persist.

The Reynolds number of a given flow in a water tunnel can be varied by changing either the tunnel velocity or the water temperature (or both). The 1.22 m tunnel provides (as part of the bypass system) for water heating. The useful temperature range of 10 to 49 deg. C permits the Reynolds number to be changed by a factor of two (2). This capability has proved useful when measuring drag coefficients on small bodies. The small bodies are necessary to minimize spurious measurement errors due to wall effects, but by operating at high temperatures, the Reynolds number range of a larger body is preserved. Higher than normal water temperatures have also been used in transition research where the demonstration of high transition Reynolds numbers by wall heating or suction was the goal. The higher water temperature effectively raises the unit Reynolds number range of the facility.

2.2 Turbulence Control

Low freestream turbulence levels in the working section of a water tunnel are absolutely necessary for many studies in hydrodynamics. This is particularly true when studying high-Reynolds number laminar flows and the transition process. Although screens have been used to reduce turbulence in wind tunnels, they are not particulary satisfactory in water tunnels; that is, large, high-speed water tunnels. The problem with screens is strength. The wire diameter required for a given velocity of operation almost always results in a wire diameter Reynolds number range where vortex shedding occurs. This shedding causes the familiar "singing" phenomenon.

Based on the fundamental work of Lumley (1964) and Lumley and McMahon (1967), honeycombs in which the cells have a large length-to-diameter ratio have been used successfully to reduce turbulence levels to order 0.1% in the ARL Penn State tunnels. The honeycomb has two effects: it reduces the level of existing turbulence, and it creates turbulence of its own. With the right design of honeycomb, the combined effect is favorable. The honeycomb completely annihilates the transverse components (u_2 and u_3) of velocity, and so only the u_1 component remains immediately downstream of the honeycomb. However, the turbulence has a very strong tendency to rearrange itself and return to an isotropic state. This state is influenced to a significant degree by the flow within individual cells. It is advantageous to have fully-developed turbulent flow in the cells as opposed to laminar or transitional flow. This surprising conclusion (Lumley and McMahon, 1967) is based on the fact that a laminar wake exiting the honeycomb cell is actually larger than a turbulent wake and it produces more turbulence than is lost in the laminarization of the cell flow. It is therefore accepted procedure to leave rough edges on honeycomb that has been sawed to size. This roughness lowers the cell transition Reynolds number.

Placement of turbulence management honeycomb in the plenum section just upstream of the nozzle section has advantages. The turbulence at the exit of the honeycomb enters the nozzle contraction and eddies are stretched (elongated) axially. This upsets the isotropic state again and the fluctuating transverse velocities are reduced. The degree of turbulence suppression by the nozzle depends on the contraction ratio. Ramjee and Hussain (1976) report on these effects of contraction ratio, and show that the larger the contraction, the greater the turbulence suppression.

For example, the 1.22 m Water Tunnel uses honeycomb with hexagonal cells 47.6 cm long X 0.56 cm across the flats. This is placed 1.39 m upstream of the nozzle which has a 9:1 contraction ratio. Robbins (1978) measured the turbulence levels and length scales in both the plenum and test section before and after installation of the honeycomb. Figure 8 shows the axial component of turbulence intensity in the test section. The level of 0.1% for velocities greater than 10 m/s is quite acceptable for most work. The high turbulence levels at lower velocities are a result of transitional flow in the honeycomb and support the Lumley and McMahon recommendation noted above. The open circles of Fig. 8 represent plenum data extrapolated to the test section using theory developed by Lumley and McMahon (1967).

A high-Reynolds number laminar pipe flow facility is described by Barker and Gile (1981). They used a honeycomb and screens upstream of a 35:1 contraction ratio nozzle to achieve test section turbulence levels of order 0.01% which is better than most wind tunnels. It demonstrates that turbulence levels in water tunnels can be made extremely low provided one can justify the large contraction ratio nozzle and entrance flow control. With very large contraction ratios, the boundary layer separates at the test section entrance which then feeds new turbulence into the working volume. Barker and Gile needed to provide wall suction in this critical region to control the separation.

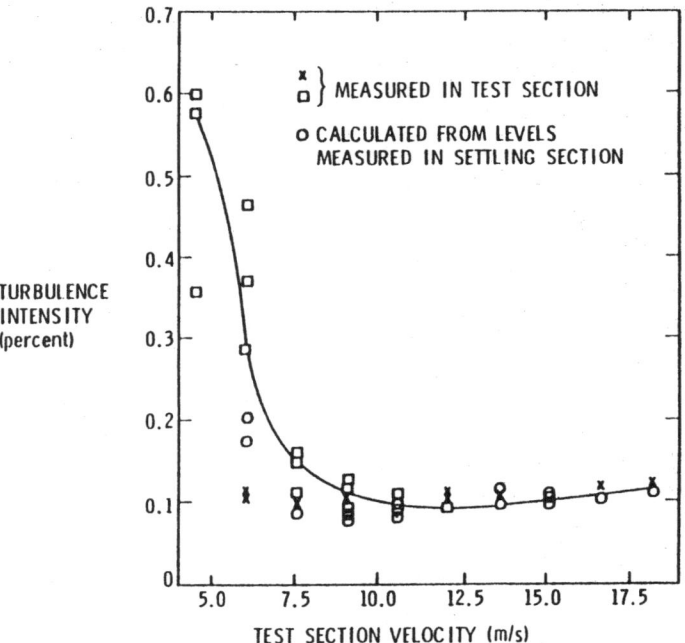

Figure 8.   Turbulence Intensity (axial component) in 1.22 m Diameter
Water Tunnel Test Section

2.3 Noise Control

Unfortunately, the control of background noise in a water tunnel is at least an
order of magnitude more difficult than the control of turbulence, water quality,
pressure, or velocity. The reason is hydroelastic in nature. The metallic materials
used in water tunnel construction have a density comparable to that of water.
Conventional methods of noise path attenuation become severely limited because there
are two paths of transmission: one through the water and the other through the
structure. One can place an acoustic baffle in the liquid that attenuates the water-
borne path, but the energy on the source side of the baffle "jumps" into the structure
at that point, bypasses the baffle, and then re-enters the water and continues to
propagate through the water. The same thing happens if one baffles the structure, say
with a rubber gasket at a joint. Structure-borne energy leaves the structure at the
gasket, travels through the water, and then back into the structure. Thus the only
way to attenuate the duct modes of acoustic energy flow is by blocking both water and
structure paths simultaneously and at the same point. Attempts at doing this by many
organizations to existing facilities have been generally unsuccessful. However,
several planned facilities have attempted to incorporate these ideas in their design
(Holl and Billet, 1987).

Water tunnels that are relatively quiet, have relatively quiet sources of noise.
The primary source is the pump. Secondary sources are separation zones in turns and

wall turbulent boundary layers. A typical tunnel can be quieted by 15 to 30 dB through complete suppression of pump cavitation. The CALTECH High-Speed Water Tunnel (Barker, 1974), the NUSC, New London Acoustic Water Tunnel (Schloemer, 1974), and the ARL Penn State 1.22 m Water Tunnel (Lauchle, 1977) are all capable of operating cavitation free; thus, they are fairly quiet tunnels. The NUSC, New London tunnel also uses fire hose in parts of the circuit to help suppress vibration.

Even with a quiet pump and well-designed turns, the turbulent boundary layer in the test section and diffuser radiate significant acoustic energy. Typical spectral data for this clear tunnel noise are shown for the CALTECH and ARL Penn State tunnels in Fig's. 9 and 10, respectively. Even though the 1.22 m tunnel is larger than the CALTECH tunnel by a factor of about four (4), it is quieter. The reason is due to the physics of turbulent boundary layer (TBL) noise, e.g. Tam (1975). The noise spectrum peaks when $\omega\delta^*/u \approx 8 \times 10^{-2}$ (ω – radian frequency, δ^*–TBL displacement thickness, u–velocity). The smaller water tunnel supports a thinner TBL at the end of the test section than does the larger tunnel at the same velocity. Consequently, the peak in TBL noise occurs at a higher frequency; the roll-off portion of the TBL noise spectrum is at a higher level and is within the frequency range shown on Fig's. 9 and 10. At ultra-low frequencies, the larger water tunnel would be the noisier of the two.

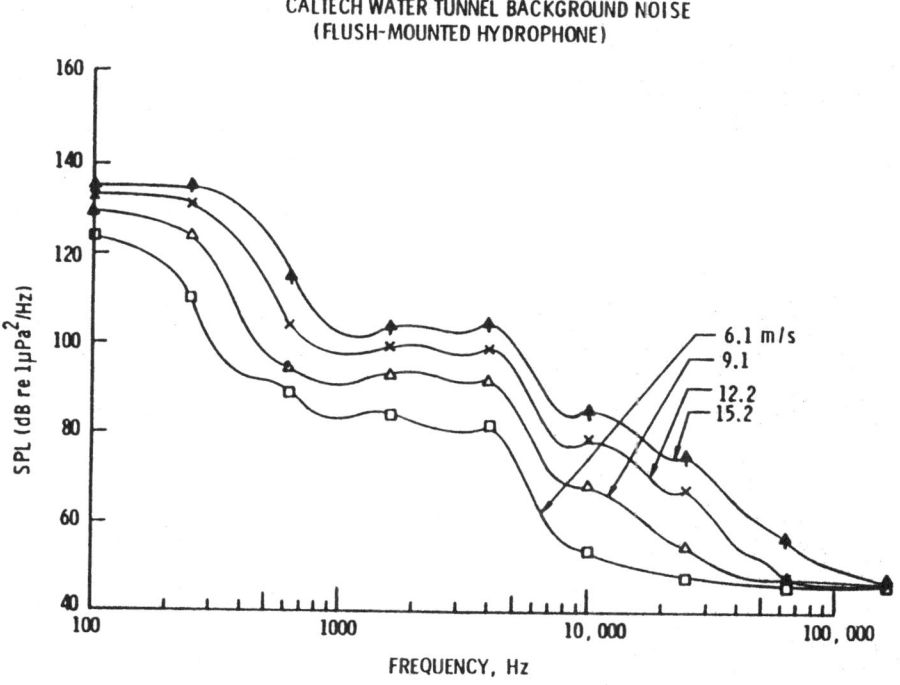

Figure 9. Clear Tunnel Background Noise of CALTECH Water Tunnel

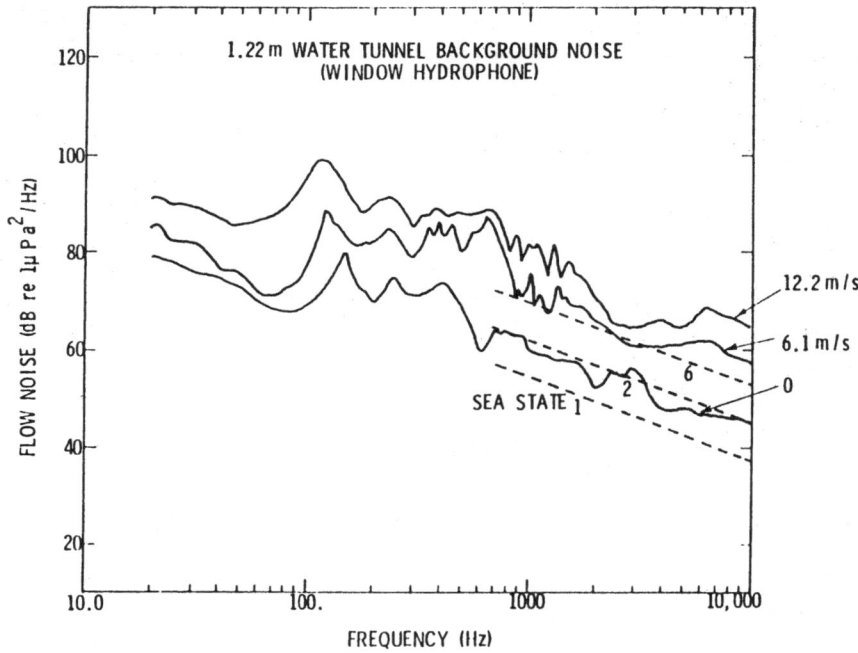

Figure 10.   Clear Tunnel Background Noise of ARL Penn State 1.22 m
Water Tunnel

3.0   BODY FORCE MEASUREMENT

Although the standard facility used to determine body forces and moments, and
powering (or propulsion) characteristics of marine-type vehicles is the towing tank
these facilities are limited to relatively low velocities (5-8 m/s) for completely
submerged bodies. This limitation is caused by surface wave effects and the water-
piercing strut interference drag. The Reynolds number range is often low and accurate
scaling of the powering or force data to larger prototype conditions is difficult.

The water tunnel is most useful for the testing of completely submerged bodies.
Higher Reynolds numbers can be achieved so that the scaling of data is more accurate.
If the body frontal area is greater than about 1/40 the test section cross sectional
area, then complications may arise from tunnel wall interference effects, e.g. Ross,
et.al. (1948). The tunnel wall causes the streamlines over the body to be modified
which changes the body pressure distribution from that observed in a free-field
environment. In addition, the skin friction and displacement thickness growth over
the body and tunnel wall give rise to a spurious axial force component called
horizontal buoyancy. Much of the potential flow blockage effect and part of the
horizontal buoyancy effect can be eliminated by using a flow correcting liner attached
to the test section wall, Fig. 11. The fundamental research on liner design was
performed by Lehman, et.al. (1958) and Peirce (1964). The basic premise in liner
design is that if the tunnel wall were to conform exactly to one of the outer stream

surfaces formed when the body operates in the free stream, then all forces acting on
the body would be unaffected by the tunnel wall. The design accuracy therefore
depends upon the accuracy that one can predict the stream surfaces over arbitrary
bodies. As computational tools and techniques improve, so do tunnel liner designs.

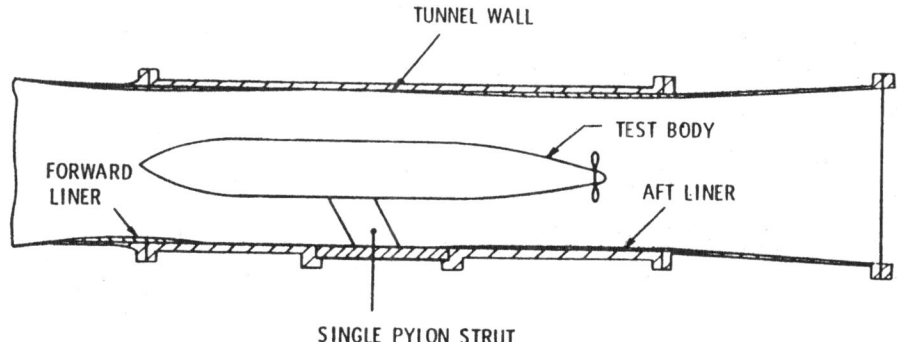

Figure 11. Installation of a Large Body in Water Tunnel Test Section

The liner design resulting from potential flow calculations can (and should) be
modified to account for the boundary layer displacement thickness growth over the
liner itself. Axisymmetric viscous flow codes can be used for this. The displacement
thickness calculated is added to the coordinates of the liner specified in the
potential flow calculation. This step accounts for some of the horizontal buoyancy
effect, but not all of it. The skin friction on both the body and tunnel wall add to
the tunnel interference and cannot be treated easily in the liner design. It is
usually handled empirically.

As an example of the empirical procedure, consider the measurement of net axial
force on a large axisymmetric body operating in a cylindrical test section with liner.
(The force sensors are described in Section 3.1). This body is large enough that
tunnel interference effects are present. The body is presumed to have an operating
propulsor. A standard propulsor powering test entails measuring the net body axial
force, C_x, as a function of the propulsor advance ratio, e.g., Fig. 12 (a). At the
self propulsion point, $C_x = 0$. However, $C_x =$ (Thrust) minus (Bare Body Drag) (and
bare body drag is affected by tunnel interference) so the true self propulsion point
cannot be determined from the data measured. One must remove the propulsor and
measure the bare body drag coefficient as a function of Reynolds number, Fig. 12(b).
These data must then be compared to the drag coefficient data measured on a very small
body of the same geometry. Usually, the Reynolds number range can be maintained
between the two tests by operating the small body at much higher speeds and by heating
the tunnel water. The difference between the two data sets is the empirical tunnel
correction. It is applied, as shown in Fig. 12(a), to the powering data. The primary
reasons why the entire test is not conducted at the very-small body scale is that
propulsors cannot be manufactured accurately at those scales. Also, the small

electric motors needed to drive the propeller shaft do not supply sufficient power to cover the advance ratio and Reynolds number range of interest. Additional details on this procedure are given by Hall (1973).

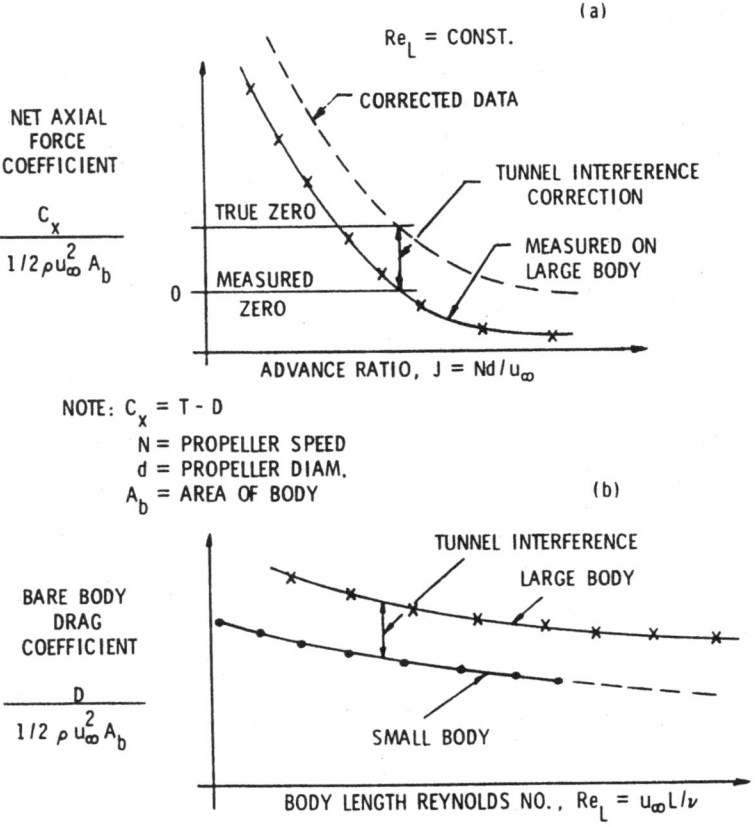

Figure 12. Tunnel Interference Correction Scheme; (a) Net Axial Force Coefficient for Body with Propeller; (b) Bare Body Drag Characteristics

Tunnel wall and horizontal buoyancy effects have a strong influence on the body static pressure distribution. Figure 13, for example, shows a predicted potential flow pressure distribution (which accounts for solid blockage effects) over a large streamlined body whose maximum diameter is 26% of the tunnel diameter (r_w – tunnel wall radius). Also shown on this figure are data obtained at various Reynolds numbers. The discrepancy between data and theory is the horizontal buoyancy effect which is viscous in origin. Lauchle (1979, 1983) has analyzed this problem and derives a closed-form solution for its correction. Figure 14 shows a result of the correction procedure applied to the potential flow solution. The agreement with data is good. One could use this approach, which entails calculating the boundary layer and shear stress distributions over the body and tunnel wall, to estimate accurate pressure distributions. Accurate pressure distributions are required when studying laminar boundary layer stability and transition, e.g. Lauchle, et.al. (1980).

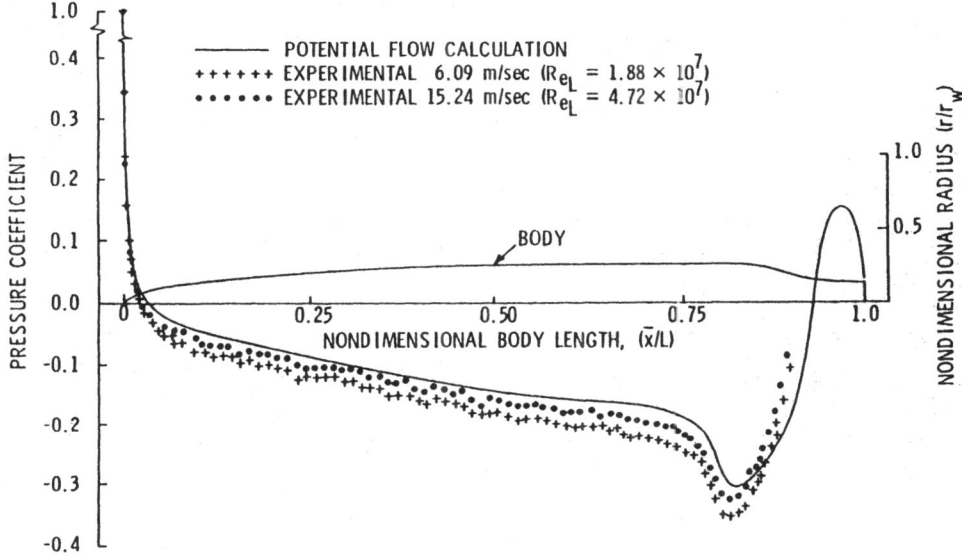

Figure 13. Measured and Predicted Body Pressure Distribution for a Large Body in the 1.22 m Water Tunnel

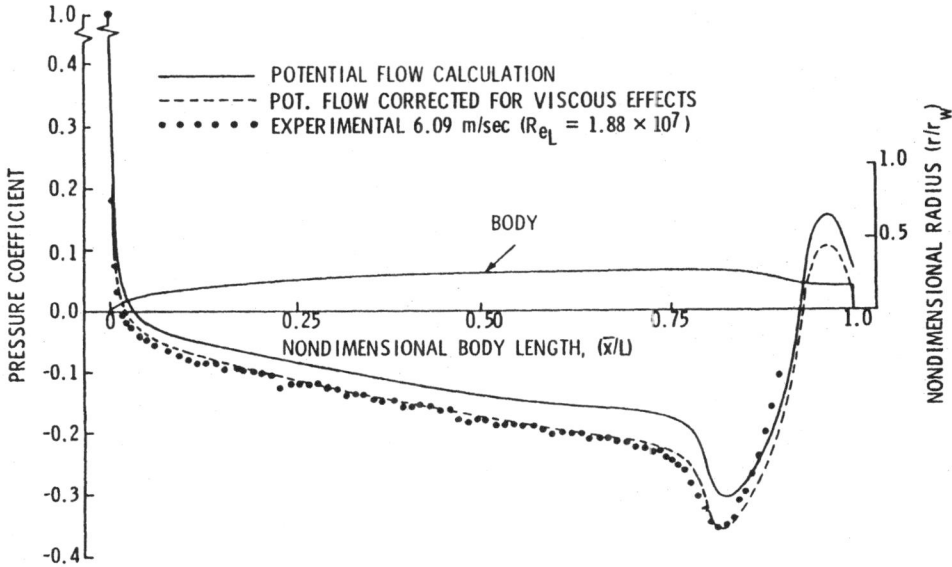

Figure 14. Corrected In-Tunnel Pressure Distributions Accounting for Horizontal Buoyancy Effects

3.1 Force Sensors

Measuring body forces and moments on water tunnel test bodies requires devices that are stiff and insensitive to changes in tunnel pressure. These devices are called balances. The basic element in balance design is the tension member, torque tube, or shear beam. Each balance is designed to meet the objective of the given test program.

A typical force cube with a tension number (Fig. 15) is instrumented with strain gages. The very small elongation of the member, caused by the hydrodynamic force being measured, is sensed directly by the gages. The force and strain are parallel to each other which minimizes cross coupling with other forces. The thin webs that connect the upper tension member side to the base (or ground) side of the cube are called flexure members; they act as double cantilever beams. Notice that the device is extremely rigid in the vertical plane while weak in the horizontal plane. However, the pre-tensioned tension members restore rigidity to the horizontal motion. In operation, the center of the tension member remains fixed to a rigid portion of the model strut system, while the ends are fastened to the body skin subject to the hydrodynamic load of concern. The load causes a microscopic translation of the cube in the left/right direction. One side of the tension member elongates while the other relaxes. The four-arm bridge becomes unbalanced and a voltage is created in the wheatstone bridge circuit that is proportional to load.

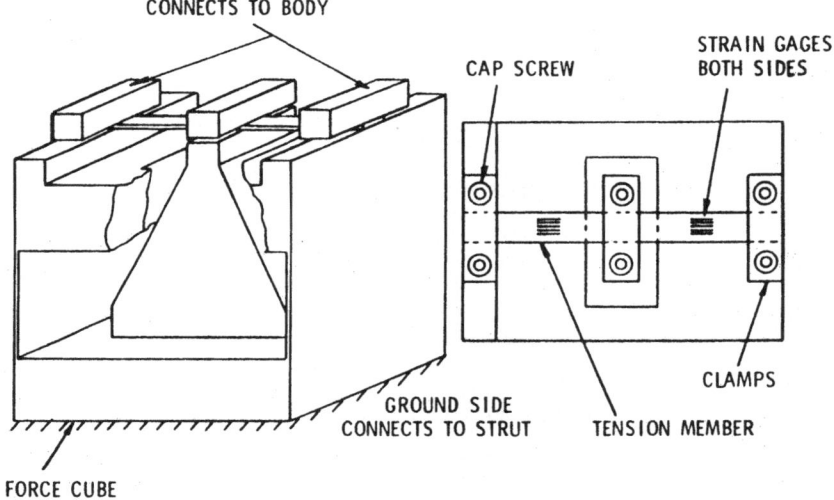

Figure 15. Force Cube - Tension Member Design Concept

Shear beam balances are schematically shown in Fig. 16. In these devices, a web replaces the tension member and the applied force causes shear in the web. The strain gages are placed in a 45° direction to the force so that the principal stress is sensed. The modified shear balance has a reduced area web to increase the shear stress which improves sensitivity further.

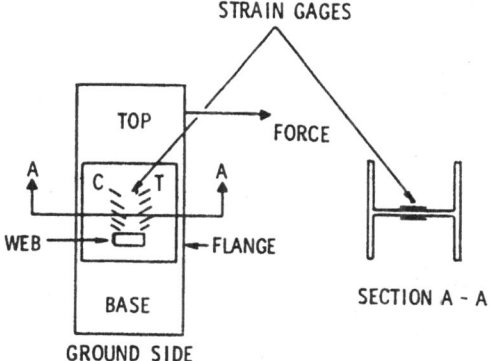

SHEAR BEAM BALANCE

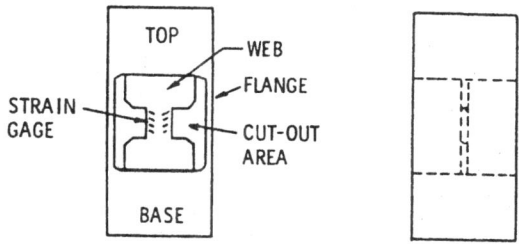

MODIFIED SHEAR BEAM BALANCE

Figure 16.   Shear Beam Balance Design Concepts

A torque tube (or cell) is used to measure moments about its longitudinal axis. This includes hinge moments on control surfaces and torque on propulsor shafts.   The typical torque cell is shown in Fig. 17.   In rotating systems, slip rings are required to bring the strain gage signals out of the test apparatus.

In any given test program, the test body may employ several steady-state force sensors simultaneously.   Figure 18 shows a typical experimental setup in which body axial force, propulsor shaft torque, thrust, and control surface hinge moment are all measured.

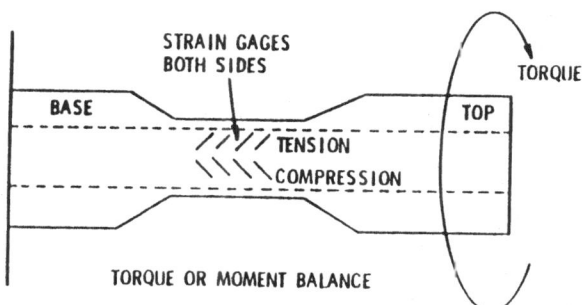

Figure 17.   Torque or Moment Balance Design Concept

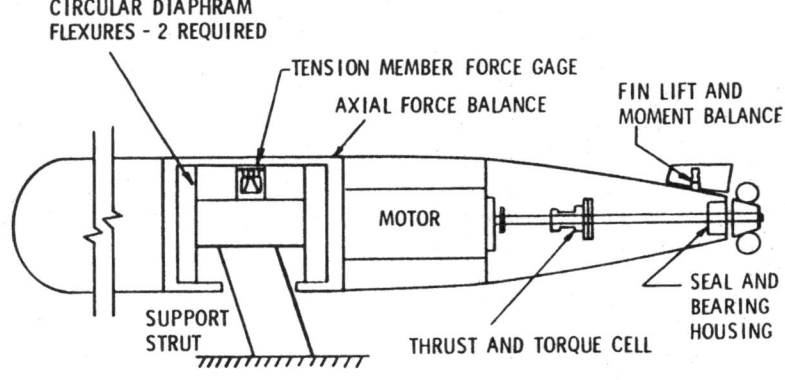

CIRCULAR DIAPHRAM
FLEXURES - 2 REQUIRED

TENSION MEMBER FORCE GAGE

AXIAL FORCE BALANCE

FIN LIFT AND
MOMENT BALANCE

MOTOR

SEAL AND
BEARING
HOUSING

SUPPORT
STRUT

THRUST AND TORQUE CELL

TYPICAL SINGLE PROPULSOR MEASURING SYSTEM

Figure 18. Typical Application of Force and Moment Balance in Water Tunnel
Model

The frequency response of strain-gaged force balances is generally poor compared
to other dynamic sensors. However, for frequencies less than about 250 Hz, they can
provide reliable time-dependent force data. Figure 19 shows an example of this. A
large, electrically-heated body (Lauchle and Gurney, 1984) was equipped with a drag
force balance in the sting mount. The body could be made to have intermittent laminar
flow by adjusting the heating level (heat stabilizes laminar layers in water) to a
critical level for the velocity of flow considered. Turbulent spots were created
randomly over the body surface and these cause time-random changes in skin friction.
A hot film senses the spots where a conditioned hot film output is shown as I(t) on
Fig. 19. Here, I(t) is equal to one when a spot is present. The simultaneous drag
signal is also shown. The data show clearly the correlation between the presence of
turbulent spots in a laminar boundary layer and the instantaneous increases in body
drag. Some spikes in drag are not correlated with the indicator function because some
spots occur on the opposite side of the body from where the hot film is located.

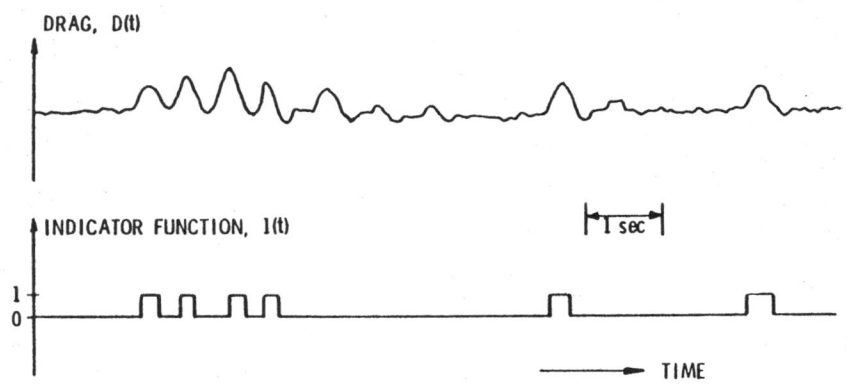

DRAG, D(t)

INDICATOR FUNCTION, I(t)

1 sec

1
0

TIME

Figure 19. Time Dependent Drag Data Compared with Occurrence of Turbulent
Spots on a Laminar Flow Body

Sensors designed specifically for unsteady force measurements usually use a piezoelectric crystal rather than strain gages. This is due primarily to resonance problems with strain gage beams. An application of this technique to the measurement of time-dependent propeller shaft forces is given by Thompson (1976). Non-steady lift measurement techniques are reviewed by Sevik (1964), and Lauchle (1974) presents a technique for measuring unsteady hinge moments.

4.0 BOUNDARY LAYER MEASUREMENTS

4.1 Turbulent Boundary Layers

Turbulent boundary layer measurements in high-speed water tunnels are in general made difficult by the high dynamic pressures involved, by the conductivity and general uncleanliness of the water, by cavitation and by the very small inner boundary layer scales. Of these potential problems, all but the last can be circumvented (more or less simply) in some way, and moreover, are not problems unique to boundary layer measurements. We shall concentrate, therefore, on the problem of very small inner boundary layer scales, and the constraints that they put on the types of measurements we can make and how we have learned to work within these constraints. As in recent years much of the boundary layer work done at ARL Penn State has been driven by the need to understand the mechanics of turbulent drag reduction--a phenomenon dominated by near wall effects--these constraints have been keenly felt.

Typical values of the momentum thickness Reynolds number, for fully developed turbulent boundary layers, which have been measured in our laboratory range from 3000 to 15000. Using approximate relationships found in White (1974), for example, we find sublayer thicknesses ($y^+ - 5$) which vary from 5 microns at 18 m/sec to 40 microns at 3 m/sec. Some measured data are shown in Table 1, which is reprinted from Deutsch and Castano (1986). The friction velocity, u_τ, is found for the data shown in Table 1 by requiring a least square fit of the data to the law of the wall (Coles and Hirst, 1968).

We have been most successful, over the past decade, with techniques that measure at the surface of the flow and with LDV techniques. Surface measurement techniques have included skin friction balances, flush mounted hot film gauges and pressure probes. Pressure probes are discussed further in Section 6.3, so that we shall concentrate here on the other three techniques.

Central to all the skin friction balance measurements that we have made, have been floating elements supported by strain-gauged force members. Early designs used the force member in bending, and although this allowed for excellent sensitivity, motion of the balance with increasing velocity was fairly large, typically on the order of several hundred microns for a speed change of 15 m/sec. This movement of the balance section made installation exceedingly difficult. In addition, it raised questions about the alignment of the balance with speed. Subsequent balance sections used an I-beam force member in shear. It was found that sensitivity could be maintained, while the stiffer balance sections were much easier to align and to maintain aligned. Movement of the shear beam balances is on the order of 2.5 microns.

TABLE 1. Comparison of LDV data.

Flat Plate

Location[*] (mm)	U_∞ (m/sec)	R_θ	u^+ (m/sec)	y at y^+ =5 (μm)
272	4.68	3372	0.195	25.6
	10.90	7358	0.423	11.8
374	4.75	4215	0.190	26.3
	10.30	8862	0.392	12.7
454	4.78	5127	0.188	26.6
	10.47	10627	0.392	12.7

Axisymmetric Body

Location[**] (mm)	U_∞ (m/sec)	R_θ	u^+ (m/sec)	y at y^+ =5 (μm)
155	5.26	2635	0.21	23.8
	10.71	5062	0.405	12.3
	16.94	8837	0.607	8.24
277	5.25	3491	0.21	23.8
	10.77	6764	0.406	12.3
	16.86	10144	0.617	8.10
403	5.30	3775	0.210	23.8
	10.90	8191	0.413	12.11
	16.88	14197	0.594	8.42

[*] From Virtual Origin
[**] From Trip Wire

Typically, we have used balance sections with surface areas of 2.5 x 10^{-3} m^2 in fully developed turbulent boundary layers. The state of the boundary layer is determined beforehand through LDV measurements. On a skin friction balance of this size the force will vary from 0 to about 10 Newtons over the speed range of 0 to 17 m/s. Some appreciation for the linearity and sensitivity of the balance can be obtained by considering a sample output, which is shown in Fig. 20[1].

Problems with balance measurements are generally caused by pressure gradient effects, balance misalignment and, in water, waterproofing. Waterproofing of the strain gauge section is certainly more art than science and we merely note that with too thick a layer of waterproofing one will trade linearity for hystersis, while too

thin a coating will lead to electrical shorting and balance failure. Uncertainty in
the measured pressure gradient is magnified in the skin friction measurement. For
example, a nominally zero, measured, pressure gradient is shown in Fig. 21. If the
variation of this pressure gradient, which is about 0.7% of the dynamic head, is
treated as a pressure gradient across the balance, it would lead to a 10% uncertainty
in the skin friction measurement. We have not been successful in measuring the
pressure within our balance gaps, perhaps because of our relatively small balance
sections, but this measurement has been successful elsewhere and is certainly worth
the effort.

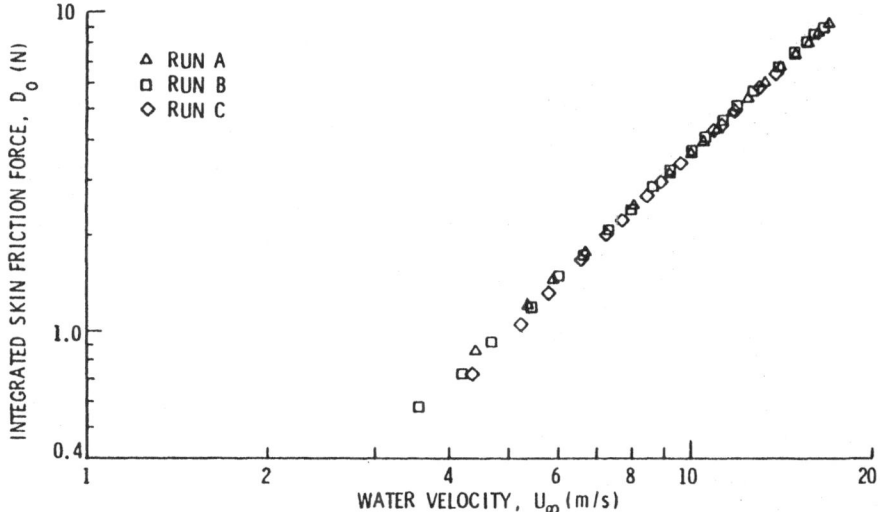

Figure 20.   Shear Stress Balance Performance Showing Sensitivity and
            Linearity

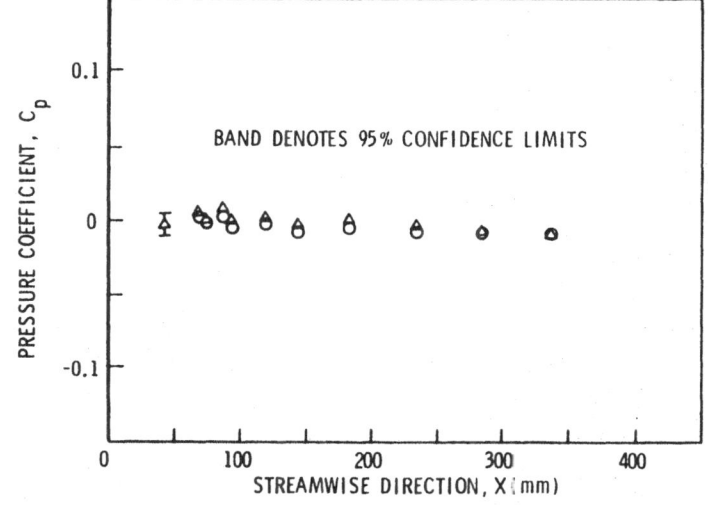

Figure 21.   Typical Pressure Gradient Over a Flat Plate Mounted in a
            Water Tunnel

Because we have used our balance sections in nominally zero pressure gradient turbulent boundary layers, a check on their performance is available by a comparison with existing data correlations. A comparison of some of our data against a correlation given in White (1974) is shown in Fig. 22. The fit to the correlation depends on a choice of virtual origin for the experimental data. This is often fixed by a trip wire or a leading edge condition or by an extrapolation of the data to zero displacement or momentum thickness. For our results, it is comforting to know that a ten percent error in virtual origin results in only a 0.8% error in the skin friction.

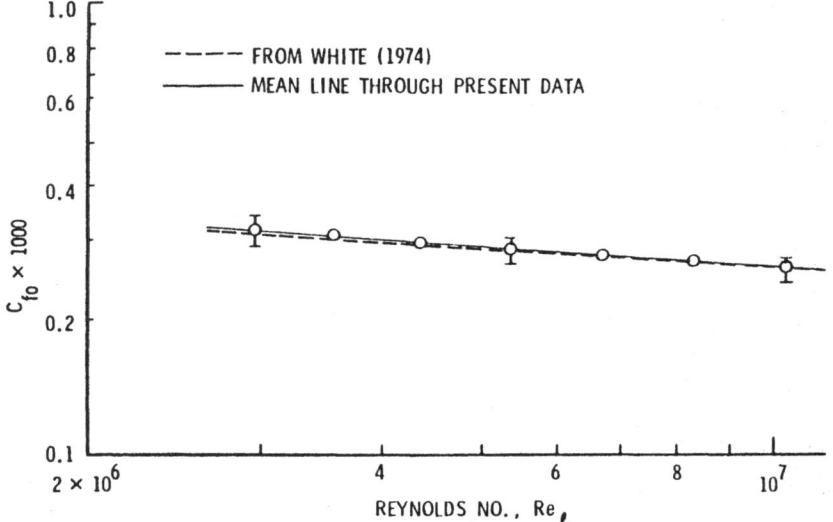

Figure 22. Skin Friction Coefficient on a Flat Plate Showing Comparison of Water Tunnel Data with Theory

Misalignment of the balance can cause extraneous pressure effects and significant errors in skin friction. In boundary layers with very small inner regions these effects can be seen for even very small misalignments, i.e. on the order of 25 microns. Results from a systematic study are shown in Fig. 23. Clearly it is better to err in the direction of a recessed element.

To obtain information about the local time dependent behavior of these turbulent boundary layers we have generally relied on flush-mounted hot film probes and LDV. For opaque microbubble flows, for example, the flush mounted films provide our only means of measuring the time dependent nature of the flow. The theory of operation of these probes can be traced back to the work of Ludweig (1950), Ludweig and Tillmann (1954) and Liepmann and Skinner (1954) and shows that the wall shear stress (τ_w) is related to the measured voltage (E) through

$$E^2 = A \, \tau_w^{1/3} + B, \qquad\qquad (4.1.1)$$

---

[1] Figures 20-27 have been reprinted from the Ph.D. thesis of Madavan (Penn State University, 1984).

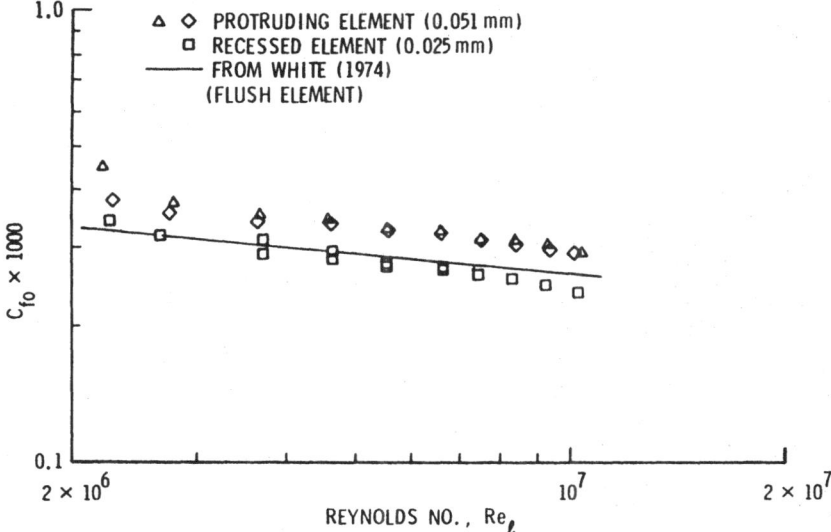

Figure 23. Effect of Shear Balance Misalignment on Skin Friction
Coefficient Measured on a Flat Plate Mounted in a Water
Tunnel

where the coefficients A and B must be found through calibration against a known flow.
Experimental evidence shows (for example, Bellhouse and Schultz, 1966) that a
relationship of the same form exists for turbulent flows provided that the thermal
boundary layer which grows on the probe lies entirely within the viscous sublayer. Of
course, for a turbulent flow the equation relates instantaneous values of the shear
stress and the measured voltage.

For all our water tunnel experiments we have been fortunate enough to have
sufficient boundary layer data to permit us to calibrate these hot film probes in
place. Our experience in the use of these probes in other types of measurements
indicates that with the small spatial scales involved in the flush-alignment, that
calibration outside the facility used would lead to only order of magnitude results.
Calibration in place implies calibration in a turbulent boundary layer and Sandborn
(1979) and Hanratty and Campbell (1982) have recommended an interactive technique
toward the determination of the constants A and B. In our measurements we have found
the initial estimate of the constants from the mean calibration data to be
sufficiently accurate. A typical hot film calibration curve is shown in Figure 24.
The fit of two straight lines, with a higher slope to the data at lower tunnel speed,
is usual and is probably a result of free convection effects at low speeds.

The "miniature" hot film probes we employ are not small when compared to our
inner boundary layer scales. Active sensor elements have a streamwise width of 400
microns by a cross-stream dimension of 900 microns. The probes then are some 10 to 80
sublayer thicknesses in streamwise extent with an L/D ratio of 2.25. With streak
spacing estimated as roughly 100 wall units (20 sublayer thicknesses) the probes may
be averaging over as many as 4 wall layer streaks. One might suppose that the effect

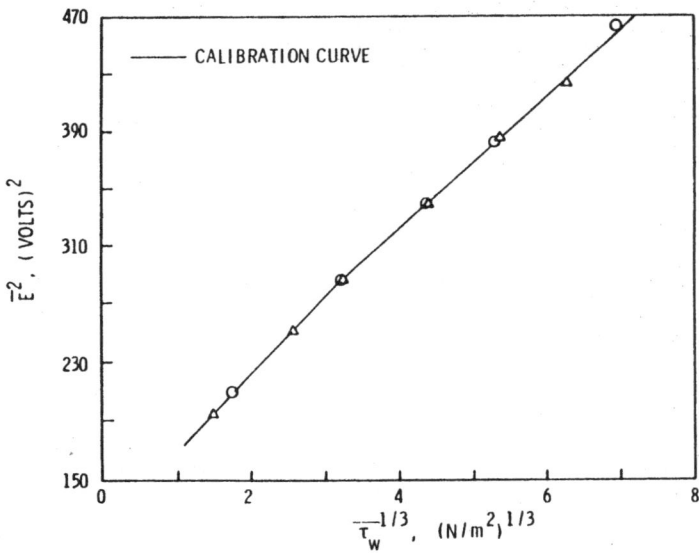

Figure 24. Typical Calibration of a Flush-Mounted Hot Film Probe

of this averaging would appear in plots of the intensity of shear stress fluctuations, which we show versus Reynolds number in Fig. 25. In Fig. 25, the dotted line represents data taken from Eckelmann's study in the thick viscous sublayer of an oil channel. Scatter in the measurements is typical of results taken from several different probes and may be related to the relative success in flush mounting the individual probes. Eckelmann (1974) notes that his value of 0.24 agrees well with the extrapolation of Laufer's hot wire measurements to the wall. Others, most notably Hanratty and his colleagues, find values closer to 0.32-0.36. While the current paper was in review, Alfredsson et. al. (1988) published an attempt to explain the large variation of intensities obtained by various researchers. They conclude that the intensity value should be close to 0.4, and that the results presented in Fig. 25 do indeed suffer from spatial averaging.

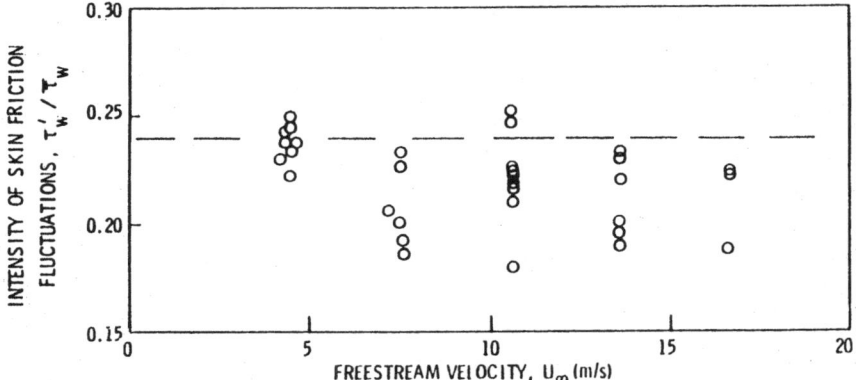

Figure 25. Typical Measurements of Fluctuating Wall Shear Stress Using Hot Film Probes

LDV measurements in high Reynolds number liquid boundary layers are also restricted by problems of small inner-scale size. Using standard backscatter optics, an argon-ion laser at 488 nm (or 512 nm) a 3.75x beam expander, and a 480 mm lens, our ellipsoidal measuring volume is 80 microns in the vertical and streamwise direction and 580 microns in the cross-stream direction. Our measurement volume then is 10 to 80 wall units in the vertical direction with an L/D of 6. Typical plots of mean velocity and turbulence intensity are shown in Fig.'s 26 and 27, respectively. Even with tilting the laser beams at some small angle to the plate to decrease reflections, y+ values of less than 50 are quite unusual to obtain[2]. Turbulence production, by way of comparison, may peak near a y+ of 15.

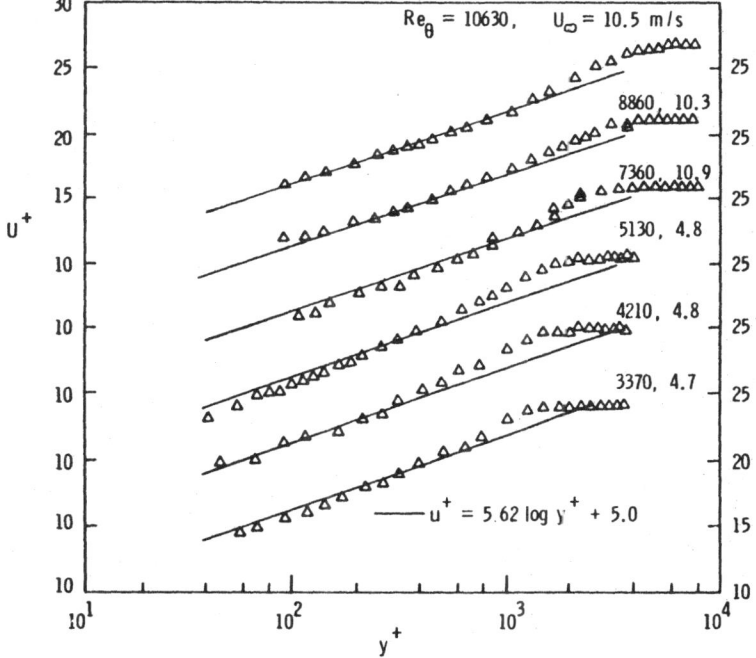

Figure 26. Laser Doppler Velocimetry (LDV) Measurements of Turbulent Boundary Layers on a Flat Plate

In summary, we have described some of our more successful measurement techniques for high-velocity turbulent water boundary layers. These experiments have the singular difficulty of very small inner scales. Because of these small scales, the measurements we would most often like to make are simply not obtainable; some measurements we do make are often not as accurate as we would like.

4.2 Transitional Boundary Layers

Much of our recent experience in transitional boundary layers has come from our

---

[2] That is, for a flat plate boundary layer. We have managed to come significantly closer on axisymmetric bodies by exploiting the curvature of the body to reduce reflections.

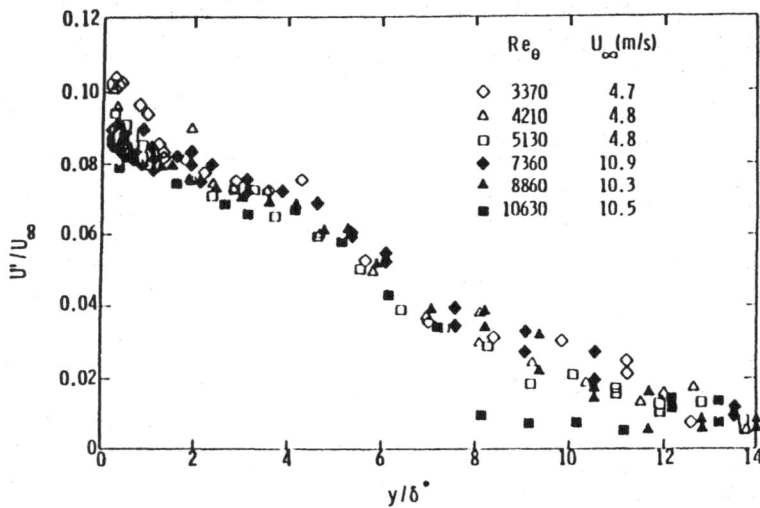

Figure 27. LDV Measurements of Turbulence Intensity (Streamwise Component)

work in heated laminar flow on underwater vehicles, again at large unit Reynolds
numbers (Lauchle and Gurney, 1984). The object of the research was to achieve laminar
flow at very large length Reynolds numbers by using surface heating to stabilize the
boundary layer. The body employed was a 3.05 m long X 0.32 m diameter axisymmetric
body and was tested in the large (1.22 m diameter) water tunnel. We found that with
sufficient heating we could increase the length Reynolds number for transition from a
cool wall value of 4.5×10^6 to a value near 37×10^6.

The laminar boundary layers that were encountered on this body were typically too
small to permit velocity profile measurements, particularly for the heated cases. At
1.92 m from the nose, for example, the displacement thickness is about 2.33 mm, while
at the same location with high heating (75 kW) the displacement thickness is only 100
microns. Our chief instrumentation has been flush-mounted hot film probes and
injected dye-flow visualization. We found early on in the experiments that injection
of fluid, at a low rate, did not interfere with the transition process, so that we
could, by using dye, visualize it. This was very useful, particularly in determining
at what position on the body the initiation of turbulent spots began. We found this
position to be consistently at the predicted critical point of the body for the given
level of heating. Heating increases the distance to the critical point. A typical
visualization, which illustrates the spot formation is shown in Fig. 28.

We used flush mounted hot film probes to provide quantitative data on
intermittency and burst rates. This is an ideal use for these probes as they may be
run uncalibrated. We used probes at a fixed location (2.12 m downstream of the nose)
so that the Reynolds number was varied by changing the mean tunnel speed.

---

[3] Typically we filter at 50 Hz. This is important if one wishes to have an
intermittency detector with long time stability.

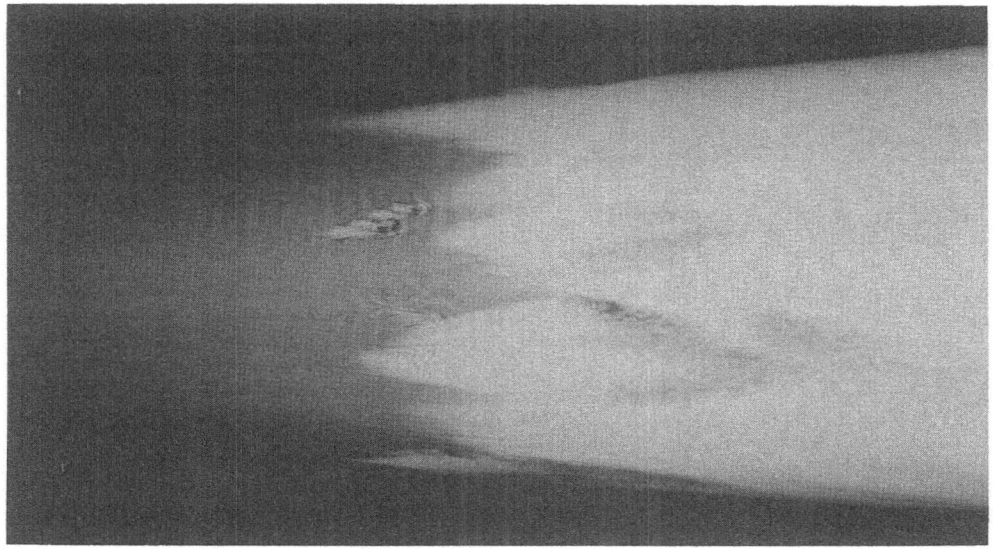

Figure 28. Flow Visualization of Natural Transition on a Heated Body
Operated in a Water Tunnel

Transitional flow is intermittently turbulent and is generally described statistically
through the intermittency function (γ) and the burst formation rate (N). We use an
indicator function, I(t), which is set to zero for laminar flow and to one for
turbulent flow, and define the intermittency as the time average value of the
indicator function. An intermittency detector is used to generate the indicator
function. First the hot-film signal is high passed filtered to remove low frequency
information[3]; second it is differentiated to highlight the remaining high frequency
fluctuations; third it is integrated with a short-time-constant low pass filter to
smooth the fluctuations within the turbulent burst; and fourth the conditioned signal
is passed through a variable threshold Schmidt trigger. The output of the Schmidt
trigger is the indicator function. The intermittency is obtained by passing the
indicator function through an integrating voltmeter, while the burst frequency is
obtained by passing I(t) through a counter that only triggers on the leading edge of a
step function. The threshold level of the Schmidt trigger is set by observing
simultaneously the raw hot film signal and indicator function on a storage
oscilloscope. Adjustments are made until I(t) tracks with the bursts sensed by the
film. Typical intermittency and burst rate plots for no surface heating and
substantial surface heating are shown in Fig. 29. Here, the arc length Reynolds
number is based on the fixed arc length location of the probe (s = 2.12m).

Ladd and Hendricks (1985) have measured boundary layer transition on an
ellipsoidal-shaped heated body which operated in the Naval Ocean Systems Center (NOSC)
0.305 in diameter water tunnel. This tunnel is capable of speeds up to 14 m/s with
turbulence intensities reported to be 0.16 percent. The maximum transition Reynolds

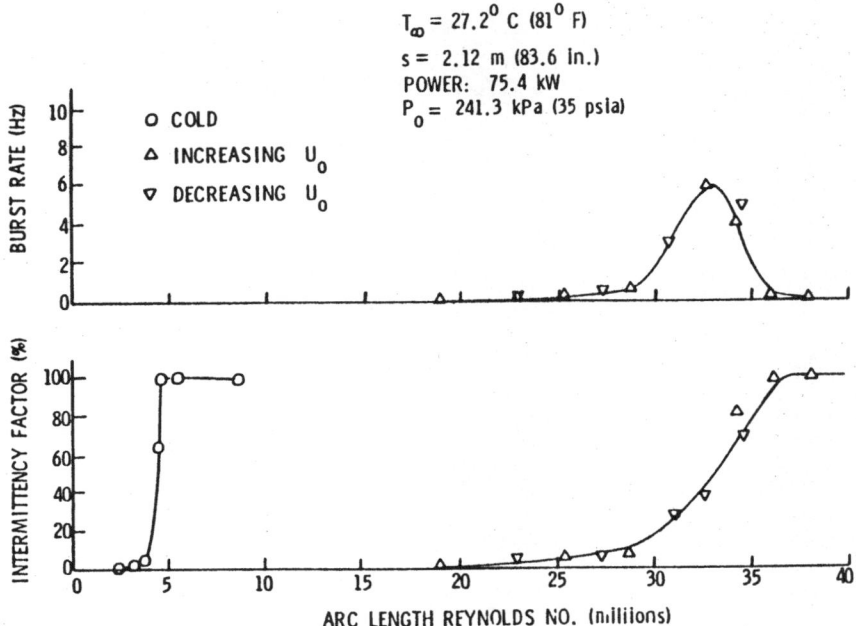

Figure 29. Transitional Boundary Layer Data Showing Spot Formation Rate
and Intermittency on a Heated Underwater Body

number considered in their experiments is 7×10^6 (based on axial distance to
transition). Intermittency in the transition zone was measured by locating the center
of an LDV measuring volume at a distance of 5.8 θ away from the model wall, where θ is
the computed laminar boundary layer momentum thickness. This distance was selected
because it assures that the LDV measuring volume is outside of the region of "sharp"
velocity gradient. They found this necessary to avoid the effects of the finite
measuring volume diameter (0.12 mm) and the sporadic characteristic of LDV signals,
which would produce a fluctuating velocity output signal indistinguishable from
turbulence, even in a fully laminar flow.

The laser scattering seeds used in the NOSC experiments were 1.5 μm silicon
carbide particles. The LDV photodetector signal was processed with a counter-type
processor that yields a voltage proportional to velocity. This signal was processed
on a computer where software was provided to form the intermittency function, $I(t)$.
The threshold velocity selected to indicate a turbulent event was 10 percent of the
maximum velocity fluctuation observed in the transitioning boundary layer. This is
consistent with the Lauchle and Gurney (1984) experience using flush-mounted hot films
and analog processing.

In transition experiments where intermittency is measured, 50 percent
intermittency is usually selected to define the transition point or transition
velocity.

5.0 MULTIPHASE FLOW MEASUREMENTS

5.1 Multiphase Flows

There are many different techniques associated with the measurement of multiphase
flows (i.e. liquid-solid-gas flows) due to the great variety of flows themselves.
Accurate measurements of the fundamental physical quantities are not only an essential
part in an understanding of multiphase flows but also in the measurement process
itself. Almost all flow visualization techniques involve measurement processes having
multiphases. Even the early work of Osborne Reynolds, a 19th-century English
scientist, utilized dye to visualize pipe flows. Today, optical techniques such as
laser velocimetry and pulsed velocimetry (Dybbs and Pfund, 1985) involve the
measurement of solids in a fluid and problems of particle velocity fidelity arise as
well as statistical particle bias problems.

There have been many international symposia and journal articles directed toward
flow visualization techniques and multiphase flow measurements. For example, the
American Society of Mechanical Engineers-Fluids Engineering Division sponsors annually
The Cavitation and Multiphase Flow Forum (see Furuya 1987) and Fluid Measurements and
Instrumental Forum (see Bajura and Billet, 1986) that serve as a source of information
and provide valuable insite into multiphase flow measurements. Although the
applications are very diverse, the use of optical measurement techniques does appear
to be one common denominator in multiphase flow measurements in high-Reynolds number
liquid flow facilities.

Multiphase flow measurements in the flow facilities at ARL Penn State have for
the most part been developed to understand the basic fluid mechanics of cavitation.
Cavitation is the gas-liquid region created by a localized pressure reduction produced
by the dynamic action of the fluid in the interior and/or the boundaries of a liquid
system. Cavitation will occur in the lowest pressure regions of the flowfield.
Figure 30 shows cavitation occurring in a pump stage. A first order measure of
cavitation is determined from the cavitation number defined as

$$\sigma = \frac{P_\infty - P_v}{1/2 \; \rho \; V_\infty^2} \qquad\qquad (5.1.1)$$

where P_∞ is the reference static pressure, P_v is the vapor pressure at the bulk
temperature of the liquid, ρ is the liquid density at the bulk temperature of the
liquid and V_∞ is a reference velocity. Thus from a classical point of view,
cavitation inception occurs when the cavitation number is equal to the absolute value
of the local minimum pressure coefficient. At higher values of cavitation number no
cavitation can occur and at lower values of cavitation number developed cavitation
will occur.

In order to determine cavitation inception, the cavitation number is obtained by
either varying the static pressure at constant velocity or varying the velocity at

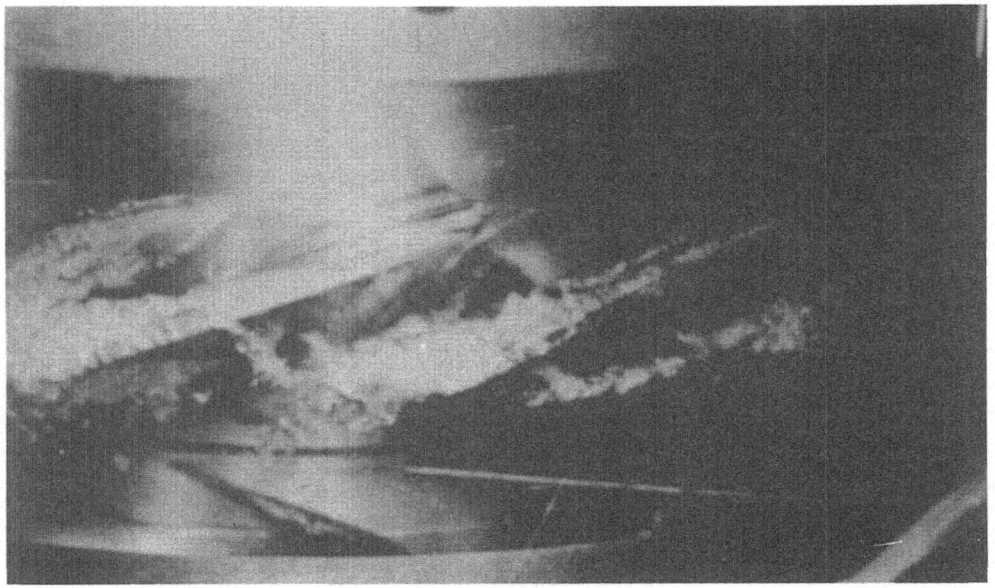

Figure 30.   Photograph of Cavitation.

constant pressure.   In either case the condition is identified by the observation of
large bubbles or by acoustic sensors through the noting of an increase in noise level.
Figure 31 shows the characteristics of cavitation noise, and cavitation inception can
be noted where a rapid increase in noise level occurs for a small variation in
cavitation number.   Broadband noise is created as the bubbles collapse in a high
pressure region.   A description of cavitation noise theories can be found for example
in Blake (1987) and Hamilton, Thompson and Billet (1986).

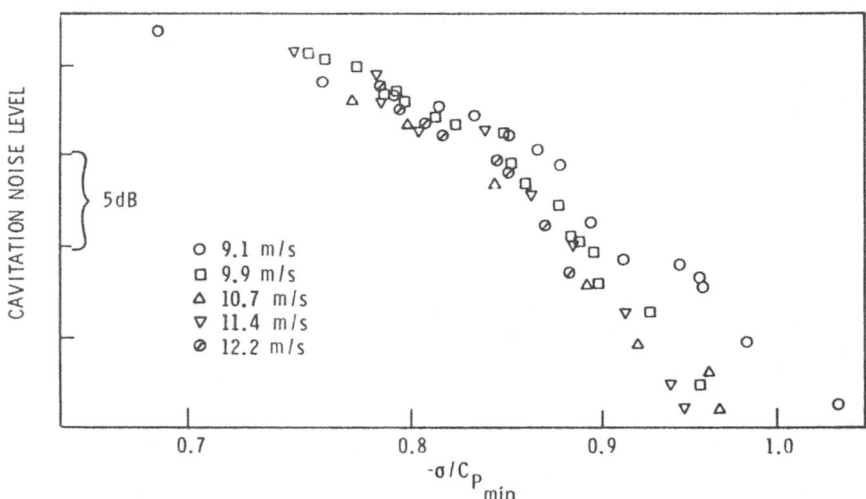

Figure 31.   Cavitation Noise.

Departures from the classical viewpoint of cavitation are due to so-called scale effects (see Holl, 1969) and are (1) due to variations in the local pressure field from the perfect fluid model, and (2) bubble dynamics which cause the pressure at inception of the microbubble to vary from equilibrium vapor pressure. Thus, cavitation measurements generally involve fluid measurements, cavitation bubble measurements, and cavitation nuclei measurements.

Past research studies have included such diverse measurement techniques such as oil-paint film, tufts, dyes, bubble tracing, photography, and Schlieren photography. Now X-rays, gamma rays, radio and lightwaves, and computer technology are combined in complex systems for these measurements. Emphasis at the ARL/Penn State has been on developing electro-optical techniques such as laser velocimetry, laser light-scattering, and holography. The application of these techniques is essentially due to two fundamental properties; it is non-intrusive, and in general, gives directly the measurement of the flow quality. As an example, the instantaneous components of the local velocity vector can be obtained with a laser velocimeter and holography gives directly the size and distribution of bubbles.

A summary of laser velocimeter and its application to water tunnel measurements is given in Billet (1987). Obviously, the major problem with these systems is optical access to the measurement location which is not always easy in a high-speed water tunnel.

5.2 Cavitation Nuclei Measurements

Cavitation nuclei is a general term used to refer to the impurities that cause weak spots in liquids and thus prevent the liquid from supporting higher liquid tensions. The importance of cavitation nuclei has been noted as early as a century ago by Besant (1859). A relationship between nuclei size and cavitation growth has been established by a bubble dynamics model started by Lord Rayleigh (1917) almost half a century ago followed by the work of Plesset (1949) and others. The importance of nuclei size and distribution has also been shown by many experimental efforts. Recently, the results of experiments conducted by members of the International Towing Tank Conference (ITTC) (see Lindgram and Johnsson, 1966 and Acosta and Parkin, 1970) have provided impetus for attempts to quantify a relationship between nuclei and cavitation. As a direct result, many different measurement techniques have been developed and a summary of many techniques is given by Billet (1986) and in Fig. 32.

Two significantly different approaches to measuring cavitation nuclei have been developed. One is to measure the particulate/microbubble distributions by utilizing acoustical (Schiebe and Killen, 1971 and Medwin, 1977), electrical (Hammitt, et.al., 1974 and Oba, 1981) or optical techniques (Keller, 1972, Farmer, 1976, Gates and Bacon, 1978 and Gowing and Ling, 1980). The other approach measures a cavitation event rate for a liquid under various tensions and establishes a cavitation susceptibility (Oldenziel, 1979, Lecoffre and Bonnin, 1979 and d'Agostino and Acosta, 1983).

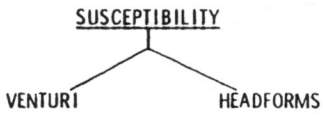

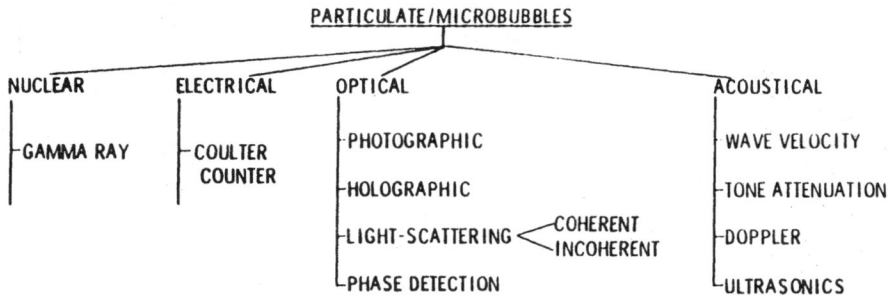

Figure 32. Cavitation Nuclei Measurement Techniques.

The most widely used method of establishing cavitation susceptibility utilizes a
venturi system. Measurements of cavitation susceptibility are based on bubble events
that occur in a prescribed pressure field. A small venturi tube made of glass
(Oldenziel, 1979) or steel (Lecoffre and Bonnin, 1979) is used and the minimum
pressure can be varied by a adjusting the volume flow rates. The number of cavitation
events can be determined optically or by recording the noise generated by their
collapse in a downstream diffuser. The corresponding concentrations of unstable
nuclei can be estimated from the pressure at the throat and the count of cavitation
bubbles. A schematic showing the principles of operation is given in Fig. 33.

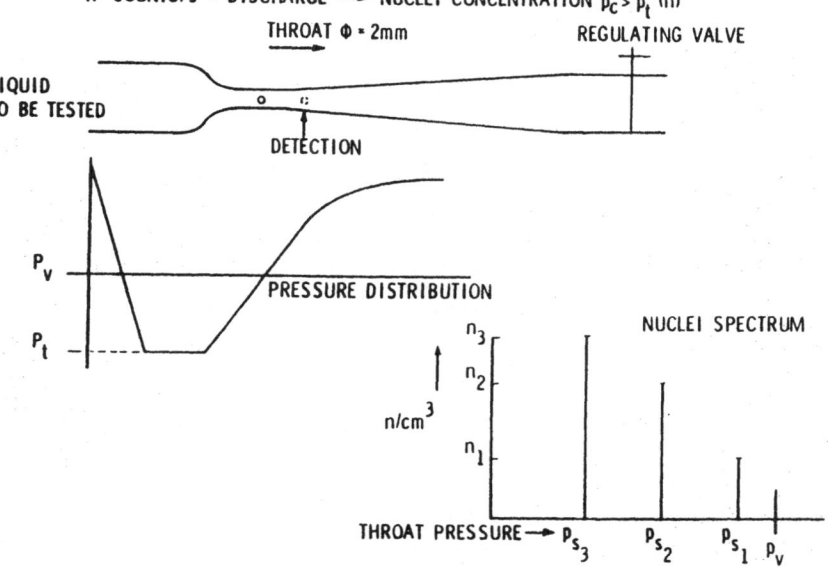

Figure 33. Venturi System.

The venturi system can be severely limited by nuclei concentration larger than $10/cm^3$; however, it can provide a direct measure of both the liquid critical tension and nuclei concentrations over a very large size range. Shen and Gowing (1985) and Chahine and Shen (1985) discuss the calculation of bubble dynamics in the throat of a venturi. A more general discussion of cavitation susceptibility meters is given by d'Agositino and Acosta (1983).

Light-scattering techniques are the most widely used methods to determine nuclei distributions. In some cases, the amplitude of the scattered light is related to the size. Other methods are based on laser doppler techniques in which particulate visibility or phase of the doppler frequency is correlated with size. One of the first applications of a scattered-light method to measure nuclei distributions in a water tunnel was done by Keller (1972) and utilized a laser as a light source.

A light-scattering system developed for water tunnel applications at ARL/Penn State utilizes the laser light-scattering principles and is shown schematically in Fig. 34. This system is based on the relationship between the radius of a scattering sphere and the scattered intensity. A comparison between experimental data and Mie theory is given in Fig. 35.

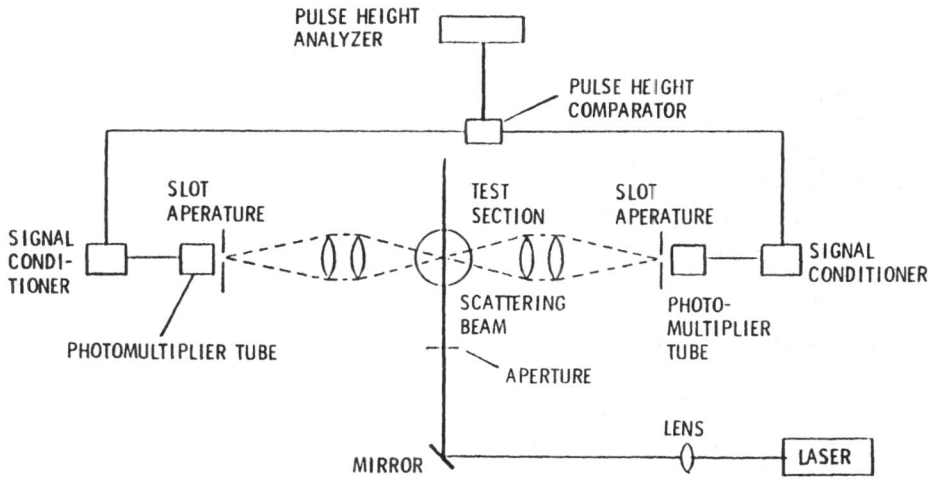

Figure 34. Light-Scattering System.

An on-line method was developed to discriminate between the light scattered from microbubbles and particulates. An analytical investigation by Kohler and Billet (1981) showed that the asymmetry of a particulate relative to the postulated spherical symmetry of a microbubble results in a scattered field that is quite different. Yungkurth (1983) conducted a series of tests in a water tunnel that showed that discrimination using this theory is only possible when the particulate population is less than the microbubble population.

The largest classifying error of this technique is the tendency for spherical microbubbles of a given size to register a number in not only the classification channel corresponding to that size, but also in all channels lower than that size.

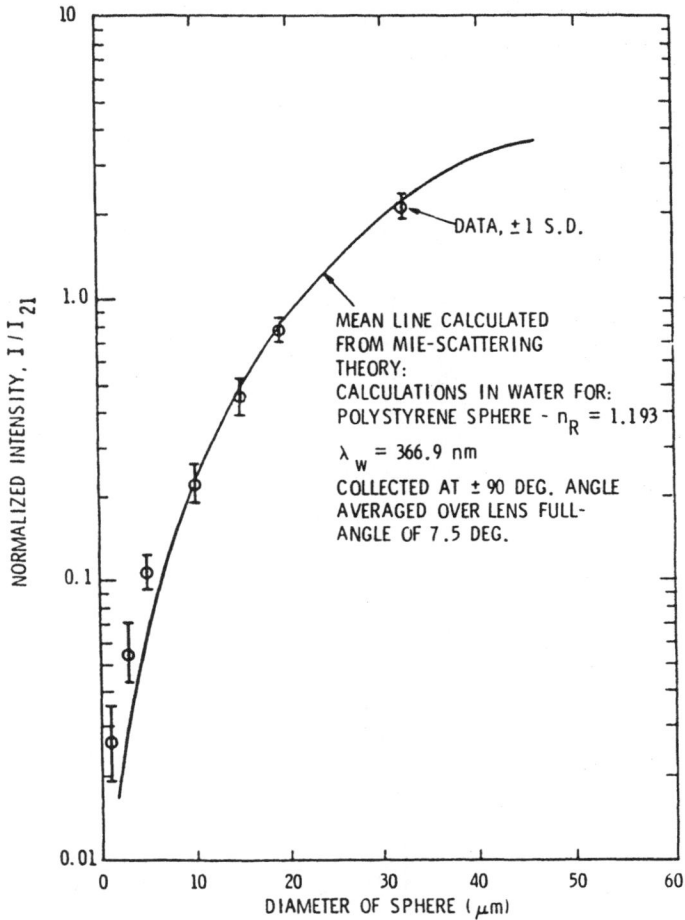

Figure 35. Comparison Between Calculated and Measured Scattered
Intensities for Spheres in Water.

This is due to the probe volume having a nonuniform light intensity. Fig. 36 shows
the different light scattering characteristic of the same size bubble located at
different locations within the laser beam.

This experimental result shows that the count histogram does not reflect directly
the actual distribution of sizes present. The amplitude of the voltage pulse (A)
produced by a photomultiplier as a result of a microbubble entering the measuring
volume can be expressed as

$$A - G \ \frac{\lambda^2}{4\pi^2 r^2} \ I(x,y) \ F(\alpha, \Omega, \eta_{rel}) \quad , \tag{5.2.1}$$

where G — gain of associated electronics

λ — light wavelengths

r = effective optical distance

I(x,y) = intensity of light beam in probe volume

$F(\alpha, \Omega, n_{rel})$ = scattering response function.

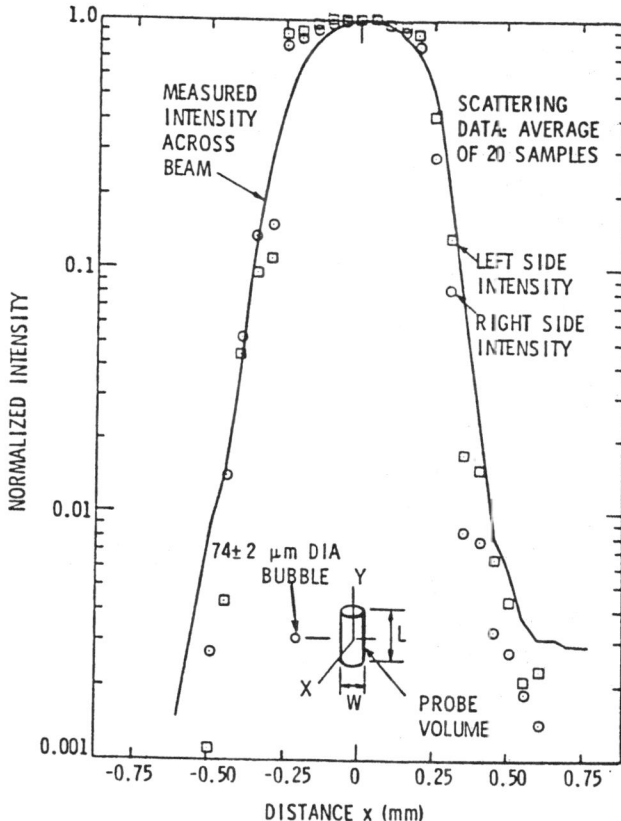

Figure 36. Normalized Maximum Scattered Intensity of a Bubble Along the Axis Normal to the Light Beam.

To correct for this, an inversion scheme that accounts for the nonuniform light intensity has been developed. Thus the above equation can be generalized in terms of discrete classification channels as

$$C_i = U \; \Delta S_{ij} \; N_j \qquad . \qquad\qquad\qquad (5.2.2)$$

where U is the bulk velocity of the flow, C_i is the column matrix of detected count rates ΔS_{ij} is the equivalent cross-sectional area of the probe volume for the microbubble response function F_j, and N_j is the column matrix of detected count rates.

It is important to note that the S_{ij} elements are simply probabilities based on the light intensity distribution and the width of the classification channel. The actual data received in an experiment are the C_i collected in the various channels. The information desired is the microbubble density, N_j, thus the above equation can be written as

$$N_j = \frac{1}{U} C_i \, \Delta S_{ij}^{-1} \quad , \qquad\qquad (5.2.3)$$

where ΔS_{ij}^{-1} is the inverse of the matrix ΔS_{ij}. Thus the microbubble density for one channel is actually a weighted sum of the count rates from every channel.

Even utilizing an inversion procedure to account for the nonuniformity of the light intensity, the question of system accuracy cannot be answered.. The system requires an appropriate statistical sample before the inversion procedure can account properly for the nonuniform light intensity.

A computer model of the probe volume - microbubble distribution - inversion procedure interaction has been developed and details are given in Billet (1986). The results clearly show the importance of the size width of the classification channels and of the sample size. A comparison of microbubble distributions obtained with the light-scattering system and by using holographic techniques is given in Fig. 37.

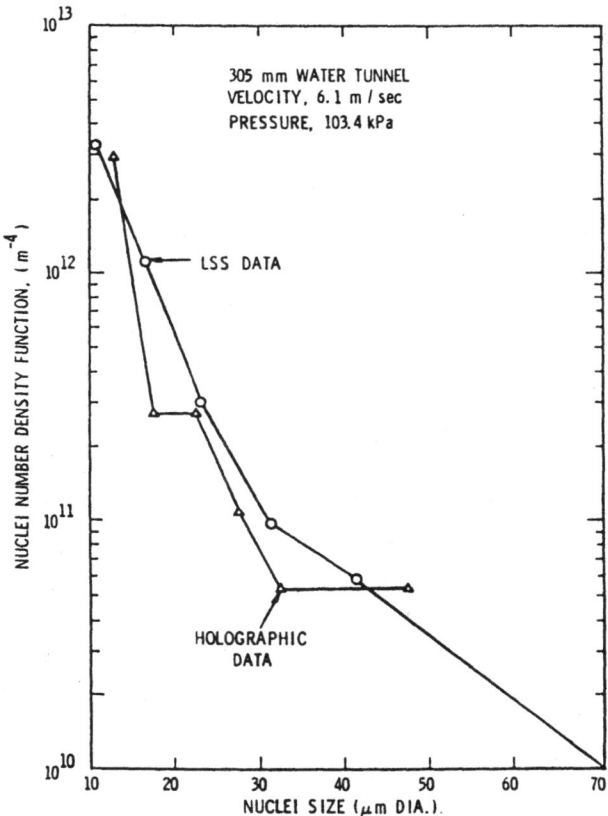

Figure 37. Number Density Data for $V_\infty = 6.1$ m/sec, P = 103.4kPa in 30.5 cm Water Tunnel.

5.3 Cavitation Bubble Measurements

Holography is utilized to record the growth of cavitation bubbles, the trajectory
of microbubbles and the size and concentration of cavitation nuclei. Holography is a
complete measurement technique that makes possible the recording of color, scale and
three-dimensional images of the multiphase flow. Unlike an ordinary photograph or a
Schlieren photograph, which record only two-dimensional images, a hologram preserves
the three-dimensional information by recording the wave front phase differences as
well as light intensity. When a coherent light beam encounters a gas bubble or some
other change in the refractive index of the medium, the phase of the light waves is
modulated. Scattered coherent light encodes all the optical properties of the object.

In the holographic process, a laser beam is split into two separate beams - one
that illuminates the object and the other which is a reference beam superimposed on
the light field scattered from the object, Fig. 38a. The interference pattern
generated by these beams is recorded in a transparent light-sensitive emulsion.
Development of the exposed emulsion produces a hologram that contains the complete
optical information.

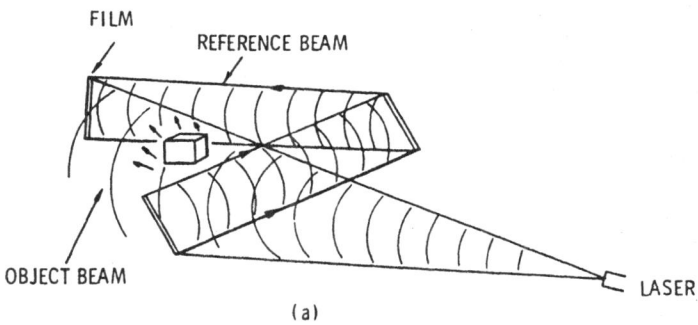

(a)

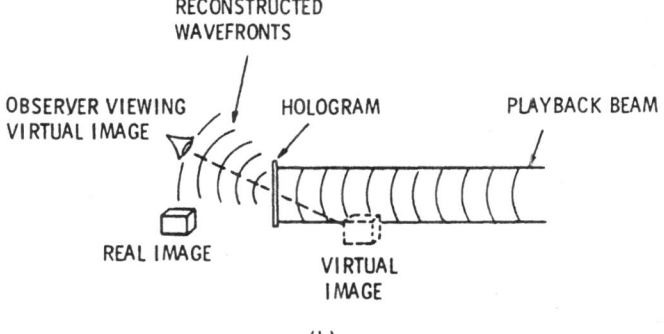

(b)

Figure 38.   (a)   Hologram Recording.
              (b)   Wavefront Reconstruction for Image Playback.

The hologram is composed of a series of light and dark fringes on the transparent substrate. When a laser beam illuminates it at the same incidence angle as the original reference beam, the light is diffracted by the fringes. Converging and diverging wavefronts, similar to those scattered by the object are reconstructed (Fig. 38b). The diverging wavefronts appear to come from a virtual image, visible from behind the hologram. This image, seen through the hologram, appears to be at the location of the original object. The converging wavefronts form a real image of the object on the opposite side of the hologram.

Two holographic configurations are used for cavitation research in the water tunnel. An off-axis system has spatially separated object and reference beams. The term off-axis indicates that the beams are not coincident along a single axis. An in-line system used one beam as both the object beam and the reference beam.

In-line holography is a very well established technique for the study of microbubbles in a dynamic situation. Although the technique is almost as old as modern holography, it is still one of the most reliable measurement methods. Consequently, research to improve different aspects of the method are continuing, e.g., Vikram and Billet (1982), (1983), (1984).

In-line holography has a number of advantages for cavitation measurements. The limited window area and working space around a water tunnel test section provide little off-axis space for a separate reference beam. The in-line system has been used to successfully measure microbubble concentration as high as $200/cm^3$ over a size range of 20-200 microns in the tunnel. Also, it is capable of obtaining bubble growth information via multiple exposures for bubbles having wall velocities as high as 100 m/sec. However, when more than 20% of the reference beam is distorted, an off-axis system is necessary and has been utilized to measure concentrations as high as $1500/cm^3$ in the water tunnel.

The holographic system has three main components: the holocamera, a hologram recording system and a reconstruction system. Schematics are shown in Fig.'s 39 and 40. The holocamera utilized at ARL/Penn State is a double-pulse ruby laser Q-switched

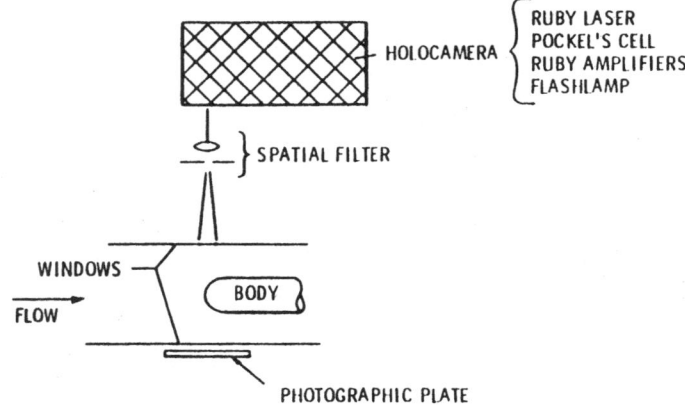

Figure 39. Three Main Components of an In-Line System.

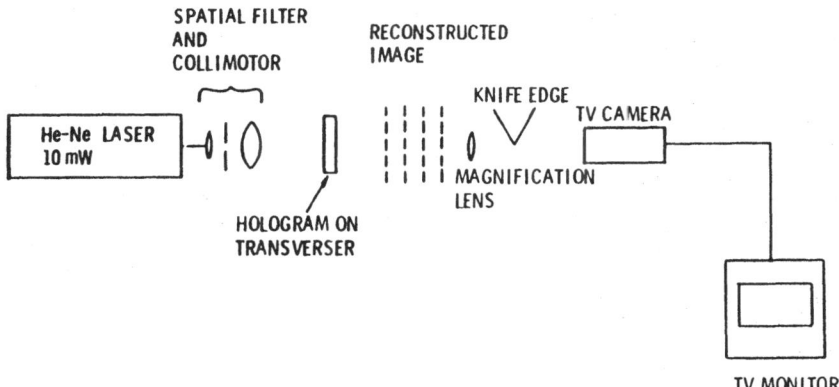

Figure 40. Holographic Reconstruction System.

with a Pockels cell. The power of the system is up to 10 joules in a single 20-msec
pulse or two 20-msec pulses of 5 joules each. The pulse-repetition rate can be varied
from 400 msec to about 1 msec. A resolution of 15 μm can be obtained when the central
maximum fringe of the diffraction pattern is recorded with at least three side lobes.
This also requires an allowable fringe movement of less than 1/40 of the fringe
spacing.

6.0 HYDROACOUSTICS MEASUREMENTS

As noted previously, radiated noise generated by turbulent stresses depends on a
high power of the Mach number. Even in a high-speed water tunnel the Mach number may
be as low as 0.01; the flow is nearly incompressible, and the level of sound radiation
is very low compared to air at the same Reynolds number. Nevertheless,
hydrodynamically-generated sound does occur under water and its physical understanding
and control is an important subject of past and current research projects.

The most efficient sources of hydrodynamic noise are cavitation and other
multiphase flows. These sources are monopole and the sound level increases with the
fourth power of Mach number. Such flows create noise that is usually more intense
than the facility background noise which results in a good signal-to-noise ratio.
Examples of cavitation noise measured in water tunnels may be found in Blake, et.al.
(1977), Marboe (1982), Barker (1974), Gavigan, et.al. (1974), and Hamilton (1986).

Flow noise in single-phase flow is generated in both free and bounded shear
layers. Turbulent boundary layer noise obeys a fifth to sixth power dependence on
Mach number and is hence less efficient than cavitation noise. Its directivity is
that of dipole radiation. Free-shear flows are quite inefficient sound radiators at
low Mach numbers; the intensity scales with the eighth power of the Mach number and
the directivity is described by quadruples. Boundary-layer noise can, under some
instances, be measured in a water tunnel environment (Skudrzyk and Haddle, 1960,
Lauchle, 1977 and Greshilov and Mironov, 1983), but difficulties in signal-to-noise
ratio often invalidate the data. The problem of signal-to-noise ratio is compounded

in free-shear flows and no known noise data are reported.

The above discussion is concerned with the radiated noise associated with unsteady flows. The subject of turbulent wall pressure fluctuations is a different matter. These fluctuations are generally quite intense since they scale with the dynamic head and depend only secondarily on Mach number. Their measurement, at a point, in liquid flows have been somewhat successful and several studies are reported; for example, see Willmarth (1975) for a review, and Bakewell (1968), Barker (1973), Lyamshev, et. al. (1984), Kadykov and Lyamshev (1970), Skudrzyk and Haddle (1960), Lauchle and Daniels (1987), and Horne and Hansen (1981) for more specific studies involving liquid flow tunnels.

The radiation field from liquid stress fluctuations is non-dispersive and depends on only one wavenumber, the acoustic wavenumber. On the other hand, turbulent wall pressure fields are dispersive, being dependent on a wide range of characteristic velocities, and hence on a wide range of subsonic wavenumbers. Therefore, in measuring wall pressure fluctuations, one must be concerned with both spatial and temporal information.

A complete survey of such fields requires that measurements be made at many locations, simultaneously. These measurements are made using flush-mounted pressure transducers, the size of which is important. Large transducers tend to filter out the high wavenumber components, while small transducers give an average of the field over a large wavenumber range. In order to measure a turbulent fluctuation within some pre-determined wavenumber band, one must use an array of transducers, shaded appropriately to give the desired result. These arrays are called wavevector filters and are described by Maidanik and Jorgensen (1967). Several wavevector filters, each of a different design, must be used to cover a given broad wavenumber range. The difficulties of this approach are significant; successful measurements of the entire wavevector/frequency spectrum of turbulent boundary layer wall pressure fluctuations have yet to be reported for any medium. A predominant problem is facility background noise. This noise is capable of leaking into the side lobes of a wavevector filter and hence contaminating the desired data. The problem remains unsolved.

In the remainder of this section, methods for measuring the flow-induced radiation field and turbulent boundary layer wall pressure fluctuations in water tunnel testing will be high-lighted. The examples discussed will show new measurement procedures.

6.1 Radiated Noise Measurements

The water tunnel is not considered to be an ideal acoustic environment for making radiated noise measurements from a test object. The hard-wall construction of a typical tunnel results in internal sound fields that are highly reverberant. One can approach the problem by treating the tunnel as a reverberation chamber, calibrating it for its reverberation characteristics, and then using it to determine sound power (spatially-averaged sound intensity). This is the approach routinely used by Blake, et.al. (1977) in the DTRC 0.914 m water tunnel. The method does suppress the effects

of standing waves in the tunnel, but precludes the use of focused arrays or
directional receivers to localize suspected noise sources. A second approach, used in
the ARL Penn State 1.22 m tunnel, is to use individual receivers calibrated in situ.
The receiver response is thus influenced significantly by the tunnel reverberation and
one must be extremely careful is setting up the calibration. This requires some
anticipation as to where the hydroacoustic sources of noise are situated in the actual
test.

To illustrate the procedure, consider the directional hydrophone system shown in
Fig. 41. The reflector is a pair of copper spinnings, welded at the edges, and filled
with air. The air reflects sound underwater. A commercial hydrophone is placed at
the focus of the ellipsoid forming the shape of the reflector. The other focus is in
line with the tunnel centerline. Sound originating within the tunnel test section
passes through the acoustically-transparent plexiglass viewing window of the tunnel
hatch cover (Fig. 42), into the water-filled external tank, and is then sensed by the
reflecting hydrophone. The reflector can be traversed up and down the length of the
test section to help in isolating different sources of sound.

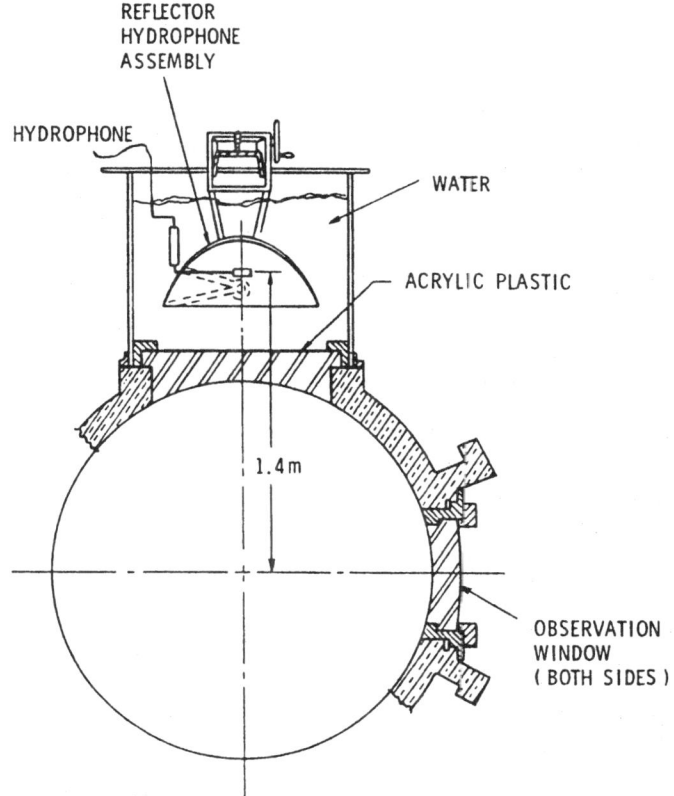

Figure 41.   Schematic of the Reflecting Hydrophone System used on the
1.22 m ARL Penn State Water Tunnel

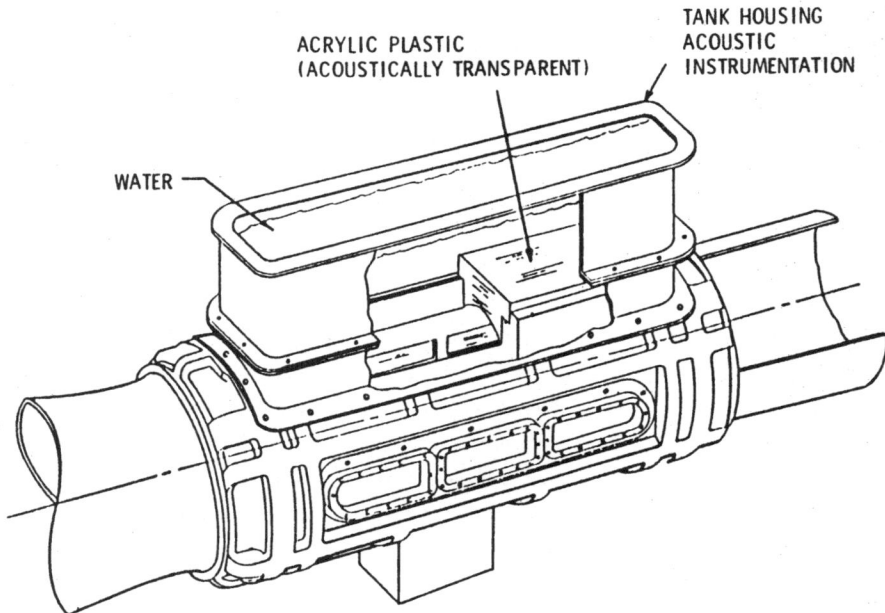

ACRYLIC PLASTIC
(ACOUSTICALLY TRANSPARENT)

TANK HOUSING
ACOUSTIC
INSTRUMENTATION

WATER

Figure 42. Overall View of the Acoustic Tank Mounted on the 1.22 m Water
Tunnel.

The calibration procedure is to place a standard projector (piezoelectric
hydrophone) at the anticipated location of the hydroacoustic source of interest. The
transmitting frequency response of the standard must be know and is given as dB_S
(referenced to 1 μPa - m/input volts). Typically, the sound output per unit input
voltage at a fixed distance increases with frequency at 12 dB/octave. The projector
is driven with broadband noise from a random noise generator. The mean-square level
of this voltage is arbitrary but it must be measured by passing it through a spectrum
analyzer. This level is called dB_X (re 1 volt). The raw output voltage from the
receiver (dB_V re 1 volt) is also passed through the spectrum analyzer. The ordinate
of this spectrum is then converted to sensitivity by:

$$dB_{M_o} = dB_V - dB_S - dB_X + 20 \log d, \qquad\qquad (6.1.1)$$

where d is the separation distance between source and receiver expressed in meters.

A typical sensitivity curve for the reflecting hydrophone is shown in Fig. 43.
The tunnel was free of a test body and the projector was at the geometric center.
Radial standing waves created in the test section are quite apparent in this
sensitivity curve. The free-field sensitivity does not exhibit the rapid oscillation
with frequency characteristic of the in-tunnel sensitivity. These oscillations are
due to radial standing waves. The same waves are created when a broadband
hydrodynamic source of sound (say a spot of cavitation) is situated at the same

location as was the projector in the calibration test. Then, through application of the sensitivity curve, these standing wave resonances (and anti-resonances) cancel; the reduced spectrum will be broadband and quite accurate.

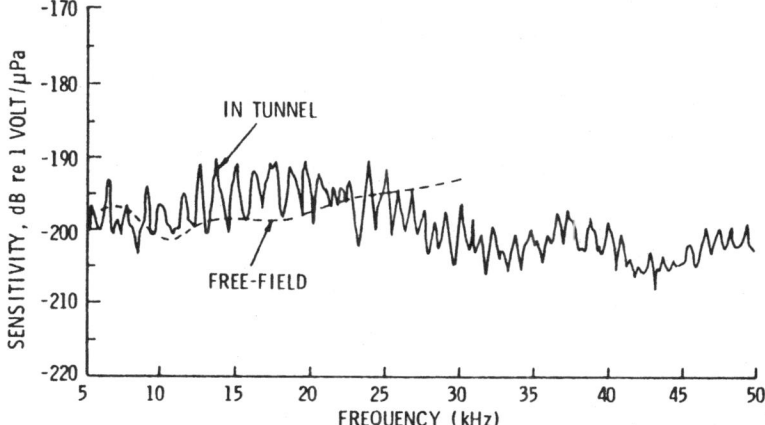

Figure 43. Reflecting Hydrophone Receiving Sensitivity

In many instances the hydrodynamic source of noise occupies a significant volume; for example, cavitating blades on a propeller. In these situations, the projector is placed at several locations within the source volume. The sensitivity curves measured for each location are then averaged together. The averaging eliminates much of the radial standing wave response and a fairly smooth sensitivity curve results.

With the projector at the test section center, a scan of the reflecting hydrophone results in data typified in Fig. 44. Based on the broadband data shown, the half-power beamwidth of the system is about 40 cm.

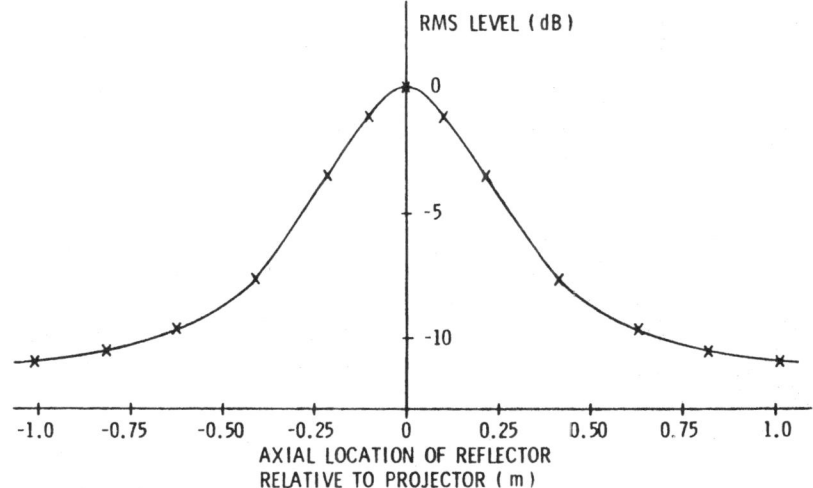

Figure 44. Typical Directivity of Reflecting Hydrophone

6.2 Volume Source Strength Measurements

Cavitation and other two-phase flows create omni-directional noise that is characterized by a random collection of point monopole sources. The time history of the acoustic pressure at distance r from a free-field monopole source is given by:

$$p(r,t) = \frac{\rho \, \ddot{V}(t-r/c)}{4\pi r} \quad , \tag{6.2.1}$$

where ρ is the water density and $\ddot{V}$ is the volume acceleration of the source region evaluated at retarded time (c is the sonic velocity). The volume source strength is of fundamental importance in many radiation and scattering problems because it can be used in theoretical models employing various geometries and known Greens functions. Therefore, its measurement is of practical interest.

In principle, one could estimate the source strength of a cavitation source occurring in a water tunnel by the techniques given in Section 6.1. But those techniques break down at low frequencies where the acoustic wavelength becomes greater than the tunnel diameter. Then the back reaction of reflected waves on the known projector used during calibrations affects adversely the output of the projector (Waterhouse, 1958). This is why the data of Fig. 43 go no lower than 5kHz.

A method which alleviates this problem, while giving directly the acoustic source strength of cavitation in reverberant tunnel test sections, is now presented. The method is based on acoustic reciprocity and permits the determination of low-frequency volume source strength. Wolde (1973) first used the method for propeller cavitation studies, and later Bistafa (1984) used it for free shear flow cavitation noise produced by orifice plates. The method has an added benefit in that the actual acoustic measurements are performed with a microphone (in air) outside of the tunnel. This eliminates the use of hydrophones in the flow, or mounted in the walls of the tunnel, which can be influenced by other sources of noise. The cavitating orifice plate of Bistafa (1984) and Bistafa, Lauchle, and Reethof (1987) is used as an example.

The basis of the reciprocity technique is to establish a tunnel transfer function. Here, a sound projector is placed at some convenient location near the tunnel and the sound pressure is measured inside the tunnel (filled with water but with no flow) at the location where cavitation will occur during the test. A requirement is that the sound projector be reciprocal so that it can be used as the receiver during the cavitation test. A loudspeaker meets this requirement. Figure 45 depicts the two experiments, where the "direct" experiment is when cavitation occurs and $\dot{V}_2'$ is the volume velocity desired. The open-circuit voltage, e_1' of the speaker is measured in the direct experiment. In the "reciprocal" experiment, the speaker is driven with known current i_1'' and the pressure inside the tunnel is measured with an omni-directional hydrophone, p_2'' .

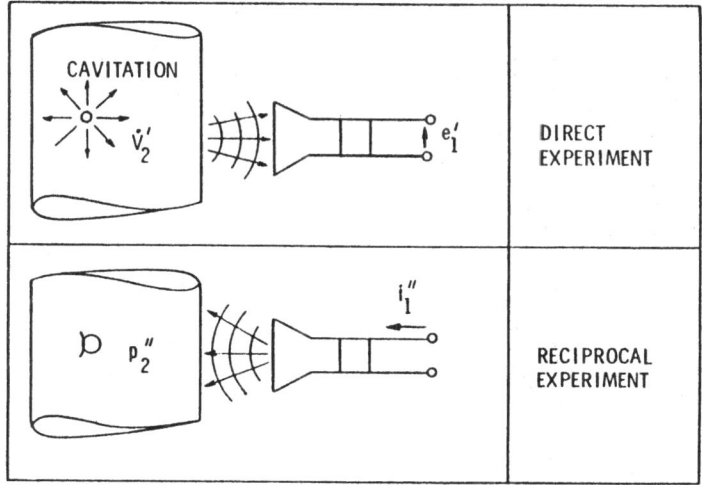

Figure 45. Determination of the Volume Velocity of a Point Source in a
Water Tunnel

The volume velocity is then calculated from

$$\dot{V}' = \frac{i_1'' e_1'}{P2''} .$$
(6.2.2)

At this point it has been assumed that the electro-acoustic transducers are
reciprocal. One must check further to see that the entire system (which includes the
water tunnel) is also reciprocal. This is accomplished by placing a projector in the
tunnel and listening to it with the loudspeaker outside the

tunnel. The current driving the projector is i_1' and the speaker output voltage

is e_2' . Then, the direction of energy flow is reversed by driving the speaker

with i_2'' and measuring the sound inside the tunnel by monitoring e_1'' ; the output

of the projector (now a hydrophone). Transfer functions

$$\frac{i_1'}{e_2'} \quad \text{and} \quad \frac{i_2''}{e_1''}$$

are calculated and compared. If they are equal, the total system is reciprocal.
Figure 46 shows these transfer functions for both the 15.3 cm and 30.5 cm water
tunnels at ARL Penn State. It is clear that the direct and reciprocal experiments
yield the same transfer functions, so both tunnels (and electro-acoustic transducers)
are reciprocal for the frequency range shown.

Figure 47 shows a schematic view of an orifice plate mounted in the 30.5 cm water
tunnel test section. Cavitation occurs in the highly turbulent mixing zone downstream
of the orifice. The bubble collapse zone is where most of the noise occurs so it was

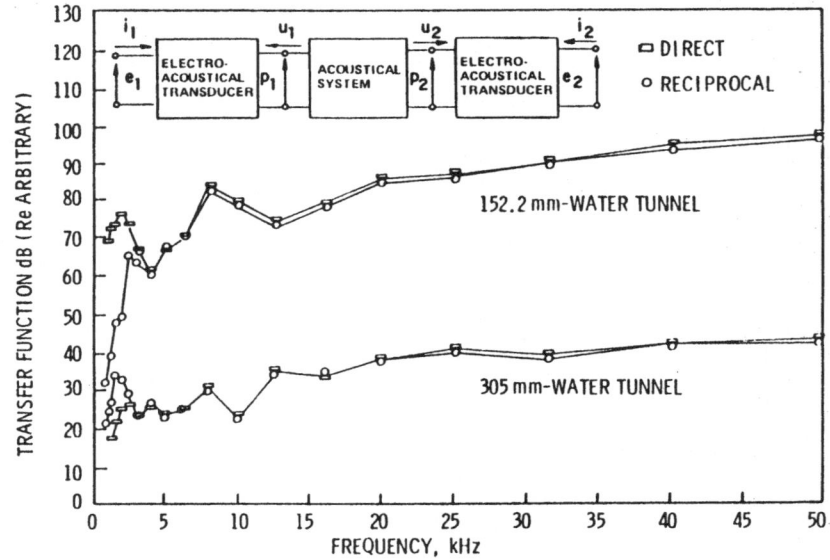

Figure 46. Direct and Reciprocal Transfer Functions for the 15.2 cm and
30.5 cm Water Tunnels at ARL Penn State

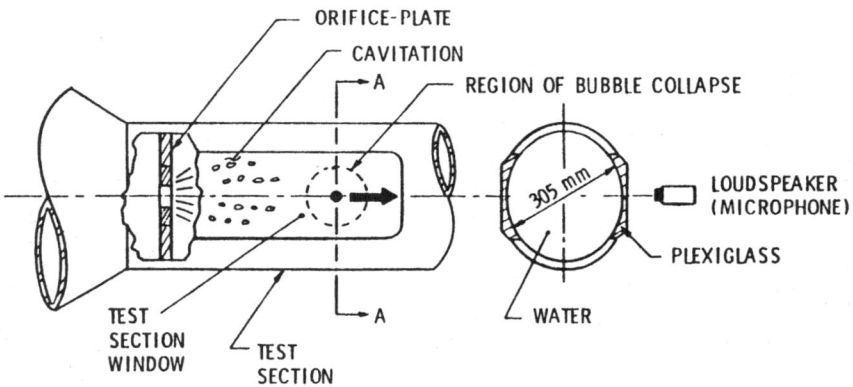

Figure 47. Schematic of Experiment on Cavitation Noise Induced by
Circular Orifice Plate

there where the reciprocal transducer was placed during the calibration experiments.
Figure 48 shows some typical volume velocity spectra for this type of cavitation
noise. The cavitation number, K, for an orifice plate is defined by

$$K = \frac{P_d - P_v}{P_u - P_d} \,, \qquad\qquad (6.2.3)$$

where P_d is the static pressure downstream of the plate, P_u is the upstream pressure,
and P_v is the vapor pressure. The reader is referred to Bistafa (1984) for additional
data and cavitation noise data scaling.

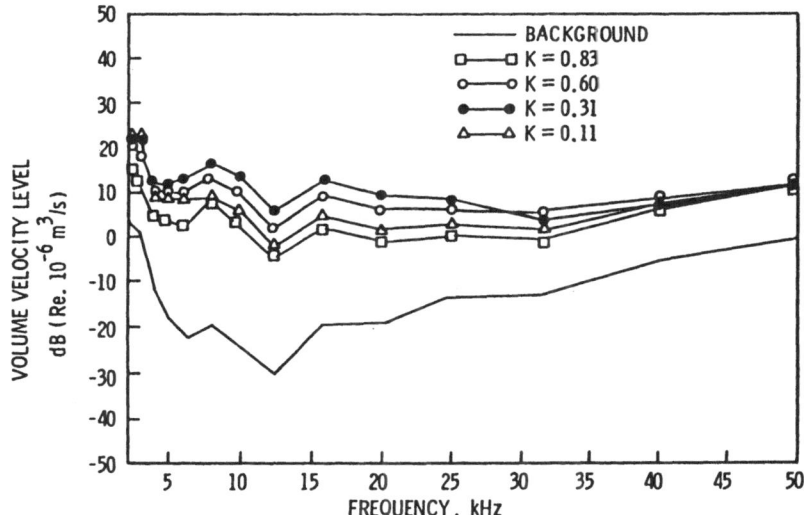

Figure 48. Typical Volume Velocity Spectra Measured Using the Reciprocity Technique for a Cavitating Orifice Plate

6.3 Background Noise Cancellation

In some instances, particularly at low frequencies, the acoustic background noise of a water tunnel can be removed from the desired noise data by a cancellation technique. The method requires that the acoustic background noise, usually caused by the pump, propagates around the tunnel circuit in plane waves. The plane wave cut-off frequency for ducts of circular cross section is accurately predicted from (Skudrzyk, 1971)

$$f_0 \approx 0.586 \ c/D \ , \qquad\qquad\qquad (6.3.1)$$

where c is the sound velocity and D is the duct inside diameter. Below this frequency, acoustic noise propagates in plane waves which are phase coherent in planes normal to the duct axis.

Suppose one is interested in measuring the wall pressure statistics under a turbulent boundary layer (TBL), say formed on the inside wall of the tunnel test section. These statistics are usually determined using flush-mounted hydrophones (in water). However, the hydrophones, when used individually, sense both the TBL noise and the tunnel background noise. If the frequency range of interest is less than f_0 (given by Eq. 6.3.1) then one can use a pair of hydrophones mounted in a plane perpendicular to the test section axis. The circumferential spacing of the two hydrophones must be large in comparison to the spanwise correlation length of the TBL wall pressure fluctuations. This correlation length is typically less than the TBL displacement thickness (Willmarth, 1975), so the two pressure sensors need not be very far apart. With this arrangement, one subtracts (in real time) the two signals. The TBL contributions (the desired contributions) contained in the two signals are statistically independent, so the subtraction does little to the statistics of the turbulence signal. However, the background noise is phase coherent and the differencing operation cancels it completely out. The difference signal thus contains only TBL information.

Lauchle and Daniels (1987) successfully used this background noise cancellation technique in the glycerine tunnel of Fig. 5. Others who have used the method include Wilson, et.al. (1979), Horne and Hansen (1981), and Simpson, et.al. (1987). It is to be noted that these latter three studies considered the cancellation of acoustic noise only, while the former study extended the technique to include vibration-induced background noise as well.

To highlight the method of Lauchle and Daniels (1987) consider Fig. 49 which shows a schematic of a co-planar three-sensor array mounted in the glycerine tunnel. Signals $a(t)$, $b(t)$, and $c(t)$ are hydrophone outputs while signals $v_a(t)$ through $v_c(t)$ are accelerometer outputs. A typical hydrophone signal is decomposed; for example,

$$a(t) = a_T(t) + a_A(t) + a_V(t) + a_E(t) \quad . \qquad (6.3.2)$$

Here, one has

$a_T(t)$ = TBL wall-pressure component,

$a_A(t)$ = acoustic background noise caused by the facility,

$a_V(t)$ = vibration-induced pressure caused by the facility,

$a_E(t)$ = electronic noise component from instruments.

There are two other equations, identical to Eq. (6.3.2) for $b(t)$ and $c(t)$. At the outset, the electronic noise spectrum was measured with the facility off. This spectrum for a_E (or b_E, or c_E) was found to be more than 60 dB below the spectra measured with the facility on; thus, no further reference will be made to the electronic noise components.

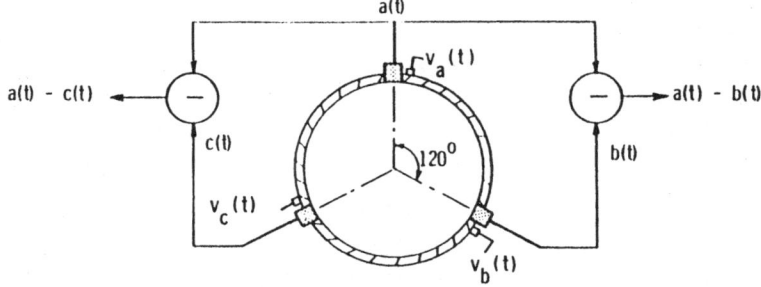

Figure 49.   Schematic of a 3-Sensor Coplanar Hydrophone Array in a Tunnel Test Section

Following the spectral notation of Bendat and Piersol (1986), we see that the autospectrum of a difference signal is given by:

$$G_{a-b,a-b} - G_{a-b}$$

$$- \lim_{T_o \to \infty} (2/T_o)E\{(A_T^* - B_T^* + A_V^* - B_V^* + A_A^* - B_A^*)$$

$$\times (A_T - B_T + A_V - B_V + A_A - B_A)\} , \qquad (6.3.3)$$

where upper-case letters denote finite Fourier transforms of the corresponding time signal, E { } is the expectation operation, and T_o is the record length.

Because the acoustic components (A_A and B_A) are in phase they cancel in Eq. (6.3.3). We are left with

$$G_{a-b} - \lim_{T_o \to \infty} (2/T_o)E\{(A_T^* - B_T^*)(A_T - B_T)$$

$$+ (A_V^* - B_V^*)(A_V - B_V)$$

$$+ (A_V^* - B_V^*)(A_T - B_T) \qquad (6.3.4)$$

$$+ (A_T^* - B_T^*)(A_V - B_V)\} .$$

The first product can be expanded to yield

$$\lim_{T_o \to \infty} (2/T_o)E\{A_T^*A_T - A_T^*B_T - B_T^*A_T + B_T^*B_T\} .$$

The first and fourth terms of this expansion are the autospecta of the wall-pressure fluctuations at locations a and b, respectively. Because of axisymmetry, they are equal. The middle two terms of the expansion are cross spectra of the TBL wall pressure in the spanwise (transverse) direction. These should be zero because the spacing between a and b is of order D, which is larger than the displacement thickness. Equation (6.3.4) reduces to:

$$G_T - G_{a_T} - G_{b_T}$$

$$- \frac{1}{2} G_{a-b} - \frac{1}{2} G_{a_V-b_V} - RE (G_{a_V-b_V, a_T-b_T}). \qquad (6.3.5)$$

Now, if the vibration components are correlated completely over the circumference of the pipe, the second two terms of Eq. (6.3.5) will be zero and the TBL pressure spectrum is equal to one-half the auto spectrum of a difference signal.

Equation (6.3.4) is essentially described by the simple single-input, single-output model of Fig. 50. The frequency response function H(f) relates the pressure response of the transducer to normal acceleration. The broken-line path indicates

that the turbulent pressures can cause the tunnel wall to vibrate and hence contribute
to the machinery-induced wall vibrations. This contribution is expected to be small
because the small-area pressure transducers are dominated by the TBL energy at the
convective wavenumber ($k_c - \omega/u_c$). The free-bending wavenumber of the pipe is at
least an order of magnitude lower than k_c for the frequency range of measurement; thus
there is not the good wavenumber matching required for there to be a strong coupling
between the TBL and pipe vibration. Nevertheless, the last term of Eq. (6.3.5)
describes this potential coupling.

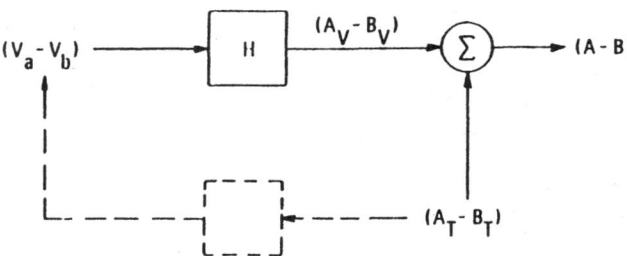

Figure 50. Schematic of a Single-Input, Single-Output Model with
Extraneous Signal at the Output

In the absence of the broken-line path of Fig. 50 one can infer the relative
contribution of (V_a - V_b) and (A_T - B_T) to (A - B) by measuring a coherent output
power (COP) spectrum between the accelerometer difference signal and the pressure-
transducer difference signal. That is, the COP spectrum is

$$G_{a_V-b_V} - \gamma^2_{a-b,v_a-v_b}\, G_{a-b} \qquad\qquad (6.3.6)$$

and the TBL contribution is

$$G_{a_T-b_T} - (1 - \gamma^2_{a-b,v_a-v_b})\, G_{a-b} \; . \qquad\qquad (6.3.7)$$

Here, the ordinary coherence function

$$\gamma^2_{xy} - |G_{xy}|^2/G_x G_y \qquad\qquad (6.3.8)$$

is utilized for the measured-difference signals. Figure 51 shows typical measured
spectra for G_{a-b} and the COP spectrum. Because there is a large spread between these
two spectra, except at one frequency near 340 Hz, we conclude that the vibration
components in the difference signal is very weak and that (a_T - b_T) dominates (a-b);
Eq. (6.3.5) reduces to

$$G_T \sim \frac{1}{2}\, G_{a-b} \; . \qquad\qquad (6.3.9)$$

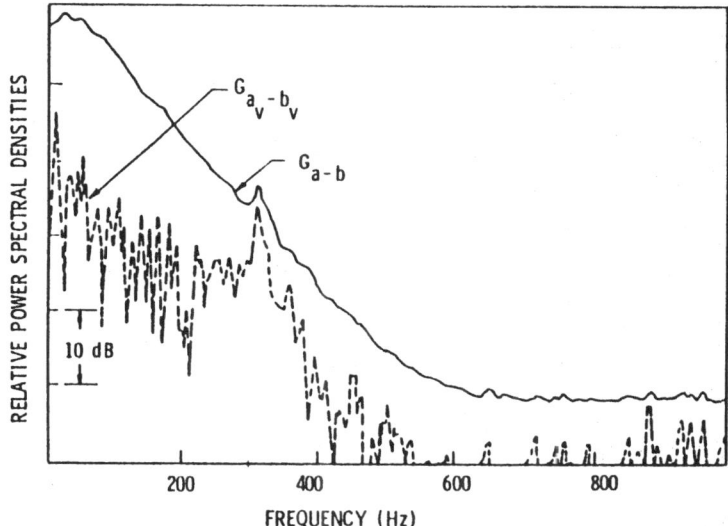

Figure 51.　Spectrum of a(t)-b(t) and the Coherent Output Power Spectrum
Between a(t)-b(t) and $v_a(t)$-$v_b(t)$

Figure 49 shows a third hydrophone and difference signal between it and a(t).
Without showing the details (Lauchle and Daniels, 1987) it also follows that the TBL
spectrum is equal to the cross spectrum between two difference signals, i.e.,

$$G_T \simeq G_{a-b,a-c} \simeq \frac{1}{2} G_{a-b} \quad . \tag{6.3.10}$$

A verification of this is shown in Fig. 52 where it is clearly seen that
$G_{a-b,a-c}$ and G_{a-b} are separated by 3 dB which is what Eq. (6.3.10) predicts.

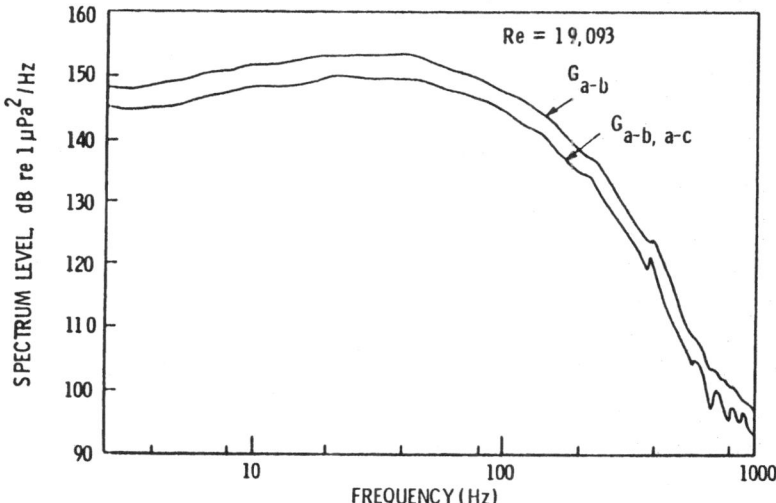

Figure 52.　Spectrum of a(t)-b(t) Compared to the Cross-Spectrum Between
a(t)-b(t) and a(t)-c(t)

The hydrophone differencing technique described above has enabled us to measure the TBL point spectrum in glycerine ($C_3 H_8 O_3$) pipe flow. A summary of the results are shown in Fig. 53. The spectrum level is normalized on mean flow velocity ($\overline{U}$), wall shear stress (τ_w) and inside diameter (D). The frequency is normalized on the viscous time scale (ν/u_τ^2). The importance of these data, which cover a pipe Reynolds number range of 10,000 to 26,500, lies in the value of transducer diameter (d) relative to the viscous sublayer thickness. The sublayer approaches 1 mm in this unique facility so it is a fairly easy matter of achieving values of $d^+ - u_\tau d/\nu$ of order one. It is agreed that $d^+ \simeq 1.0$ for a measurement to be absent of transducer spatial averaging effects. The spectrum of Fig. 53 is not only void of transducer spatial averaging effects, but is also void of tunnel background noise.

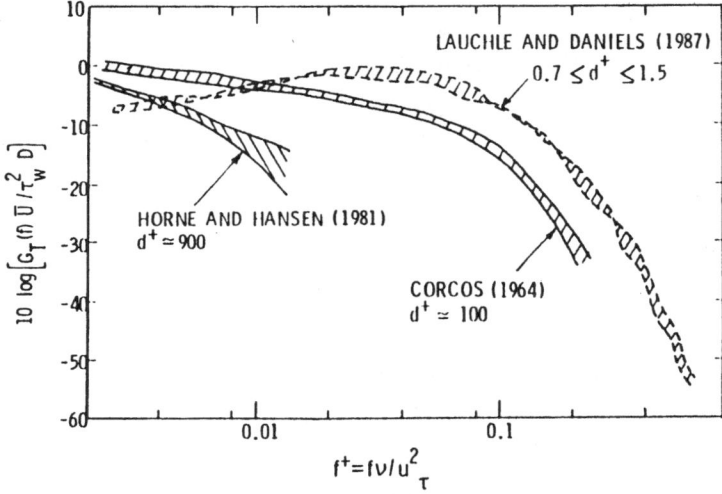

Figure 53. Normalized Wall Pressure Fluctuation Spectra Measured Under Fully-Developed Turbulent Pipe Flow (d^+-900 is in Water, d^+-100 is in Air, and $d^+ \approx 1.0$ is in Glycerine)

7.0 INTERNAL FLOW MEASUREMENTS

7.1 Internal Flow Measurement Facility

A High Reynolds Number Pump (HIREP) has been specifically designed to investigate high Reynolds number flow associated with internal flows of fluid handling machinery. The facility shown in Fig. 54 consists of a 1.07-m diameter pump stage driven by a 1.22-m diameter downstream turbine. The two units rotate on a common shaft and operate in the 1.22-m diameter test section of the Garfield Thomas Water Tunnel.

The facility was designed to make measurements to quantify the effects of rotor tip/end-wall clearance and rotor blade geometry on hydrodynamic loads, rotor

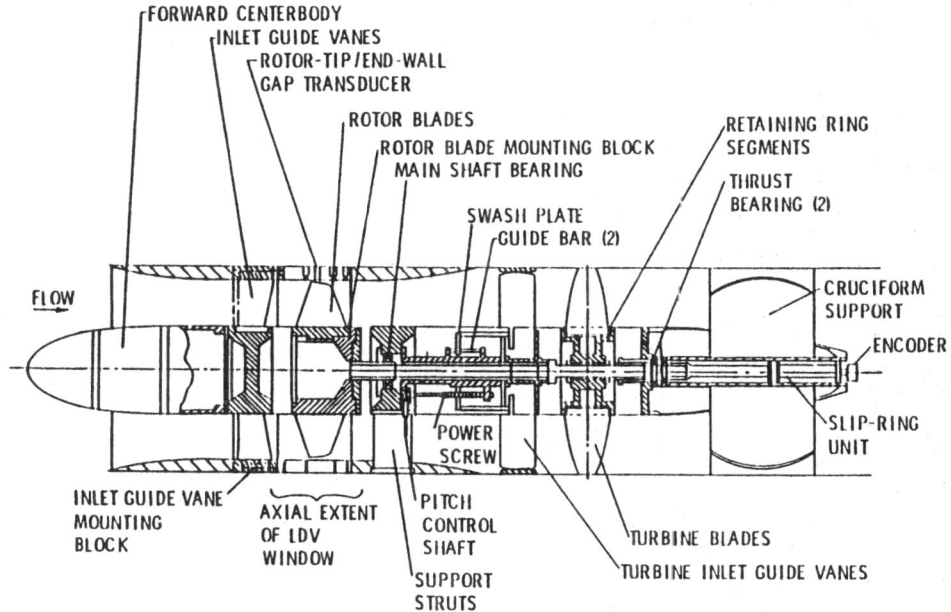

FORWARD CENTERBODY
INLET GUIDE VANES
ROTOR-TIP/END-WALL
GAP TRANSDUCER
ROTOR BLADES
ROTOR BLADE MOUNTING BLOCK
MAIN SHAFT BEARING
SWASH PLATE
GUIDE BAR (2)
RETAINING RING SEGMENTS
THRUST BEARING (2)
FLOW
CRUCIFORM SUPPORT
ENCODER
SLIP-RING UNIT
POWER SCREW
INLET GUIDE VANE MOUNTING BLOCK
AXIAL EXTENT OF LDV WINDOW
PITCH CONTROL SHAFT
SUPPORT STRUTS
TURBINE BLADES
TURBINE INLET GUIDE VANES

Figure 54. Schematic of HIREP Facility.

efficiency, cavitation, blade boundary layers and wakes, rotor dynamic response, and rotor-stator interactions. The flow coefficient of the pump is variable by adjusting the pitch of the turbine inlet guide vanes which are continuously variable. The operation of the pump/turbine is remarkably stable for an inlet velocity of 1.5 to 15 m/s with rotational speeds of 40-400 rpms. This range corresponds to a blade chord Reynolds number at the rotor tip of one-half million to six million.

The facility was designed to accommodate laser velocimetry measurements in the pump stage, radially traversing five-hole probes in every stage, a number of steady and unsteady pressure transducers in the rotating frame of reference, force and torque cells, and accelerometers. In addition, several advanced instrumentation systems for blade static pressure measurements, downstream stator blade unsteady pressure measurements, rotor-tip/end-wall gap measurement and cavitation viewing were developed.

A photograph of the facility installed in the test section is given in Fig. 55. Details of the hydrodynamic and mechanical design, and the operating characteristics are discussed by Farrell, McBride and Billet (1987).

7.2 Instrumentation/Measurement System

The facility contains provisions for a wide range of fluid dynamics measurements and methods. A 120-channel, low noise slip-ring unit accommodates many measurements in the rotating frame. The large, hollow chamber in the rotor hub houses an assembly of insulation displacement connectors which form the termination of the leads exiting

Figure 55. Photograph of HIREP Facility in the Test Section of the
1.22-m Water Tunnel.

the rotating end of the slip ring unit as shown in Fig. 56. From this termination,
four ribbon cables with high flexing capability carry signals to the rotating end of

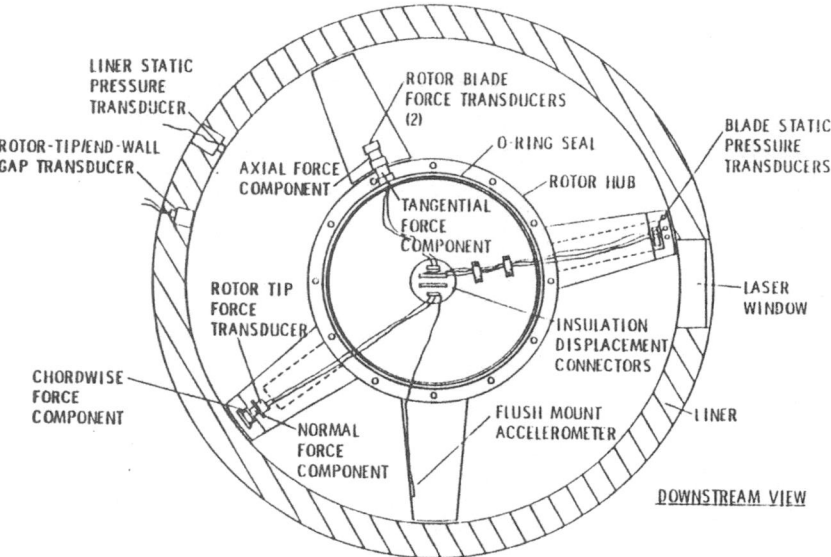

Figure 56. Rotor Hub Showing Instrumentation Capabilities.

the drive shaft. Ribbon cables form the connection between the stationary end of the slip-ring unit and the data rack and power sources outside of the facility. The signals from the transducers are sampled and processed through an extensive data analysis and reduction system consisting of the driving software program, a low pass filter, an integrating voltmeter, a multiplexer, and a VAX computer system. A schematic of the instrumentation block diagram for the performance test is shown in Fig. 57.

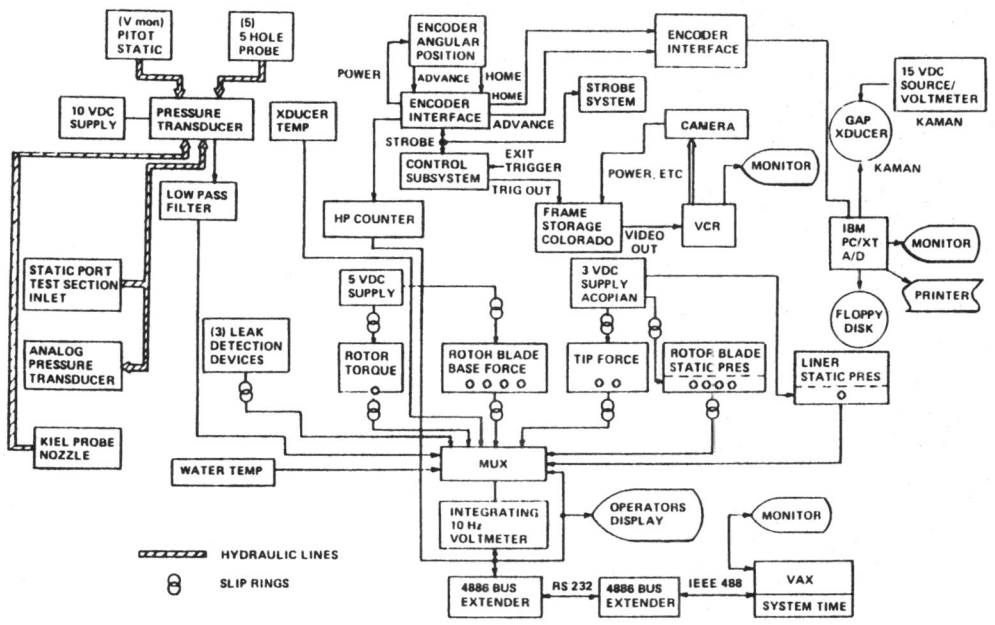

Figure 57. Instrumentation Block Diagram.

7.2.1 Velocity and Pressure Measurement

Five-hole probes can be located at the inlet and exit of the rotor, at the inlet of the turbine inlet guide vanes, and at the turbine exit. A pitot-static probe can be placed upstream of the rotor inlet guide vanes where the nose and liner are parallel and is used to calculate an axial velocity through the pump. A total pressure probe in the nozzle upstream of the test section and the static pressure port in the liner can also used to determine the axial velocity through the pump. This measurement is used as the reference axial velocity and for data normalization. The tunnel pressure control system is connected to the static pressure port. The hydraulic lines connect to a bank of differential pressure transducers located at a height equal to that of the water tunnel centerline.

The standard viewing window adjacent to the rotor is replaced by an optical window suitable for laser measurements of the rotor inlet, exit, tip-gap, and passage flows. A three-component backscatter system with a 5-watt laser is utilized for make of these measurements. The optics/laser package is positioned by a three-axis traversing mechanism. A field point velocity measurement technique discussed by Billet (1987) is used that incorporates the instantaneous laser data with an optical shaft encoder signal. An incremental optical encoder is located on the downstream end of the shaft. In this system an angular position is associated with each instantaneous velocity measured at a point. A computer then processes this digital information.

7.2.2 Force and Torque Measurement

Forces and torque are measured on individual rotor blades and drive shaft, respectively. The rotor blade force is measured by two two-component shear beams which are mounted in the base and extend radially outward to support the lower portion of the blade span. On another blade, the outer ten percent of the blade span is instrumented with a two-component force balance to measure the normal and chordwise forces. Finally, the inner diameter of the drive shaft is strain gaged to measure torque. The excitation voltages for the strain gages are remotely sensed to correct for line loss.

7.2.3 Blade Pressure Measurement

Several pressure transducers are located in the rotor blade tip region, in the wall liner and in some cases on a downstream stator blade. For the transducers that were mounted on the rotor blade, the transducer leads are channeled through milled slots on the surface to a center hole which is drilled through the pressure and suction surfaces to a center cavity. The leads are then soldered to a small printed circuit board and ribbon cable is connected to this board. The transducers are mounted beneath a Helmholtz cavity to protect the diaphragm from high, localized pressures caused, for example, by the collapse of cavitation bubbles. During the assembly of the instrumented blade, petroleum jelly was placed into the cavity so that no trapped air would be present in the cavity.

7.2.4 Dynamic Rotor-Tip/End-Wall Measurement

A dynamic measuring system is available to measure the rotor-tip/end-wall gap. A variable impedance transducer is placed in the wall liner surrounding the rotor such that the face of the transducer is flush with the inner diameter of the liner. The impedance variation is caused by the occurrence of eddy currents in the conductive, metallic rotor tip. The coupling between the coil in the sensor and the rotor tip is dependent upon their relative displacement. Thus, the gap of any particular blade can be measured by sampling the signal from the transducer during the appropriate angular positions of the rotor as triggered by a conditioned signal from the shaft encoder. These data are processed through an A/D data acquisition system. Then software analyzes these data to determine the minimum voltage which occurs when the blade-tip is directly over the top of the sensor face. This voltage is directly related to the gap.

7.2.5 Cavitation Inception Measurement

Cavitation inception cannot be viewed by the naked eye even with the aid of stroboscopic lighting due to the low rpm of the high Reynolds number pump. Thus, a low-rpm video viewing system is used. The low-rpm video viewing system captures a stroboscopically illuminated image and maintains a display of this image on a TV monitor until the next image has been digitized and placed in memory.

8.0 SUMMARY

A description of several important techniques used in the measurement of steady and time-dependent fluid dynamics quantities in high-Reynolds number liquid flows has been given in this Chapter. Our emphasis has been on the measurement of hydrodynamically-induced force, velocity, stress, and acoustic noise in both single and multi-phased liquid flow situations. Because water tunnels are commonly used to address high-Reynolds number liquid flows, and because they offer an extremely stable and time-invariant environment in which to conduct basic and applied fluid dynamics research, we have emphasized their use in making these measurements.

A brief introduction to the construction and operation of the typical water tunnel was given. Of particular importance is the experimenter's ability to maintain constant values of free-stream velocity and ambient pressure. Also, with appropriate sub-systems, the water (or other liquid) temperature and quality (free and dissolved gas content and particulate matter) can be controlled. Temperature control is particularly important if one wishes to raise (or lower) the unit Reynolds number range of the facility. Turbulence in the test section can also be controlled by placing the appropriate screens and honeycomb far upstream in the tunnel settling section. It was pointed out that the control of acoustic background noise in water tunnels is quite difficult, but that the elimination of all sources of facility-generated cavitation is the first most important step to effective noise control.

Body force and moment measurements are routinely conducted at high Reynolds numbers in water tunnels and towing tanks. The sensors used in such tests are usually strain-gaged balance devices. We emphasized that when large bodies (relative to the test section diameter) are tested, spurious forces and moments can be created by blockage effects and by horizontal buoyancy (viscous) effects. Methods to account for these problems were discussed.

There are many problems associated with making detailed measurements in laminar and turbulent liquid boundary layers at high speed. One of the more serious of these problems is the very small size of the inner scales of the turbulent boundary layers. Optical techniques, such as Laser Doppler Velocimetry (LDV), or techniques which employ surface measurement devices, such as flush-mounted hot film probes have proven most effective. Often, however, the spatial scales of the measurement devices

are so large that significant averaging of the results is unavoidable. In addition,
the very near-wall regions of a turbulent boundary layer are often not accessible
to measurement. Thus, while water tunnels are the proper choice for the study of
phenomena associated mainly with high Reynolds number liquid flows, such as many drag
reduction mechanisms, they are generally not the proper choice for fundamental studies
of the turbulent boundary layer.

Measurements in multiphased flows are difficult to generalize since each given
flow situation usually requires a unique measurement technique. One of the most widely
used techniques, however, is some form of flow visualization. This includes the use of
tufts, oil paint films, injected dyes, bubble tracking, Schlieren photography, light
scattering techniques, and holographic photography. The use of some of these methods
in cavitation research was discussed with emphasis on the measurement of bubble
distributions using scattered laser light, and on the measurement of cavitation bubble
dynamics using holography.

Flow-induced acoustic noise can be measured in some water tunnels using
reverberation room acoustic measurement techniques or special techniques developed for
the specific tunnel and tunnel test. The methods usually involve the use of single
hydrophones or arrays of hydrophones. Hydrophone directivity is important because it
can be used to improve the signal-to-noise ratio; the more directive is the receiver,
the more it can be used to discriminate the desired source of noise from the
background sources of noise. We presented a discussion on the calibration and use of a
reflector-type hydrophone in radiated noise measurements. A non-intrusive method of
making acoustic measurements in a water tunnel (or any other closed flow facility)
based on the theory of acoustic reciprocity was also presented. The method is quite
new and provides the hydroacoustics experimenter a method whereby the source strength
of an acoustic source can be measured directly. Wall pressure fluctuations under
turbulent boundary layers is a subject of importance, but many measurements are
inaccurate because of transducer spatial averaging problems, facility background noise
(particularly at low frequencies), and the paramount difficulties of implementing wide
area transducer arrays for mapping the space/time statistics of these pressure
fluctuations. The issues of background noise and transducer resolution were described
in detail with a specific measurement example given to verify that both problems can
be eliminated. This example used a passive background noise cancellation technique
together with sub-miniature pressure transducers under a glycerine turbulent boundary
layer.

The measurement of cavitation inception, cavitation noise, local pressure and
velocity fields, and body (surface) forces and moments within complex liquid handling
turbomachines under realistic high-Reynolds number operation is an extremely difficult
undertaking. In this Chapter we have described one way of approaching this important
problem. The method virtually models an axial-flow turbomachine outer casing by a
water tunnel test section. The viewing windows in the test section permit almost total
visual access to the internal hydrodynamics of the turbomachine; thus, flow

visualization, conventional LDV, and specialized internal probing can be achieved. The Reynolds number range of such experiments can be very large if large tunnels coupled with high-speed operation are used.

REFERENCES

Alfredsson, P. H., Johansson, A. V., Haritonidis, J. H. & Eckelmann, H. 1988 The fluctuating wall-shear stress and velocity field in the viscous sublayer. Phys. Fluids 31, 1026-1033.

Bajura, R. A. & Billet, M. L., ed. 1986 Fluid Measurements and Instrumentation Forum - 1986, ASME, (Atlanta, Georgia).

Bakewell, H. P. & Lumley, J. L. 1967 Viscous sublayer and adjacent wall region in turbulent pipe flow. Phys. Fluids 10, 1880-1889.

Bakewell, H. P. 1968 Turbulent wall-pressure fluctuations on a body of revolution. J. Acoust. Soc. Am. 43, 1358-1363.

Barker, S. J. 1974 Measurements of radiated noise in the CALTECH high-speed water tunnel. Graduate Aeronautical Lab Report NR 062-462.

Barker, S. J. & Gile, D. 1981 Experiments on heat stabilized laminar boundary layers in water. J. Fluid Mech. 104, 139-158.

Barker, S. J. 1973 Radiated noise from turbulent boundary layers in dilute polymer solutions. Phys. Fluids 16, 1387-1394.

Bellhouse, B. J. & Schultz, D. L. 1966 Determination of mean and dynamic skin friction, separation and transition in low speed flow with a thin film heated element. J. of Fluid Mech. 24, 379.

Bendat, J. S. & Piersol, A. D. 1986 Random Data, 2nd Ed. (Wiley, New York) p. 132.

Besant, W. 1859 Hydrostatics and Hydrodynamics, Cambridge University Press, Cambridge, England.

Billet, M. L. 1987 Wake measurments using a laser doppler velocimeter system. Proceedings of The International Towing Tank Conference, Kobe, Japan.

Billet, M. L. 1986 Cavitation nuclei measurments with an optical system. J. of Fluids Engr., 108, 366-372.

Bistafa, S. R. 1984 An experimental study on the noise generated by vaporous cavitation in turbulent shear flows produced by confined orifice plates. Ph.D. Thesis, The Pennsylvania State University.

Bistafa, S. R., Lauchle, G. C. & Reethof, G. 1987 Noise generated by cavitation in orifice plates. To appear in J. of Fluids Engr.

Blake, W. K. 1986 Mechanics of Flow-Induced Sound and Vibration, Vol. I and II (Academic Press, Orlando).

Blake, W. K., Wolpert, M. J. & Geib, F. E. 1977 Cavitation noise and inception as influenced by boundary-layer development on a hydrofoil. J. Fluid Mech. 80, 617-640.

Chahine, G. & Shen, Y. T. 1985 Bubble dynamics and cavitation inception in cavitation susceptibility meters. Proc. Int. Sym. on Fundamental Aspects of Gas-Liquid Flows (Miami Beach, Florida).

Coles, D. E. 1968 A young person's guide to the data. In _Proceedings of the AFOSR-IFP Stanford Conference on Computation of the Turbulent Boundary Layers_, (D. E. Coles and E. A. Hirst, Eds.), Stanford University Press, Stanford, CA.

Corcos, G. M. 1964 The structure of the turbulent pressure field in boundary layer flows. _J. Fluid Mech._ 18, 353.

d'Agostino, L. & Acosta, A. J. 1983 On the design of cavitation susceptability meters (C.S.M.'s). _Proc. 20th American Towing Tank Conference_ (Hoboken, New York).

Deutsch, S. & Castano, J. 1986 Microbubble skin friction reduction on an axisymmetric body. _Phys. of Fluids_ 29, 3590 .

Dybbs, A. & Pfund, P. A. 1985 _International Symposium of Laser Anemometry_, ASME, (Miami Beach, Florida).

Eckelmann, H. 1974 The structure of the viscous sublayer and adjacent wall region in a turbulent channel flow. _J. Fluid Mech._, 65, 439.

Farmer, W. M. 1976 Sample space for particle size and velocity measuring interferometers. _Applied Optics_, 15, No. 8.

Furuya, O. 1987 _Cavitation and Multiphase Flow Forum_ - 1987, ASME (Cincinnati, Ohio).

Farrell, J. J., McBride M. W. & Billet, M. L. 1987: High Reynolds number pump facility for cavitation research. _Proc. Int. Sym. on Cavitation Research Facilities and Techniques - 1987_ (Boston, Massachusetts).

Gates, E. M. & Bacon, J. 1978 Determination of cavitation nuclei distributions by holography. _J. Ship Res._, 22, 29-31.

Gad-el-Hak, M. 1987 The water towing tank as an experimental facility, an overview. _Exp. in Fluids_ 5, 289-297.

Gavigan, J. J., Watson, E. E. & King, W. F. 1974 Noise generation by gas jets in a turbulent wake. _J. Acoust. Soc. Am._ 56, 1094-1099.

Gowing, S. & Ling, S. C. 1980 Measurements of microbubbles in a water tunnel, _Proc. 19th American Towing Tank Conference_ (Ann Arbor, Michigan).

Greshilov, E. M. & Mironov, M. A. 1983 Experimental evaluation of sound generated by turbulent flow in a hydrodynamic duct. _Sov. Phys. Acoust._ 29, 275-280.

Hall, W. R. 1973 Tunnel wall interference for bodies-of-revolution in non-steady motion. Master of Science Thesis, The Pennsylvania State University.

Hamilton, M. F., Thompson, D. E. & Billet, M. L. 1986 An experimental study of travelling-bubble cavitation noise. _J. of Fluids Engr._ 108, 241-247.

Hammitt, F. G., Keller, A., Ahmed, O., Pyun, J. J. & Tilmaz, E. 1974 Observation and measurement of cavitation nuclei using Coulter counter and laser light scattering in cavitation and multiphase flow laboratory at the University of Michigan. Report No. VMICH 03157-24-I.

Hanratty, T. J. & Campbell, J. A. 1982 Measurement of wall shear stress. In _Fluid Mechanics Measurements_, (R. J. Goldstein, Ed.), Hemisphere Publishing Corporation, New York.

Henderson, R. E. & Parkin, B. R. 1982 Hydrodynamic test facilities at ARL Penn State. _Proc. 22nd Defense Research Group Seminar on Adv. Hydro. Test Facilities_, NATO DS/A/DR 83252.

Holl, J. W. 1970 Nuclei and cavitation. _J. of Basic Engr._ 92, 681-688.

Holl, J. W.; Billet, M. L. & Weir, D. 1975 Thermodynamic effects on developed cavitation. _J. of Fluids Engr._ 97, 507-514.

Holl, J. W. & Billet, M. L. 1987 _International Symposium on Cavitation Research Facilities and Techniques - 1987_. ASME, Boston, Massachusetts, December 1987.

Horne, M. P. & Hansen, R. J. 1981 Minimization of farfield acoustic effects in turbulent boundary layer wall pressure fluctuation experiments. _Proc. 7th Symp. on Turb._ (Univ. Missouri, Rolla) p. 139.

Kadykov, I. F. & Lyamshev, L. M. 1970 Influence of polymer additives on the pressure fluctuations in a boundary layer. _Sov. Phys. Acoust._ 16, 59-63.

Keller, A. P. 1972 The influence of the cavitation nucleus spectrum on cavitation incpetion, investigated with a scattered light counting method. _J. of Basic Engr._, 94, 917-925.

Kohler, R. A. & Billet, M. L. 1981 Light scattering by a nonspherical particle. _Cavitation and Dolyphase Flow Forum - 1981_ (Boulder, Colorado).

Ladd, D. M. & Hendricks, E. W. 1985 The effects of background particulates on the delayed transition of a heated 9:1 ellipsoid. _Exp. in Fluids 3_, 113-119.

Lauchle, G. C. 1977 Noise generated by axisymmetric turbulent boundary-layer flow. _J. Acoust. Soc. Am._ 61, 694-703.

Lauchle, G. C. 1979 Horizontal buoyancy effects on the pressure distribution of a body in a duct. _J. Hydronautics_ 13, 61-67.

Lauchle, G. C. 1983 In-duct viscous flow pressure distribution on a body with boundary-layer control. _J. Ship Res._ 27, 34-38.

Lauchle, G. C., Eisenhuth, J. J. & Gurney, G. B. 1980 Boundary-layer transition on a body of revolution. _J. Hydronautics_ 14, 117-121.

Lauchle, G. C. & Gurney, G. B. 1984 Laminar boundary-layer tranistion on a heated underwater body. _J. Fluid Mech._ 144, 79-101.

Lauchle, G. C. 1974 Dynamic hinge moment of a low aspect ratio control surface. _J. Hydronautics_ 8, 119-120.

Lauchle, G. C. & Daniels, M. A. 1987 Wall-pressure fluctuations in turbulent pipe flow. _Phys. of Fluids_ 30, 3019-3024.

Lecoffre, Y. & Bonnin, J. 1979 Cavitation tests and nucleation control. _Proc. Int. Sym. Cavitation Inception_ (New York, New York).

Lehman, A. F. 1959 The Garfield Thomas Water Tunnel. Ordnance Research Laboratory, The Pennsylvania State University Tech. Report NOrd 16597-56.

Lehman, A. F., Light, J. H. & Peirce, T. E. 1958 Elimination of water-tunnel interaction with a coaxial test body by a flow correcting liner. Ordnance Research Laboratory, The Pennsylvania State University, Report NOrd 16597-59.

Liepmann, H. & Skinner, G. 1954 Shearing-stress measurements by use of a heated element. NACA TM 1824.

Ludwieg, H. 1950 Instrument for measuring the wall shearing stress of turbulent boundary layers. NACA TM 1284.

Ludwieg, H. & Tillmann, W. 1950 Investigation of the wall shearing stress in turbulent boundary layers. NACA TM 1285.

Lumley, J. L. 1964 Passage of a turbulent stream through honeycomb of large length-to-diameter ratio. Trans. A.S.M.E., J. Basic Engr. D86, 218-220.

Lumley, J. L. & McMahon, J. F. 1967 Reducing water tunnel turbulence by means of a honeycomb. Trans. A.S.M.E., J. Basic Engr. 89, 764-770.

Lyamshev, L. M., Chelnokov, B. I. & Shustikov, A. G. 1984 Pressure fluctuations in a turbulent boundary layer under the conditions of injection of a Continous medium through a permiable boundary. Sov. Phys. Acoust. 30, 394-397.

Maidanik, G. & Jorgensen, D. W. 1967 Boundary wavevector filters for the study of the pressure field in a turbulent boundary layer. J. Acoust. Soc. Am. 42, 494-501.

Marboe, R. C. 1982 Bubble dynamics and resulting noise from traveling bubble cavitation. M.S. Thesis, Massachusetts Inst. Tech. (Also, ARL TM No. 82-94, The Pennsylvania State University).

Medwin, H. 1977 Acoustical determination of bubble size spectra. J. of Acoust. Soc. Am. 62, 1041-1044.

Oba, R. 1981 Cavitation nuclei measurments by a newly made Coulter counter without adding salt in water. Rep Inst. High Speed Mech. (Tohoku University) 43, 340.

Oldenziel, D. M. 1979 New instrument in cavitation research, Proc. Int. Sym. Cavitation Inception (New York, New York) 111-124.

Oosterveld, M. W. C. & Johnson, B. 1985 16th International Towing Tank Conference Catalog of Facilities (Contact Prof. Bruce Johnson, U.S. Naval Academy, Annapolis, MD 21402, U.S.A.).

Parkin, B. R. 1983 Hydrodynamics and fluid mechanics. Naval Research Rev. 35, 20-34.

Peirce, T. E. 1964 Tunnel wall interference effects on drag and pitching moment of an axisymmetric body. Ph.D. Thesis, The Pennsylvania State University.

Ramjee, V. & Hussain, A.K.M.F. 1976 Influence of the axisymmetric contraction ratio on free-stream turbulence. Trans. A.S.M.E., J. Basic Engr. 98, 506-515.

Robbins, B. E. 1978 Water tunnel turbulence measurements behind a honeycomb J. Hydronautics 12, 122-128.

Ross, D., Robertson, J. M. & Power, R. B. 1948 Hydrodynamic design of the 48-inch water tunnel at the Pennsylvania State University. S.N.A.M.E. Trans. 56, 5-29.

Sandborn, V. A. 1979 Surface shear stress fluctuations in turbulent boundary layers. In Proceedings of the Second Symposium on Turbulent Shear Flow, Imperial College, London.

Schiebe, F. R. & Killen, J. M., 1971 An evaluation of acoustical techniques for measuring gas bubble size distributions in cavitation research. St. Anthony Falls Hydraulic Lab No. 120. (University of Minnesota).

Schloemer, H. H. 1974 Installation of a rectangular test section for acoustic water tunnel studies of flow-induced noise. NUSC Tech. Report 4763.

Sevik, M. 1964 Measurement of unsteady thrust in turbo-machinery. A.S.M.E. Paper 64-FE-15.

Shen, Y. T.; Gowing, S. 1985 Scale effects on bubble growth and cavitation inception in cavitation susceptability meters. Cavitation and Multiphase Flow Forum - 1985 (Albuquerque, New Mexico) 14-16.

Simpson, R. L., Ghodbane, M. & McGrath, B. E. 1987 Surface pressure fluctuations in a separating turbulent boundary layer. J. Fluid Mech. 177, 167-186.

Skudrzyk, E. J. & Haddle, G. P. 1960 Noise radiation from a turbulent boundary layer by smooth and rough surfaces. J. Acoust. Soc. Am. 32, 19-34.

Skudrzyk, E. 1971 The Foundations of Acoustics, Basic Mathematics and Basic Acoustics (Springer, New York) p. 330.

Tam, C. K. W. 1975 Intensity, spectrum, and directivity of turbulent boundary layer noise. J. Acoust. Soc. Am. 57, 25-34.

Thompson, D. E. 1976 Propeller time-dependent forces due to non-uniform inflow. Ph.D. Thesis. The Pennsylvania State University.

Vikram, C. S. & Billet, M. L. 1983 Gaussian beam effects in far-field in-line holography. Applied Optics. 22, 2830-2835.

Vikram, C. S. & Billet, M. L. 1984 Optimizing image-to-background irradiance in far-field in-line holography. Applied Optics 25, 1995-1998.

Vikram, C. S. & Billet, M. L. 1984 In-line Fraunhofer holography of a few far fields. Applied Optics 23, 3091-3094.

Vikram, C. S. & Billet, M. L. 1982 Holographic image formation of objects inside a chamber. Optik 61, No. 4, 427-432.

Vikram, C. S. & Billet, M. L. 1983 Magnification with divergent beams in Fraunhofer holography of object inside a chamber. Optik 63, No. 2, 109-114.

Waterhouse, R. V. 1958 Output of a sound source in a reverberation chamber and other reflecting environments. J. Acoust. Soc. Am. 30, 4-13.

White, F. M. 1974 Viscous Fluid Flow, (McGraw-Hill, New York).

Willmarth, W. W. 1975 Pressure fluctuations beneath turbulent boundary layers. Ann. Rev. Fluid Mech. 7, 13-37.

Wilson, R. J., Jones, B. G. & Roy, R. P. 1979 Measurement technique of stochastic pressure fluctuations in annular flow. Proc. 6th Symp. on Turb. (University Missouri, Rolla), p. 4-1.

Wolde, T. T. 1973 Reciprocity experiments on the transmission of sounds in ships. Ph.D. Thesis, Delft University, The Netherlands.

Yungkurth, C. B. 1983 A light-scattering system to measure cavitation nuclei analysis and calibration. M.S. Thesis, The Pennsylvania State University.

The turbulent boundary layer

K.R. Sreenivasan

Mason Laboratory, Yale University, New Haven, CT 06520

Abstract

This article is intended as a brief review of the dynamics of the turbulent boundary layer on a smooth flat wall. The emphasis is on physical arguments, and the contents offer the minimum ground that an advanced graduate student (who has some familiarity with the turbulence problem but not much with the boundary layer itself) ought to cover before embarking on his own research. But it is believed that the perspective is relatively new at places, and that the article is therefore of interest also the specialist. In section 2, classical notions are presented in a way that their strengths and weaknesses become transparent. This is followed in section 3 by a description of the structural features of the boundary layer. In both sections, some emphasis is given to the inner/outer interactions. The article ends with a discussion of a few open ended issues concerning the problem.

1. Introduction

A major practical reason for interest in turbulence is that it enhances mixing as well as transport of energy and matter; yet another is that it is responsible for a significant fraction of energy loss in internal and external flows. The understanding and control of turbulent boundary layers is an especially important problem technologically, and vast sums of money and effort have been spent on it. While a wealth of information now exists on the description of its various facets, a complete picture is yet to emerge, and our predictive and control capabilities based on first principles have remained far less adequate than practical needs would suggest.

We restrict attention largely to the fully-developed, two-dimensional turbulent boundary layer on a smooth, semi-infinite flat plate with no imposed pressure gradient, and the flow is assumed to be isothermal and incompressible. It is impossible to realize these conditions strictly in a laboratory, but close approximations have been produced. In practice, one encounters the turbulent boundary layer in the presence of three-dimensionalities, roughnesses, pressure gradients, curvature, heat transfer, compressibility, etc., some of which can have profound effects on the flow development. It is therefore useful to comment on the rationale for the excessive emphasis on this paradigmatic case. First, it is the simplest possible wall-bound turbulent flow: There is little hope of understanding other complex effects without the benefit of understanding this 'simple' case. Secondly, the properties of this special case are useful even in the presence of these complex effects – quantitatively if they are small and qualitatively otherwise.

The turbulent boundary layer possesses both 'universal' and case-specific properties; by definition, the universal aspects do not depend on the precise flow configuration (at least to a good approximation). Here, we focus attention on those aspects which make the turbulent boundary layer unique, and dwell less on the universal aspects.

Although the velocity field at each instant of time is believed to be governed exactly by the Navier-Stokes equations, their complexity makes it impossible to obtain the required information by directly solving them. (See section 4.1 for comments on the role of computer simulations.) Following Reynolds (1894), traditional turbulence studies reject the need for detailed instantaneous quantities, and proceed by decomposing the motion into its time mean (which is steady in the flow we are considering) and fluctuations around it; only the evolution of the mean and perhaps a first few moments of fluctuations are sought. We assume that the reader is familiar with the resulting closure problem (see, for example, Monin & Yaglom 1971), and the necessity for heuristic modeling. Alternatives to this so-called Reynolds decomposition have been proposed (e.g., Hussain 1983), especially motivated by the awareness that the coherent component of the turbulent motion needs to be explicitly built into a workable decomposition scheme, but their advantages are unclear at the moment. We shall follow tradition and adopt the Reynolds decomposition, keeping in mind that we must be willing to do away with this artifact under sufficiently compelling reasons.

The presence of the wall makes the turbulent boundary layer structure and dynamics profoundly different from those of unbounded (or 'free') shear flows. One of the primary differences is that there are in the turbulent boundary layer two distinct scales which are highly disparate at large Reynolds numbers. This makes true self-preservation impossible (except in the special case of the flow between two converging planes where the two scales can be forced to be proportional to each other at all Reynolds numbers). Unlike other fully turbulent flows in which *direct* viscous effects are negligible, they are *always* present in a small region near the wall of the turbulent boundary layer. The turbulent boundary layer can maintain large fluctuation levels close to the wall, successfully countering viscous and turbulent diffusion. The rate of turbulent energy production and dissipation peak very near the wall (figure 1). More than about a third of the total production and dissipation occurs within a distance from the wall which occupies no more than 2% of the flow thickness at typical laboratory Reynolds numbers (and an even smaller fraction at higher Reynolds numbers). This feature has no analog in free shear flows where the rate of turbulent energy production is the largest near the inflection point of the mean velocity profile, and is spread over the entire flow width; the energy dissipation is even more diffusely distributed (figure 2).

The wall-region of the boundary layer is thus perceived to serve an important function in maintaining the flow. As we shall see in section 2.1, the characteristic Reynolds number of this region is a small number of the order 30. Boundary layers at such low Reynolds numbers cannot be maintained 'turbulent' unless buffetted constantly by strong disturbances, and this necessary function can be expected to be served by the outer layer. At least in this sense, the maintenance of the turbulent boundary layer must owe itself to the interaction of the outer layer fluctuations with the near-wall region. Unfortunately, the precise way in which this interaction comes about is not fully understood: The complexity of the problem is such that there are differing views even on the kinematic description of the region near the wall, and on the very importance of the events occurring there. Section 2 provides a description based on classical notions (albeit with different emphasis at places), while section 3 summarizes the work on the boundary layer structure. In each of these sections, we first present descriptions of the wall-dominated region and the rest of the flow separately, and then examine the more complex problem of interactions; this latter is our emphasis

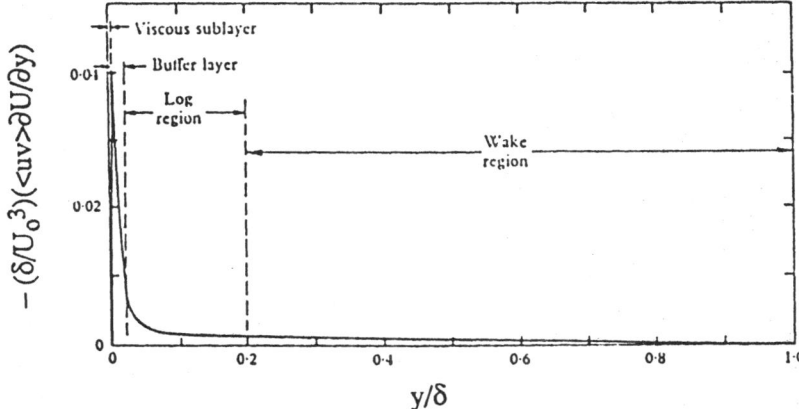

Figure 1. The turbulent energy production rate per unit volume is the product of the Reynolds stress -<uv> and the local velocity gradient $\partial U/\partial y$. This quantity normalized by the freestream velocity U_0 and the boundary layer thickness δ is plotted against y/δ, y being the distance measured from the wall normal to it. The curve, computed from Klebanoff's data, is adapted from Kline et al. (1967). The various regimes in the boundary layer will be explained at appropriate places in the text; so will other details.

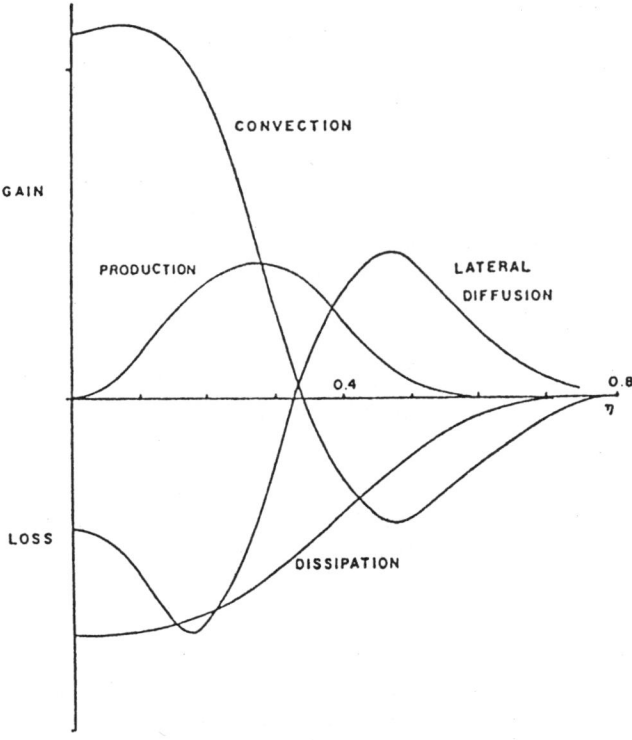

Figure 2.   The turbulent energy balance in the plane wake behind a circular cylinder. Data of Townsend (1956). Here, η is the appropriately normalized distance in the direction of the largest shear. $\eta = 0$ is the wake centerline.

but, unfortunately, the task will have to be left unfinished at the present state of knowledge.

In the following section, we occasionally invoke similarities in the near-wall region between the boundary layer on the one hand, and pipe and channel flows on the other. The conditions in the outer region can be expected to be different in these two classes of flows because there is usually a turbulence-free uniform outer stream in the boundary layer while the flow in fully developed pipes and channels is turbulent right up to the geometric center. Furthermore, flow variables (except for the pressure) in fully-developed pipes and channels do not depend on the streamwise coordinate while such a dependence, even if a weak one, exists in the boundary layer.

2. The mean velocity and Reynolds stresses

In the Reynolds decomposition, the most important dynamical quantity affecting the mean motion is the so-called Reynolds or turbulent shear stress, $\tau = -<uv>$, where u and v are the fluctuation velocities in the streamwise direction x and the direction y normal to the flat plate (see figure 3); the angular brackets indicate the time mean. (Here and elsewhere, unless explicitly remarked otherwise, capital letters indicate mean values and the corresponding lower case letters indicate fluctuations around the mean; for example, U and u are respectively the mean and fluctuating velocities in the direction x.) Since the Reynolds stress represents the significantly larger momentum transport accomplished by turbulence, modeling its behavior is one of the prime concerns of various prediction schemes.

A fully turbulent flow is characterized by the inequality that the turbulent stress is much larger than the viscous stress. As already mentioned, this condition is *not* satisfied everywhere in the boundary layer. Figure 4 shows the Reynolds shear stress distribution at four different Reynolds numbers. (Some details of how these distributions were obtained are given in the figure caption, and more can be found in section 2.3.) Also shown in figure 4 is the viscous shear stress distribution. For reasons that will soon become clear, the distance from the wall is normalized by the so-called wall variables, which are the kinematic viscosity coefficient v of the fluid and the friction velocity $U^* = \tau_w^{1/2}$, τ_w (= $v\partial U/\partial y$ at the wall) being the kinematic wall shear stress.

One can identify several fairly distinct regions in figure 4. Very near the wall, there is a small region in which the viscous stress is overwhelmingly large compared to the turbulent shear stress. For a certain distance beyond this region, neither the viscous stress nor the turbulent shear stress is negligible. This region merges with another within which τ is nearly a constant and equal to τ_w at high Reynolds numbers. (We shall return to the low Reynolds number cases later.) Even further away from the wall, τ drops off to zero gradually. These various regions are described in greater detail below.

2.1. The viscous layer: The region closest to the wall in which the viscous shear stress is dominant is called the *viscous sublayer*. To a first approximation, it is reasonable to think that the entire shear stress in this layer is the result of viscous action, which implies that the only quantities of relevance are U^* and v. It follows from the definition of the shear stress that the velocity variation in the sublayer is linear with the distance from the wall, and is given by

$$U/U^* = yU^*/v. \tag{1}$$

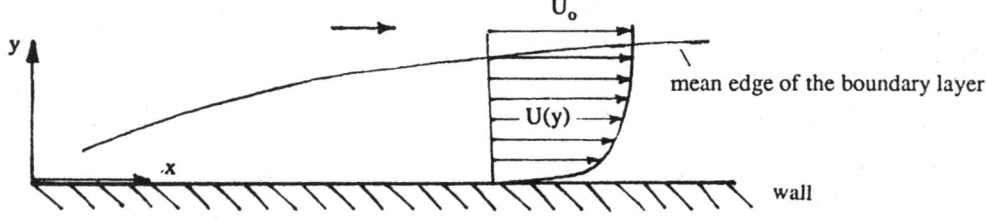

Figure 3.   The schematic of the boundary layer defining coordinates.

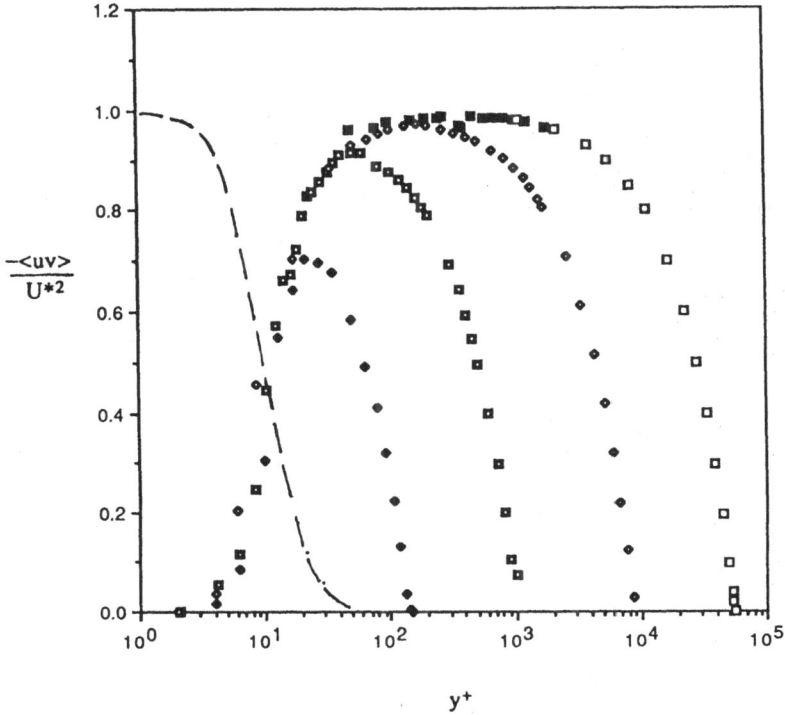

Figure 4. The viscous and turbulent shear stresses in pipe flows. The region up to and including the constant-stress layer is similar to that in the boundary layer. We chose pipe data because the shear stress can be computed from mean velocity using equation (18), section 2.3. The dashed line is the viscous stress. From left to right, the calculated Reynolds stresses correspond to: Nikuradse (1932), $R^* = U^*a/v = 140$; Laufer (1954), $R^* = 1050$: Laufer, $R^* = 8600$; Nikuradse, $R^* = 5.54 \times 10^4$; a is the pipe radius. For the last flow, the closest velocity measurement is at y^+ of 1110, and points plotted at smaller y^+ are based on a reasonable extrapolation of the measured mean velocity. They may therefore be somewhat uncertain.

It is convenient to introduce the superscript + which refers to the normalization by wall variables. In this notation, equation (1) becomes $U^+ = y^+$. It is known from empirical data that the height of the viscous sublayer is on the order of $5\nu/U^*$, where the combination ν/U^* is called the wall or viscous unit of length.

As already noted, the viscous action extends substantially beyond the viscous sublayer. If we define, for definiteness, this layer to be one in which 95% of the direct viscous dissipation occurs, its thickness, to within experimental uncertainties, is on the order of 30 wall units. (This number may vary somewhat with the various flow conditions, but it is not much.) We shall designate this layer as the *viscous layer*. The thickness of the viscous layer as a function of the momentum thickness Reynolds number (R_θ) is shown in figure 5. Although it occupies about 25% of the total thickness at the lowest Reynolds number that turbulent boundary layers are known to exist (R_θ of about 350, see Preston 1958), it is a small fraction of the total boundary layer thickness at high Reynolds numbers. But its importance far exceeds its size, with more than a third of the turbulent energy being produced here. (This last statement is the result of integrating the distributions of turbulent energy production at several Reynolds numbers up to the highest value for which reliable measurements are available. Whether this is in principle true in the limit of infinitely large Reynolds number cannot be decided simply; we indicate in section 2.4 a crucial point that needs to be settled for a satisfactory resolution of the problem.)

The region of the viscous layer outside of the viscous sublayer is often called the *buffer layer* ($5 \leq y^+ \leq 30$, approximately). The peak production and dissipation of turbulent energy, already mentioned in section 1, occur in the middle of the buffer layer where the viscous and turbulent stresses are approximately equal. To within experimental uncertainty, this height corresponds to y^+ of about 12 (see figure 4). A close examination of the available data indicates that this result is independent of Reynolds number; again, the asymptotic validity of this statement is related to the issue raised in the preceding paragraph.

In the close vicinity of the wall, all flow quantities can be expanded by the Taylor's series, equation (1) for the mean velocity being only the first term in such an expansion. From considerations of the no-slip condition, and continuity and dynamical equations, Monin & Yaglom (1971) show that the appropriate expressions for the mean and fluctuating velocity are of the form

$$U^+ = y^+ - a y^{+4} + b y^{+5} + ... \tag{2a}$$

$$u'^+ = a_1 y^+ + b_1 y^{+2} + ... \tag{2b}$$

$$v'^+ = a_2 y^{+2} + b_2 y^{+3} + ... \tag{2c}$$

$$w'^+ = a_3 y^+ + b_3 y^{+2} + ... \tag{2d}$$

$$-\langle uv \rangle^+ = a_4 y^{+3} + b_4 y^{+4} + ... \tag{2e}$$

Primes here and elsewhere denote root-mean-square *(rms)* values. Note the absence of the second and third order terms in the expansion for the mean velocity, and of the first order term in (2c) for normal velocity fluctuations; also, the leading term in the Reynolds shear stress expansion is of order y^{+3}. Since in the viscous layer the sum of the viscous stress and the Reynolds shear stress is a constant — this being readily apparent, to the lowest order, from the equations of motion — it follows that $a_4 = 4a$, and $b_4 = -5b$. There is no first-principles theory that gives these coefficients, and it is difficult to determine them *accurately* from experimental data. Both the constants in the mean velocity can be determined more reliably than those for fluctuations (although their *absolute* accuracy is not very high), and a reasonable fit to data from several experiments turn out to be a = 10^{-4}, b = 1.6×10^{-6}. Equation (2a) with these numerical values for the coefficients is valid for $y^+ <$ 20. Rough estimates for other leading coefficients provided by Monin & Yaglom are $a_1 = 0.3$, $a_2 = 0.008$ and $a_3 = 0.07$.

Figure 6 shows the *rms* intensities of the three fluctuation velocity components normalized by U^*. Careful measurements by Kreplin & Eckelmann (1979) near the wall show (figure 7) that the streamwise and spanwise fluctuations normalized by the local mean velocity have finite non-zero limiting values, whereas that in the direction y possesses a zero limit. While the fluctuations vanish at the wall, figure 7 also shows that they rise to rather high relative levels fairly quickly in the viscous layer. Although relatively large, the sublayer fluctuations are not capable of transmitting momentum (figure 4). Further, they are not responsible for significant production or dissipation of turbulence (although they are marginally more effective at dissipation than at production). A conceivable conclusion is that the viscous sublayer performs little dynamic function, merely providing the right boundary condition to the rest of the flow. A strong evidence for this comes from comparisons between rough-wall and smooth-wall boundary layers. As long as the roughness elements are confined to the viscous sublayer, their effect on the mean velocity distribution outside of the sublayer is negligible.

Expressions (2) enable us to deduce the *rms* fluctuating vorticity components at the wall. Since the derivatives of averages with respect to x and z vanish at the wall, it follows that

$$\omega'_x(y=0) = 0.07U^{*2}/\nu, \quad \omega'_y(y=0) = 0, \quad \text{and} \quad \omega'_z(y=0) = 0.3U^{*2}/\nu. \tag{3}$$

Here, ω_x, ω_y and ω_z are the fluctuating vorticity components in x, y and z directions respectively. Thus, even though the wall (except at the leading edge) is *not* a source of mean vorticity, there is a concentration of fluctuating vorticity there. The vorticity components at the wall can also be estimated from the knowledge of the limiting behavior of velocity fluctuation intensities shown in figure 7 or the fluctuating shear stress measurements right at the wall (e.g., Fortuna & Hanratty 1971, Sreenivasan & Antonia 1977). The conclusions that follow are:

$$\omega'_x(y=0) = 0.065U^{*2}/\nu, \quad \omega'_y(y=0) = 0, \quad \text{and} \quad \omega'_z(y=0) = 0.24U^{*2}/\nu. \tag{4}$$

The direct numerical simulations of Moin & Kim (1982) show (figure 8) that

$$\omega'_y(y=0) = 0, \quad \omega'_z(y=0)/\omega'_x(y=0) = 1.5. \tag{5}$$

The ratio is much smaller in experiments, and it is not clear whether   .s merely a consequence of difficulty in numerically resolving the viscous sublayer. Note that the numerical simulations show that the largest ω_x and ω_z components occur at the wall, and are diffused outwards. The largest ω_y component is produced somwhere in the middle of the viscous layer, and is diffused both inwards and outwards.

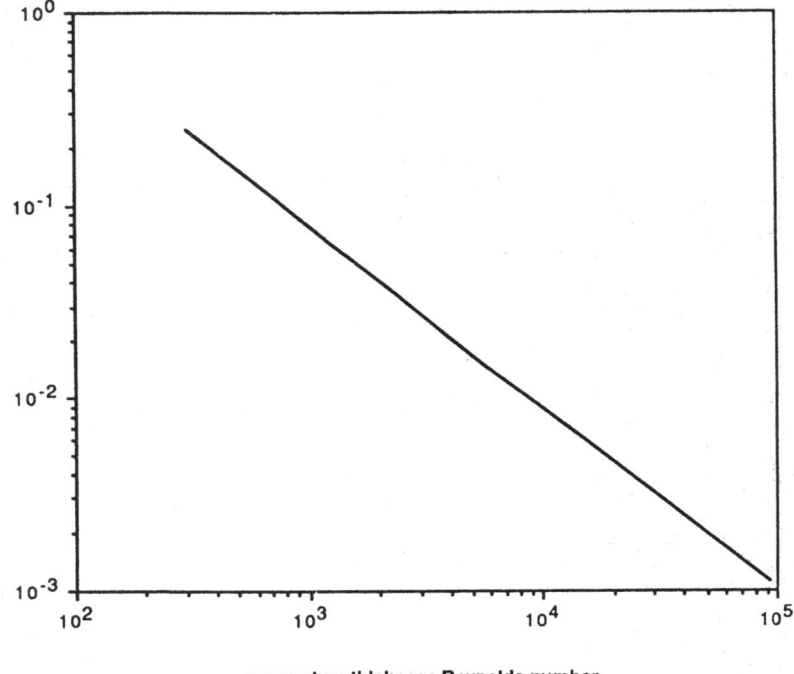

Figure 5. The thickness of the viscous layer as a fraction of the boundary layer thickness.

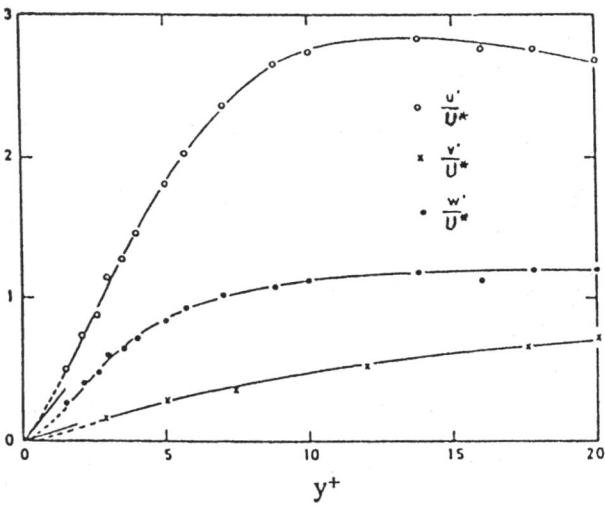

Figure 6. The distribution of u', v', and w' normalized by the friction velocity U*. Data from Kreplin & Eckelmann (1979).

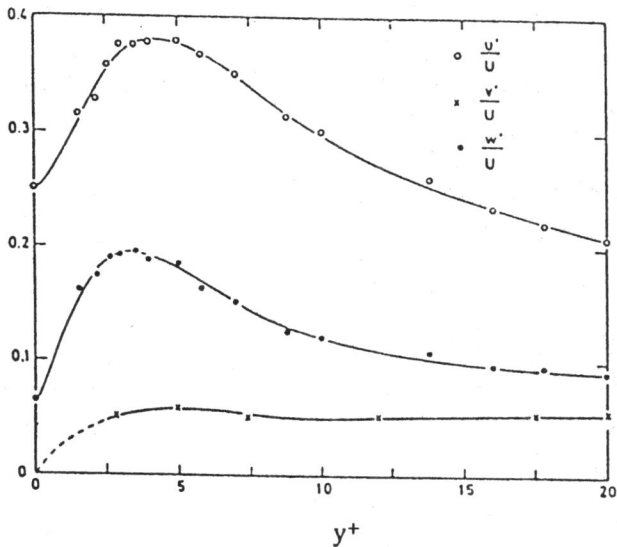

Figure 7. The distribution of u', v', and w' normalized by the local mean velocity U(y). Data from Kreplin & Eckelmann (1979).

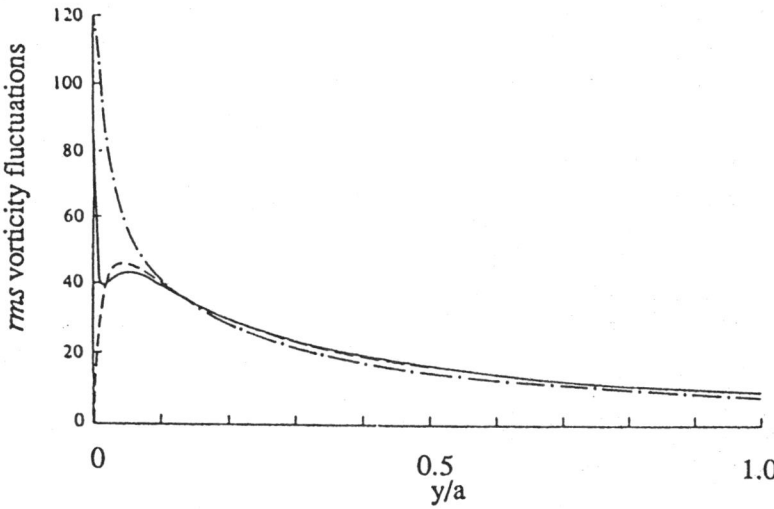

Figure 8. The *rms* vorticity fluctuations resolved in the direct numerical simulations of Moin & Kim (1982). ——— , ω'_x, – – – , ω'_y, – . – , ω'_z.

2.2.The constant-stress layer: Outside the viscous region, the turbulent shear stress is many times greater than the viscous stress, which is the same as saying that the momentum flux across layers of fluid is accomplished nearly entirely by turbulence. We see from figure 4 that this region is substantial at high Reynolds numbers, and characterized by an approximately constant Reynolds shear stress. Hence, the variation of the mean velocity in this region must not depend on fluid viscosity but only on the turbulent stress transmitted across it. We are discussing here the variation of the velocity, not its absolute value, this being so because the addition of an arbitrary constant velocity will not change the momentum flux. The precise value of the velocity in the constant-stress region will depend on the 'boundary condition' provided by the viscous layer. Dimensional analysis gives

$$\partial U/\partial y = (1/\kappa)\tau^{1/2}/y, \tag{6}$$

where the constant κ is named after von Kármán, and is presumed to be 'universal' (see section 2.4). In writing (6) we explicitly assume that the only relevant length scale in this region is the distance from the wall y; neither the viscous length v/U^* nor the boundary layer thickness δ is relevant, the former because it is very small compared to y and the latter because y/δ is very small. The validity of (6) depends on these two conditions, which can empirically be ascertained to hold at large Reynolds numbers.

It can be seen from figure 4 that the Reynolds shear stress in this region is equal to the wall stress τ_w, so that (6) can be written as

$$\partial U/\partial y = (1/\kappa)U^*/y.$$

Integrating, we have

$$U(y) = (1/\kappa)U^*\log(y) + A. \tag{7}$$

Consistent with previous remarks, the constant A should depend on boundary conditions arising from the existence of the viscous layer underneath. If we assume that the velocity at the edge of the viscous layer, defined by $\beta v/U^*$, can be expressed as αU^*, where α and β are dimensionless constants, we get

$$A = \alpha U^* - (1/\kappa)U^*\log(\beta v/U^*) = (1/\kappa)U^*\log (U^*/v) + BU^*. \tag{8}$$

We can then write (7) as

$$U^+ = (1/\kappa)\log(y^+) + B \tag{9}$$

for all $y \gg v/U^*$; for the smooth-wall case case considered here, B is expected to be another 'universal' constant. The constants κ and B have so far not been determined from first principles. Typical empirical values are $\kappa = 0.41$ and $B = 5.5$.

The mean velocity distribution given by (9) is the so-called log-law. It can be derived by other arguments also, Millikan's (1939) asymptotic arguments among them, but they all depend on the existence of the constant stress layer. (For a different view, see section 2.4.) The log-law (see figure 9) does not extend all the way to the viscous sublayer but smoothly blends with equation (2a).

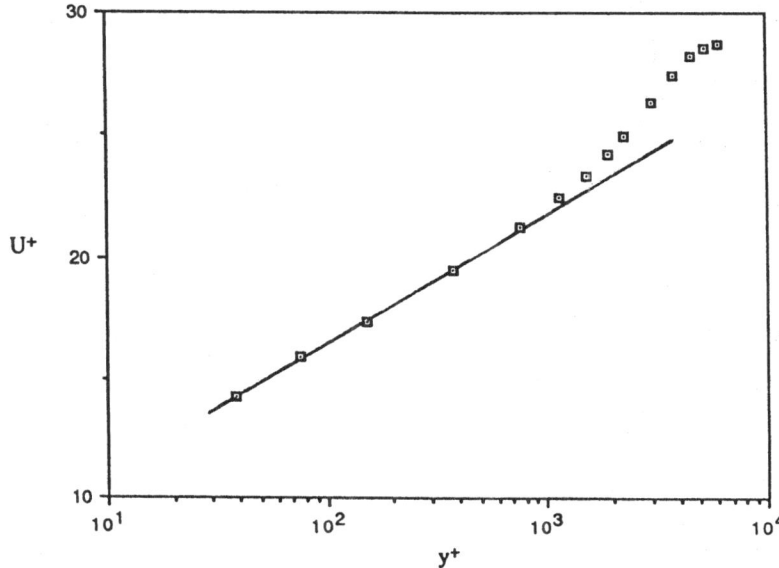

Figure 9. A typical semi-logarithmic plot, from the flat plate data of Weighardt & Tillmann (1951). Free stream speed $U_o = 33$ ms^{-1}, the momentum thickness Reynolds number $R_\theta = 15160$, the shape factor $H = 1.31$. Deviations from log-law are generally perceptible below a y$^+$ of 30 (viscous region); the measurements given here do not extend to the viscous region.

The requirement that the turbulent stress be constant for the logarithmic region to exist is not stringent because only the half power of the turbulent stress appears in the above equations. The constant-stress region can therefore be interpreted generously to include the entire region 'within the −3dB points' of the peak stress (that is, the region within which the stress value is 70.7% of the peak) without losing accuracy. Examination of all the available experimental data suggests the following bounds for the constant stress region:

$$\text{lower bound: } y^+ = 30 \tag{10a}$$

$$\text{upper bound: } y^+ = 0.2R^*. \tag{10b}$$

Here, $R^* = U^*\delta/\nu$, δ being the boundary layer thickness. It is convenient to think of the Reynolds number R^* as the ratio of the boundary layer thickness (the outer scale) to the viscous length scale ν/U^*.

Since it is the half-power of the stress (namely U^*) that is the relevant quantity, the log-law (with the same value of the constant κ) has been observed under a variety of circumstances where it is not expected to be valid *a priori*. One such example is the low Reynolds number wall flow where the constant-stress region is essentially non-existent, and the maximum Reynolds shear stress is substantially less than the wall stress (see figure 4); another is the case of mild pressure gradients whose influence on the stress distribution itself is significant but negligible on the mean velocity.

Velocity variations in viscous sublayer and constant-stress region depend only on y^+, and it is therefore useful to include the buffer layer into this description − even though neither (1) nor (9) is valid there − and write the velocity distribution in the so-called inner layer encompassing the entire region from the wall to the outer edge of the 'constant-stress' region as

$$U^+ = f_i(y^+). \tag{11}$$

This is the so-called inner law of velocity distribution, first formulated by Prandtl (1925).

The flow close to the wall (including the logarithmic part) is essentially the same in boundary layers, pipes and channel flows, and altogether encompasses about a fifth of the boundary layer thickness, pipe radius or channel half-height. (With slightly readjusted constants κ and B, it holds almost up to the centreline in pipes and channels.) This is a strong indication that the important dynamics of that region are controlled essentially by the total stress transmitted, and not by other details.

2.3. Fluctuating quantities in the constant-stress region: This last fact imposes some dynamic constraints on the flow. In a broad sense, these have been set forth by Townsend (1976), and enlarged by Perry & Abell (1975); see also Tennekes & Lumley (1972, Chapter 5) and section 2.6. Dimensional arguments similar to those above show that fluctuation intensities in the constant stress layer must be constant also (to the lowest order, to the same degree of approximation as the shear stress). This has been verified experimentally (see figure 10 for u'), and some typical values are

$$u'/U^* = 2.0, \ v'/U^* = 1 \text{ and } w'/U^* = 1.4. \tag{12}$$

It also follows that the average energy dissipation rate $<\varepsilon>$ and production rate $<p>$ must vary inversely with distance from the wall so that we can write

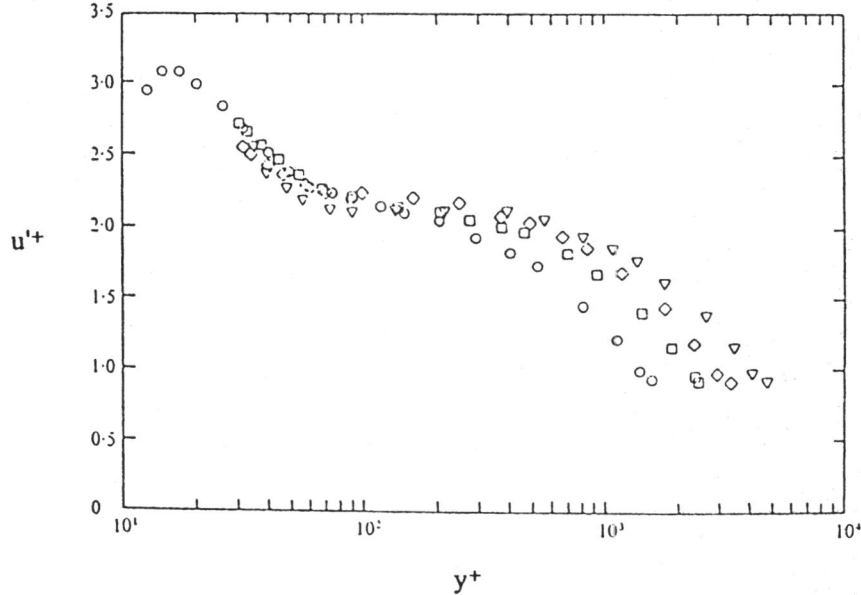

Figure 10. Streamwise turbulence intensity in the wall region of a pipe flow, from Perry & Abell (1975). O, Re = 7.8×10^4; □ 13.3×10^4; ◊, 17.3×10^4; ∇, 25.7×10^4.

$$<\varepsilon> = <p> = U^{*3}/\kappa y. \qquad (13)$$

Figure 11, in which the ordinate is $(<p>\delta/U^{*3})\kappa$ and the abscissa is y/δ, shows an extensive straight line region of slope -1, confirming (13) adequately. (There is no special significance to the use of δ as the normalizing length because it appears in both the abscissa and the ordinate.) Similarly, one can write the *rms* vorticity fluctuation in the direction i as

$$\omega_i'v/U^{*2} = C_i/y^+. \qquad (14)$$

Vorticity fluctuations from computations confirm this conclusion (figure 12). Measurements are rather sparse and it is uncertain that they support (14); also, the constant C is different in the two sets of measurements shown in figure 12. Clearly, more work is needed.

One can also obtain specific results for the spectral form in the constant-stress region. From the knowledge that there is in this region no preferred length scale other than the distance from the wall, and that the dynamics there is governed by the turbulent shear stress transmitted across different fluid layers, one can write the energy spectral density $\phi(ky)$ in the form

$$\phi(k) = DU^{*2}y \, \zeta(ky). \qquad (14)$$

Here, k is the wave number magnitude, $\int\phi(k) \, d(k) = <u^2>$, and $\zeta(ky)$ and D are expected to be a universal function and a universal constant respectively. Similar expressions hold for the other two components. If we argue that the spectral density should not depend on y (because not much distinguishes one layer from another in the constant stress region; see below), $\zeta(ky)$ has to be of the form $(ky)^{-1}$, and we are left with the result

$$\phi(ky)/U^{*2} = D/ky. \qquad (15)$$

Figures 13 and 14 show that this is valid over an extended wave number regime $ky < 1$. It is therefore clear that small changes in y will not affect its dynamics; this is the *a posteriori* justification for expecting the spectral density in that region to be independent of y. From the experimental data, one can estimate D to be of the order unity. Note that figure 14 is for fluctuations in wind speed over sea, and is measured 200 m above the mean water level in the sea; we therefore emphasize that, even when the flow underneath the log-region is vastly different in the two cases, the same relation is valid at the low wave number end of the spectrum (which, for convenience, will be denoted the energy containing range).

Obviously, (15) is valid only for the stress- (or energy-) carrying scales, and is not expected to hold either on the low end which depends on eddies characteristic of the flow width, or on the high end where viscous effects are important. This latter fact can be seen explicitly in figures 13 and 14: At large enough wave numbers, we have the usual $-5/3$ law of Kolmogorov; even at moderate Reynolds numbers, the $-5/3$ region extends all the way to the high end of (15). The first feature is not evident in figures 13 and 14 because the data do not extend to low enough wave numbers. But it can be seen clearly in figure 15 taken from Perry & Abell (1975).

Closer to the wall where viscous effects are also important, we expect to have:

$$u'/U^* = f_1(y^+)$$
$$v'/U^* = f_2(y^+)$$

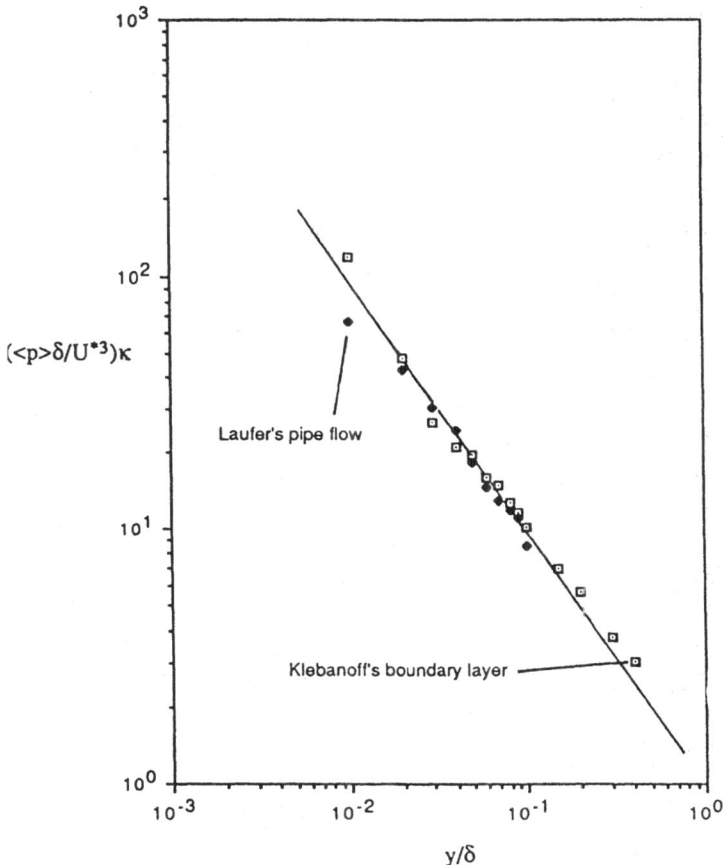

Figure 11. Non-dimensional turbulent kinetic energy production in the constant-stress region. Data from Laufer's (1954) pipe flow and Klebanoff's (1955) boundary layer.

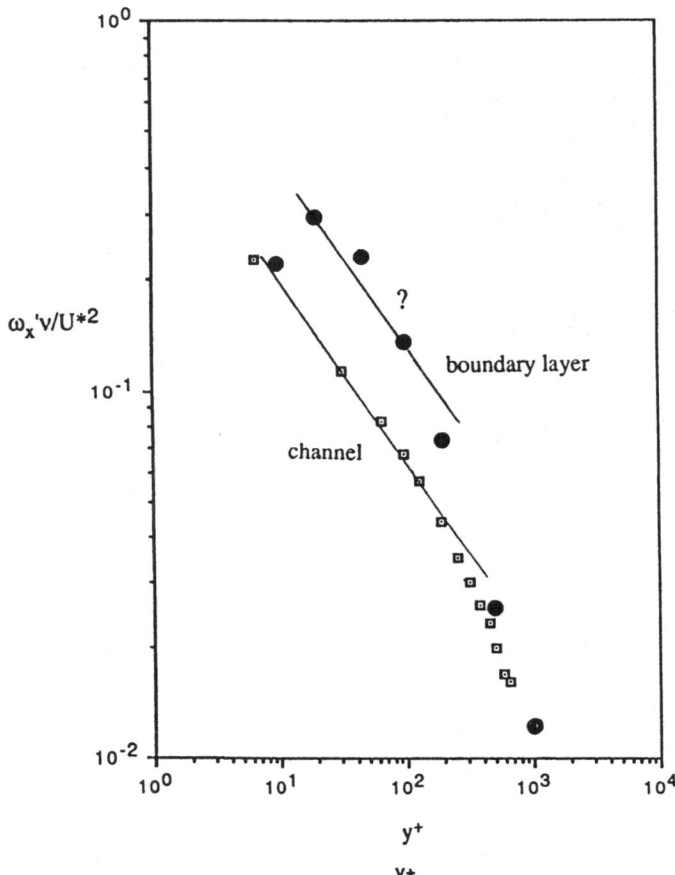

Figure 12. Non-dimensional vorticity fluctuation data from measurement (Wallace 1986, boundary layer), and direct numerical simulations (Moin & Kim 1982, channel).

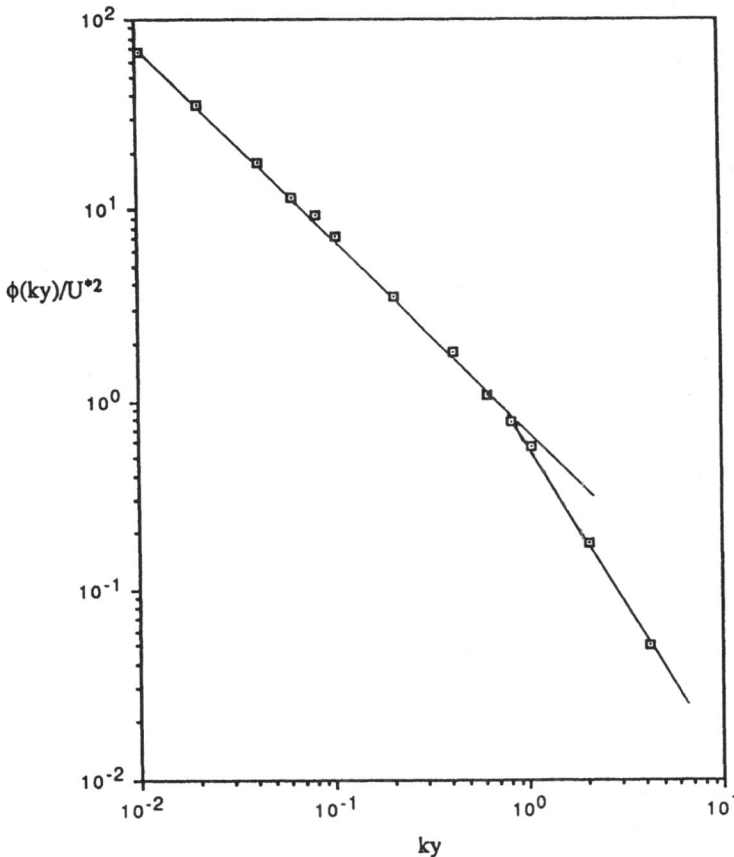

Figure 13. The −1 power law displayed in the pipe data of Laufer (1954), The power law at higher wave numbers is the Kolmogorov's −5/3 law. There is no discernible range of intermediate scales for which neither is valid.

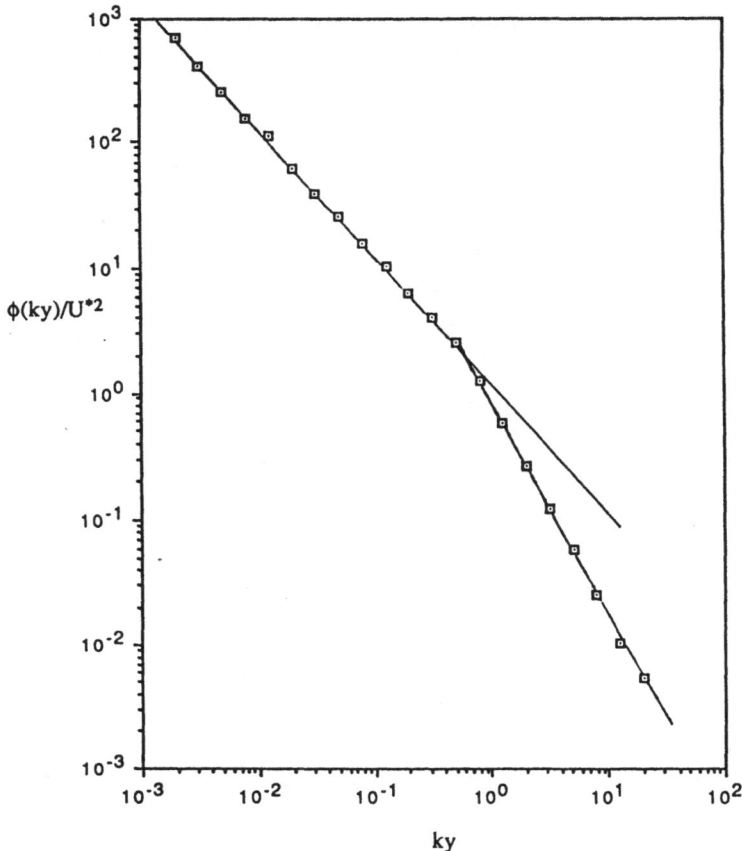

Figure 14. The wind data over the ocean (Pond et al. 1966) displaying the −1 and −5/3 power laws.

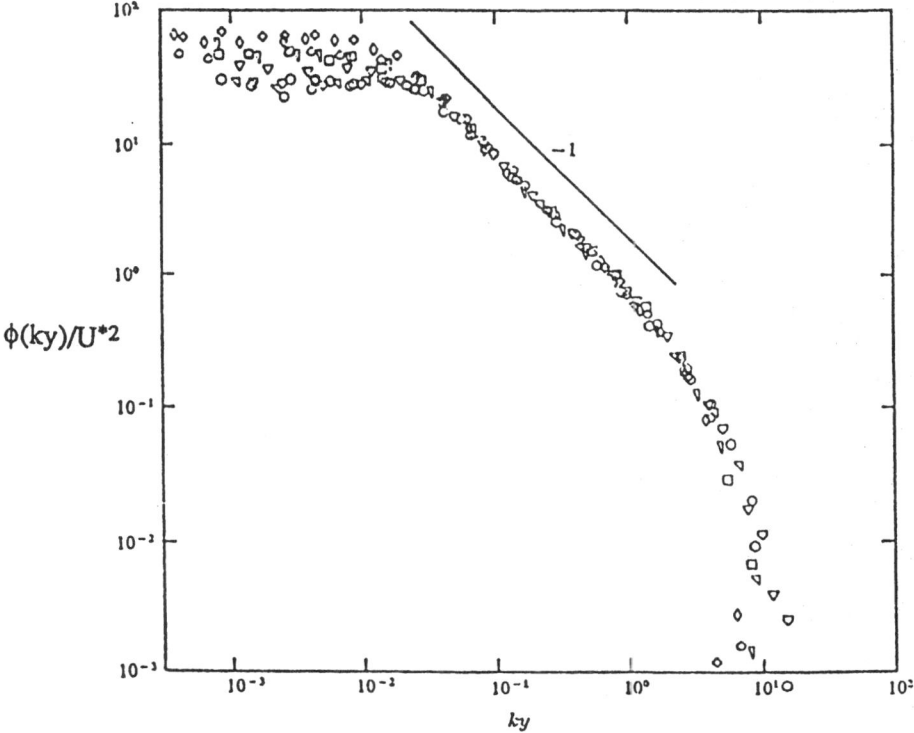

Figure 15. Pipe data of Perry & Abell (1975) show that there are very large scale fluctuations outside of the equilibrium hypothesis; these constitute the so-called inactive motion. The small scale features lying outside the equilibrium range may either obey the usual −5/3 law (figures 13 and 14) if the Reynolds number is sufficiently high, or simply be viscosity-dependent, at low Reynolds numbers.

	$Re \times 10^{-3}$	y^+	y/R	U/u_τ
○	80	150	0·0934	17·6
□	120	100	0·043	16·6
▽	120	150	0·0645	17.59
▽	120	200	0·086	18·3
◇	180	104	0·03	16·6
◁	180	138	0·04	17·31
◇	180	275	0·08	19·2
✢	260	148	0·03	17·6
△	260	246	0·05	18·8
▽	260	444	0·09	20·4

$$w'/U^* = f_3(y^+)$$

$$(<p>/U^{*3})\kappa y = f_4(y^+)$$

$$(<\varepsilon>/U^{*3})\kappa y = f_5(y^+)$$

$$\phi(k^+) = \phi(k^+, y^+).$$

Not enough work has gone into determining these empirical functions.

The arguments used in deriving the spectral form (15) are qualitatively similar to those employed by Kolmogorov (1941) in discussing the energy cascade. We recall that in this theory the spectral shape in the inertial subrange is completely detemined at high Reynolds numbers by the energy flux across the wave number domain. The analogy here is the following: The primary boundary condition on the boundary layer dynamics is that the wall provides a sink of momentum, similar to the energy sink at fine scales in the Kolmogorov scenario. This loss has to be replenished constantly by the outer stream *via* the momentum flux across different 'fluid layers' in the turbulent boundary layer, qualitatively analogous to the energy flux across the wave number domain. Just as in Kolmogorov's spectral theory, it is plausible to expect this process of momentum flux in physical space to determine, to a first approximation, the average dynamics in the constant-stress region.

Another point can be brought to bear by examining the locations in the boundary layer where the largest *rms* fluctuations occur. In figure 16 is plotted the distance from the wall where the streamwise fluctuation peaks. It appears that this height in wall variables is sensibly independent of the Reynolds number. Similar data (figure 17) on the normal component show larger scatter, perhaps reflecting the difficulty in measuring it accurately in the wall region and the fact that the peak in the v' distribution is rather flat; in spite of the scatter, it is clear that the location of this peak is a strong function of the Reynolds number. A rough fit to the data is given by the relation

$$y^+_{v'max} = R^{*0.75}, \tag{16}$$

suggesting that v-fluctuations are essentially an outer layer (and therefore inviscid) phenomenon. Similar data on the spanwise component are rather sparse (figure 18). Even though the position of peak v' does not scale on wall variables, and the conclusion on w' is uncertain, the total fluctuation energy does so because, near the wall, it is essentially overwhelmed by the streamwise component.

The peak Reynolds shear stress occurs at increasingly higher y^+ at increasing Reynolds numbers (figure 19a); as a fraction of the boundary layer thickness, however, the peak location moves closer to the wall. A simple equation fitting the data is

$$y_p^+ = 2R^{*0.5}. \tag{17}$$

Since accurate measurement of the Reynolds shear stress near the wall is beset with several problems including the probe size, it would be valuable to corroborate (17) by some other means. As already noted, the wall layer dynamics in the boundary layer is the same as that in pipe and channel flows; for the latter two in the fully developed state, the Reynolds stress distribution is related exactly to the mean velocity distribution U(y) by

$$-uv = -v(du/dy) + U^{*2}(1-y/a), \tag{18}$$

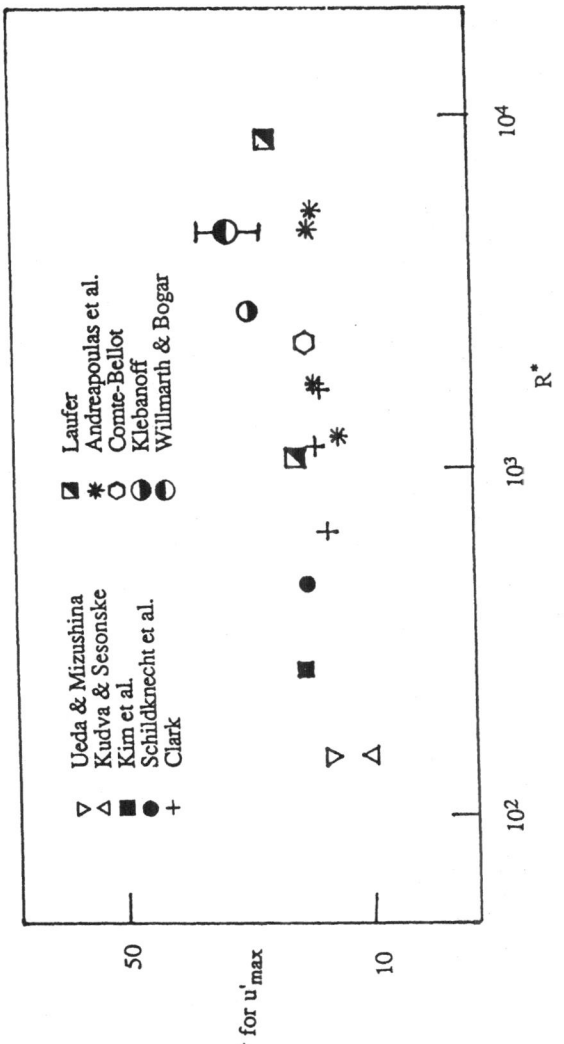

Figure 16. The location of the peak in u', plotted in wall units, as a function of the Reynolds number R*.

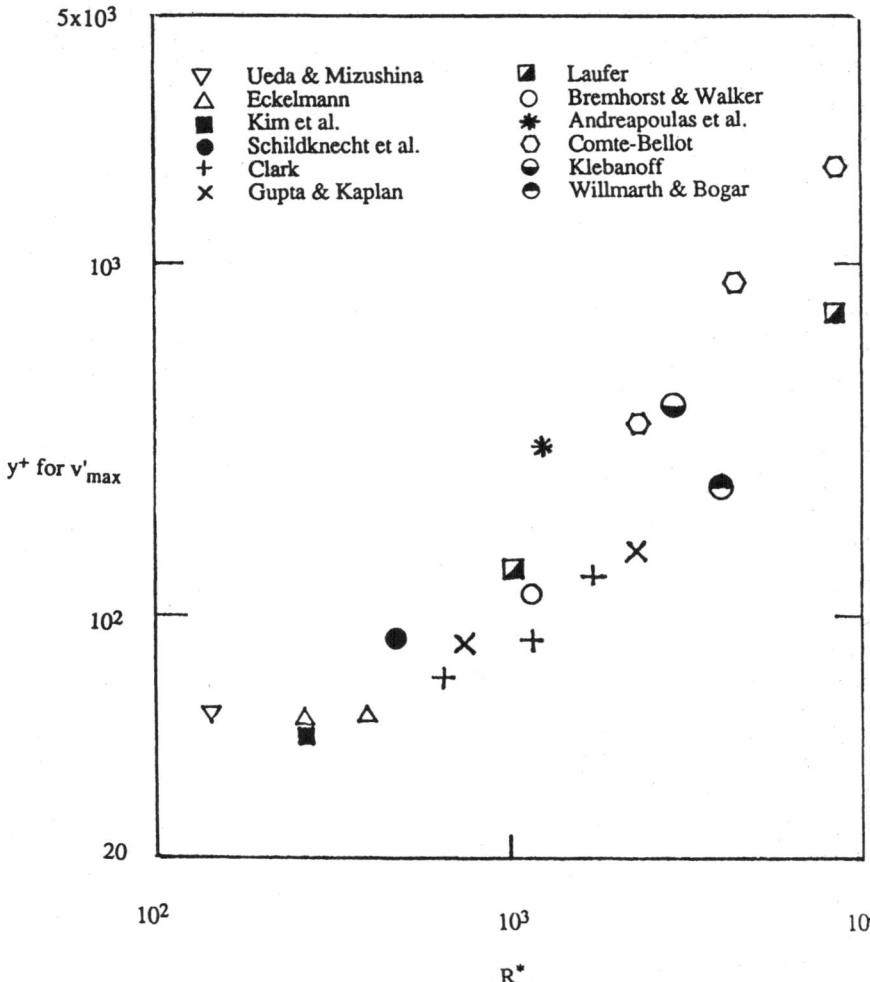

Figure 17. The location of the peak in v', plotted in wall units, as a function of the Reynolds number R^*.

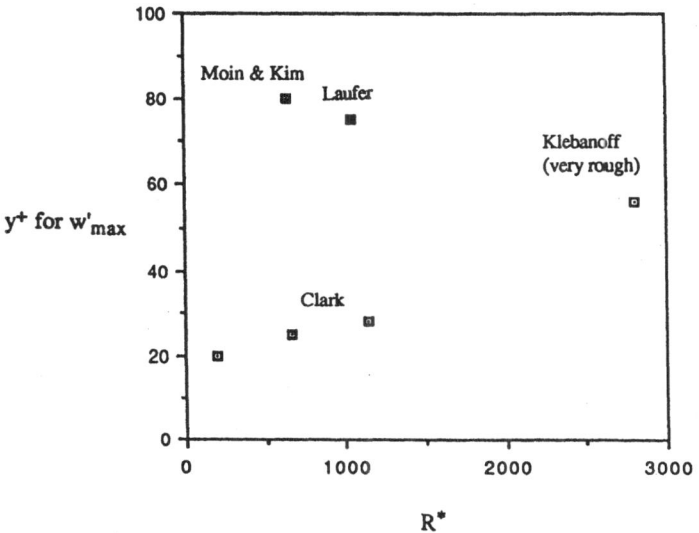

Figure 18. The location of the peak in w', plotted in wall units, as a function of the Reynolds number R^*.

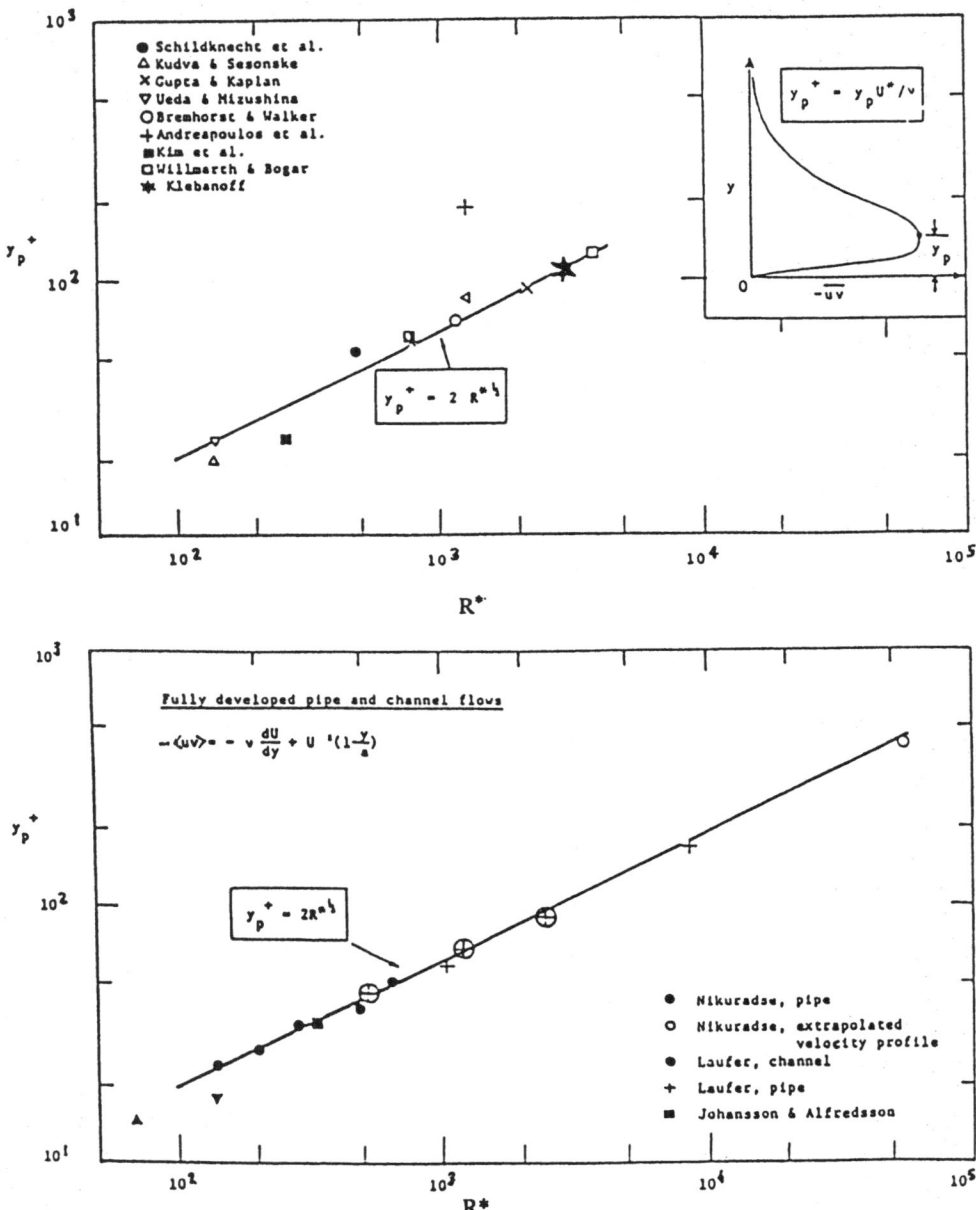

Figure 19. The peak Reynolds shear stress location in boundary layer, pipe and channel flows. (a) Data from various experiments. (b) Data computed from measured mean velocity distributions at various Reynolds numbers. See equation (18).

where a is the channel half-height or pipe radius. Mean velocity measurements are generally accurate, and so the computed shear stress distribution can be expected to be quite reliable. The results obtained in several flows (figure 19b) substantiate (17) satisfactorily. (For a more detailed commentary on the experimental data, see Sreenivasan 1987.) It is clear that at high Reynolds numbers the peak stress occurs substantially outside the viscous region.

The fact that the peak Reynolds shear stress follows (17) in the turbulent boundary layer as well as pipe and channel flows is interpreted to mean that, in all these flows, the motion producing the Reynolds shear stress does not primarily reside at constant y^+. Taken together with the earlier remark that the position of the maximum kinetic energy scales on wall variables, this leads to the conclusion that the energy and stress carrying eddies are quite different. It was primarily to obviate the need for this awkward conclusion that Townsend (1961) and Bradshaw (1967) invoked the so-called inactive motion, the part of the energy containing motion that does not scale on the Reynolds shear stress.

2.4. The power-law variation of the mean velocity distribution: We have tried to describe the boundary layer structure in terms of self-similarity arguments that make the underlying assumptions rather transparent, but a little closer examination is helpful. In general, the state of the motion at some point in the inner layer can be expected to depend on the turbulent shear stress τ, the viscosity coefficient ν, the distance from the wall y, and the boundary layer thickness δ. Thus

$$\partial U/\partial y = f(\tau, \nu, y, \delta). \qquad (19)$$

We can write this non-dimensionally as

$$y(\partial U/\partial y)/U^* = g(y^+, R^*). \qquad (20)$$

If the function g asymptotes to a constant in the limit of $y^+ \to \infty$ (which clearly implies $R^* \to \infty$), we can write

$$y(\partial U/\partial y)/U^* = \text{constant}$$

and recover the log-law discussed earlier.

In general, there is no special reason to anticipate that the limiting form of the function g is well-behaved. It could, for instance, diverge in some power-law fashion on the parameter y^+. We can then write (20) alternatively as

$$y(\partial U/\partial y)/U^* = (y^+)^\lambda h(R^*),$$

where λ is a parameter that might depend on R^*. Barenblatt (1979) has called this the incomplete self-similarity in the parameter y^+, and has explored its consequences. Following him, we integrate the above equation and write

$$U^+ = (y^+)^\lambda/(\lambda \kappa(R^*)) \qquad (21)$$

where $\kappa(R^*)$ is $1/h(R^*)$. This shows that the variation of the mean velocity occurs according to a power law. Power laws, although in use in the turbulence literature for a long time, have come to be discredited since Millikan (1939) derived the log-law from asymptotic arguments. However, the basis for them is *a priori* as sound as for the log-law, especially at low Reynolds numbers.

Barenblatt has also presented a preliminary anlaysis of the experimental data in pipe flows of Nikuradse (1932), and shown that κ is essentially independent of R*; while, on the other hand, λ does vary as shown in figure 20. Although it is hardly possible to distinguish power-laws with such small exponents from logarithmic variation, and the differences are immaterial from an experimental point of view, it makes a fundamental difference to our understanding of the boundary layer asymptotics. Of particular interest is the limiting behavior of the constant λ as $R^* \rightarrow \infty$. If λ goes to zero, we recover the log-law as before. On the other hand, if the limiting value is a non-zero constant, the log-law does not strictly hold, and the dependence on viscosity of the velocity distribution in the boundary layer will be nontrivial even at infinitely large Reynolds numbers.

The log-law scenario has been questioned in the past by Long & Chen (1981). A moment's reflection with respect to figure 4 shows that the constant-stress region (and therefore the log-law) does not obtain if the lower limit for the effective constant-stress region, namely equation (10a), is replaced by $R^{*\alpha}$, where α is any small positive number. Unfortunately, it is not possible from experiments either to dismiss such a possibility or support it. The chief difficulty is the paucity very near the wall of experimental data on the Reynolds stress at truly high Reynolds numbers. One might think that in pipe and channel flows the turbulent shear stress can be inferred from the mean velocity data using (18), but sufficiently accurate mean velocity data is also not available close enough to the wall at truly high Reynolds numbers. This important problem is therefore unresolved at present, but it is this that bears strongly on whether the earlier statements in section 2.1 about turbulence production in the viscous region are valid asymptotically.

2.5. The outer layer: The mean velocity profile in the outer part of the boundary layer is characterized by the velocity defect $U_0 - U(y)$, where U_0 is the free-stream velocity. The previous assertion (section 2.3) that the momentum flux is the primary governing factor suggests that the wall is felt entirely through the friction velocity U^*, from which it follows that the only quantities on which the defect velocity can depend are U^*, U_0, y, and δ. We can then write

$$(U_0 - U(y))/U^* = \text{fn}(y/\delta, U^*/U_0). \tag{22}$$

Experiments show that the dependence of the defect velocity ratio on U^*/U_0 is rather weak (if at all) so that one can write

$$(U_0 - U(y))/U^* = f_0(y/\delta). \tag{23}$$

This so-called defect law, first formulated by von Kármán (1930), is well confirmed by experiment (figure 21). The defect law applies even to the logarithmic region.

Coles (1956) has combined the defect law and the inner law by writing the velocity distribution at any height as

$$U(y)/U^* = f_i(y^+) + (\Pi/\kappa)\, w(y/\delta), \tag{23}$$

where Π is a constant, and w(y/δ) is the so-called wake function. The constant Π depends strongly on the Reynolds number for small Reynolds numbers, and may asymptote (Coles 1962) or vary only little (Crocco 1965) for $R_\theta > 6000$. The emphasis in this formulation is that, in the outer layer,

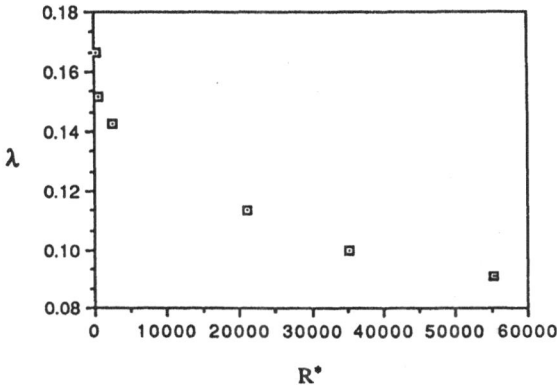

Figure 20. The Reynolds number variation of the power-law exponent from Nikuradse's pipe flows. Nikuradse's own power-law exponent is 0.1 for the highest two Reynolds numbers, but our fit to his data shows that, while a range of exponents is possible, the best value for the highest Reynolds number is somewhat different as shown; other values are Nikuradse's.

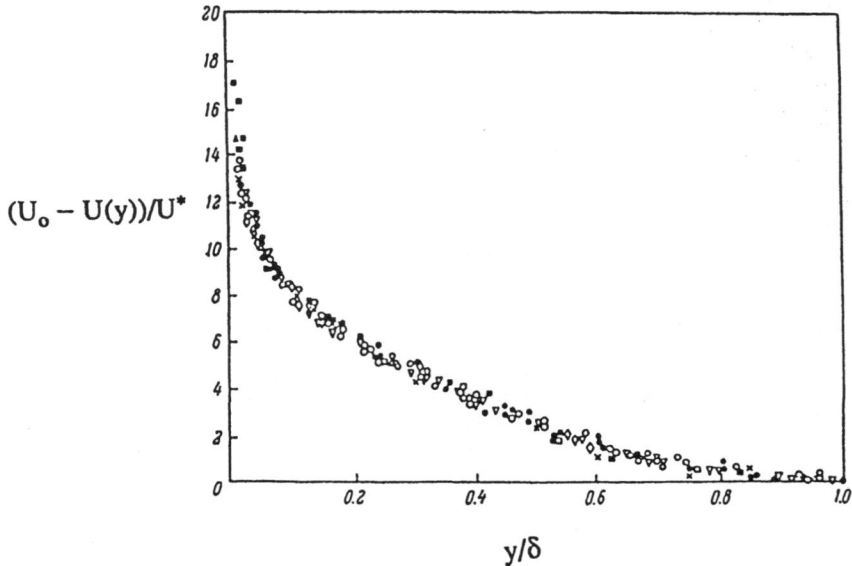

Figure 21. Verification of the defect law for the turbulent boundary layer, according to data from different authors (different symbols). Figure taken from Monin & Yaglom (1971).

the departure of the measured velocity distribution from the inner law is similar to that in the wake of a bluff body; for example, the wake function can be represented adequately by $2\sin^2[(\pi/2)(y/\delta)]$, which is reminiscent of the wake profile. (For this reason, the outer part of the boundary layer is called the wake region.) This observation emphasizes that the interaction between the turbulent flow and the free stream is similar to that in wakes and, by inference, in other free shear flows.

Such a similarity is reflected, among other things, by the so-called outer layer intermittency. A point probe located in the outer layer is sometimes in the turbulent region and sometimes outside of it, and therefore shows a roughly on-off characteristic (Corrsin & Kistler 1955). The simplest characterization of this intermittent behavior is the so-called intermittency factor (Townsend 1947), which is merely the average time spent by a point probe in the turbulent region. The intermittency factor varies with the distance from the wall, being unity nearly all the way in the constant stress region, decreasing further outwards roughly as an error function – a feature shared by all commonly studied free-shear flows. Much of this type of work beginning with Corrsin & Kistler (see also Klebanoff 1955) perceived that a sharp contiguous interface separates the turbulent regions from the non-turbulent ones. Later, the notion of such an interface was incorporated (for example, by Kovasznay, Kibens & Blackwelder 1970) into special forms of conditional measurements such as averages in the turbulent and non-turbulent zones, as well as the 'fronts' and 'backs' of the interface. Measurements of the same type in wakes (Thomas 1973) as well as other free shear flows support this similarity.

Similar interfaces are thought to exist for scalar quantities such as temperature in a heated boundary layer, or for a dye injected upstream in a liquid flow. Although turbulence does spread the scalar very effectively, there is no reason to expect that a scalar interface will in general be the same as the vorticity interface.

Now we know from visualization by optical techniques of thin sections of the flow and subsequent image processing that such interfaces are far more fragmented (see, for example, Prasad & Sreenivasan 1989), and probably not even contiguous, and is reasonably well represented by a fractal-like entity (Sreenivasan & Meneveau 1986); it takes a contiguous form essentially when some coarse-graining is applied. (We do not imply that large structures are absent; see section 2.6 and 3.3). Unlike mathematical fractals (Mandelbrot 1982) where scale-similarity extends over an infinite range of scales, the self-similar scale range is bounded here on both sides: The outer cut-off is of the same order as the boundary layer thickness, while the inner cut-off is on the order of the Kolmogorov scale. Incorporating this notion of cut-offs, ideas have been advanced (Sreenivasan, Ramshankar & Meneveau 1988) for successfully calculating the scaling laws for the mixing and entrainment of the ambient fluid into the turbulent region.

Outer layer intermittency in vorticity fluctuations is not possible in pipe and channel flows. This does not necessarily preclude a thermal interface between boundary layers if one of the walls of a channel gets heated, but we do not know of any such experiments.

2.6. A first approximation to interaction dynamics: We have already noted that a potentially useful point of view to take in explaining the boundary layer dynamics is to consider the wall as a sink of momentum, and that the inward flux of momentum governs the boundary layer dynamics to a first approximation; we further assume that only neighboring fluid layers interact, and that distant layers of fluid, for example a fluid layer in the viscous region and another in the outer region, do not interact directly. This 'local momentum transfer theory' also neglects possible 'intermittency effects' concerning large variability in space and time of quantities of primary interest such as the turbulent shear stress. As already remarked, this description of the boundary layer dynamics is analogous to Kolmogorov's (1941) theory of local energy transfer across wave number space, ignoring both non-local and intermittency effects. Spectral theory in turbulence research has since been embroiled over questions of whether this picture is correct, and a similar line of enquiry (although in a different guise, see below and section 4) has been much on the horizons of research in turbulent boundarys layers also.

Before we criticize, modify or discard this point of view, it is useful to remind ourselves of its success, paralleling the similar success in the Kolmogorov-type spectral theories. It predicts that the inner layer dynamics is of the quasi-equilibrium type, governed entirely by the parameters ν, τ_w and y; in the equilibrium layer, ν becomes irrelevant, predicting the log-law, the -1 power law in spectral space, etc. These predictions have been verified amply by experiments. Although intermittency effects can be included, it is not clear that they will be profound.

In this view, the inward momentum flux constitutes a hierarchical 'momentum cascade' process. Recall that although the Kolmogorov energy cascade proceeds on the average from the low wave number to the high, there is no restriction that the energy flux cannot reverse its direction instantaneously. In a similar fashion, momentum can flow from the wall region to the outside region locally in time and space. In contrast to Kolmogorov's theory in which the significance of the anticascade of energy is unclear, the analogous anticascade here should clearly be important. The experiments of Uzkan & Reynolds (1967) seem to suggest that the wall plays a greater role than being merely a sink of momentum; the precise role of the wall is not fully understood, however, and we can only attempt a roughly self-consistent description.

Lighthill (1963) argued that the main effect of a solid surface on the incoming turbulent vorticity is to stretch it in the spanwise direction, and briefly set forth the basis for his arguments. Even though some caution is in order because vortex stretching is not the only possible nonlinear action possible (as can be seen from the equations for the mean square vortcity), such arguments are of some value. This can be seen from the fluctuating velocity and vorticity data (section 2.1) which show that the largest vorticity component is indeed in the spanwise direction; measurements of Blackwelder & Eckelmann (1979) also support Lighthill's picture. That the stretching is primarily in the spanwise direction (although streamwise stretching no doubt plays a part) is consistent with the observation that the principal effect of increasing only the streamwise stretching (as in highly accelerating boundary layers) is to kill turbulence rather than enhance it. The inward flux of momentum must accompany the stretching effect which gets stronger as one nears the wall until, in the buffer layer, the viscous effects balance the nonlinear stretching.

One can obtain a feel for near-wall vorticity generation by measurements of the skewness of the velocity derivative. In homogeneous and isotropic turbulence, this quantity signifies the inertial transfer of energy across the wave number domain; it is also proportional to the production of mean square vorticity by stretching. Both these effects are due to the nonlinearity of the equations of motion. It is not clear if these interpretations hold for the wall-bound anisotropic turbulence, but we shall assume that it is valid in a qualitative sense. There are no data on the skewness of the space derivative of velocity fluctuations, but comparable data on the time derivative do exist. Although, because of the high shear near the wall, converting time to space by invoking Taylor's frozen flow hypothesis is questionable, the expectation is again that qualitative interpretations will be possible.

The derivative skewness measurements have been made, among others, by Comte-Bellot (1963, channel), Ueda & Hinze (1975, boundary layer), and Elena & Dumas (1978, pipe flow). Data from Elena & Dumas shown in figure 22 are representative, and suggest that the derivative skewness peaks at around $y^+ = 12$ where it is of order unity, and drops off to both sides. The value in the fully turbulent region is on the order of 0.4, not very different from the usual value in isotropic and homogeneous turbulence. The behavior closer to the wall is somewhat different in different experiments, but the collective evidence seems to favor the view that it drops off to zero at the wall. Whatever the details, it seems clear that the nonlinear effects are prominent around $y^+ = 12$, and more effective in that vicinity than elsewhere.

Lighthill viewed that most of the vorticity fluctuations thus generated constantly battle against turbulent and viscous diffusion, but that some of it will escape outwards. This escaped vorticity is facilitated in its upward mobility by the v-fluctuation, itself generated by the stretching action. This, in turn, is responsible for entraining the outer inviscid fluid, which begins its long route towards replenishing the lost momentum at the wall. The story is, in some sense, thought to be complete.

While this view explains some observed features, it does not account for two important aspects. First, the *energy containing motion* in the constant-stress region is independent of the details of the viscous region; as pointed out already, these aspects in the log-region of the atmospheric surface layer above the ocean are not different from those in the smooth-wall boundary layer. Second, this does not explain the observed structural features – that is, those aspects of the motion which can be given some identity either in visual or probe measurements. The sense that there is more to the boundary layer dynamics is motivated by flow visualization observations which call to question the basic premise of local interactions. One thought, for example, is that bursts occurring in the buffer layer travel far into the logarithmic region, and, if low-Reynolds number experiments are any guide, also into the outer region. Another thought is that the outer layer large eddies promote nonlocal interactions. It is possible that our visual perception is burdened by low-Reynolds number experience, and detailed experiments at high Reynolds numbers are therefore expected with much eagerness. It may well be that no new structures are likely to arise at high Reynolds numbers, but their interaction and relative importance is almost certain to undergo a change. Although high Reynolds number experiments have profound experimental difficulties, it is clear that dynamical views based exclusively on structural observations at low Reynolds numbers must be tempered by caution. It is useful to remind ourselves that an important aspect of the turbulent boundary layer, namely the separation of the inner and outer scales, does not obtain at low Reynolds numbers, and that any physical picture derived only from low Reynolds number measurements must be examined for consistency with asymptotic arguments mentioned already at different places.

3. The structural elements

Experiments have identified a multiplicity of structural elements, and the first comprehensive compilation is now being made by Robinson & Kline (1988). The value of the 'structure work',

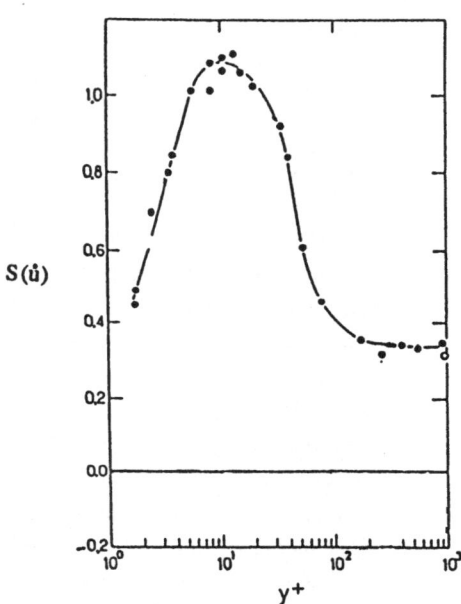

Figure 22. The skewness of the velocity derivative du/dt in the inner region of a pipe flow. Data from Elena & Dumas (1978).

which inquires into the mechanisms of various transport processes occurring in the boundary layer, cannot be overemphasized, especially in the context of boundary layer control.

A significant part of the early work relied heavily on measurements at a single point or a few small number of points in the flow. From appropriate correlation measurements, deductions about 'average' turbulent eddies were made (Townsend 1956, 1976, Grant 1956, Tritton 1967); note must also be made of the outer layer intermittency measurements of Corrsin & Kistler (1955) which brought forth the notion of an outer layer large eddy. The space-time correlation measurements (Favre et al. 1957, Willmarth & Wooldridge 1962, Kovasznay et al. 1970, to name but a few) have been especially valuable. Later work has emphasized flow visualization work, and an example of how it can be incorporated as an integral tool of quantitative point measurements can be found in Falco (1977, 1983). Low Reynolds number flows have now been simulated directly on supercomputers without invoking any specific turbulence models (Moin & Kim 1982, Spalart 1987); from these data bases one can deduce many flow features normally inaccessible to measurement. A number of kinematic details have been extracted already (e.g., Kim 1983, Brausser & Lee 1987, Robinson & Kline 1988), and it is clear that rapid advances in this direction will continue to be made. Our efforts here are better spent on the dynamical interpretation of some generic aspects. A useful complementary reference is Cantwell (1981).

3.1. The viscous region: The evidence appears incontrovertible (Kline et al. 1967) that a streaky structure in the velocity distribution near the wall is an ubiquitous feature of the turbulent boundary layer, at least at low Reynolds numbers; see figure 23. (In this figure one is looking down into the flow, and the flow is from top to bottom.) The streaks pervade the entire viscous region, but seem most pronounced around a y^+ of 9. They bear some resemblance to Townsend's (1956) countercurrent attached eddies, but details are quite different. The streaks occur randomly in space and time, but any given realization preserves definite spatial coherence. One significant conclusion of the work of Kline et al. concerns the average spanwise spacing of streaks. Figure 24 shows the collection of all the data available on the spanwise spacing $\Lambda^+ = \Lambda U^*/\nu$ of the streaks. There is no discernible trend with Reynolds number, and the mean spacing of around 100 wall units, borne out by nearly all subsequent investigators, is remarkably close to the early values given by Kline et al. Further measurements have refined the specifics of this picture. The sublayer streaks are believed to be of the order of 1000 wall units long (Blackwelder 1978, Praturi & Brodkey 1978, Blackwelder & Eckelmann 1979), but the vortical structures associated with them are more like a tenth as long (Kim 1983): The streaks, being the product of the accumulated dye, hydrogen bubble or other markers, are longer. They are centered at about 10 wall units from the wall (e.g., Blackwelder & Eckelmann 1978).

The streaks are believed to be significant because they form the precursor to the so-called bursting event (figure 25). As more and more observations have become available, refinements of the definition of bursting have been introduced; however, the essence is contained in Kline's (1978) proposal: It consists of a sequence of quasiperiodic events comprising of the formation of streaks, their lift-up, oscillation and breakdown. The bursting event itself is perceived to be significant because it presents a possible scenario for the interaction between the inner and outer regions of the flow; further, it is believed to be responsible for a significant fraction of the turbulent energy

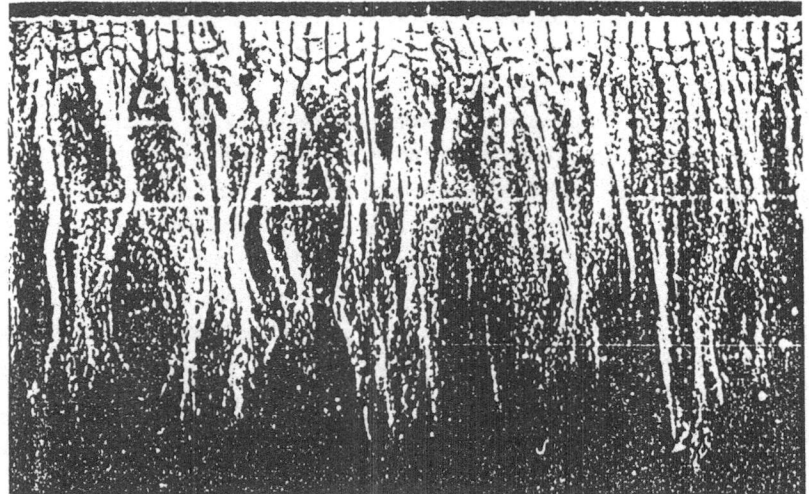

$y^+ = 2.7$

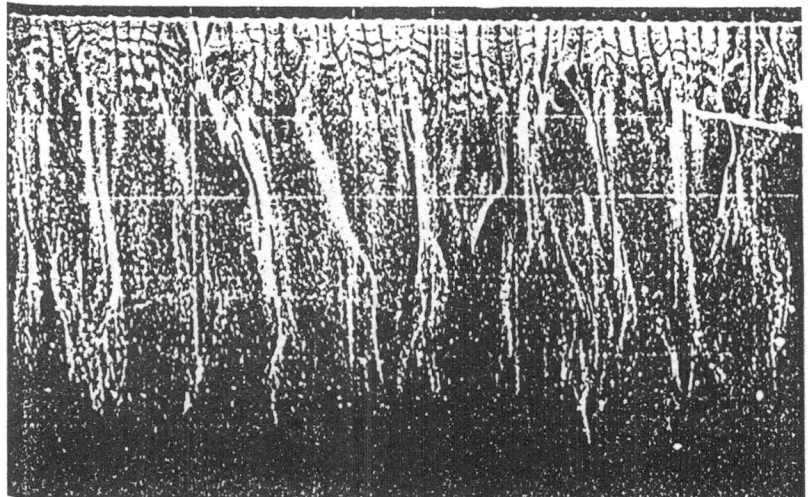

y+ = 4.5

Figure 23. Photographs showing the accumulation of hydrogen bubbles in the form of streaky structures in the wall region of the boundary layer. Pictures from Kline et al. (1967).

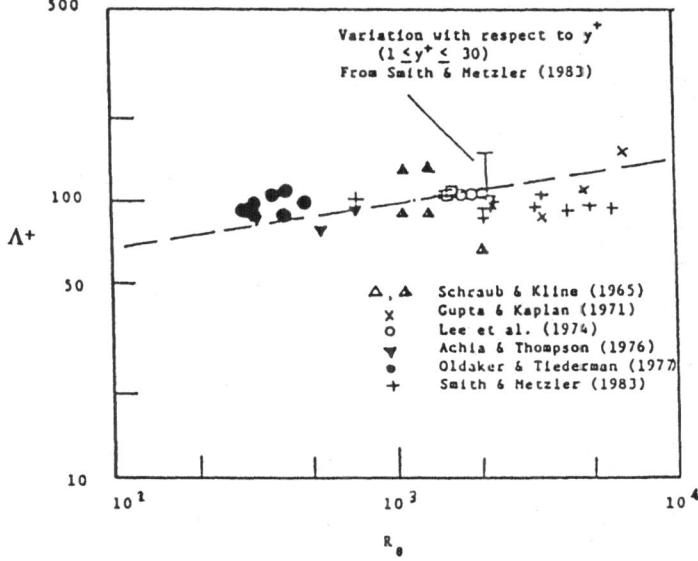

Figure 24. The spanwise spacing of streaks, normalized by wall variables. All measurements are in the viscous region, but measurement techniques differ from one author to another. The line drawn through the data will be explained in section 3.3. From Sreenivasan (1987).

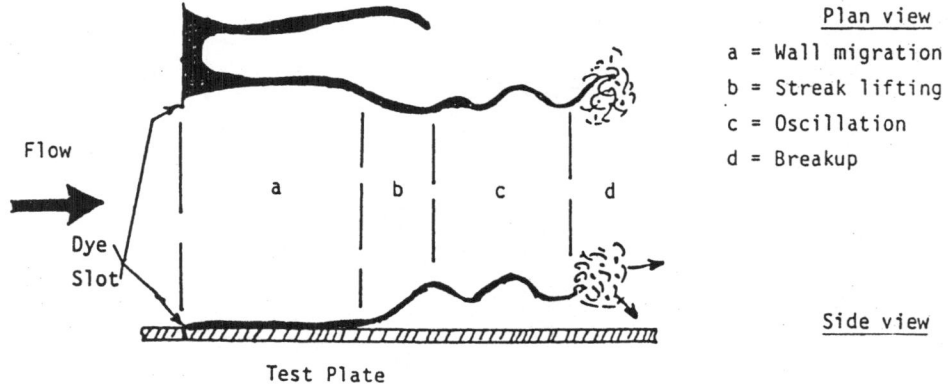

Figure 25. Schematic view of the near-wall bursting as seen by dye injection through the wall, from Kline (1978).

production (Kim, Kline & Reynolds 1971, later elaborated also by Wallace, Eckelmann & Brodkey 1972). In the experiments of Kline et al. (1967) the streaks of hydrogen bubbles drifted away from the wall as they move slowly, corresponding essentially to viscous diffusion effects. When a streak reaches a y^+ of about 10, it begins to oscillate and, upon reaching sufficient amplification, terminates in an abrupt break-up in the buffer region. These experiments showed that hydrogen bubbles after the break-up move along in a trajectory at least up to the lower portion of the log-region. To this picture, Corino & Brodkey (1969) added the so-called sweep event consisting of the streamwise movement of the upstream fluid sweeping away the previously ejected fluid – an event essentially demanded by continuity. The notion is that the sweep stabilizes the wall region, until a new cycle of these same events repeats. Other suggestions include pockets and ring vortices (Falco 1987).

Although remarkable amount of information is now available about streaks – only gross details have been presented here – their importance in the interaction dynamics remains unclear. In an interesting set of experiments, Grass (1971) has shown that the nature of interaction between the inner and outer regions remains quite similar whether the wall is smooth or covered with roughness elements protruding up to about 80 wall units, well outside the viscous region. This may be interpreted as minimizing the importance of streaks. On the other hand, smooth-wall data in drag reduction by polymer addition (e.g., Donohue, Tiederman & Reischman 1972, Oldaker & Tiederman 1977, Tiederman, Luchik & Bogard 1985) or by stable stratification (e.g., Kasagi & Hirata 1976) show that there is a direct correlation between increased streak spacing and skin friction reduction.

Even if it were true that streaks play a lesser role than has often been believed, a proper explanation for their existence has remained a challenge. The earliest explanation due to Kline et al. followed Lighthill's (1963) suggestion that a spanwise variation of the streamwise velocity will result when the stretched vortex elements in the viscous layer relax and contract. Another simple explanation (Acarlar & Smith 1987a, b) is that they mark the tracks of the haippin eddies in the outer layer (see later) that are being dragged along. Coles (1978) and, following him, Brown & Thomas (1977) have tried to explain them by invoking some kind of centrifugal instability of the viscous layer, it being generated when a 'big eddy' (which is two-dimensional on the scale of the streak spacing) from the outer layer passes over the sublayer region. The notion is essentially that when the big eddy moves with a convection velocity different from the local velocity, the particle paths underneath assume a curvature that is concave upwards – leading to the Görtler instability. For the notion to be useful, it is necessary to show that, among a host of other possible instabilities, it is indeed the Görtler kind that dominates; it should also be shown that the Görtler number exceeds the critical value substantially, and that the corresponding growth rates are large enough for the perturbations to grow to saturation amplitudes before encountering regions of convex curvature – which also occur on either side of the concave one. The situation is hopeless in a strict sense, but some plausible details were filled in by Sreenivasan (1987), who noted that the eddies driving the centrifugal instability were not of the boundary layer scale, but of the intermediate scale equal to the geometric mean of the inner and outer scales. His estimate for the streak spacing was in close agreement with the experimental data of figure 24. Yet another explanation is due to Phillips (1988) who proposed that they are merely a result of the back effect of the oscillatory disturbances on the mean flow.

If the primary effect of the wall is to produce a large magnitude of shear, it is reasonable to examine whether the uniform shear of the right magnitude can account for the observed streaky structure. The numerical simulations of Lee, Kim & Moin (1987) seem to favor this view, although, to see the right structure, Lee et al. had to use a substantially larger shear than is found experimentally. They also tried to explain their numerical results by means of the rapid distortion theory. But, as we saw in section 2.6 (see also section 4.2), the largest nonlinear effects in the boundary layer are to be found in exactly the region where the rapid distortion calculations were being made, and so the qualitative correspondence between observations and rapid distortion results must be considered tentative. It is a correct but impotent conclusion that completely satisfactory explanation is yet to emerge.

There is nontrivial difficulty in quantifying the bursting event by probe measurements, even though, at least at low Reynolds numbers, it appears to be well-defined visually. One of the most controversial results concerns the mean period of the bursting event. Part of the problem in measuring this quantity is that bursting occurs randomly in space and time, and a few point probes would be unable to determine their occurrence unambiguosly. A fuller discussion can be found in Offen & Kline (1974) and Bogart & Tiederman (1986,1987). Earlier conclusions of Kline et al. (1967), substantiated by several follow-up studies (for example, Blackwelder & Haritonidis 1983), is that the appropriate scaling is on wall variables. This is a natural conclusion consistent with the notion that wall phenomena are governed by wall variables. Interest surged when, on the basis of hot-wire measurements around the edge of the viscous layer, Rao, Narasimha & Badri Narayanan (1971) concluded that the mean period $<T>$ between bursts scaled on the outer variables U_o and δ, and that $U_0<T>/\delta$ is about 5. There is evidence in favor of this observation also (Laufer & Badri Narayanan 1971, Willmarth & Lu 1972, Blackwelder & Kaplan 1976). Bandyopadhyay (1982) examined all the available data, and concluded that neither the inner scaling nor the outer scaling explains all observations. Mixed scaling consisting of the geometric mean of the inner and outer scales has also been proposed (Johansson & Alfredsson 1982). There is some approximate rational basis for expecting this scaling to work (Sreenivasan 1986 – see also section 3.5).

It is difficult to see how a frequently occurring event in the viscous layer can be controlled by the boundary layer thickness, but it is not hard to rationalize that some rare event can in fact be so. We also note that even some gross characteristics of the boundary layer (see section 2.5) show some Reynolds number dependence at least up to R_θ of the order of 6000, and there are very few clean experiments at higher Reynolds numbers. The concept of incomplete similarity introduced in section 2.4 can be invoked to support the notion that an inner layer event should scale on outer layer variables (see Barenblatt 1979), but a more likely parameter on which incomplete similarity may occur is y^+ itself: Except below a y^+ of about 5, there is no reason to think that the only controlling parameters are ν and U^* (see section 2.3). A study of the details of measurements, not merely of the final results reported in publications, suggests that an equally likely explanation for the conflicting measurements is that different authors were in fact measuring different facets of the same event, or even different events altogether. This, however, is not the place for further elaboration of this controversial point.

3.2. The outer layer: Some evidence has been put forth (Townsend 1966; Kovasznay et al. 1970) to suggest that the outer structure of the boundary layer is dominated by the so-called large eddies of size comparable to δ. Kovasznay et al. (1970) determined that these bulges are of the order of 3δ in the streamwise direction, and δ in the spanwise direction. They conjectured that the large eddies are passive so that new turbulence originating in the wall region is responsible for producing the Reynolds stress, and that the bursting initiates a process that eventually leads to the above large scale structure. Kovasznay (1970) speculated that 'bursting' of some kind occurs at all intermediate scales, and that the large structure is the culmination of a sequence of amalgamations occurring in the constant stress region (see also Offen & Kline 1974). On the other hand, Head & Bandyopadhyay (1981) suggest that the large eddy apparent in flow visualization studies is only the slow overturning motion of a collection of smaller scale eddies of the hairpin type, each of which is inclined, over a major part of the thickness, at around 45° to the plane of the flat plate (see figure 26). Among other things, they emphasize that hairpin structure is found unambiguously only for R_θ > 7000, and that it is Reynolds number dependent below this R_θ and so untypical. Their hairpin structures are similar to those hypothesized many years ago by Theodorsen (1955). One of their claims is also that Falco's (1977) typical eddies are tips of the hairpin eddies. Motivated by these observations, Perry & Chong (1982) and Perry, Henbest & Chong (1986) hypothesized that the essential structure of the outer flow is a hierarchy of hairpin eddies, and proposed a model for the near-wall dynamics on that basis.

Other structural elements such as the double roller-type eddies have also been identified (Townsend 1956, Kim & Moin 1986, Nagib & Guezennec 1986). This structure has been deduced in the latter two papers from conditionally averaged measurements (although the details of the conditional averaging are different in the two cases). The primary conclusion is that this structure has a characteristic length of the order δ in the streamwise direction, and is responsible (see Nagib, Naguib & Wark 1988) for the production of the Reynolds stress – even though it is quite weak. A possible dynamical explanation for the appearance of this structure was given by Sreenivasan (1987). Earlier, Praturi & Brodkey (1978) had identified the so-called transverse structure, which they linked to outer layer bulges as well as wall layer dynamics. In spite of the similarity to the double roller structure, the scales in the latter two cases appear to be different.

3.3. The inner-outer interactions: It is clear that the boundary layer structure is many-faceted and complex. In describing them, we have clearly not done full justice to every aspect discussed in the literature, partly because some of these structural elements are not independent, and partly because we do not understand where they all fit into the broader scheme of things. Even so, we are left with a number of them, and the puzzle is to decide which among them are dynamically significant, and how they arise and interact. If, for the sake of the argument, we assume that the large structure is indeed the dominant and driving mechanism, the question arises as to how it originates and maintains itself. We can rule out laminar-turbulent transition as the source, because their occurrence has been noted in heavily tripped flows where the usual transition details have been bypassed. It is also not hard to argue that the bursting process is unlikely to accomplish this, merely on the basis that its scales and the large eddy are quite disparate at high Reynolds numbers; to our knowledge, no one has observed the hierarchical amalgamation of scales required to make the point

200

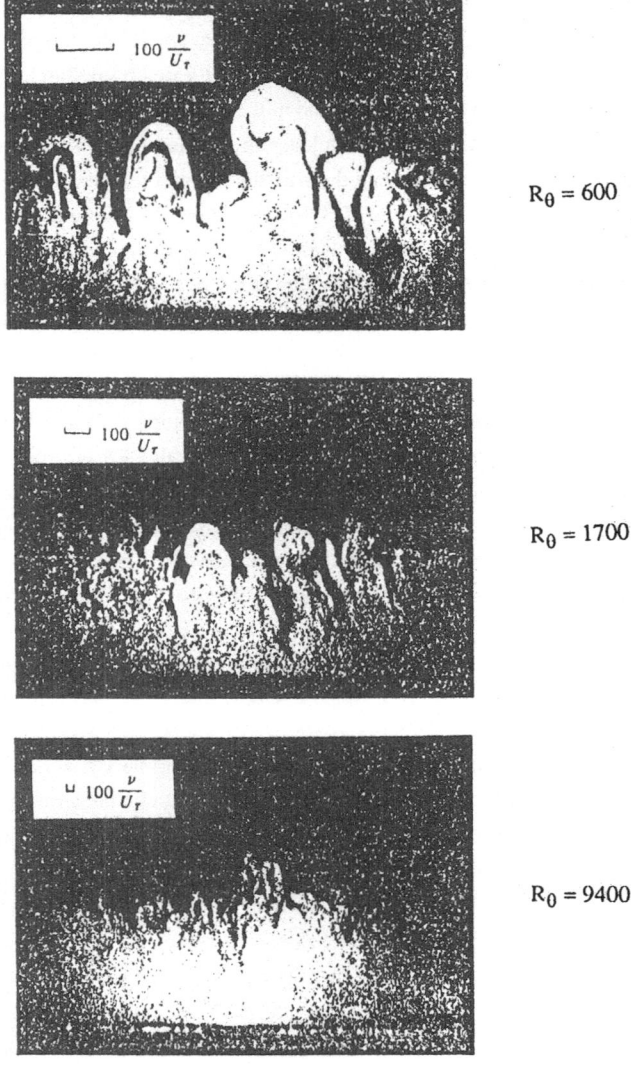

$R_\theta = 600$

$R_\theta = 1700$

$R_0 = 9400$

Figure 26. Sections of the boundary layer with 45° downstream plane illumination showing hairpin structures. Scales of v/U^* are indicated. Picture from Haed & Bandyopadhyay (1981).

of view viable. Alternatively hypothesizing that bursting is the dominant aspect shifts our concern to *its* origin and sustained maintenance, and difficulties still remain. If we argue (e.g., Blackwelder 1978) that bursting arises purely from local instabilities in the viscous region, it is tantamount to saying, at high Reynolds numbers, that an extremely small fraction of the entire thickness (see figure 5) drives the rest of the flow having no active role – a sentiment that has been called to question. If we propose instead that bursting is driven by the outer structure (Rao et al. 1971), the nature of the inner/outer interaction has to be understood, and we are back to asking how the outer structure which drives the phenomenon came into being in the first place. If we take the view that the outer structure is of the hairpin-type (Theodorsen 1955, Head & Bandyopadhyay 1981, Wallace 1984, Lynn 1987), we are faced with difficulties in explaining their origin. The same can be said of the the origin of the double rollers and transverse structures. Few explanations in currency are quantitative and self-consistent – that is, almost no existing explanation satisfactorily accounts for each aspect of the boundary layer in relation to every other aspect.

One prevalent view is that many observed structures are the result of some kind of instability. It is useful to bring to focus the qualitative meaning, usually unspoken, of the deterministic stability calculations in explaining turbulence structure; for convenience, our discussion is focused on the large eddy. When all turbulence scales in the boundary layer are resolved, it is immediately clear, even at moderate Reynolds numbers, that a wide spectrum of scales is present, ranging from the Kolmogorov scale to a multiple of an integral scale. There is in turbulence theory a long-held belief (Richardson 1922) that these scales are statistically self-similar. Together, these two features lead to the expectation that the turbulence structure is fractal-like (Sreenivasan & Meneveau 1986); the value of this notion has now been demonstrated at least in some cases (e.g., Sreenivasan et al. 1988, Gouldin 1987). When some coarse graining is performed on these scales (either by instrument smoothing, conditional averaging, or some other means), one begins to perceive the large scale contiguous structures. The general belief is that the degree of coarse-graining beyond a certain point is not crucial to conclusions on the geometry and topology of the large structures. One may therefore view the large structure as the result of a judicious coarse-graining performed on the spatially and temporally resolved three dimensional flow field. An alternative point of view – whose basis is yet to be explored in detail, but whose usefulness becomes obvious as soon as it is stated – is that these large structures are the result of instability of a caricature flow which itself is the result of applying a suitable coarse-graining to the real flow. Whether the structures deduced by coarse graining the full solution are the same as those deduced by the instability of its caricature deserves a lot of thought but, if correct, the benefits of this notion are likely to be many.

As already remarked, Sreenivasan (1987) argued that the appropriate caricature of the turbulent boundary layer is a fat vortex sheet located at a distance from the wall given by (17), representing the location of the maximum Reynolds shear stress. It was shown that two and three dimensional instability, both inviscid and viscous, of this coarse-grained flow leads to an explanation for a number of observed features such as the double-roller structures, hairpin eddies, and the wall streaks; the line drawn through the data in figure 24 is from this theory. One of the primary conclusions of this analysis is that a dynamic reason exists for expecting energetic motions on the scale equal to the geometric mean of the inner and outer ones. One can infer that such scales have been seen in a number of experiments (Kline et al. 1967, Dinkelacker et al. 1977, Brown &

Thomas 1977, Falco 1977). Another interesting point is that the convection speed of these eddies, deduced by Sreenivasan, was $0.65U_o$, consistent with conclusions of the experimental work just cited. Further, the theory explains one of the peaks in the space-time correlation measurements of the wall pressure (Willmarth & Wooldridge 1962): Their correlation surface $R_{pp}(x,t)$ showed that pressure-producing small eddies travel at a speed of $0.69U_o$, quite close to $0.65U_o$ mentioned above.

It is worth emphasizing that in 'concentrating' all the mean flow vorticity in a fat 'sheet' of vorticity we are no longer concerned about the details of the mean velocity distribution; the implication is that the large-scale instabilities of the two are the same. It follows that one of the *a priori* requirements for the reasonableness of this approach is that the mean velocity distribution itself be unstable. This requirement is not necessarily satisfied by the boundary layer at the linear stage of instability, but the perturbation environment is in general *not* linear.

4. Some concluding remarks

We have presented a brief review of the turbulent boundary layer in some consistent way. It must be noted that the literature on the problem is so undauntingly large (and some of it is undigested) that it has not been possible to do justice to every paper on the subject, even all worthy ones. We have chosen to present what in our view constitute the essential aspects, sometimes at the expense of lacking elaborate references to fine details. We now discuss some isolated aspects deserving of further attention.

4.1 The role of direct numerical simulations and experiments: Much can now be computed, and the future looks even brighter because of the likely progress in computing technology, especially parallel architecture and special purpose machines. The power of such computations is that almost any quantity of interest can be extracted, many of them inaccessible to experimentalists in spite of the tremendous progress made in instrumentation in the last decade. Even so, it is clear that direct simulation of turbulent flow fields will be impossible at all but 'moderate' Reynolds numbers. Unfortunately, reasonable extrapolation of available experimental techniques based on laser diagnostics and other optical methods does not suggest that they can do much better.

An eventual goal of detailed scientific study is to build successful working models. By definition, even a highly successful model does not account for every detail, but two of its hallmarks are wide applicability and well-understood limitations. In this regard, a useful role of computations is that they should unravel the proper physics, which can then be used in modeling high Reynolds number flows. A useful role of experiments is that they should determine the proper scaling relations for the dynamically important events. Fortunately, useful scaling relations can be determined without requiring full three-dimensional flow field data, as long as we understand what we are measuring – say, with a single probe.

In computations as well as modern experiments dealing with turbulent phenomena, the crucial problem is often one of reducing the massive bulk of data in ways that can be comprehended efficiently. Here, a lot of help is being offered by formal advances in nonlinear dynamics (e.g., Aubry, Holmes, Lumley & Stone 1988), proper orthogonal decomposition (e.g., Sirovich, Maxey & Tarman 1988), multifractals (e.g., Meneveau & Sreenivasan 1987a) as well as graphical display capabilities. Unfortunately, the information extracted to-date has largely been kinematic in nature. Kinematic description, however interesting, is only the beginning: One must graduate beyond merely showing that a complex phenomenon like turbulent boundary layer is indeed complex.

4.2. Linear modeling of the inner layer: It is often thought (e.g., Einstein & Li 1956, Hanratty 1956, Sternberg 1962, Landhal 1967, Schubert & Corcos 1967) that the viscous layer can be modeled by linear equations. While the details are different from one work to another in this list, the commonality among them justifies bunching them together, at least for present purposes.

The behavior of the ratio of a typical nonlinear term to the linear term is shown in figure 27. (In these computations, the linear term was estimated by the zero-crossing measurements of Sreenivasan, Prabhu & Narasimha 1983, and the nonlinear term was estimated by the measured stress gradients.) This suggests that strong nonlinearities are essentially confined to a narrow region centered around a y^+ of 15 or so (the precise value seems to depend marginally on the Reynolds number). This feature should be explored in the predictive modeling of the turbulent boundary layer, but we know of no such effort. One thought is that the rest of the boundary layer can be taken to be a linear problem driven essentially by strong random forcing in this region. The statistics of the random forcing must, of course, be provided by experiment. At around a y^+ of 15, the probability densities of all three velocity components are essentially symmetric but non-Gaussian. It is not clear, whether for present purposes, there is a need to consider these non-Gaussian effects.

4.3. The role of intermittent phenomena in turbulent boundary layers: It has long been felt that many important processes in turbulence are highly intermittent in nature. For example, the Reynolds shear stress (Willmarth & Lu 1972), the turbulent energy production (Kline et al. 1967), turbulent energy dissipation (Batchelor & Townsend 1949), etc. An extreme view would be that a small number of rarely occurring large amplitude events account for a major share of the dynamics. It is not clear if this is so, and needs to be quantified in detail. (If this is *not* true, the enormous emphasis in the recent turbulence literature on special and rare events is more or less misplaced.) At any rate, it is already clear that turbulence dynamics is not governed by Central Limit type statistics where a large number of marginally dissimilar events control the statistics. Such processes are best modeled by multiplicative processes in which some a simple dynamical step repeats over and over leading to the observed intermittency. This view has led to some remarkable success in modeling the energy cascade (Meneveau & Sreenivasan 1987b).

4.4. Non-canonical boundary layers: As remarked early in section 1, the study of the canonical case of the flat-plate smooth-wall boundary layer, we believe, will help us understand the boundary layer dynamics in the presence of other complex effects. On the other hand, there is some merit to the view that the study of the non-canonical boundary layers will help us understand the canonical case better. A case in point is the rough-wall boundary layer: By eliminating the viscous layer as we know it to be in the smooth-wall case, it can help us understand the role of the viscous layer. The Stratford boundary layer (Stratford 1959) in which the skin friction is zero is another example because it eliminates one of the driving forces in the turbulent boundary layer, namely the wall-ward momentum flux, and can therefore help us understand better its role. A study of pressure

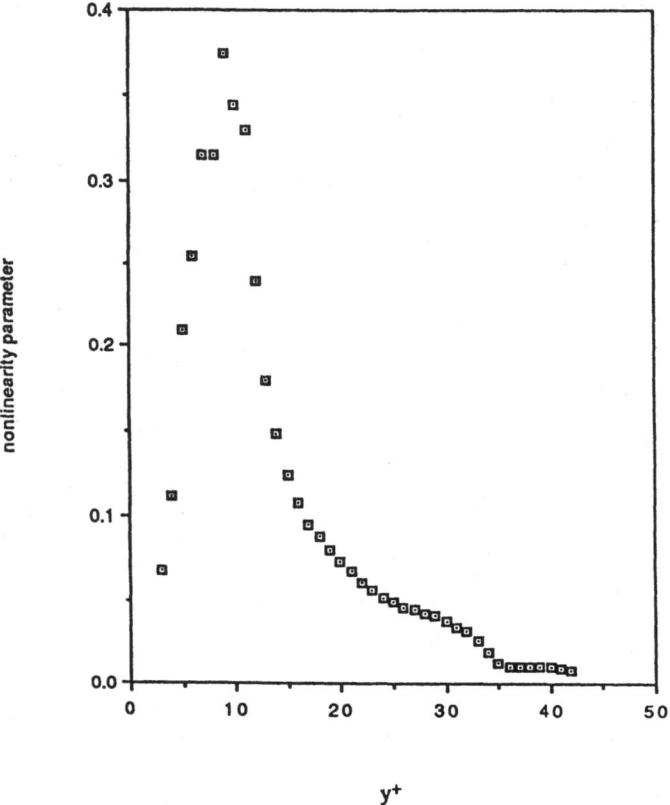

Figure 27. A typical ratio of the nonlinear to linear terms in the equations of motion computed in the viscous region.

gradient effects, curvature effects, compressibility effects, etc. can all be very helpful. The suggestion is not that data be generated to the same level of detail in all these non-canonical flows, but that some effort be devoted to them within a certain framework, with specific questions in mind.

Acknowledgements

This article would not have been written without the persistent efforts of Professor M. Gad-el-Hak, who constantly propped my flagging will by cajoling telephone calls and reminders. I have benefitted from discussions with a number of colleagues; to them, and particularly to B.-T. Chu, C. Meneveau, P. Kailasnath and K. Sreenivasan who commented on the draft, I am grateful. The work was supported by a grant from the Air Force Office of Scientific Reserach managed by Dr. James McMichael.

References

Acarlar, M.S. & Smith, C.R. 1987 J. Fluid Mech. **175**, 1.

Acarlar, M.S. & Smith, C.R. 1987 J. Fluid Mech. **175**, 43.

Achia, B.V. & Thompson, D.W. 1976 J. Fluid Mech. **81**, 439.

Andreopoulos,T., Durst, F., Zaric, Z. & Jovanic, J. 1984 Exp. in Fluids **2**, 7.

Aubry, N., Holmes, P., Lumley, J.L. & Stone, E. 1988 J. Fluid Mech. **192**, 115.

Badri Narayanan, M.A., Rajagopalan, S. & Narasimha, R. 1977 J. Fluid Mech. **80**, 237.

Batchelor, G.K. & Townsend, A.A. 1949 Proc. Roy. Soc. **199**, 238.

Blackwelder, R.F. 1978 In Lehigh Workshop on Coherent Structure in Turbulent Boundary
   Layers, ed. C.R. Smith & D.E. Abbot, p 211.

Blackwelder, R.F. & Eckelmann, H. 1979 J. Fluid Mech. **94**, 577.

Blackwelder, R.F. & Haritonodis, J.H. 1976 J. Fluid Mech. **132**, 87.

Blackwelder, R.F. & Kaplan, R.E. 1976 J. Fluid Mech. **107**, 89.

Bogard, D.G. & Tiederman, W.G. 1987 J. Fluid Mech. **162**, 389.

Bogard, D.G. & Tiederman, W.G. 1987 J. Fluid Mech. **179**, 1.

Bradshaw, P. 1967 J. Fluid Mech. **30**, 241.

Brausser, J.G. & Lee, M.J. 1987 NASA Rep. CTR-S87, p.165.

Bremhorst, K. & Walker, T.B. J. Fluid Mech. **61**, 173.

Brown, G.L. & Thomas, A. W. 1977 Phys. Fluids **20**, s243.

Cantwell, B.J. 1981 Ann. Rev. Fluid Mech. **13**, 457.

Clark, A.J. 1968 J. Basic Engg. **D90**, 455.

Coles, D.E. 1956 J. Fluid Mech. **1**, 191.

Coles, D.E. 1962 Rand Rep. R-403-PR.

Coles, D.E. 1978 In Lehigh Workshop on Coherent Structure in Turbulent Boundary Layers, ed.
   C.R. Smith & D.E. Abbot, p 462.

Comte-Bellot, G. 1963 Ph.D. Thesis, Univ. Grenoble. (Translated into English by P. Bradshaw as
   ARC31609, FM 4102, 1969.)

Corino, E.R. & Brodkey, R.S. 1969 J. Fluid Mech. **37**, 1.

Corrsin, S. & Kistler, A.L. 1955 NACA Tech. Rep. 1244.

Crocco, L. 1965 J. Soc. Ind. Appl. Math. **13**, 206.

Dinkelacker, A., Hessel, M., Meier, G. & Schewe, G. 1977 Phys. Fluids **20**, s216.

Donohue, G.L., Tiederman, W.G. & Reischman, M.M. 1972 J. Fluid Mech. **56**, 559.

Einstein, H.A. & Li, H. 1956 ASCE Proc. **82**, no. EM 2.

Elena, M. & Dumas, R. 1978 Paper 78-HT-4 presented at the ASME-AIAA Thermophysics & Heat
Transfer Conference, Palo Alto, CA.

Falco, R.E. 1977 Phys. Fluids **20**, s124.

Falco, R.E. 1983 AIAA paper 83-0377.

Falco, R.E. 1987 Presentation at the Workshop on Turbulent Boundary Layer, Austin,Texas.

Favre, A.J., Gaviglio, J.J. & Dumas, R. 1957 J. Fluid Mech. **2**, 313.

Fortuna, G. & Hanratty, T.J. J. Fluid Mech. **53**, 575.

Gouldin, F. 1987 Rep. E-87-01 Cornell University; to appear in AIAA J.

Grant, H.L. 1958 J. Fluid Mech. **4**, 149.

Grass, A.J. 1971 J. Fluid Mech. **50**, 223.

Gupta, A.K & Kaplan, R.E. 1972 Phys. Fluids **15**, 981.

Gupta, A.K., Laufer, J. & Kaplan, R.E. 1971 J. Fluid Mech. **50**, 493.

Hanratty, T.J. 1956 AIChE J. **2**, 359.

Head, M.R. & Bandyopadhyay, P. 1981 J. Fluid Mech. **107**, 297.

Hussain, A.K.M.F. 1983 Phys. Fluids **26**, 2816.

Johansson, A.V. & Alfredsson, P.H. 1982 J. Fluid Mech. **122**, 295.

von Kármán, Th. 1930 Mechanische Ähnlichkeit und Turbulenz, Nachr. Ges. Wiss. Göttingen. Math. Phys. K1. p. 58.

Kasagi, M. & Hirata, N. 1976 In Proc. ICHMT Seminar on Turbulent Buoyant Convection

Kim, J. 1983 Phys. Fluids **26**, 2088.

Kim, J. & Moin, P. 1986 J. Fluid Mech. **162**, 339.

Kim, H.T., Kline, S.J. & Reynolds, W.C. 1971 J. Fluid Mech. **50**,133.

Klebanoff, P.S. 1954 NACA Rep. 1247.

Kline, S.J., Reynolds, W.C., Schraub, F.A. & Runstadler, P.W. 1967 J. Fluid Mech. **30**, 741.

Kline, S.J. 1978 In Lehigh Workshop on Coherent Structure in Turbulent Boundary Layers, ed. C.R. Smith & D.E. Abbot, p 1.

Kolmogorov, A.N. 1941 C.R. Acad. Sci. U.R.S.S. **30**, 301.

Kovasznay, L.S.G. 1970 Ann. Rev. Fluid Mech. **2**, 95.

Kovasznay, L.S.G., Kibens, V. & Blackwelder, R.F. 1970 J. Fluid Mech. **41**, 283.

Kreplin, H.-P. & Eckelmann, H. 1979 Phys. Fluids **22**, 1233.

Kudva, A.K. & Sesonske, A. 1972 Int. J. Heat Mass Tr. **15**, 127.

Landahl, M.T. 1967 J. Fluid Mech. **29**, 441.

Laufer, J. 1951 NACA Rep. 1053.

Laufer, J. 1954 NACA Rep. 1174.

Laufer, J. & Badri Narayanan, M.A. 1971 Phys. Fluids **14**, 182.

Lee, M.J., Kim, J. & Moin, P. 1987 In Sixth Symposium on Turbulent Shear Flows, Toulouse, Sept. 7-9.

Lee, M.K., Eckelman, L.D. & Hanratty, T.J. 1974 J. Fluid Mech. **66**, 17.

Lighthill, M.J. 1963 In Laminar Boundary Layers, ed. L. Rosenhead, Oxford University Press, p.48.

Long, R.R. & Chen, T-C. 1981 J. Fluid Mech. **105**, 19.

Lu, S.S. & Willmarth, W.W. 1973 J. Fluid Mech. **60**, 481.

Lynn, T.B. 1987 Ph. D. Thesis, Yale Univ.

Mandelbrot, B.B. 1982 The Fractal Geometry of Nature, W.H. Freeman.

Meneveau, C. & Sreenivasan, K.R. 1987a Nucl. Phys. B. (Proc. Suppl.) **2**, 49..

Meneveau, C. & Sreenivasan, K.R. 1987b Phys. Rev. Lett. **59**, 1424.

Millikan, C.B. 1939 In Proc. Fifth Cong. Appl. Mech. Cambridge, MA. p. 386.

Moin, P. & Kim, J. 1982 J. Fluid Mech. **118**, 341.

Monin, A.S. & Yaglom, A.M. 1971 Statistical Fluid Mech. vol. I, M.I.T. Press.

Nagib, H.M. & Guezennec, Y.G. 1986 Paper presented at the tenth Symp. on Turbulence, Univ. Missouri-Rolla.

Nagib, H.M., Naguib, A. & Wark, C. 1988 Bull. Amer. Phys. Soc. **33**, 2262 (abstract only).

Nakagawa, H. & Nezu, I. 1981 J. Fluid Mech. **104**, 1.

Nikuradse, J. 1932 Forsch. Arbeiten Ing.- Wesen, no. 356.

Offen, G.R. & Kline, S.J. 1974 J. Fluid Mech. **62**, 223.

Oldaker, D.K. & Tiederman, W.G. 1977 Phys. Fluids. **20**, s133.

Perry, A.E. & Chong, M.S. 1982 J. Fluid Mech. **67**, 257.

Perry, A.E. & Chong, M.S. 1982 J. Fluid Mech. **119**, 173.

Perry, A.E., Henbest, S. & Chong, M.S. 1986 J. Fluid Mech. **165**, 163.

Phillips, W.C.R. 1988 In Near-Wall Turbulence, ed. S.J. Kline, Hemisphere.

Pond, S., Smith, S.D., Hamblin, P.F. & Burling, R.W. 1966 J. Atmos. Sci. **23**, 376.

Prandtl, L. 1925 ZAMM **5**, 136.

Prasad, R.R. & Sreenivasan, K.R. 1989 Exp. in Fluids (in press).

Praturi, A.K. & Brodkey, R.S. 1978 J. Fluid Mech. **89**, 251.

Preston, J.H. 1958 J. Fluid Mech. **3**, 373.

Rao, K.N., Narasimha, R. & Badri Narayanan, M.A. 1971 J. Fluid Mech. **48**, 339.

Reynolds, O. 1894 Phil. Trans. Roy. Soc. Lond. **186**, 123.

Richardson, L.F. 1922 Weather Prediction by Numerical Process, Cambridge University Press.

Robinson, S. & Kline, S.J. 1988 In Near-Wall Turbulence, ed. S.J. Kline, Hemisphere.

Schildknecht, M., Miller, J.A. & Meir, G.E.A. 1979 J. Fluid Mech. **90**, 67.

Schraub, F.A. & Kline, S.J. 1965 Stanford Univ. Rep no. MD-12.

Sirovich, L., Maxey, M. & Tarman, H. 1987 In Sixth Symp.Turb.Shear Flow, Toulouse, p. 6-12.

Smith, C.R. & Metzler, S.P. 1983 J. Fluid Mech. **129**, 27.

Spalart, P. 1987 NASA TM 89407.

Sreenivasan, K.R. 1987 In Turbulence Management and Relaminarization, eds. H.W. Liepmann & R. Narasimha, Springer, p. 37.

Sreenivasan, K.R. & Antonia, R.A. 1977 J. Appl. Mech. **44**, 389.

Sreenivasan, K.R. & Meneveau, C. 1986 J. Fluid Mech. **173**, 357.

Sreenivasan, K.R., Prabhu, A. & Narasimha, R. 1983 J. Fluid Mech. **137**, 251.

Sreenivasan, K.R., Ramshankar, R. & Meneveau, C. 1988 Proc. Roy. Soc. Lond. (in press).

Sternberg, J. 1962 J. Fluid Mech. **13**, 241.

Stratford, B.S. 1959 J. Fluid Mech. **5**, 17.

Theodorsen, T. 1955 In 50 Jahre Grenzschichtforschung, ed. Goertler & Tollmien, Vieweg & Sohn

Thomas, R.M. J. Fluid Mech. **57**, 549.

Tennekes, H. & Lumley, J.L. 1972 A First Course in Turbulence, M.I.T. Press.

Tiederman, W.G., Luchik, T.S. & Bogard, D.G. 1985 J. Fluid Mech. **156**, 419.

Townsend, A.A. 1956 The Structure of Turbulent Shear Flow, Cambridge Univ. Press.

Townsend, A.A. 1961 J. Fluid Mech. **11**, 97.

Townsend, A.A. 1966 J. Fluid Mech. **26**, 689.

Townsend, A.A. 1976 The Structure of Turbulent Shear Flow, Cambridge Univ. Press, second ed.

Tritton, D.J. 1967 J. Fluid Mech. **28**, 439.

Ueda, H. & Hinze, J.O. 1975 J. Fluid Mech. **67**, 125.

Ueda, H. & Mizushina, T. 1977 In Proc. 5th Biennial Symp. on Turb. Univ.Missouri-Rolla.

Uzkan, T. & Reynolds, W.C. 1967 J. Fluid Mech. **28**, 803.

Wallace, J.M. 1982 In 11th Sotheast Conf. on Theo. and Appl. Mech. Huntsville, Alabama.

Wallace, J.M. 1986 Exp. Fluids. **4**, 61.

Wallace, J.M., Eckelmann, H. & Brodkey, R.S. 1972 J. Fluid Mech. **54**, 39.

Wieghardt, K. & Tillmann, W. 1951 NACA TM 1314.

Willmarth, W.W. & Bogar, T.J. 1977 Phys. Fluids **20**, s9.

Willmarth, W.W. & Lu, S.S. 1972 J. Fluid Mech. **55**, 65.

Willmarth, W.W. & Wooldridge, R. 1962 J. Fluid Mech. **14**, 187.

THE ART AND SCIENCE OF FLOW CONTROL

Mohamed Gad-el-Hak

Department of Aerospace & Mechanical Engineering
University of Notre Dame
Notre Dame, IN 46556

ABSTRACT

The ability to actively or passively manipulate a flow field to effect a desired change is of immense technological importance. In this article, methods of control to achieve transition delay, separation postponement, lift enhancement, drag reduction, turbulence augmentation, or noise suppression are considered. Emphasis is placed on external boundary-layer flows although applicability of some of the methods reviewed for internal flows will be mentioned. Attempts will be made to present a unified view of the different methods of control to achieve a variety of end results. Performance penalties associated with a particular method such as cost, complexity, or trade-off will be elaborated.

CONTENTS

1. INTRODUCTION AND SCOPE

As defined by Flatt (1961), the term boundary layer control includes any mechanism or process through which the boundary layer of a fluid flow is caused to behave differently than it normally would were the flow developing naturally along a smooth straight surface. Prandtl (1904) pioneered the modern use of flow control in his epoch-making presentation to the third International Congress of Mathematicians held at Heidelberg, Germany. In just 8 pages (as required for acceptance by the Congress) of a paper entitled "On the Motion of a Fluid with Very Small Viscosity," Prandtl introduced the boundary layer theory, explained the mechanics of steady separation, opened the way for understanding the motion of real fluids, and described several experiments in which the boundary layer was controlled.

In its broadest sense, the art of flow control probably has its roots in prehistoric times when streamlined spears, sickle-shaped boomerangs, and fin-stabilized arrows evolved empirically by archaic Homo sapiens (Williams, 1987). Modern man also artfully applied flow control methods to achieve certain technological goals. Relatively soon after the dawn of civilization and the establishment of an agriculture way of life 8,000 years ago, complex systems of irrigation were built along inhabited river valleys to control the water flow, thus freeing man from the vagaries of the weather. For centuries, farmers knew the value of windbreaks to keep top soil in place and to protect fragile crops.

For the purpose of this article, we narrow our focus considerably to discuss a particular kind of flow control; namely that of manipulating an external boundary-layer flow, such as that developing on the exterior surface of an aircraft or a submarine, to achieve transition delay, separation postponement, lift enhancement, drag reduction, turbulence augmentation, or noise suppression. These objectives are not necessarily mutually exclusive. For example, by maintaining as much of a boundary layer in the laminar state as possible, the skin-friction drag and the flow-generated noise are reduced. However, a turbulent boundary layer is in general more resistant to separation than a laminar one. By preventing separation, lift is enhanced and the form drag is reduced. An ideal method of control that is simple, inexpensive to build and operate, and does not have any trade-off's does not exist, and the skilled engineer has to make continuous compromises to achieve a particular goal.

Prandtl (1904) used artificial control of the boundary layer to show the great influence such a control exerted on the flow pattern. He used suction to delay boundary-layer separation from the surface of a cylinder. Notwithstanding Prandtl's success, aircraft designers in the three decades following his convincing demonstration were accepting lift and drag of airfoils as predestined characteristics with which no man could or should tamper (Lachmann, 1961). This predicament changed mostly due to the German research in boundary layer control pursued vigorously shortly before and during the Second World War. In the two decades following the war, extensive research on laminar flow control, where the boundary layer formed along the external surfaces of an aircraft is kept in the low-drag laminar state, was conducted in Europe and the United States, culminating in the successful flight test program of the X-21 where suction was used to delay transition on a swept wing up to a chord Reynolds number of 4.7 x 10^7. The oil crisis of the early 1970's brought renewed interest in novel methods of flow control to reduce skin-friction drag even in turbulent boundary layers. In the 1980's, the need for supermaneuverable fighter planes, faster and quieter underwater vehicles, and hypersonic transport (e.g., the U.S. 'National Aerospace Plane') provides new challenges for researchers in the field of flow control.

The major goal of this article is to present a unified view of the different control methods to achieve a variety of end results. Emphasis is placed on external boundary-layer flows. The same vorticity considerations brilliantly employed by Lighthill (1963) to place the boundary layer correctly in the flow as a whole is used to explain many of the flow control techniques reviewed in here. The paper is organized according to the desired goal of applying the boundary layer control. The governing equations are developed in the following section. In Section 3 methods to delay laminar-to-turbulent transition are discussed. Available techniques to postpone separation and to enhance lift are reviewed in Section 4. In Section 5, methods to reduce drag are discussed. Turbulence augmentation and noise control are reviewed in Sections 6 and 7, and finally concluding remarks are given in Section 8. As noted before, a particular flow control method may serve more than one end result or may help one goal while adversely affecting another. Such situations will be elaborated as they occur throughout this review.

2. GOVERNING EQUATIONS

The principles of conservation of mass, momentum and energy govern all fluid motions. In general, a set of partial, nonlinear differential equations expresses these principles, and together with appropriate boundary and initial conditions constitute a well-posed problem. It is of course beyond the scope of this article to derive these equations and the reader is referred to any advanced textbook in Fluid Dynamics (Landau and Lifshitz, 1963; Batchelor, 1967; Hinze, 1975; Kays and Crawford, 1980; Panton, 1984). The equations will be first recalled in as general a form as possible. This approach will become particularly useful when discussing surface heating/cooling (viscosity varies spatially), drag-reducing polymers (non-Newtonian fluid), and other nonconventional situations.

In Cartesian tensor notation, the equation of conservation of mass reads:

$$\frac{\partial \rho}{\partial t} + \frac{\partial}{\partial x_i}(\rho U_i) = 0, \qquad (2.1)$$

and Newton's second law is:

$$\rho \left(\frac{\partial U_i}{\partial t} + U_j \frac{\partial U_i}{\partial x_j} \right) = \frac{\partial}{\partial x_j} \tau_{ji} + F_i \ , \tag{2.2}$$

where U_i is an instantaneous velocity component, ρ is the density, F_i is the body force and τ_{ji} is the second-order stress tensor which must be related to the deformation tensor in order to reduce the number of unknowns to be equal to the number of equations. The independent variables are time t and the three spatial coordinates x_1, x_2, and x_3 (or x, y, and z). For a Newtonian fluid a linear relation between the stress and the rate of strain is assumed. If this fluid is further assumed to be isotropic and to have a negligible bulk viscosity (Stokes's hypothesis), its constitutive relation reads:

$$\tau_{ji} = - P \delta_{ji} + \mu \left(\frac{\partial U_i}{\partial x_j} + \frac{\partial U_j}{\partial x_i} - \frac{2}{3} \frac{\partial U_k}{\partial x_k} \delta_{ji} \right) , \tag{2.3}$$

where P is the hydrostatic pressure, μ is the dynamic coefficient of viscosity, and δ_{ji} is the unit second-order tensor (Kronecker delta). While (2.3) is valid for air and water under most practical conditions, different constitutive relations must be sought for non-Newtonian fluids such as polymer solutions.

For a Newtonian fluid, (2.3) is substituted in (2.2) to yield the momentum equation:

$$\rho \left(\frac{\partial U_i}{\partial t} + U_j \frac{\partial U_i}{\partial x_j} \right) = - \frac{\partial P}{\partial x_i} + \frac{\partial}{\partial x_j} \left[\mu \left(\frac{\partial U_i}{\partial x_j} + \frac{\partial U_j}{\partial x_i} \right) \right]$$

$$- \frac{2}{3} \frac{\partial}{\partial x_i} \left(\mu \frac{\partial U_k}{\partial x_k} \right) + F_i \ . \tag{2.4}$$

For a compressible, Newtonian fluid, (2.1) and (2.4) must be complemented by an equation of state and the energy equation to form six equations for the six unknowns U_i, ρ, P and T, where $T(x_i, t)$ is the temperature field.

If the flow is incompressible, then $\dfrac{\partial U_k}{\partial x_k} = 0$, density is assumed given, and (2.4) reads:

$$\rho \left(\frac{\partial U_i}{\partial t} + U_j \frac{\partial U_i}{\partial x_j} \right) = - \frac{\partial P}{\partial x_i} + \frac{\partial}{\partial x_j} \left[\mu \left(\frac{\partial U_i}{\partial x_j} + \frac{\partial U_j}{\partial x_i} \right) \right] + F_i \ . \tag{2.5}$$

Note that in (2.5), μ has not been assumed constant, a useful generalization when surface heating/cooling is used.

The familiar Navier-Stokes equation is obtained from (2.2) by assuming a Newtonian, incompressible, constant-viscosity fluid:

$$\rho \left(\frac{\partial U_i}{\partial t} + U_j \frac{\partial U_i}{\partial x_j} \right) = - \frac{\partial P}{\partial x_i} + \mu \frac{\partial^2 U_i}{\partial x_j \partial x_j} + F_i \ . \qquad (2.6)$$

In this case, the momentum and continuity equations form four equations for the four unknowns U_i and P.

All the above equations are valid for nonturbulent as well as turbulent flows. However, in the latter case all the dependent variables are in general random functions of space and time. No straightforward method exists for solving stochastic, nonlinear partial differential equations. The recent attempts to use dynamical systems theory to study turbulent flows has not yet reached fruition especially at Reynolds numbers far above transition. The brute-force numerical integration of the equations using the supercomputer is prohibitively expensive at practical Reynolds numbers. For the present at least, a statistical approach, where a temporal, spatial or ensemble mean is defined and the equations of motion are written for the various moments of the fluctuations about this mean, is the only route available to get meaningful engineering results. Unfortunately, the nonlinearity of the Navier-Stokes equations guarantees that the process of averaging to obtain moments results in an open system of equations, where the number of unknowns is always more than the number of equations, and more or less heuristic modeling is used to close the equations.

In (2.6), let $U_i = \overline{U_i} + u_i$ and $P = \overline{P} + p'$, where $\overline{U_i}$ and $\overline{P}$ are temporal averages for the velocity and pressure, respectively, and u_i and p' are the velocity and pressure fluctuations about the respective averages. The equation governing the mean velocity for a Newtonian, incompressible, constant-viscosity, turbulent flow becomes:

$$\rho \left(\frac{\partial \overline{U_i}}{\partial t} + \overline{U_j} \frac{\partial \overline{U_i}}{\partial x_j} \right) = - \frac{\partial \overline{P}}{\partial x_i} + \frac{\partial}{\partial x_j} \left(\mu \frac{\partial \overline{U_i}}{\partial x_j} - \rho \overline{u_i u_j} \right) + F_i \ . \qquad (2.7)$$

This equation is written in a form that facilitates the physical interpretation of the turbulent stress tensor (Reynolds stresses), $-\rho \overline{u_i u_j}$, as additional stresses on a fluid element to be considered along with the conventional viscous stresses and pressure. An equation for the components of this tensor may be derived but it will contain third-order moments such as $\overline{u_i u_j u_K}$, and so on. The equations are closed by expressing the second- or third-order quantities in terms of the first- or second-moments, respectively. For a review of these first- and second-order closure schemes see Lumley (1983; 1987).

For external flows at high Reynolds number, viscous forces are confined to a relatively thin layer along the surface of a body, although this layer's thickness increases in the downstream direction. Outside the boundary layer, the flow could be computed using the potential flow theory. Within the viscous region, the classical boundary-layer approximations apply (Rosenhead, 1963; Schlichting, 1979). Ignoring body forces, the continuity equation and the streamwise and normal momentum equations along a two-dimensional (or axisymmetric) surface of small curvature are obtained from (2.1) and (2.4), respectively:

$$\frac{\partial \rho}{\partial t} + \frac{\partial}{\partial x_1} (\rho U_1) + \frac{\partial}{\partial x_2} (\rho U_2) = 0 \; , \tag{2.8}$$

$$\rho \left(\frac{\partial U_1}{\partial t} + U_1 \frac{\partial U_1}{\partial x_1} + U_2 \frac{\partial U_1}{\partial x_2} \right) = -\frac{\partial P}{\partial x_1} + \frac{\partial}{\partial x_2} \left(\mu \frac{\partial U_1}{\partial x_2} \right) , \tag{2.9}$$

$$-\frac{\rho}{R} U_1^2 = -\frac{\partial P}{\partial x_2} \; . \tag{2.10}$$

These equations are valid for variable properties ρ and μ. In (2.8) through (2.10), x_1 is in the main-flow direction along the body, x_2 is normal to the surface, and R is the radius of curvature of the two-dimensional surface or the radius of revolution of the axisymmetric body. Equation (2.10) gives the pressure gradient required to balance the centrifugal effect of flow round a curved wall (Milne-Thomson, 1968). For a plane wall $(R \to \infty)$ and within the boundary-layer approximation, the pressure is constant in the normal direction and its value is determined by the inviscid flow at the outer edge of the boundary layer.

For steady flow, the above equations can be integrated in the x_2 direction resulting in the momentum integral equation:

$$\frac{C_f}{2} = \frac{d \delta_\theta}{d x_1} + \delta_\theta \left[\left(2 + \frac{\delta^*}{\delta_\theta} \right) \frac{1}{U_\infty} \frac{d U_\infty}{d x_1} + \frac{1}{\rho_\infty} \frac{d \rho_\infty}{d x_1} + \frac{1}{R} \frac{d R}{d x_1} \right] - \frac{\rho_0 v_0}{\rho_\infty U_\infty} \; . \tag{2.11}$$

In (2.11), C_f = local skin-friction coefficient $\equiv \dfrac{\tau_0}{1/2 \, \rho_\infty U_\infty^2}$, $\tag{2.12}$

$$\delta^* = \text{displacement thickness} \equiv {}_0\!\!\int^\infty \left(1 - \frac{\rho U_1}{\rho_\infty U_\infty} \right) d x_2 \; , \tag{2.13}$$

$$\delta_\theta = \text{momentum thickness} \equiv {}_0\!\!\int^\infty \frac{\rho U_1}{\rho_\infty U_\infty} \left(1 - \frac{U_1}{U_\infty} \right) d x_2 , \tag{2.14}$$

ρ_∞ and U_∞ are the density and velocity of the freestream, respectively, ρ_0 and v_0 are the density and normal velocity of fluid injected through the surface, and τ_0 is the shear stress at the wall.

Since the skin-friction coefficient in the momentum integral equation is defined in terms of the shear stress and not in terms of the velocity gradient at the wall, (2.11) is, in fact, valid for both laminar and turbulent flows as well as for both Newtonian and non-Newtonian fluids; the only assumptions being that the boundary-layer flow is steady and two-dimensional in the mean. For an incompressible fluid, $\rho = \rho_\infty = $ constant. In case of a turbulent flow, the mean streamwise velocity, $\overline{U_1}(x_1, x_2)$, is used in the definition of δ^* and δ_θ. For a Newtonian fluid,

$$\tau_0 = \mu \frac{\partial U_1}{\partial x_2} \bigg|_{wall} \quad . \tag{2.15}$$

3. TRANSITION DELAY

3.1 INTRODUCTORY REMARKS

Delaying laminar-to-turbulent transition of a boundary layer has many obvious advantages. Depending on the Reynolds number, the skin-friction drag in the laminar state can be as much as an order of magnitude less than that in the turbulent condition (Figure 1). For an aircraft or an underwater body, the reduced drag means longer range, reduced fuel cost/volume, or increased speed. Flow-induced noise results from the pressure fluctuations in the boundary layer and, hence, is virtually nonexistent in the laminar case. Reducing the boundary layer noise is crucial to the proper operation of an underwater sonar. On the other hand, turbulence is an efficient mixer and rates of mass, momentum and heat transfer are much lower in the laminar state, so early transition may be sought in some applications as for example when enhanced heat transfer rates are desired or when rapid mixing is needed.

The routes to transition are many; some are more understood than others. On a semi-infinite, smooth flat plate placed in a clean, uniform, incompressible flow with as little external disturbances as possible, the laminar flow (somewhat downstream of the leading edge) is in the form of a Blasius profile and exists at low enough Reynolds number, typically $R_x < 6 \times 10^4$. This laminar shear layer is, however, unstable to small perturbations that invariably exist in any flow. Squire's (1933) theorem shows that two-dimensional travelling waves (Tollmien-Schlichting waves) are the most dangerous for incompressible-flow instability and become unstable when the Reynolds number exceeds a critical value. However, as soon as the T-S waves are amplified, gain a certain amplitude and nonlinear effects take place, three-dimensional disturbances can no longer be excluded (Itoh, 1987). Following the linear step, the originally two-dimensional waves inherently acquire a nearly periodic spanwise modulation and three-dimensional structures evolve as a result of a secondary instability. Hairpin vortices develop (tertiary instability?) and finally breakdown to turbulence occurs (Klebanoff et al., 1962). Not surprisingly, these nonlinear processes are far less understood than the linear step. The linear amplification step is, however, the slowest of the successive multiple steps in the transition process and, hence, factors that affect the linear amplification determine the magnitude of the transition Reynolds number.

Reshotko (1976; 1985; 1987) asserts that transition is a consequence of the nonlinear response of the laminar shear layer (a very complicated oscillator) to random forcing disturbances that result from freestream turbulence, radiated sound, surface roughness, surface vibrations, or combination of these environmental factors. If the initial disturbances are small, transition Reynolds number depends upon the nature and spectrum of these disturbances, their signature and excitation of the normal modes in the

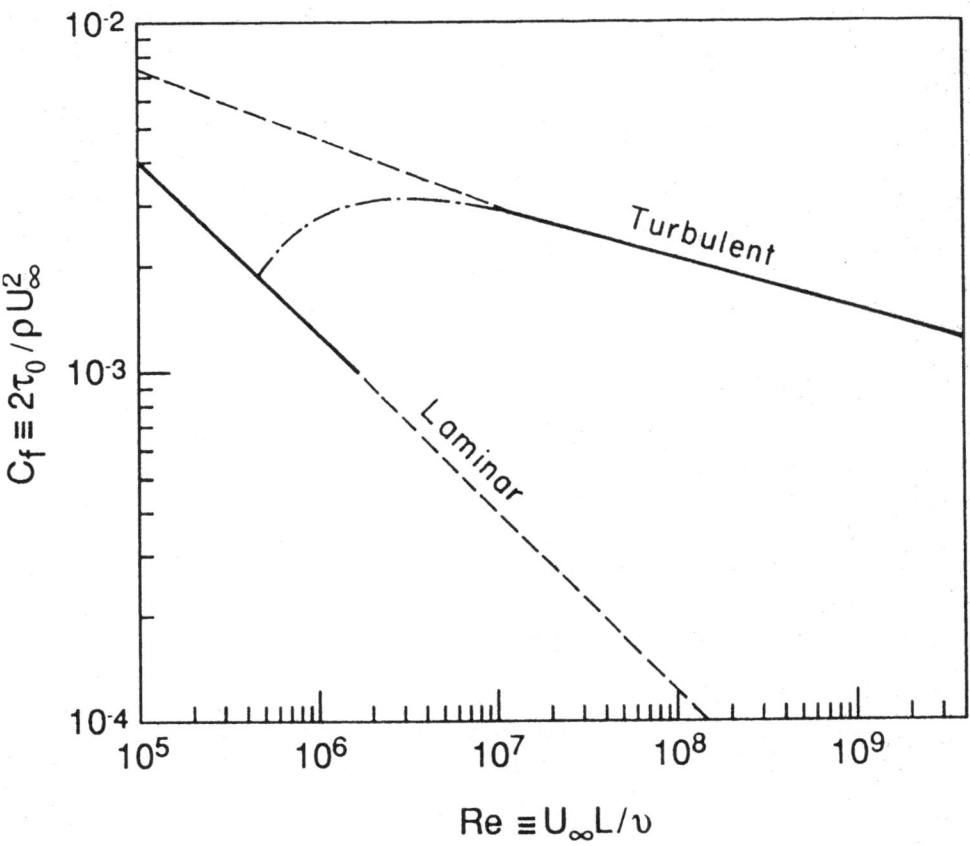

Figure 1. Skin - Friction Coefficient for a Smooth Flat Plate at
Zero Incidence to a Uniform, Incompressible Flow.

boundary layer (receptivity), and the linear amplification of the growing normal modes. Once wave interaction and nonlinear processes set in, transition is quickly completed. If the initial disturbance levels are large enough, the relatively slow linear amplification step mentioned above is bypassed (Morkovin, 1984; 1988) and transition can occur at much lower Reynolds numbers. In fact, a sufficiently violent disturbance, $u_{rms} / U_\infty \sim 10\%$, can cause transition of a laminar boundary layer to advance to the position upstream of which perturbations of all wave numbers decay (Klebanoff et al., 1955).

Other routes to transition include the Taylor-Görtler vortices forming on a concave surface as a set of counter-rotating, streamwise standing eddies (Stuart, 1963; DiPrima and Swinney, 1985), the cross-flow instabilities occurring on a swept wing or a rotating disk (Reed and Saric, 1987; 1989), and any other situation where an instability will occur as a result of disturbing the equilibrium of body forces, inertia and surface forces (Drazin and Reid, 1981). Once again, these instabilities will generally occur at Reynolds numbers lower than the critical Reynolds number for the growth of Tollmien-Schlichting waves.

To delay transition to as far downstream position as possible, the following steps may be taken. First, since factors that affect the linear amplification of Tollmien-Schlichting waves determine the magnitude of the transition Reynolds number, these waves should be either inhibited or cancelled. In the former method of control, the growth of the linear disturbance is minimized using any or a combination of the so-called stability modifiers which alter the shape of the velocity profile and include increased length of favorable pressure gradient, wall transpiration, wall motion, and surface heating/cooling. Wave cancellation of the growing perturbation is accomplished through exploiting but not altering the stability characteristics of the flow. Secondly, the forcing disturbances in the environment in which the laminar shear layer develops must be reduced. This is accomplished by using smooth surfaces, reducing the freestream turbulence and the radiated sound, minimizing body vibration, and ensuring a particulate-free incoming flow or, in case of a contaminated environment such as the ocean, using a particle-defense mechanism. Practically achieved surface smoothness and levels of radiated noise place an upper limit for unit Reynolds number required for a successful laminar flow control system. For aircraft, this typically translates into a requirement for high altitude operation (above 10 Km). Thirdly, provide a flow where other kind of instabilities, e.g., Taylor-Görtler vortices or cross-flow instabilities, will not occur or at least will not grow at a rapid rate. This is done by avoiding as much as possible concave surfaces or concave streamlines, minimizing the sweep on lifting surfaces, etc. In the next section, the elements of linear stability theory are briefly recalled. This is followed by a discussion of the first of the three steps outlined above, stability modifiers and wave cancellation methods.

3.2 LINEAR STABILITY THEORY

The linear stability of a laminar flow is governed by the Orr-Sommerfeld equation and appropriate boundary conditions:

$$(U_1 - c)\,(\phi'' - \alpha^2 \phi) - U_1'' \phi = \frac{i}{\alpha Re}\,(\phi'''' - 2\alpha^2 \phi'' + \alpha^4 \phi). \qquad (3.1)$$

This fourth-order, linear, ordinary differential equation is derived from the Navier-Stokes equation $(2.6)^*$ by assuming a two-dimensional small disturbance superimposed upon a steady, unidirectional mean flow, $U_1(x_2)$. The stream function for the perturbation is given by:

$$\psi(x_1, x_2, t) = \phi(x_2)\, e^{i\alpha(x_1 - ct)}, \tag{3.2}$$

where α is the wave number, c is the wave speed, ϕ is the amplitude function of the fluctuation, the superscript ' denotes the derivative with respect to x_2, and Re is the Reynolds number based upon the freestream velocity and the boundary layer thickness. The Orr-Sommerfeld equation has been nondimensionalized using the boundary layer thickness as a length scale and the freestream velocity as a velocity scale.

Either the temporal or the spatial growth of instability waves is considered as an eigenvalue problem. In the former case, a disturbance oscillates in space but either grows or decays exponentially with time. A complex eigenvalue $c = c_r + i c_i$ is determined for each pair of values of the real parameter α and the Reynolds number. The real part of c is the phase velocity of the prescribed disturbance and the sign of the imaginary part determines whether the wave is temporary amplified ($c_i > 0$) or temporary damped ($c_i < 0$). The more realistic spatial stability problem involves disturbances that oscillate in time but either grow or decay exponentially with downstream distance. In this case, a complex eigenvalue $\alpha = \alpha_r + i \alpha_i$ is determined for each pair of values of the real circular frequency αc and the Reynolds number. The real part of α is the wave number and the sign of α_i determines whether the wave is spatially amplified ($\alpha_i < 0$) or spatially damped ($\alpha_i > 0$). In either case, the major difficulty in numerically integrating the Orr-Sommerfeld equation lies in the fact that it is highly stiff and unstable (Lin, 1945), which makes the use of conventional numerical schemes virtually impossible. Explicit numerical methods with a step size that is commensurate with the global behavior of the solution cannot be used to integrate (3.1) because of the numerical instabilities that characterize this ordinary differential equation. The papers by Orszag (1971) and Scott and Watts (1977) give examples of special codes to handle stiff eigenvalue problems.

In a spatially growing boundary layer, transition to turbulence involves the integrated effect of the growth of unstable waves as they are convected downstream by the mean flow. If A represents the amplitude of a wave, then its local spatial growth is given by:

$$\frac{d\,|A|}{d\,x_1} = -\,\alpha_i\,|A|. \tag{3.3}$$

The net growth of a wave with a given real frequency, from the location x_1^0, where it first becomes unstable, to the location x_1^{00}, where transition to turbulence occurs, is given by:

---

*Note that (2.6) is the equation of motion for a Newtonian, incompressible, constant-viscosity fluid. If viscosity is not constant, additional terms containing the first and second derivatives of viscosity with respect to x_2 result and one obtains the so-called modified Orr-Sommerfeld equation (see, for example, Wazzan et al., 1968).

$$\ln \left[\frac{|A|}{|A_0|} \right] = -\int_{x_1^0}^{x_1^{00}} \alpha_i \, dx_1 = n. \tag{3.4}$$

Emperically, a modal amplification n of between 8 and 11 is found at transition to turbulence (e.g., Smith and Gamberoni, 1956; Van Ingen, 1956; Jaffe et al., 1970). Despite the apparent success of the amplitude-ratio method to predict transition, Reshotko (1976) maintains that the method is defective in principle and perhaps also in practice. Transition location depends strongly on the freestream turbulence levels and other environmental factors. Moreover, nonlinear effects that are physically significant in the transition process cannot be accounted for by procedures based on linear stability theory. Notwithstanding these drawbacks, the e^n rule provides a practical scheme for estimating the effectiveness of a transition-delay method of control (Bushnell et al., 1988; Bushnell, 1989).

The Reynolds number below which perturbations of all wave numbers decay is termed the critical Reynolds number or the limit of stability. For a given velocity, $U_1(x_2)$, the critical Reynolds number and the rate of growth of perturbations depends strongly on the shape of the velocity profile. A profile with an inflectional point $(\partial^2 U_1 / \partial x_2^2 = 0)$ above the wall provides a necessary and sufficient condition for inviscid instability (Rayleigh, 1880; Tollmien, 1935). Such profile must have a positive curvature at $x_2 = 0$, since $\partial^2 U_1 / \partial x_2^2$ is negative at a large distance from the wall (Figure 2). Even when viscous effects are included (right-hand side of the Orr-Sommerfeld equation $\neq 0$), a velocity profile becomes more stable as its second derivative near the wall becomes negative, $[\partial^2 U_1 / \partial x_2^2]_0 < 0$. The profile is then said to be more full, having a smaller ratio of displacement thickness to momentum thickness than, for example, an inflectional velocity profile. In the former case, the critical Reynolds number is increased, the range of amplified frequencies is diminished and the amplification rate of unstable waves is reduced.

3.3 STABILITY MODIFIERS

Stability modifiers are those methods of laminar flow control which alter the shape of the velocity profile to minimize the linear growth of unstable waves. Recall the streamwise-momentum equation along a two-dimensional surface of small curvature, (2.9). At the wall, $x_2 \to 0$, and after some rearranging the equation, valid for variable properties ρ and μ, becomes:

$$\left[\rho \, \frac{\partial U_1}{\partial t} \right]_0 + \left[\rho \, U_1 \, \frac{\partial U_1}{\partial x_1} \right]_0 + \left[\rho \, U_2 \, \frac{\partial U_1}{\partial x_2} \right]_0 + \frac{dP}{dx_1} - \frac{d\mu}{dT} \left[\frac{\partial T}{\partial x_2} \frac{\partial U_1}{\partial x_2} \right]_0$$

$$= \left[\mu \, \frac{\partial^2 U_1}{\partial x_2^2} \right]_0 , \tag{3.5}$$

where the subscript $[\ ]_0$ indicates flow quantities computed at the wall and T is the temperature field. For a two-dimensional laminar boundary layer, vorticity is only in the spanwise direction and is equal to $-\partial U_1 / \partial x_2$. The right-hand side of (3.5), therefore, represents the flux of vorticity at the surface, as

222

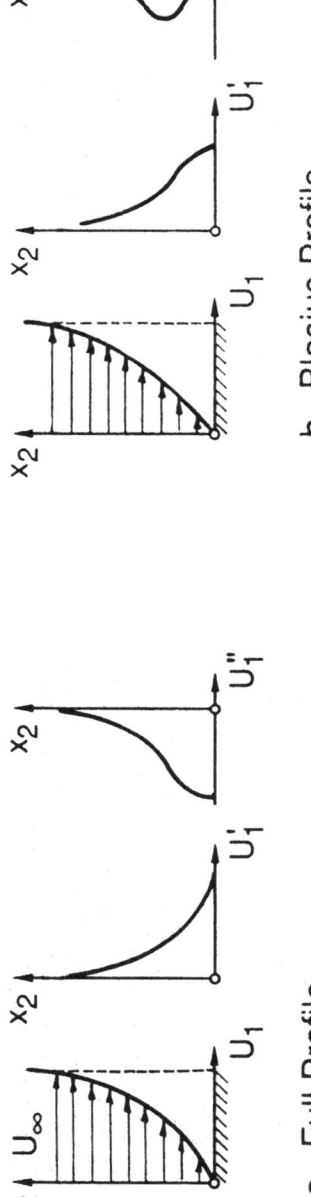

a. Full Profile.

b. Blasius Profile.

c. Inflectional Profile.

d. Profile at the Point of Steady, Two-Dimensional Separation.

Figure 2. Normal Velocity Profile in a Boundary Layer.

brilliantly demonstrated by Lighthill (1963). Any of the terms on the left-hand side of (3.5) can affect the sign of the second derivative of the velocity profile (or the direction of the vorticity flux) at the wall and, hence, the flow stability. Stability modifiers do just that and include wall motion such that the first, second or third term on the left-hand side of (3.5) is negative, wall suction ($[U_2]_0 < 0$), favorable pressure gradient ($dP/dx_1 < 0$), surface cooling in gases ($d\mu/dT > 0$; $[\partial T/\partial x_2]_0 > 0$), or surface heating in liquids ($d\mu/dT < 0$; $[\partial T/\partial x_2]_0 < 0$). Any or a combination of these methods will cause the curvature of the velocity profile at the wall to become more negative and, hence, increase the lower critical Reynolds number and reduce the spatial or the temporal amplification rates of unstable waves.

Boundary layers which are stabilized by extending the region of favorable pressure gradient are known as natural laminar flow (NLF), while the other methods to modify the stability of the shear flow are termed laminar flow control (LFC). It is clear from (3.5) that the effects of these methods are additive. The term hybrid laminar flow control normally refers to the combination of NLF and one of the LFC techniques.

3.3.1 WALL MOTION

Wall motion can be generated by either actively driving the surface or by using a flexible coating whose modulus of rigidity is low enough so that surface waves are generated under the influence of the stress field in the fluid. In the former case, the wall motion is precisely controlled and can be made to affect the shape of the velocity profile in a desired manner. This method of control is, however, quite impractical and is used mainly to provide controlled experiments to determine what type of wave motion is required to achieve a given result. The more practical passive flexible walls can be broadly classified into two categories: truly compliant coatings which have extremely small damping and modulus of elasticity and can therefore respond with little phase lag to the boundary layer flow; and the more readily available resonant walls in which vibration modes in the solid, acting as an active vibration damper, are excited by the flow-disturbance forcing function (Bushnell et al., 1977). The flow stabilization in this case may be a result of altering the phase relation in the viscous region rather than changing the curvature of the velocity profile at the wall.

Theoretical studies of boundary-layer stability in the presence of a flexible wall started in the early 1960's, stimulated in part by the pioneering experimental work of Krämer (1960). In both the classical work (e.g., Benjamin, 1960; Landahl, 1962; Kaplan, 1964) and the more recent research (e.g., Carpenter and Gerrad, 1985; Willis, 1986; Yeo and Dowling, 1987), steady-state stability theory has been used with an assumed velocity profile which is not allowed to change as the wall moves. It is clear, however, that the wall displacement will induce a travelling pressure signal which in turn will modulate the mean velocity profile. Either a quasi-steady stability analysis of the modulated flow or a true time-dependent calculations would be preferable to conventional linear-stability theory, but neither has been attempted yet due to the obvious complexities of the problem. Notwithstanding the shortcoming of present stability calculations, Willis (1986) obtained a very impressive agreement with his carefully conducted experiments. His results are depicted in Figure 3. Eigenvalue calculations were performed to predict the amplification factors for a

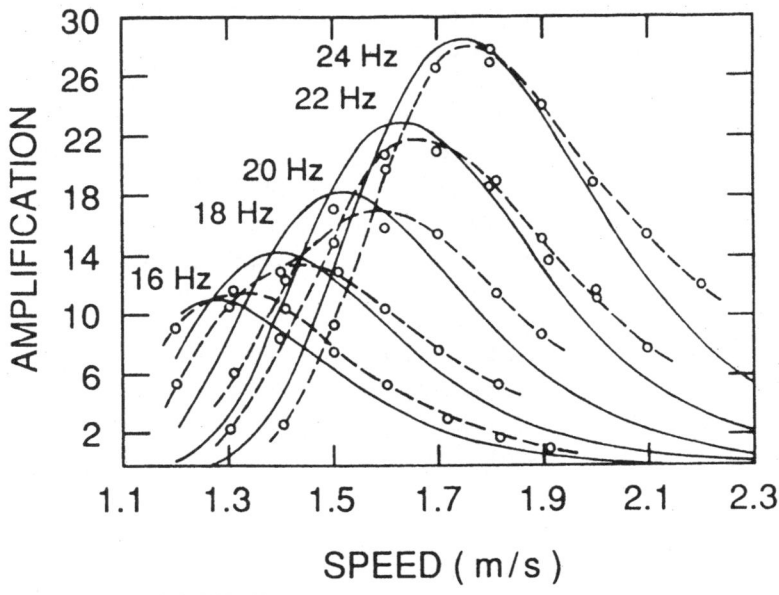

a. Rigid Wall.

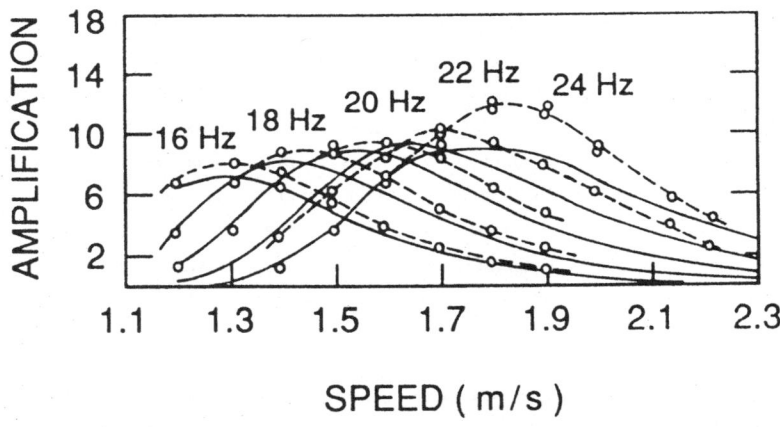

b. Flexible Wall. Modulus of Rigidity = 5000 N/m².

Figure 3. Growth Curves of Artificially Introduced Tollmien - Schlichting Waves. Solid Lines are Theoretical Prediction and Circles are Experimental Data. (from Willis, 1986).

range of modal frequencies. The flexible coating was a silicon-rubber/silicon-oil mix covered by a thin latex rubber skin stretched across the surface. The experiments were conducted in a water towing tank using a flat plate, and controlled, harmonic two-dimensional disturbances were introduced upstream of the compliant surface. As indicated in Figure 3, a reduction of the wave growth by an order of magnitude is feasible when this rather simple coating is used, almost eliminating transition due to Tollmien-Schlichting type of instability. The flexible wall itself is, however, susceptible to other kinds of instability and care must be taken to ensure that these surface waves will not grow to an amplitude that will promote transition through a roughness-like effect (see the recent review article by Riley et al., 1988).

Although the original Kramer's (1960) experiments were discredited up until a few years ago, new theoretical and experimental evidence confirm Kramer's results (Carpenter and Gerrard, 1985). Passive flexible coatings with density the order of the fluid density appear to be capable of considerable transition postponement. The density requirement makes this method of control suitable for water applications only. Transitional Reynolds numbers (based on distance from the leading edge) that are 5-10 times those for a rigid surface seem to be readily achievable with a simple method that does not require energy expenditure, slots, ducts, or internal equipment of any kind.

3.3.2 SUCTION

A second method for postponing transition is the application of wall suction. As seen from (3.5), small amounts of fluid withdrawn from the near-wall region of the boundary layer change the curvature of the velocity profile at the wall and can dramatically alter the stability characteristics of the boundary layer. Additionally, suction inhibits the growth of the boundary layer, so that the critical Reynolds number based on thickness may never be reached.

Although laminar flow can be maintained to extremely high Reynolds number provided that enough fluid is sucked away, the goal is to accomplish transition delay with the minimum suction flow rate. Not only this will reduce the power necessary to drive the suction pump but also the momentum loss due to suction, and hence the skin friction, is minimized. This latter point can easily be seen from the momentum integral equation. Rewriting (2.10) for an incompressible flow ($\rho_0 = \rho_\infty = $ constant) over a flat plate ($d\, U_\infty / d\, x_1 = 0$; $d\, R / d\, x_1 = 0$) with uniform suction through the wall (v_0 negative), the equation reads:

$$\frac{C_f}{2} = \frac{d\, \delta_\theta}{d\, x_1} + \frac{|v_0|}{U_\infty} \ . \tag{3.6}$$

The second term on the right-hand side is the suction coefficient, C_q, and although withdrawing the fluid through the wall leads to a decrease in the rate of growth of the momentum thickness, C_f increases directly with C_q . Fluid withdrawn through the wall has to come from outside the boundary layer where the streamwise momentum per unit mass is at the relatively high level of U_∞. The second term is proportional to the rate of momentum loss due to withdrawing a mass per unit time and area of $\rho\ |v_0|$. Note that this term does not exist for pipe flows because of the mass flow constraint. Hence, this momentum penalty is not paid for channel flows with wall transpiration, an important distinction between internal and external flows.

Although Prandtl (1904) used suction to prevent flow separation from the surface of a cylinder near the beginning of this century, the first experimental demonstration that boundary-layer transition can be delayed by withdrawing near-wall fluid did not take place until about four decades later. Holstein (1940), Ackeret et al. (1941), Ras and Ackeret (1941), and Pfenninger (1946) used carefully shaped, single and multiple suction slits to demonstrate the decrease in drag associated with delaying transition. Braslow et al. (1951) used continuous suction through a porous wall to maintain laminar flow on an airfoil to chord Reynolds number of 2.0×10^7. In the early 1960's, test flights of two X-21 aircrafts (modified U.S. Air Force B-66's) indicated the feasibility of maintaining a laminar flow on a swept wing to chord Reynolds number as high as 4.7×10^7 (Whites et al., 1966). The wing surfaces contained many thin and closely spaced spanwise suction slots, and the total airplane drag was reduced by 20% as compared to the no suction case.

Although discrete suction slots were used first, because of the unavailability of suitable porous surfaces in the early 1940's, the theoretical treatment of the problem is considerably simplified by assuming continuous suction through a porous wall where the characteristic pore size is much smaller than a boundary layer thickness. In fact, the case of a uniform suction from a flat plate at zero incidence is an exact solution of the Navier-Stokes equation (2.6). Assuming weak enough suction that the potential flow outside the boundary layer is unaffected by the loss of mass at the wall (sink effects), the asymptotic velocity profile in the viscous region is exponential and has a negative curvature at the wall:

$$U_1(x_2) = U_\infty [1 - \exp(-|v_0| x_2 / \upsilon)] . \tag{3.7}$$

The displacement thickness has the <u>constant</u> value $\delta^* = \upsilon / |v_0|$, where υ is the kinematic viscosity and $|v_0|$ is the absolute value of the normal velocity at the wall. In this case, (3.6) reads:

$$C_f = 2 C_q . \tag{3.8}$$

Bussmann and Münz (1942) computed the critical Reynolds number for the above asymptotic velocity profile to be $R_{\delta^*} \equiv U_\infty \delta^* / \upsilon = 70,000$. From the value of δ^* given above, the flow is stable to all small disturbances if $C_q \equiv |v_0| / U_\infty > 1.4 \times 10^{-5}$. The amplification rate of unstable disturbances for the asymptotic profile is an order of magnitude less than that for the Blasius boundary layer (Pretsch, 1942). This treatment ignores the development distance from the leading edge needed to reach the asymptotic state. When this is included into the computation, a higher C_q (1.18×10^{-4}) is required to ensure stability (Iglisch, 1944; Ulrich, 1944). Wuest (1961) presented a summary of transpiration boundary layer computations up to the early 1960's.

The more complicated analysis for the stability of a boundary layer with suction through discrete spanwise strips was only carried out satisfactorily very recently. Reed and Nayfeh (1986) conducted a numerical-perturbation analysis of a linearized, triple-deck, closed-form basic state of a flat plate boundary layer with suction through a finite number of spanwise porous strips. Their results were compared to interacting boundary layer calculations (Ragab and Nayfeh, 1980) as well as to the carefully conducted experiments of Reynolds and Saric (1986). Suction applied through discrete strips can be as effective as suction applied continuously over a much longer streamwise length. Reed and Nayfeh suggested a scheme for optimizing the strip configuration. Their results showed that suction should be concentrated nearer the leading edge (branch I of the neutral stability curve) when disturbances are still small in amplitude.

Suction can be applied through porous surfaces, perforated plates, or carefully machined slots. It is of course structurally impossible to make the whole surface of an aircraft's wing or the like out of porous material and often strips of sintered bronze or steel are used. A relatively inexpensive woven stainless steel, Dynapore, is now available and provides some structural support (Reynolds and Saric, 1986). Superior surface smoothness and rigidity are obtained by drilling microholes in tatanium using the recently developed electron-beam technology. The lower requirement for a pressure drop in the case of a perforated plate translates directly into pumbing-power saving. However, outflow problems may result from regions of the wing having strong pressure gradients (Saric and Reed, 1986). Outflow in the aft region of a suction strip can cause large destabilizing effects and local three-dimensionality.

While structurally a surface with multiple slits is more rigid than a porous surface, slots are more expensive to fabricate accurately. Moreover, the higher mass flow rates associated with them may result in high Reynolds number instabilities such as separation and backflow, which adversely affect the stability of the basic flow. The rule of thumb is that the Reynolds number based on slot width (or hole diameter in the case of a perforated plate) and the local suction velocity should be kept below 10 to avoid adverse effects on the boundary layer stability, although Saric and Reed (1986) claim a hole Reynolds number an order of magnitude higher than that without destabilization of the basic flow.

Delaying transition using suction is a mature technology, most of the remaining problems are in the maintainability and reliability of suction surfaces and the optimization of suction rate and distribution. To protect the delicate suction surfaces on the wing of an aircraft from insect impacts and ice formation at low altitudes, special leading edge systems are used (Wagner and Fischer, 1984; Wagner et al., 1984). Suction is less suited for underwater vehicles because of the abundance of suspended ocean particulate that can clog the suction surface as well as destabilize the boundary layer.

3.3.3 SHAPING

The third method of control to delay laminar-to-turbulent transition is perhaps the simplest and involves the use of suitably shaped bodies to manipulate the pressure distribution. In (3.5), the pressure gradient term can affect the sign of the curvature of the velocity profile at the wall and, hence, change the stability characteristics of the boundary layer. According to the calculations of Schlichting and Ulrich (1940), the critical Reynolds number based on displacement thickness and freestream velocity changes from about 100 to 10,000 as a suitably nondimensionalized pressure gradient (the shape factor, Λ , defined in Section 4.1) varies from $\Lambda = -6$ (adverse) to $\Lambda = +6$ (favorable). Moreover, for the case of a favorable pressure gradient, no unstable waves exist at infinite Reynolds number. In contrast, the upper branch of the neutral stability curve in the case of an adverse pressure distribution tends to a non-zero asymptote so that a finite region of wavelengths at which disturbances are always amplified remains even as Re → ∞.

Streamlining a body to prevent separation and reduce form drag is quite an old art, but the stabilization of a boundary layer by pushing the longitudinal location of the pressure minimum to as far back as possible dates back to the 1930's and led to the successful development of the NACA 6-Series NLF airfoils. Newer, low-Reynolds-number lifting surfaces used in sailplanes, low-speed drones and executive business jets have their maximum thickness point far aft of the leading edge. The recent success of the Voyager's nine-day, unrefueled flight around the world was due in part to a wing design employing natural laminar flow to approximately 50% chord. Application of NLF technology to underwater vehicles is feasible but somewhat more limited (Granville, 1979).

The favorable pressure gradient extends to the longitudinal location of the pressure minimum. Beyond this point, the adverse pressure gradient becomes steeper and steeper as the peak suction is moved further aft. For an airfoil, the desired shift in the point of minimum pressure can only be attained in a certain narrow range of angles of incidence. Depending on the shape, angle of attack, Reynolds number, surface roughness and other factors, the boundary layer either becomes turbulent shortly after the point of minimum pressure or separates first then undergoes transition. One of the design goals of NLF is to maintain attached flow in the adverse pressure gradient region and some method of separation control (Section 4) may have to be used there.

Factors that limit the utility of NLF include crossflow instabilities and leading edge contamination on swept wings, insect and other particulate debris, high unit Reynolds numbers at lower cruise altitudes, and performance degradation at higher angles of attack due to the necessarily small leading edge radius of NLF airfoils. Improvement of surface waviness and smoothness of modern production wings, special leading edge systems to prevent insect impacts and ice formation, higher cruise altitudes of newer airplanes, and higher Mach numbers all favor the application of NLF (Runyan and Steers, 1980). To paraphrase a recent statement by Holmes (1988), an NLF airfoil is no longer as finicky as Morris the Cat. It is true that a boundary layer that is kept laminar to extremely high Reynolds numbers is very sensitive to environmental factors such as roughness, freestream turbulence, radiated sound, etc. However, the flow is durable and reliable within certain conservative design corridors which must be maintained by the skillful designer and eventual operator of the vehicle. Current research concentrates on understanding the achievability and maintainability of natural laminar flow, expanding the practical applications of NLF technology, and extending the design methodology to supersonic aviation (Bushnell and Malik, 1988; Bushnell, 1989).

3.3.4   WALL HEATING/COOLING

The last of the stability modifiers is the addition or removal of heat from a surface, which causes the viscosity to vary with distance from the wall. In general, viscosity increases with temperature for gases, while the opposite is true for liquids. Thus, if heat is removed from the surface of a body moving in air, the fifth term on the left-hand side of (3.5) is negative. In that case, the velocity gradient near the wall increases and the velocity profile becomes fuller and more stable. The term containing the viscosity derivative will also be negative if the surface of a body moving in water is heated. With heating in water or cooling in air, the critical Reynolds number is increased, the range of amplified frequencies is diminished and the amplification rate of unstable waves is reduced. Substantial delay of transition is feasible with a surface that is only a few degrees hotter (in water) or colder (in air) than the freestream.

The first indirect evidence of this phenomenon was the observation that the drag of a flat plate placed in a wind tunnel increases by a large amount when the plate is heated (Linke, 1942). Both Frick and McCullough (1942) and Liepmann and Fila (1947) showed that the transition location of a flat-plate boundary layer in air at low subsonic speeds is moved forward as a result of surface heating. The stability calculations of Lees (1947) confirmed these experiments and, moreover, showed that cooling has the expected opposite effects. The critical Reynolds number based on distance from the leading edge increases from 10^5 to 10^7 when the wall of a flat plate placed in an air stream is cooled to 70% of the absolute ambient temperature. Even a modest cooling of the wall to $0.95T_\infty$ results in doubling of the critical Reynolds number (Kachanov et al., 1974). With cooling, the range of amplified frequencies is diminished and the growth rate of T-S waves is reduced resulting in a substantial decrease in transition Reynolds number. These same trends were dramatically confirmed in subsonic and supersonic flights[*] by Dougherty and Fisher (1980) who studied the transition on an airborne cone at the Mach number range of 0.55 - 2.0. They reported a transition Reynolds number that varied approximately as T_0^{-7}, where T_0 is the wall temperature. For aircraft, this method of transition delay is feasible only for a vehicle which uses a cryo-fuel such as liquid hydrogen or liquid methane. In that case, a sizeable heat sink is readily available. The idea being that the fuel is used to cool the major aerodynamic surfaces of the aircraft as it flows from the fuel tanks to the engines. Reshotko (1979) examined the prospects for the method and concluded that, particularly for a hydrogen-fueled aircraft, substantial drag reductions are feasible. His engineering calculations indicated that the weight of the fuel saved is well in excess of the weight of the required cooling system.

The above effects are more pronounced in water flows due to the larger Prandtl number (good thermal coupling) and the stronger dependence of viscosity on temperature.[**] In a typical low-speed situation , a surface heating of 1 °C in water has approximately the same effect on the curvature of the velocity profile at the wall as a surface cooling of 20°C in air or a suction coefficient (in air or water) of 0.0003 (Liepmann et al., 1982). Wazzan et al. (1968;1970) used a modified fourth-order Orr-Sommerfeld equation combined with the e^9 method of Smith (1957) and confirmed that wall heating can produce large increases in the transition Reynolds number of water boundary layers. They predicted a transition Reynolds number, based on freestream velocity and distance from the leading edge of a flat-plate, as high as 2×10^8 for wall temperatures that are only 40°C above the ambient water temperature.

---

[*] In hypersonic flows, a different mode of stability, Mack's Second Mode, dominates the transition process and cooling, in fact, promote earlier transition to turbulence.

[**]For water at room temperature, $Pr \simeq 7$ and absolute viscosity is decreased by approximately 2% for each 1°C rise in temperature. For room temperature air, $Pr \simeq 0.7$ and absolute viscosity is decreased by approximately 0.2% for each 1°C drop in temperature.

Lowell and Reshotko (1974) refined Wazzan et al.'s (1968; 1970) calculations by introducing a coupled sixth-order system of vorticity and energy disturbance equations. The predicted critical Reynolds numbers for wall overheats of up to 2.8 °C were confirmed by the experiments of Strazisar et al. (1977) who measured the growth rates of small disturbances generated by a vibrating ribbon in a heated flat-plate in water. These experiments did not yield data on transition or on stability at higher overheats.

The transition predictions of Wazzan et al. (1968; 1970) at higher overheats were partially confirmed by the very carefully conducted experiments of Barker and Gile (1981) who used the entrance region of an electrically heated pipe. The displacement thickness was much smaller than the pipe radius and, thus, the boundary-layer development was approximately the same as that of a zero-pressure gradient flat-plate. Barker and Gile reported a transition Reynolds number of 4.7×10^7 for a wall overheat of 8 °C. No further increase in $Re_{transition}$ was observed as the wall was heated further, in contradiction to the computations of Wazzan et al. (1968; 1970). Barker and Gile (1981) investigated possible causes of this discrepancy including buoyancy effects, wall roughness, effects of geometry, flow asymmetries, and suspended particulate matter. Their analysis and numerous subsequent work (e.g., Chen et al., 1979; Kosecoff et al., 1976; Hendricks and Ladd, 1982; Lauchle and Gurney, 1984) indicate that increased concentration and size of suspended particulate diminish the stabilizing effect of surface heating until at some point surface heating no longer stabilizes the boundary layer but is in fact a destabilizing influence.

On a heated body of revolution in a high-speed water tunnel, Lauchle and Gurney (1984) observed an increase in transition Reynolds number from 4.5×10^6 to 3.6×10^7 for an average overheat of 25 °C. Clearly, surface heating in water can be an extremely effective method of transition delay and, hence, drag reduction for small, high-speed underwater vehicles where the reject heat from their propulsion system is used to increase the surface temperature along the body length. The detrimental effects of freestream particulate alluded to above are, however, a major obstacle at present for a practical implementation of this method of control. Suspended particulate having a wide-band concentration spectra are abundant in the oceans and "particle-defense" mechanisms must be sought before using any of the transition delay methods in a contaminated environment.

In addition to surface heating, several other techniques are available to lower the near-wall viscosity ($\partial\mu / \partial x_2 > 0$) in a water boundary layer and, thus, favorably affect the stability of the flow. These include film boiling, cavitation, sublimation, chemical reaction, or wall injection of a gas or lower-viscosity liquid. Finally, a shear-thinning additive could be introduced into the boundary layer. Since the shear increases as the wall is approached, the effective viscosity of the non-Newtonian fluid decreases there and $\partial\mu / \partial x_2$ becomes positive.

3.4 WAVE CANCELLATION

An alternative approach to increase the transition Reynolds number of a laminar boundary layer is wave cancellation. If the frequency, orientation and phase angle of the dominant element of the spectrum of growing linear disturbances in the boundary layer is detected, a control system and appropriately located disturbance generators may then be used to effect a desired cancellation or suppression of the detected disturbances. In this case, the stability characteristics of the boundary layer are exploited but not altered (Reshotko, 1985). Wave cancellation is feasible only when the disturbances are still relatively small, their growth is governed by a linear equation, and the principle of superposition is still valid.

The first reported use of wave cancellation is that due to Schilz (1965/66). He used a vibrating ribbon to excite a T-S wave on a test plate which had a flexible surface. A unique wall-motion device flush mounted into the plate moved the flexible wall in a transverse, wavelike manner with a variety of frequencies and phase speeds. Significant amount of cancellation resulted when the flexible wall motion had the opposite phase but the same frequency and phase speed as the T-S wave. Both Milling (1981) and Thomas (1983) used two vibrating wires, one downstream of the other, to generate and later cancel a single frequency T-S wave. Thomas (1983) observed that interaction between the primary disturbance and background excitations prevented complete cancellation of the primary wave. To further study the consequences of wave interactions, Thomas applied the same method of control to eliminate two interacting waves of different frequency. Although the primary waves were behaving linearly, a nonlinear interaction gave rise to a low-amplitude difference frequency that could only be partially reduced and ultimately led to transition. Thomas (1983) concluded that it is not possible to return the flow completely to its undisturbed base state because of wave interactions and that it is perhaps more appropriate to describe this control method as wave superposition rather than wave cancellation.

The same principle of wave superposition could be applied using wall heating/cooling (Liepmann et al., 1982; Liepmann and Nosenchuck, 1982; Ladd and Hendricks, 1988), plate vibration (Gedney, 1983), compliant wall (McMurray et al., 1983), or periodic suction/blowing (Biringen, 1984). Liepmann and Nosenchuck (1982) used flush-mounted hot-film probes to sense natural T-S waves in a flat-plate boundary layer in a water tunnel. A feed-forward control loop was then used to synthesize and introduce disturbances of equal amplitude but of opposite phase via flush-mounted wall heaters. Quite recently, Ladd and Hendricks (1988) performed their experiment in a water tunnel on a 9:1 fineness-ratio ellipsoid. Strip heaters were again used to create and actively attenuate T-S waves. They applied digital filtering techniques to synthesize the attenuation signal. The filter was able to actively adapt the attenuation signal to changes in amplitude and frequency of the artificially introduced instability wave with no loss in attenuation downstream. A sample of Ladd and Hendrick's results is shown in Figure 4. Time records of a flush-mounted hot-film probe are shown for the natural flow, artificially introduced T-S wave using a heating ring, and canceled wave using downstream heating ring.

The transition delay achieved by active wave cancellation is modest, typically a factor of two or less increase in the transition Reynolds number based on distance from the leading edge. Reshotko (1985) maintains that to achieve significant delay in transition using this technique would require an extensive array of disturbance detectors and generators as well as prohibitively complicated control system that could cancel both the primary and residual disturbance spectra. Significant delay in transition is more readily achieved via the stability modifiers summarized in Section 3.3.

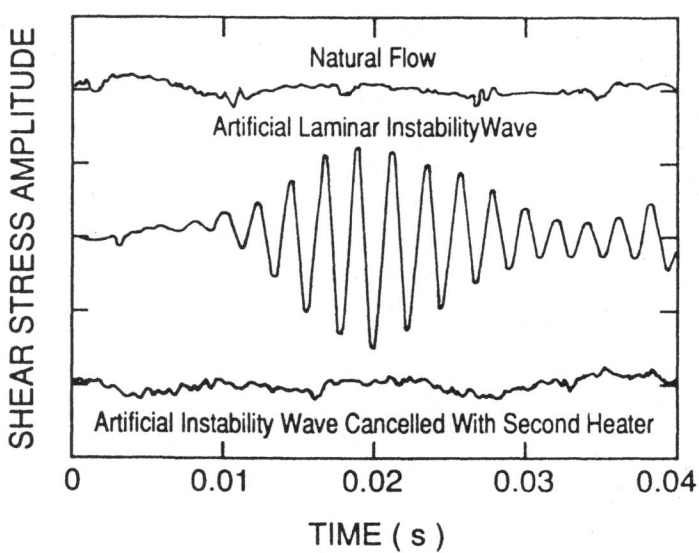

Figure 4. Cancellation of Artificial Instability Waves Using
Adaptive Heat. (from Ladd and Hendricks, 1988).

4. SEPARATION POSTPONEMENT AND LIFT ENHANCEMENT

Fluid particles in a boundary layer are slowed down by wall friction. If the external potential flow is sufficiently retarded, for example due to the presence of an adverse pressure gradient, the momentum of those particles will be consumed by both the wall shear and the pressure gradient. At some point (or line), the viscous layer departs or breaks away from the bounding surface. The surface streamline nearest to the wall leaves the body at this point and the boundary layer is said to separate (Maskell, 1955). At separation, the rotational flow region next to the wall abruptly thickens, the normal velocity component increases, and the boundary-layer approximations are no longer valid.

Due to the large energy losses associated with boundary-layer separation, the performance of many practical devices is often controlled by the separation location. For example, if separation is postponed, the pressure drag of a bluff body is decreased, the circulation and hence the lift of an airfoil at high angle of attack is enhanced, and the pressure recovery of a diffuser is improved.

Prandtl (1904) was the first to explain the mechanics of separation. He provided a precise criterion for its onset for the case of a steady, two-dimensional boundary layer developing over a fixed wall. In case such a flow is retarded, the near-wall fluid may have insufficient momentum to continue its motion and will be brought to rest at the point of separation. Fluid particles behind this point move in a direction opposite to the external stream and the original boundary-layer fluid passes over a region of recirculating flow. Since the velocity at the wall is always zero, the gradient $[\partial U_1 / \partial x_2]_0$ will be positive upstream of separation, zero at the point of separation, and negative in the reverse flow region. The velocity profile at separation must then have a positive curvature at the wall. However, $[\partial^2 U_1 / \partial x_2^2]$ is negative at a large distance from the wall, which means the velocity profile at separation must have a point of inflection somewhere above the wall as shown in Figure 2d. Since $[\partial^2 U_1 / \partial x_2^2]_0 > 0$ is a necessary condition for a steady, two-dimensional boundary layer to separate, the opposite, i.e., a negative curvature of the velocity profile at the wall, is a sufficient condition for the boundary-layer flow to remain attached.

4.1 SEPARATION CONTROL

The above arguments naturally lead to several possible methods of control to delay separation. Namely, the object is to keep $[\partial^2 U_1 / \partial x_2^2]_0$ as negative as possible, in other words to make the velocity profile as full as possible. In this case, the spanwise vorticity decreases monotonically away from the wall and the surface vorticity flux is in the positive x_2 direction. Not surprisingly, then, methods of control to postpone separation are similar to those used to delay transition. Rewrite (3.5) for a steady flow and fixed wall:

$$\left[\rho U_2 \frac{\partial U_1}{\partial x_2}\right]_0 + \frac{dP}{dx_1} - \frac{d\mu}{dT}\left[\frac{\partial T}{\partial x_2}\frac{\partial U_1}{\partial x_2}\right]_0 = \mu\left[\frac{\partial^2 U_1}{\partial x_2^2}\right]_0 . \qquad (4.1)$$

Separation control methods include the use of wall suction ($[U_2]_0 < 0$), favorable pressure gradient ($dP/dx_1 < 0$), surface cooling in gases ($d\mu/dT > 0$; $[\partial T/\partial x_2]_0 > 0$), or surface heating in liquids ($d\mu/dT < 0$; $[\partial T/\partial x_2]_0 < 0$). Obviously any or a combination of these methods could be used in a particular situation. For example, beyond the point of minimum pressure on a streamlined body the pressure gradient is adverse and the boundary layer may separate if the pressure rise is sufficiently steep; however, enough suction could be applied there to overcome the retarding effects of the adverse pressure and to prevent separation. Each of these control methods is covered in more details in the following.

For any two-dimensional, closed-surface body, adverse pressure gradient must always occur in subsonic flow. Streamlining, quite an old art, can greatly reduce the steepness of the pressure rise. In supersonic flows, pressure always rises across a shock wave and a boundary layer may separate as a result of the wave interaction with the viscous flow (Young, 1953; Lange, 1954). Laminar boundary layers can only support very small adverse pressure gradients without separation. In fact, if the ambient incompressible fluid decelerates in the streamwise direction faster than $U_\infty \sim x_1^{-0.09}$, the flow separates (Schlichting, 1979). On the other hand, a turbulent boundary layer, being an excellent momentum conductor, is capable of overcoming much larger adverse pressure gradient without separation. In this case, separation is avoided for external flow deceleration up to $U_\infty \sim x_1^{-0.23}$. The efficient momentum transport that characterizes turbulent flows provides the mechanism for mixing the slower fluid near the wall with the faster fluid particles further out. The forward movement of the boundary-layer fluid against pressure and viscous forces is facilitated and separation is, thus, postponed. According to the experimental results of Schubauer and Spangenberg (1960), a larger total pressure increase without separation is possible in the turbulent case by having larger adverse pressure gradient in the beginning and continuing at a progressively reduced rate of increase.

The second method to avert separation involves withdrawing the near-wall fluid through slots or porous surfaces. Prandtl (1904) applied suction through a spanwise slit on one side of a circular cylinder. His flow visualization photographs convincingly showed that the boundary layer adhered to the suction side of the cylinder over a considerably larger portion of its surface. By removing the decelerated fluid particles in the near-wall region, the velocity gradient at the wall is increased, the curvature of the velocity profile near the surface becomes more negative, and separation is avoided. In the following, an approximate method to compute the amount of suction needed to prevent laminar separation is briefly recalled.

For a laminar boundary layer, the ratio of pressure forces to viscous forces is proportional to the shape factor, Λ :

$$\Lambda = \frac{\delta^2}{\upsilon} \frac{dU_\infty}{dx_1} = -\frac{dP}{dx_1} \frac{\delta^2}{\mu U_\infty} , \qquad (4.2)$$

where δ is the boundary layer thickness, U_∞ is the freestream velocity, υ is the kinematic viscosity, μ is the dynamic viscosity, and dP/dx_1 is the pressure gradient. At separation,

$[\partial U_1 / \partial x_2]_0 = 0$ and Equation (4.1) reads:

$$\frac{dP}{dx_1} = \mu \left[\frac{\partial^2 U_1}{\partial x_2^2} \right]_0 .$$ (4.3)

The shape factor at the point of separation of a laminar boundary layer is $\Lambda = -12$, and from (4.2) and (4.3) the expressions for the curvature of the velocity profile at the wall and the boundary-layer thickness become, respectively:

$$\left[\frac{\partial^2 U_1}{\partial x_2^2} \right]_0 = \frac{12 \, U_\infty}{\delta^2} ,$$ (4.4)

$$\delta = \sqrt{\frac{-12 \, \upsilon}{d \, U_\infty / d \, x_1}} .$$ (4.5)

The freestream velocity distribution, $d \, U_\infty / d \, x_1$, is determined from the potential flow solution. As an example, suppose we wish to compute the suction coefficient, C_q, which is just sufficient to prevent laminar separation from the surface of a cylinder. By assuming that the velocity profiles along the whole length are identical with that at the point of separation and computing $d \, U_\infty / d \, x_1$ at the downstream stagnation point, Prandtl (1935) used the momentum integral equation, (2.10), and the above results to make a simple estimate of the required suction:

$$C_q = 4.36 \, Re^{-0.5} ,$$ (4.6)

where Re is the Reynolds number based on the cylinder diameter and the freestream velocity.

Several researchers used similar approximate methods to calculate the laminar boundary layer on a body of arbitrary shape with arbitrary suction distribution (see, e.g., Chang, 1970; Schlichting, 1979). A particularly simple calculations is due to Truckenbrodt (1956). He reduces the problem to solving a first-order ordinary differential equation. As an example, for a symmetrical Zhukovskii airfoil with uniform suction, Truckenbrodt predicts a suction coefficient just sufficient to prevent separation of:

$$C_q = 1.12 \, Re^{-0.5} ,$$ (4.7)

where Re is the Reynolds number based on the airfoil chord and the freestream velocity.

For turbulent boundary-layers, semi-empirical methods of calculations are inevitably used due to the closure problem alluded to in Section 2. Suction coefficients in the range of $C_q = 0.002 - 0.004$ are sufficient to prevent separation on a typical airfoil (Schlichting, 1959; Schlichting and Pechau, 1959). Optimally, the suction should be concentrated on the low-pressure side of the airfoil just a short distance behind the nose where, at large angles of attack, the largest local adverse pressure gradient occurs.

The third term in (4.1) points to yet another method to delay boundary-layer separation. By transferring heat from the wall to the fluid in liquids or from the fluid to the wall in gases, this term adds a negative contribution to the curvature of the velocity profile at the wall and, hence, causes the separation point to move farther aft. Although this method of control has been successfully applied to delay transition in both water and air flows (Section 3.3.4), its use to prevent separation has been demonstrated only for high-speed gaseous flows. If the surface of a body in a compressible fluid is cooled, the near-wall fluid will have larger density and smaller viscosity than that in the case with no heat transfer. The smaller viscosity results in a fuller velocity profile and higher speeds near the wall. Combined with the larger density, this yields a higher momentum for the near-wall fluid particles and, hence, the boundary layer becomes more resistant to separation. These effects are demonstrated in the analytical results of Libby (1954), Illingworth (1954), and Morduchow and Grape (1955). Experimental verification are provided by the work of Gadd et al. (1958), Bernard and Siestrunck (1958), and Lankford (1960; 1961). Excellent summaries of the problem of heat transfer effects on the separation of a compressible boundary layer are available in Gadd (1960) and Chang (1970).

4.2   TIME-DEPENDENT AND 3-D SEPARATIONS

Unlike the relatively simple case of steady, two-dimensional (or axisymmetric) boundary-layer flow over a fixed wall treated above, the point of vanishing wall shear and the occurrence of reverse flow do not necessarily indicate separation for two-dimensional flow over moving walls, two-dimensional unsteady flows, or three-dimensional steady flows, and a more complex criterion for separation must be sought (Chang, 1970; Williams, 1977; Gad-el-Hak, 1987a). A comparison of streamlines and velocity profiles for stationary and moving walls illustrate this point. In Figure 5, separation in the case of a moving wall occurs when the velocity and its gradient simultaneously vanish at some point above the wall. Separation is delayed when the wall moves downstream and is advanced when the wall moves upstream.

The moving-wall effects can be exploited to postpone separation. Prandtl (1925) demonstrated the effects of rotating a cylinder placed in a uniform stream at right angles to its axis. Separation is completely eliminated on the side of the cylinder where the wall and the freestream move in the same direction. On the other side of the cylinder separation is developed only incompletely. In fact, for high enough values of circulation, the entire flow field can be approximated by the potential flow theory. The asymmetry causes a force on the cylinder at right angle to the mean flow direction. This important phenomenon, known as the Magnus effect, is exploited in several sport balls (Mehta, 1985a) and even in an experimental device used for propelling ships, known as the Flettner's (1924) rotor. From a practical point of view, wall motion for body shapes other than circular cylinders or spheres is prohibitively complicated, although it is feasible to replace a small portion of the surface of an airfoil by a rotating cylinder thus energizing the boundary-layer and avoiding separation (Alvarez-Calderon, 1964).

For time-dependent flows, the separation point is no longer stationary but rather moves along the surface of the body. The Moore-Rott-Sears (MRS) criterion states that unsteady separation occurs when

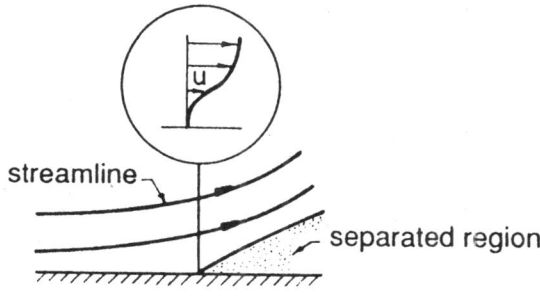

a. Fixed Wall.

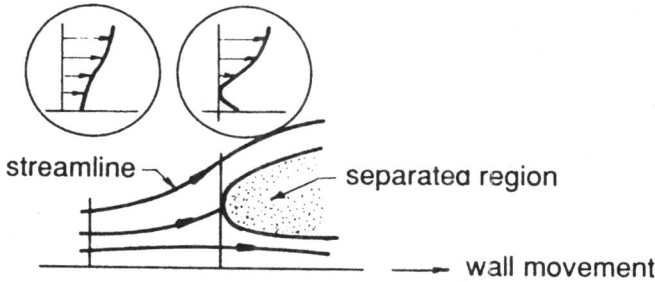

b. Wall Moving Downstream.

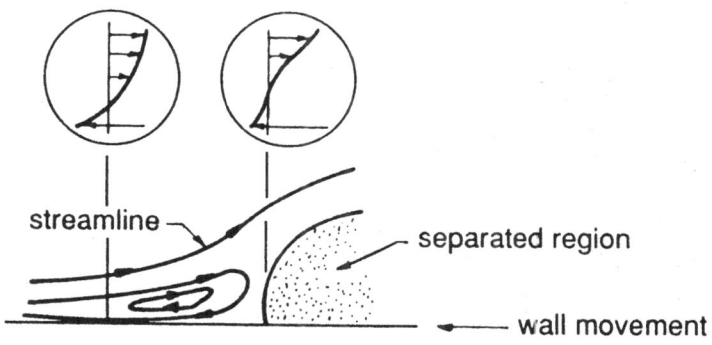

c. Wall Moving Upstream.

Figure 5. Streamlines and Velocity Profiles When a Boundary Layer over Fixed or Moving Wall Separates.

the shear and velocity vanish simultaneously and in a singular fashion at a point within the boundary layer as seen by an observer moving with the separation point (Rott, 1956; Sears, 1956; Moore, 1958). Steady separation is clearly included in this model as a special case. The main drawback of the MRS criterion is that the speed of the separation point is not known a priori, making it difficult to locate this point and forcing researchers to rely on more qualitative measures for unsteady separation.

In analogy to the moving wall case, unsteady separation is postponed when the separation point moves upstream as is the case on the suction side of an airfoil undergoing a pitching motion from small to large angles of attack. Conversely, when an airfoil is pitched from large to small attack angle, the separation point on the suction side moves downstream and separation is advanced, much the same as the case of a wall moving upstream.

An airfoil oscillating sinusoidally through high angles of attack can produce very high lift coefficients and maintain flow attachment well beyond static stall attack angles (McCroskey, 1977; 1982). During the upstroke, the separation point moves upstream and reverse flow exists in an attached and mathematically well-behaved boundary layer. The global aerodynamic properties of a pitching airfoil are strongly influenced by the local unsteady separation. Sudden changes of lift, drag, and pitching moment occur near the onset of separation. Moment stall is observed when the reverse-flow region extends over most of the airfoil and a large-scale vortex is formed near the leading edge. During this phase of the cycle, lift continues to increase. Lift stall follows moment stall and occurs when the separation vortex reaches the latter half of the airfoil. In other words, the suction on the upper surface of the airfoil continues to increase at the initial stages of separation, and a sudden decrease in suction does not occur until the leading edge is in the wake of separation. Similar phenomena are observed on three-dimensional lifting surfaces undergoing pitching motion (Gad-el-Hak, 1986a; 1988a; 1988b; Gad-el-Hak and Ho, 1985; 1986a; 1986b).

Insects, most of whom mate and eat while airborne, exploit unsteady separation effects to achieve remarkable aerodynamic characteristics. The dragonfly, in existence for approximately 250 million years, presumably survived innumerable life and death aerodynamic struggles (Luttges et al., 1984). The enviably large lift coefficients generated by the chalcid wasp during hovering suggest the existence of an efficient unsteady lift generation mechanism (Weis-Fogh, 1973; Lighthill, 1973; Maxworthy, 1979; 1981).

As mentioned earlier, the point of boundary-layer separation from a three-dimensional body does not necessarily coincide with the point of vanishing wall shear. Instead, the shear stress at the wall is equal to zero only at a limited number of points along the separation line. The number and type of these critical or singular points must satisfy certain topological laws (Lighthill, 1963; Tobak and Peake, 1982). The projection of the limiting streamlines nearest to the wall coincides with the skin-friction lines on the surface of the body. Oil-streak techniques and the like are usually used to obtain separation and attachment patterns (Maltby, 1962). A necessary condition for the occurrence of flow separation is the convergence of skin-friction lines onto a particular line. Because of the three-dimensionality of the flow, the near-wall fluid may move in a direction in which the pressure gradient is more favorable and not against the adverse pressure in the direction of the main flow as is the case for two-dimensional flows.

Consequently, three-dimensional boundary-layers are in general more capable of overcoming an adverse pressure gradient without separation.

4.3   ADDITIONAL CONTROL METHODS

In his 1976 monograph, Chang reviews several other passive and active methods to postpone separation for low- and high-speed flows. Common to all these control methods is an attempt to supplying additional energy to the near-wall fluid particles which are being retarded in the boundary layer. Passive techniques do not require auxiliary power, but do have an associated drag penalty, and include intentional tripping of transition from laminar to turbulent flow upstream of what would be a laminar separation point (Section 6), boundary-layer fences to prevent separation at the tips of swept-back wings, placing an array of vortex generators on the body to raise the turbulence level and enhance the momentum and energy in the neighborhood of the wall (Mehta, 1985b; Rao and Kariya, 1988), geometric design to avoid shock-induced separation, machining a series of lateral grooves on the surface of the body upstream of separation, or using a screen to divert the flow and increase the velocity gradient at the wall. Howard and Goodman (1985; 1987) recently investigated the effectiveness of two passive techniques to reduce the flow separation, transverse rectangular grooves and longitudinal V-grooves placed in the aft shoulder region of a bluff body. Both types of grooves were beneficial in reducing the form drag on a body at zero and moderate angles of yaw.

Active methods to postpone separation require energy expenditure. Obviously, the energy gained by the effective control of separation must exceed that required by the device. In addition to suction or heat transfer reviewed earlier, fluid may be injected parallel to the wall to augment the shear-layer momentum or normal to the wall to enhance the mixing rate. Either a blower is used or the pressure difference that exists on the aerodynamic body itself is utilized to discharge the fluid into the retarded region of the boundary layer. The latter method is found in nature in the thumb pinion of a pheasant, the split-tail of a falcon, or the layered wing feathers of some birds. In man-made devices, passive blowing through leading-edge slots and trailing-edge flaps is commonly used on aircraft wings. Although in this case direct energy expenditure is not required, the blowing intensity is limited by the pressure differentials obtainable on the body itself. In some applications, the injected fluid differs from the freestream fluid. Typically, a lighter gas is injected to reduce the rate at which heat is exchanged between the wall and the external stream and, thus, providing thermal protection at high supersonic velocities.

More recently developed active methods for controlling boundary layer separation include the use of acoustic excitations (Ahuja et al., 1983), oscillating surface flaps (Koga et al., 1984) and oscillatory surface heating (Maestrello et al., 1988). Ahuja and his colleagues successfully demonstrated that sound at a preferential frequency can postpone the turbulent separation on an airfoil in both pre- and post-stall regimes. The optimum frequency was found to be $4\,U_\infty\,/\,c$ (Strouhal number = 4), where U_∞ is the freestream velocity and c is the airfoil chord. Goldstein (1984) speculates that the delay in separation in Ahuja et al.'s (1983) experiment resulted from enhanced entrainment promoted by instability waves that were triggered on the separated shear layers by the acoustic excitation. Koga et al. (1984) used a computer-controlled spoiler-like flap in a flat-plate turbulent boundary layer with and without modelled upstream separation. They were able to manipulate the separated flow region and its reattachment length characteristics by varying the frequency, amplitude and waveform of the oscillating flap. Reynolds and Carr (1985) offer a plausible explanation, from the viewpoint of a vorticity framework, for the experimental observations of Koga et al. (1984).

On a sharp-leading-edge delta wing, the separation position is fixed and a strong shear layer is formed along the entire edge. The shear layer is wrapped up in a spiral fashion, which results in a large-bound vortex on each side of the wing. The two vortices appear on the suction surface of the wing in the form of an expanding helix when viewed from the apex. The low pressure associated with the vortices produces additional lift on the wing, often called nonlinear or vortex lift, which is particularly important at large angles of attack. The recent experiments of Gad-el-Hak and Blackwelder (1985) have indicated that small discrete vortices are shed parallel to the leading edge at a repeatable frequency, determined by the angle of attack and Reynolds number. Repeated vortex pairings result in the formation of progressively larger vortices. This process can be modulated by weak, periodic suction/injection through a leading edge slot. In particular, when the perturbation frequency is a subharmonic of the natural shedding frequency, the evolution of the bounded shear layer is dramatically altered (Gad-el-Hak and Blackwelder, 1987a).

Many of the control methods discussed so far in this section could be used for external as well as internal flows. Viets (1980) used an asymmetrical rotating cam embedded in the wall to produce large eddies in a turbulent boundary layer with zero- and adverse-pressure gradients. By using this device in a wide-angle diffuser, Viets et al. (1981) were able to postpone the natural separation and dramatically improve the diffuser's performance.

Although the object of this section is to review control methods to prevent separation, under certain circumstances the designer may wish to provoke separation or the flow may separate as a by-product of another primary goal of control, such as for example relaminarization (Section 5.4). Consider a thin and sharp-edged airfoil, suitable for good cruising performance at supersonic speeds. During takeoff and landing, a thicker airfoil is desired, however, to generate sufficient lift at low speeds. In this case, a free-stream flap may be used to provoke leading-edge separation followed by reattachment at the leading edge of the flap, thus forming a thicker pseudo-body with the desired aerodynamic shape as shown in Figure 6 adapted from Hurley (1961).

The detached bow shock forming upstream of a blunt body in supersonic flight may be changed into a weaker, attached oblique shock by placing a spike in front of the body. The pressure rise and the presence of a solid surface on which a boundary layer forms causes the flow to separate downstream of the spike tip. A properly designed spike may result in lower drag, higher lift, and corresponding change in pitching moment (Wood, 1961).

Periodic separation may also be provoked by changes in the wall geometry. Francis et al. (1979) initiated separation on an airfoil by periodically inserting and removing a spoiler at the wall. Viets et al. (1984) studied separation inducement and control by use of a cam-shaped rotor mounted on an airfoil. The cam, either driven or free-wheeling, periodically extended out into the flow causing the boundary layer to separate. Large-scale coherent spanwise structures were periodically generated and were responsible for the flow detachment.

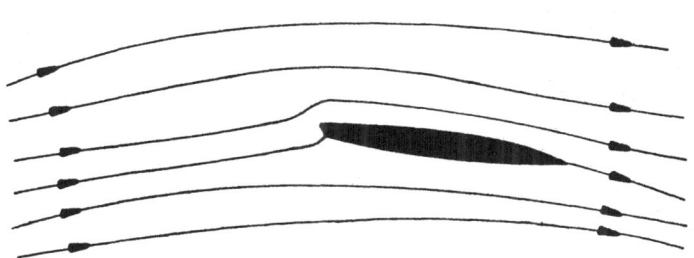

a. During Supersonic Flight

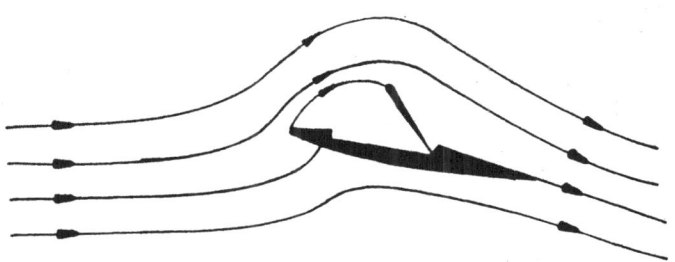

b. During Takeoff and Landing

Figure 6. Free - Stream Flap.

5. DRAG REDUCTION

5.1 INTRODUCTORY REMARKS

Whenever there is relative motion between a solid body and the fluid in which it is immersed, the body experiences a net force due to the action of the fluid. The component of this force parallel to the direction of motion is termed the drag, and the body will move at a constant speed when the thrust generated by its propulsion system is equal to the drag force. The total drag consists of skin friction, equal to the streamwise component of the integral of all shearing stresses over the body's surface, and pressure drag, equal to the streamwise component of the integral of all normal forces. Flow Separation is the major source of pressure drag with additional contributions due to displacement effects of the boundary layer, wave resistance in a supersonic flow or at an air/water interface, and drag induced by lift on a finite body. The object of this section is to review available methods to reduce the total drag. For a vehicle, the reduced drag means longer range, reduced fuel cost/volume, higher payload, or increased speed. For pipes and channels, where 100% of the drag is due to skin friction, drag reduction can result in improved throughput, reduced pumping power, or reduced duct size.

Nature provides numerous instances where drag reduction is essential for the survival of many species of air and marine animals. In here, we cite instead two examples of the importance of minimizing the drag of man-made vehicles. At present, the annual fuel bill for all commercial airlines in the United States is about $10 billion (Hefner, 1988). At subsonic cruising speeds, approximately half of the total drag of conventional takeoff and landing aircraft is due to skin friction. So, a reduction in skin-friction drag of 20% translates into an annual fuel saving of $1 billion. Not only is this sum substantial (about 60% of the entire annual budget of the National Science Foundation), but also many believe that a return of the 1973's energy crisis is inevitable (Phillips, 1979; Kannberg, 1988). While the world may have another 100-year supply of natural petroleum, it is estimated that by the year 2000, the United States supply will be virtually exhausted (Nagel et al., 1975). Fuel conservation is certainly an important tool to ward off future shortages.

The second example is from the military sector. The amount of propulsive power available for an underwater vehicle is limited by the volume allocated to its power plant and the efficiency of the various propulsive components. For these vehicles, about 90% of the total drag is due to skin friction. Accordingly, a reduction in skin-friction drag of 20% translates into an increase in speed of 6.8%. Although modest, this extra speed may be vital for the survival of a submarine being chased by another underwater vehicle.

Attempts to reduce drag goes back to antiquity as mentioned in Section 1. Streamlining and other control methods summarized in Section 4 can eliminate most of the pressure drag due to flow separation. Some drag remains, however, even when the flow remains attached to the trailing edge. Due to the displacement effects of the boundary layer, the pressure distribution around the body differs from the symmetric distribution predicted by potential flow theory. This remanent drag can be reduced by keeping the boundary layer as thin as possible.

Wave resistance and induced drag can also be reduced by geometric design. By sweeping the wings of a subsonic aircraft, drag divergence is delayed to higher Mach numbers, thus allowing the aircraft to fly at higher speeds without experiencing a sudden increase in drag. Additionally, the so-called area rule or coke-bottle effect typically leads to a factor of 2 reduction in wave drag at Mach number of 1 (Whitcomb, 1956). For surface ships, where typically half the drag is due to wave resistance, bow and stern bulbs can reduce the energy dissipated into waves at the air/water interface. An even simpler solution to reduce wave drag is to operate the ship well below the hull design speed, as is the case for supertankers. The induced drag of an aircraft's wing, about 25% of the airplane's total drag at subsonic cruising speeds, is inversely proportional to its aspect ratio and, hence, a lifting surface is typically designed with as large an aspect ratio as permissible by structural considerations and desired degree of manueverability. End plates or other vortex diffusers can also be used to further reduce the induced drag.

Most of the current research efforts are directed towards reducing the skin-friction drag, and this topic will occupy the remainder of this section. According to Bushnell (1983), the leverage in this area of research is quite considerable and justifies the study of unusual or high-risk approaches on an exploratory basis.

Let us start by estimating typical Reynolds number for the vehicles whose drag is to be minimized. A commercial aircraft travelling at a speed of 300 m/s would have a unit Reynolds number of 2×10^7 /m at sea level and 1×10^7 /m at an altitude of 10 km. Due to the much smaller kinematic viscosity of water, an underwater vehicle moving at a modest speed of 10 m/s ($\approx$ 20 knots) would have the same unit Reynolds number of 1×10^7 /m.

Three flow regimes can be identified for the purpose of reducing skin friction. First, if the flow is laminar, typically at Reynolds numbers based on distance from leading edge $< 10^6$, then methods of reducing the laminar shear stress are sought. This topic will be covered in Section 5.2. Secondly, in the range of Reynolds numbers from 1×10^6 to 4×10^7, active and passive methods to delay transition as far back as possible are sought. These techniques were reviewed in Section 3, and can result in substantial savings. As shown in Figure 1, the skin-friction coefficient in the laminar flat plate can be as much as an order of magnitude less than that in the turbulent case. Note, however, that all the stability modifiers discussed in Section 3.3 result in an increase in the skin friction over the unmodified Blasius layer. The object is, of course, to keep the penalty below the saving, i.e., the net drag will be above that of the flat-plate laminar boundary-layer but well below the viscous drag in the flat-plate turbulent flow. Thirdly, for Re $> 4 \times 10^7$, transition to turbulence cannot be delayed with any known practical method without incurring a penalty that exceeds the saving. The task is then to reduce the skin-friction coefficient in a turbulent boundary layer. This topic will be covered in Section 5.3. Relaminarization will be covered in Section 5.4, although achieving a net saving is problematic at present.

For a vehicle with a unit Reynolds number of 1×10^7 /m, the first regime exists in the first few centimeters and, hence, is of no great consequence. Transition delay is feasible on the wing and other appendages of an aircraft but not on its much longer fuselage. On the fuselage, where half of the skin-friction drag takes place, some alteration of the turbulence structure is sought. Short underwater bodies, such as torpedoes, are ideal target for applying transition control methods. On the much longer submarine, these techniques are feasible on the first few meters of its surface. Beyond that, turbulence drag reduction or relaminarization is sought.

5.2 LAMINAR BOUNDARY LAYERS

In the absence of transition promoters, such as cross flow, concave surface, adverse pressure gradient, roughness or freestream disturbances, the boundary layer is laminar to $Re = 0 [10^6]$. In this case, methods to reduce the laminar skin-friction are sought. This may be useful for some land vehicles, airborne vehicles at very high altitude, or small hang-gliders and the like, but obviously large underwater or air vehicles cannot benefit from these techniques.

Recall the definition of local skin-friction coefficient for a Newtonian fluid:

$$C_f \equiv \frac{2 \tau_0}{\rho U_\infty^2} = \frac{2 \upsilon}{U_\infty^2} \left[\frac{\partial U_1}{\partial x_2} \right]_0 , \qquad (5.1)$$

where τ_0 is the wall shear stress and $[\partial U_1 / \partial x_2]_0$ is the slope of the velocity profile (or the spanwise vorticity) at the wall. From equations (2.10) and (3.5), it is clear that any or a combination of the following techniques can be used to make the curvature of the velocity profile at the wall more positive and, thus, lower C_f ; wall motion such that the first three terms on the left-hand side of (3.5) are positive; injection of fluid normal to the wall $\left(v_0 = [U_2]_0 > 0 \right)$; adverse pressure gradient $(dP / d x_1 > 0)$; wall heating in air $(d\mu / dT > 0; [\partial T / \partial x_2]_0 < 0)$; or wall cooling in water $(d\mu / dT < 0; [\partial T / \partial x_2]_0 > 0)$. Note that any of these methods will promote flow instability (Section 3.3) and separation (Section 4.1). These tendencies have to be carefully considered when deciding how far to go with the attempt to lower C_f.

Two other techniques can be used to lower laminar skin-friction. Narasimha and Ojha (1967) considered the higher order effects of moderate longitudinal surface curvature. Their similarity solutions show a definite decrease in skin friction when the surface has convex curvature in all cases including zero pressure gradient. Narasimha and Ojha attributed the decrease in C_f to the fact that the velocity in the potential flow region tends to decrease away from the surface. The second technique is used in rarefied gas flows. Appropriate surface preparation could be used to introduce a slip velocity at the wall and, thus, lower the tangential momentum accommodation coefficient (Steinheil et al., 1977; Gampert et al., 1980).

5.3 TURBULENT BOUNDARY LAYERS

At very large Reynolds numbers, transition can no longer be postponed and the boundary layer becomes turbulent and, thus, an excellent momentum conductor. Such a flow is less prone to separation but is characterized by large skin-friction. Several techniques are available to reduce the turbulent skin-friction coefficient, but only very few are in actual use. The basic reason being that the majority of these methods are relatively new and, thus, are still in the research and development stage.

Most of these techniques alter the structure of the turbulent boundary layer. It is useful to start the discussion by recalling the governing equations, followed by a brief reminder of the different quasi-periodic structures known to exist in the inner and outer regions of the bounded shear flow. Available techniques to generate the different structures artificially will be reviewed in Section 5.3.3. The various drag reduction methods will then be grouped into three categories: techniques that reduce the near-wall momentum; methods involving the introduction of foreign substance; and techniques involving the geometry. Three excellent reviews of the various turbulent skin-friction reduction methods are available, Bushnell (1983), Bandyopadhyay (1986), and Wilkinson et al. (1988).

5.3.1 GOVERNING EQUATIONS

Applying the boundary-layer approximations to the time-averaged continuity, streamwise-momentum, and normal-momentum equations for a steady, two-dimensional turbulent flow of a Newtonian, incompressible fluid, the resulting equations read:

$$\frac{\partial \overline{U_1}}{\partial x_1} + \frac{\partial \overline{U_2}}{\partial x_2} = 0 \quad , \tag{5.2}$$

$$\rho \left(\overline{U_1} \frac{\partial \overline{U_1}}{\partial x_1} + \overline{U_2} \frac{\partial \overline{U_1}}{\partial x_2} \right) = -\frac{\partial \overline{P}}{\partial x_1} + \frac{\partial}{\partial x_2} \left(\mu \frac{\partial \overline{U_1}}{\partial x_2} - \rho \overline{u_1 u_2} \right), \tag{5.3}$$

$$0 = -\frac{\partial \overline{P}}{\partial x_2} + \frac{\partial}{\partial x_2} (-\rho \overline{u_2^2}) \quad , \tag{5.4}$$

where $\overline{U_1}$ and $\overline{U_2}$ are the time-averaged velocity in the streamwise and normal directions, respectively, $\overline{P}$ is the mean pressure, ρ is the constant density, μ is the variable viscosity, $-\rho \overline{u_1 u_2}$ is the tangential Reynolds stress, and $-\rho \overline{u_2^2}$ is the normal Reynolds stress.

In (5.3), $-\rho \overline{u_1 u_2}$ is the additional stress on a fluid element due to the turbulent fluctuations. The normal derivative of this Reynolds shear stress can be related to the vorticity-velocity cross product (see, for example, Tennekes and Lumley, 1972). Neglecting all turbulent terms that can be written as gradients of dynamic pressures, it can be shown that:

$$\frac{\partial}{\partial x_2} (- \rho \, \overline{u_1 u_2}) = \rho (\, \overline{u_2 \omega_3}) - \overline{u_3 \omega_2}), \tag{5.5}$$

where ω_i is a component of the vorticity-fluctuations vector (curl of the velocity-fluctuations vector). The first term on the right-hand side of (5.5) may be thought of as a streamwise force (per unit volume) due to the transport of spanwise vorticity ω_3 by the normal velocity u_2 in a field with a mean vorticity gradient $\partial \overline{\Omega}_3 / \partial x_2$. The second term may be called a vortex-stretching* force, since it is associated with the change of size of eddies in a flow field with a varying length scale. In a surface layer with constant stress, the mean spanwise vorticity $\overline{\Omega}_3$ is constant along streamlines because the gain of mean vorticity due to a net transport surplus is balanced by the loss of $\overline{\Omega}_3$ due to the transfer of vorticity to the turbulence by vortex stretching (Tennekes and Lumley, 1972).

Equation (5.4) can be integrated to yield:

$$\overline{P} = P_0 - \rho \, \overline{u_2^2} \, , \tag{5.6}$$

where P_0 is the pressure just outside the turbulent region, determined from the potential flow solution. Within the same order of approximation,

$$\frac{\partial \overline{P}}{\partial x_1} = \frac{d P_0}{d x_1} \, , \tag{5.7}$$

leaving (5.2) and (5.3) as two equations for the three unknowns $\overline{U_1}$, $\overline{U_2}$ and $\overline{u_1 u_2}$. Obviously, no solution can be obtained from first principles, and we must rely on more or less heuristic models to close the equations. Nevertheless, both equations can be integrated in the normal direction to yield the Von Karman integral momentum-balance equation. For a steady, incompressible turbulent flow around a two-dimensional or axisymmetric surface of small curvature, this equation reads (Kays and Crawford, 1980):

$$C_f = 2 \frac{d \delta_\theta}{d x_1} + 2 \delta_\theta \left[\left(2 + \frac{\delta^*}{\delta_\theta} \right) \frac{1}{U_\infty} \frac{d U_\infty}{d x_1} + \frac{1}{R} \frac{d R}{d x_1} \right] - 2 \frac{v_0}{U_\infty} \, , \tag{5.8}$$

where C_f is the local skin-friction coefficient, δ^* and δ_θ are the displacement and momentum thicknesses, respectively, U_∞ is the freestream velocity, R is the radius of curvature of the wall, and v_0 is the normal velocity of fluid injected through the surface (positive for injection and negative for suction). As discussed in Section 2, (5.8) is valid for both laminar and turbulent boundary layers. In the latter case, the mean streamwise velocity is used in the definition of δ^* and δ_θ.

---

*On the average, there is more turbulent vortex stretching than squeezing. Vortex stretching transfers turbulent vorticity, and the energy associated with it, from large-scale fluctuations to small-scale fluctuations.

A second useful equation is obtained from (5.3) by taking the limit $x_2 \to 0$. At a fixed wall, the equation becomes after some rearranging:

$$\rho v_0 \left[\frac{\partial \overline{U_1}}{\partial x_2} \right]_0 + \frac{d P_0}{d x_1} - \frac{d \mu}{d \overline{T}} \left[\frac{\partial \overline{T}}{\partial x_2} \frac{\partial \overline{U_1}}{\partial x_2} \right]_0 + \rho \left[\frac{\partial \overline{u_1 u_2}}{\partial x_2} \right]_0$$

$$= \left[\mu \frac{\partial^2 \overline{U_1}}{\partial x_2^2} \right]_0 , \qquad (5.9)$$

where $\overline{T}$ is the mean temperature field, and the subscript $[\ ]_0$ indicates flow quantities computed at the wall. The right-hand side of (5.9) is the flux of mean spanwise vorticity, $- \partial \overline{U_1} / \partial x_2$, at the surface. In the absence of suction/injection, pressure gradient, and surface heating/cooling, the first three terms on the left-hand side of (5.9) vanish. Information regarding $\overline{u_1 u_2}$ has to come from experiment. The Reynolds stress must be zero at the wall and at the outer edge of the boundary layer. In between, $\overline{u_1 u_2}$ is negative* reaching a minimum at a height above the wall of typically $x_2 = 0.05\delta$ or less than 150 wall units (see next section), where δ is the boundary layer thickness. To evaluate the fourth term on the left-hand side of (5.9), consider the limiting case as the wall is approached (Monin and Yaglom, 1971). Consider a Taylor series expansion in powers of x_2 in the neighborhood of the point $x_2 = 0$. As a result of the no-slip condition, the streamwise velocity fluctuations u_1 varies at least as x_2. To conserve mass, the normal velocity fluctuations must vary as x_2^2 ($\partial u_1/\partial x_1 \sim \partial u_2/\partial x_2 \sim x_2$). It follows then that very near the wall (within the viscous sublayer), the tangential Reynolds stress $\overline{u_1 u_2}$ varies at least as x_2^3 and that $\partial \overline{u_1 u_2} / \partial x_2$ varies as x_2^2. At the wall itself, $x_2 = 0$ and $[\partial \overline{u_1 u_2} / \partial x_2]_0 = 0$, although close to the wall the slope of the tangential Reynolds stress profile is quite large.

The above arguments together with equation (5.9) indicate that the streamwise mean velocity profile for the canonical turbulent boundary layer (two-dimensional, isothermal, zero pressure gradient, over an impervious, rigid surface) will have a zero curvature at the wall. Notwithstanding this common characteristic with the Blasius boundary layer, the turbulent boundary layer is quite different from the laminar one. As pointed out by Lighthill (1963), the turbulent mixing concentrates most of the mean vorticity much closer to the wall as compared to the laminar case. The mean vorticity at the wall, $(\partial \overline{U}_1 / \partial x_2)_0$, is typically an order of magnitude larger than that in the laminar case. The turbulent mixing also causes the mean vorticity to migrate away from the wall and about 5% of the total is found much farther from the surface. The flux of mean spanwise vorticity is zero at the wall itself but very large close to it reaching a maximum at about the same location where the root-mean-square vorticity fluctuations peaks (near the edge of the viscous sublayer). To quote Willmarth (1975), "It is indeed remarkable that the turbulence in the boundary layer is able to maintain large gradients of mean and

---

*The turbulent mixing transports high-speed fluid toward the wall ($u_1 > 0$; $u_2 < 0$) and moves low-speed fluid away from the surface ($u_1 < 0$; $u_2 > 0$). Both of these processes create negative Reynolds stress (Willmarth, 1975).

fluctuating vorticity near the wall despite large viscous diffusion down the gradient. The processes that accomplish this are central to an understanding of the structure of turbulence in the boundary layer." Despite rapid advances in the field, our understanding of turbulent boundary layers remains far from complete. The closure problem resulting from averaging the nonlinear equations of motion precludes solutions based on first principles.

Transpiration, shaping, or viscosity gradient at the wall affect the mean velocity profile qualitatively in the same direction as the laminar case. For example, a favorable pressure gradient yields to a negative curvature at the wall, a fuller mean velocity profile, and a higher skin friction. Those effects are complicated, however, because the modulations also influence the Reynolds stress term. Suction, favorable pressure gradient, and lower viscosity at the wall lead to appreciable stabilization of the near wall flow. Experimental observations in these cases confirm that the low-speed streaks (see next section) tend to be longer and more quiescent as compared to the canonical case, and that the spanwise and longitudinal velocity fluctuations are reduced (Kline et al., 1967; Antonia et al., 1988). Sufficiently strong modulation could result in relaminarization (Section 5.4).

5.3.2 STRUCTURE OF TURBULENT BOUNDARY LAYERS

The classical view that turbulence is essentially a stochastic phenomenon having a randomly fluctuating velocity field superimposed on a well-defined mean has been changed in the last few decades by the realization that the transport properties of all turbulent shear flows are dominated by quasi-periodic, large-scale vortex motions (Laufer, 1975; Cantwell, 1981). In a boundary layer, the turbulence production process is dominated by three kinds of eddies: the large outer structures, the intermediate Falco eddies, and the near-wall eddies. The large, three-dimensional structures scales with the boundary layer thickness, δ, and extend across the entire layer. These eddies control the dynamics of the boundary layer in the outer region, such as entrainment, turbulent production, etc. They appear randomly in space and time, and seem to be, at least for moderate Reynolds numbers, the residue of the transitional Emmons spots (Zilberman et al., 1977). The Falco eddies are also highly coherent and three dimensional. Falco (1977) named them typical eddies because they appear in wakes, jets, Emmons spots, grid-generated turbulence, and boundary layers in zero, favorable and adverse pressure gradients. They have an intermediate scale of about $100\upsilon/u_\tau$, where υ/u_τ is the viscous scale (wall unit), υ is the kinematic viscosity, and u_τ is the friction velocity $\left(\equiv \sqrt{\tau_0/\rho} \right)$. The Falco eddies appear to be an important link between the large structures and the near-wall events.

The third kind of eddies exists in the near-wall region $(0 < x_2 < 100\upsilon/u_\tau)$ where the Reynolds stress is produced in a very intermittent fashion. Half of the total production of turbulent energy $(-\overline{u_1u_2}\ \partial\overline{U_1}/\partial x_2)$ typically takes place near the wall in the first 5% of the boundary layer, and the dominant sequence of eddy motions there are collectively termed the bursting phenomenon. This process was reviewed by Willmarth (1975) and summarized by Blackwelder (1978). Qualitatively, the process begins with elongated, counter-rotating, streamwise vortices having diameters of approximately $40\upsilon/u_\tau$. The vortices exist in a strong shear and induce low- and high-speed regions between them as sketched in Figure 7. The vortices and the accompanying eddy structures occur randomly in space and time.

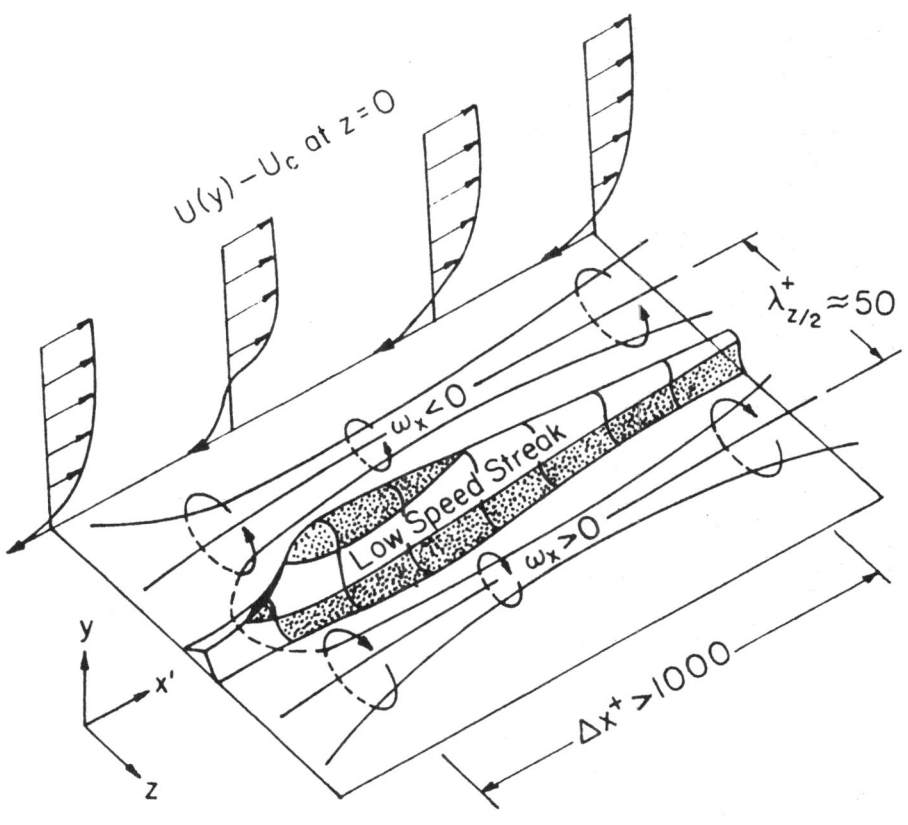

Figure 7. Model of Near - Wall Turbulent Boundary - Layer Structure. (adapted from Blackwelder, 1978).

However, their appearance is regular enough that an average spanwise wavelength of approximately 80 to $100\upsilon/u_\tau$ has been identified by Kline et al. (1967) and others. Kline et al. also observed that the low-speed regions grow downstream and develop inflectional $U_1(x_2)$ profiles. At approximately the same time, the interface between the low- and high-speed fluid begins to oscillate, apparently signaling the onset of a secondary instability. The low-speed region lifts up away from the wall as the oscillation amplitude increases, and then the flow rapidly breaks down into a completely random pattern. Since this latter process occurs on a very short time scale, Kline et al. called it a "burst". Corino and Brodkey (1969) showed that the low-speed regions are quite narrow, i.e., $20\upsilon/u_\tau$, and may also have significant shear in the spanwise direction. Virtually all of the net production of turbulent energy in the near-wall region occurs during these bursts.

Considerably more has been learned about the bursting process during the last decade. For example, Falco (1980; 1983) has shown that when a typical eddy, which may be formed in part by ejected wall-layer fluid, moves over the wall it induces a high uv sweep (positive u and negative v). The wall region is continuously bombarded by "pockets" of high-speed fluid originating in the logarithmic and possibly the outer layers of the flow. These pockets tend to promote and/or enhance the inflectional velocity profiles by increasing the instantaneous shear leading to a more rapidly growing instability. Blackwelder and Haritonidis (1983) have shown convincingly that the frequency of occurrence of these events scales with the viscous parameters consistent with the usual boundary layer scaling arguments.

5.3.3 ARTIFICIALLY GENERATED EDDIES

Naturally occurring structures in transitional and turbulent boundary layers are random in space and time. Moreover, the organized structures in the latter case are embedded in a wide-spectrum background turbulence, making even their detection extremely difficult. Single-point detection of an inherently three-dimensional random structure must produce significant smearing. However, if these structures are produced artificially at a given location and time, the smearing is substantially reduced and much can be learned about their details by phase locking the measurements to the excitation. The classical example is the use of a vibrating ribbon in a laminar boundary layer to excite Tollmien-Schlichting vorticity waves (Schubauer and Skramstad, 1947). The linear stability theory of a laminar flow was verified for the first time by these ingenious experiments. In this section, we briefly recall methods of generating Emmons spots, large outer structures, and near-wall eddies. Not only more can be learned about these eddies, but also drag-reducing techniques aimed at altering a particular structure could be more readily tried on the artificial events.

Emmons (1951) observed the random creation of "spots" of turbulence as a boundary layer underwent transition. He concluded on the basis of visual observation in a water table that the spots grow uniformly and act independently of one another as they are swept downstream by the flow. Emmons was able to trigger a spot by disturbing the thin, unstable laminar boundary layer using a falling water drop. Other methods to induce Emmons spots include electric discharge in air boundary layers (Wygnanski et al., 1976), impulsive injection of a secondary fluid from a minute hole (Gad-el-Hak et al. 1981; 1985; Riley and Gad-el-Hak, 1985), momentarily normal displacement of a small pin into the wall region (Savas and Coles, 1985), and use of acoustic horn driver through a small hole in the wall (Goodman, 1985).

By generating a row of spots, Savas and Coles (1985) modified their rate of growth by altering the spot-to-spot proximity. Goodman (1985) exploited this finding and the fact that the large outer structures in a turbulent boundary layer are the residue of the Emmons spots in an attempt to produce smaller turbulence scales. He reasoned that by triggering closely spaced spots in the spanwise and streamwise directions, smaller outer structures will result at least in the initial development region of the turbulent boundary layer. Goodman operated two acoustic horn drivers at high frequency via a spanwise row of closely spaced holes. For certain combinations of generator frequency and amplitude, hole size, and hole spacing, a reduction of the skin-friction coefficient of 15% was observed. Goodman (1985) concluded that this reduction is a result of the production of smaller turbulence scales (boundary layer appears to be at higher Reynolds number and, hence, lower C_f) and not a result of shifting the transition location.

Gad-el-Hak and Blackwelder (1987b) introduced an active flow control device to generate large outer structures in a turbulent boundary layer. A cyclic jet issuing from a spanwise slot was used to collect the turbulent boundary layer for a finite time during its on period. When the jet was turned off, all of the trapped fluid was suddenly released in one large eddy that convected downstream. These periodic events were very similar to the random, naturally occurring large eddies. This device may be used on airborne laser platforms together with adaptive lenses to reduce the optical distortion caused by the high-speed boundary layer. A drag-reducing method that uses the cyclic jet together with a large eddy breakup device will be suggested in Section 5.5.

Gad-el-Hak and Hussain (1986) artificially generated the sequence of events leading to bursting in turbulent boundary layers. One or more hairpin vortices were generated using impulsive suction through two or more minute holes separated in the spanwise direction. Bursting also resulted from sudden pitching of a miniature delta wing that was flush-mounted to the wall. The resulting sequence of events that occurred at a given location could be uniquely controlled, thus allowing detailed examination via phase-locked measurements and flow visualization. The feasibility of the selective suction technique for reducing the skin-friction drag (Section 5.5) was confirmed using artificially generated bursting events (Gad-el-Hak and Blackwelder, 1987c; 1988).

5.3.4 REDUCTION OF NEAR-WALL MOMENTUM

Methods of skin-friction drag reduction in turbulent boundary layers that rely on reducing the near-wall momentum are reviewed in this section. Recall the definition of local skin-friction coefficient:

$$C_f \equiv \frac{2\,\tau_0}{\rho\,U_\infty^2} = \frac{2\,\upsilon}{U_\infty^2}\left[\frac{\partial\,\overline{U_1}}{\partial\,x_2}\right]_0 = 2\left(\frac{u_\tau}{U_\infty}\right)^2,\qquad(5.10)$$

where $\overline{U_1}$ is the time-averaged velocity component in the streamwise direction, and $[\partial\,\overline{U_1}/\partial x_2]_0$ is the slope of the mean-velocity profile at the wall. Thus, a brute force way of achieving drag reduction is to lower the mean-velocity gradient at the wall or make the curvature of the velocity profile there as positive as possible, as was done in the laminar case (Section 5.2).

The influence of wall transpiration, shaping or heat transfer on the mean velocity profile is complicated by the additional effects of these modulations on the Reynolds stress term. However, as mentioned in Section 5.3.1, these influences are qualitatively in the same direction as in the simpler laminar case. Thus, lower skin friction is achieved by driving the turbulent boundary layer towards separation. This is accomplished by injecting fluid normal to the wall, shaping to produce adverse pressure gradient, surface heating in air, or surface cooling in water. These methods of control in general result in an increase in turbulence intensity (Wooldridge and Muzzy, 1966).

Although in the reverse flow region downstream of the separation line the skin friction is negative, the increase in pressure drag is far more than the saving in skin-friction drag. The goal of these methods of control is to avoid actual separation, i.e., lower C_f but not any lower than zero (the criterion for steady, two-dimensional separation). The papers by Stratford (1959a; 1959b) provide useful discussion on the prediction of turbulent boundary layer's separation and the concept of flow with continuously zero skin-friction throughout its region of pressure rise. By specifying that the turbulent boundary layer be just at the condition of separation, without actually separating, at all positions in the pressure rise region, Stratford (1959b) experimentally verified that such a flow achieve a specified pressure rise in the shortest possible distance and with the least possible dissipation of energy. An airfoil which could utilize the Stratford's distribution immediately after transition from laminar to turbulent flow would be expected to have a very low drag (Liebeck, 1978).

Two additional methods to reduce the near-wall momentum are tangential injection and ion wind. In the former, a low-momentum fluid is tangentially injected from a wall slot. For some distance downstream, the wall senses the lower injection velocity and not the actual freestream. Bushnell (1983) terms this situation a wall wake as opposed to a wall jet (used in high lift devices to keep the flow attached). Ion wind, generated by an electrode in the wall and discharging either to space or to a ground on the wall further downstream, causes an increase in the normal velocity component near the wall. Since there is no net mass transfer through the surface, the near-wall longitudinal momentum is again reduced and the boundary layer is driven towards separation (Malik et al., 1983). The source and amount of power to produce a measurable drag reduction are among the outstanding issues to be considered with this technique. Among the possibilities is the use of the natural buildup of static electricity on an aircraft which at present is dissipated at discharge points to avoid large-scale arcing.

When attempting to reduce drag by driving the boundary layer towards separation, a major concern is off-design conditions. A slight increase in angle of attack for example can lead to separation and consequent large drag increase as well as loss of lift. High performance airfoils with lift-to-drag ratio of over 100 utilize carefully controlled adverse pressure gradient to retard the near-wall fluid, but their performance deteriorate rapidly outside a narrow envelope (Charmichael, 1974).

In the case of tangential or normal injection, the source of fluid to be injected is important. Large ram drag penalty is associated with using freestream fluid, and reaching a break-even point is highly problematic. A lower loss source is fluid withdrawn from somewhere else to control transition or separation, for example from the wings and empennage of an aircraft. This point will be discussed further in Section 5.5.

Howard et al. (1975) conducted finite-difference calculations to show the effect of single- and multi-slot injection on skin-friction drag of a fuselage shape typical of long-haul subsonic transport. A sample of their results is shown in Figure 8 for up to 10 tangential slots. The abscissa in the figure indicates the streamwise location of each slot, ℓ, normalized with the total length of the fuselage, ℓ_T. The beneficial effect of injection is most pronounced immediately downstream of the slot exit and diminishes with distance from the slot. Performance deterioration is clearly a result of the mixing between the low-momentum slot flow and the high-momentum boundary-layer flow. Note, however, that the asymptotic level of skin friction with multi-slots is lower than that in the unperturbed flow. Without taking into account the system penalties for collecting, ducting, and injecting the secondary flow, Howard et al. (1975) computed a reduction in total skin-friction drag of the order of 50%. They maintain that further optimization may provide even greater skin-friction reductions to help compensate for system losses.

As was the case when surface heating/cooling was considered for transition delay or separation postponement, this technique should be used for skin-friction reduction only when a heat source/sink is readily available. Reject heat from the vehicle's power plant is a free source of energy, and creyo-fuel is an excellent heat sink that does not require energy expenditure.

5.3.5   METHODS INVOLVING INTRODUCTION OF FOREIGN SUBSTANCE

Turbulent skin-friction drag can be reduced by the addition of several foreign substances. Examples include long-chain molecules, surface-active agents and microbubbles in liquid flows, and small solid particles or fibers in either gases or liquids. In general, the addition of these substances leads to a suppression of the Reynolds stress production in the buffer zone and, thus, to an inhibition of the turbulent mixing and a consequent reduction in the viscous shear stress at the wall.

Drag reduction by solutions of macromolecules (molecular weight $> 10^5$) is perhaps the more mature technology with well over 1000 research papers in existence (Virk, 1975; White and Hemmings, 1976; Hoyt, 1979). Toms (1948) observed that the addition of few parts per million of polymethyl methacrylate to a turbulent pipe flow of monochlorobenzene reduced the pressure drop substantially below that of the solvent alone at the same flow rate. Successful long-chain molecules for water flow include carboxymethyl cellulose, guar gum, copolymer of polyacrylamide and polyacrylic acid, and polyethylene oxide. Skin-friction drag reduction of up to 80% is possible in both external and internal flows with 100 ppm or less of the largest molecules.

The viscosities of these dilute solutions of polymers are typically 10% higher than that of the solvents, although most polymer solutions are detectably shear thinning at higher concentrations (Berman, 1978). The polymer molecules at rest are in the form of spherical random coils in which each monomer is joined to the preceding one at a random angle. The onset of drag reduction is associated with the expansion of these coils outside the viscous sublayer. Hinch (1977) asserts that this unfolding explains

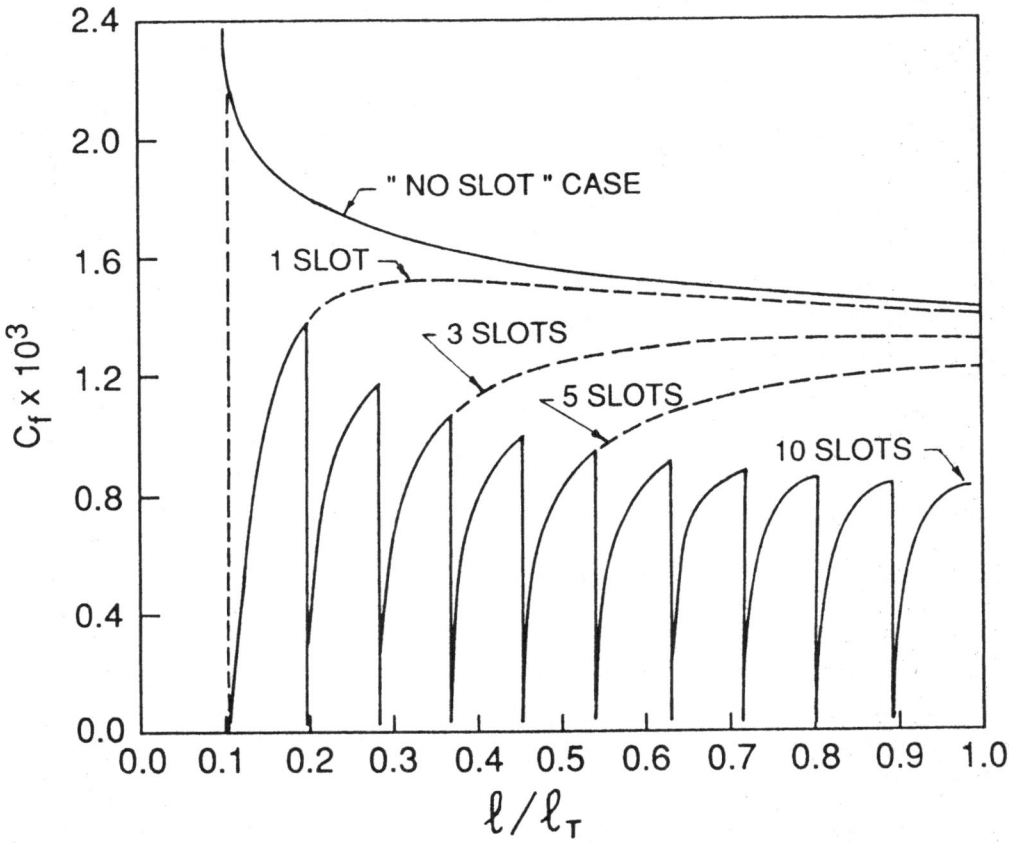

Figure 8. Computed Skin - Friction Reduction Due to Tangential Injection. (From Howard et al., 1975).

the dramatic effects of few parts per million (by weight) on the drag. The effective hydrodynamic volume fraction shows large concentration once the polymer molecules are stretched. Therefore, it is the volume fraction of spheres just enclosing the separate polymers rather than the weight which is the appropriate measure of the added substance.

When the macromolecules are coiled, the viscosity of the dilute solution changes by a few percentage, while if they are fully stretched the apparent viscosity increases by several orders of magnitude. Onset of drag reduction occurs when $D_M u_\tau / \upsilon = 0.015$, where D_M is the effective molecular diameter. This diameter is about two orders of magnitude less than the shortest dynamically significant length in a typical turbulent flow (Virk et al., 1967; Lumley, 1969). Numerous experiments have indicated that the drag reduction is associated with a thickening of the buffer layer and a corresponding shift of the logarithmic part to accommodate the larger buffer zone (Reischman and Tiederman, 1975). The effect of the additive may then be represented as a virtual slip at the wall (Virk et al., 1967).

The percentage drag reduction increases as the polymer concentration is increased and the buffer layer is thickened. This process tends to an asymptote when the buffer layer reaches the center of the pipe or the edge of the boundary layer. As the Reynolds number is increased at a given concentration, a second limit in drag reduction is reached when the macromolecules become fully stretched. Moreover, degradation or breaking of the molecules by the flow occurs at sufficiently high strain rates, with a consequent loss of the ability to reduce the drag (Berman and George, 1974). Because of this latter point, care has to be exercised when mixing the macromolecules into the solvent as well as when delivering the solution into the flow.

By injecting the polymer in different regions of the boundary layer, Wells and Spangler (1967), Wu and Tulin (1972) and McComb and Rabie (1979) have shown that the additive must be in the wall region for drag reduction to occur. Furthermore, the channel flow experiments of Tiederman et al. (1985) have clearly demonstrated that drag reduction is measured downstream of the location where the additives injected into the viscous sublayer begin to mix in significant quantities with the buffer region, i.e., $10 < x_2 u_\tau/\upsilon < 100$. At streamwise locations where drag reduction does occur, the dimensionless spanwise streak spacing increases and the average bursting rate decreases, although the decrease in bursting frequency is larger than the corresponding increase in low-speed streak's spacing. Tiederman et al. (1985) conclude that the polymers have a direct effect on the flow processes in the buffer region and that the linear sublayer appears to have a rather passive role in the interaction of the inner and outer portions of a turbulent wall layer.

Many mechanisms have been proposed to explain the experimental observations. Notable among these is the work of Lumley (1973; 1977), Landahl (1973; 1977), and the more recent quantitative theory by Ryskin (1987). According to Lumley's model, the macromolecules are expanded outside the viscous sublayer due to the fluctuating strain rate. The resulting increase in effective viscosity damps only the small dissipative eddies. The suppression of small eddies in the buffer layer leads to increased scales, a delay in the reduction of the velocity profile slope, and consequent thickening of the wall region. In the viscous sublayer, the scales remain unchanged since the effective viscosity of the dilute solution in steady shear is only slightly affected. In the buffer zone, the scales of the dissipative and energy containing eddies are roughly the same and, hence, the energy containing eddies will also be suppressed resulting in reduced momentum transport and reduced drag.

Lumley's more conservative general hypothesis is not incompatible with Landahl's specific, rather speculative mechanism (Lumley and Kubo 1984). The polymer suppresses the small eddies responsible for the inflectional profile and the secondary instability in the buffer zone. Kim et al. (1971) have shown that the greatest turbulent kinetic energy production for a Newtonian boundary layer occurs near the beginning of the buffer zone at $x_2 \approx 12\upsilon/u_\tau$. A sharp maximum in the turbulent intensity, $\sqrt{\overline{u_1^2}}$, occurs at the same location. For a dilute solution of drag-reducing polymer, the peak of $\sqrt{\overline{u_1^2}}$ moves away from the wall and becomes much broader as compared to the solvent alone. Production of Reynolds stress in this important buffer zone is, thus, diminished.

The use of drag-reducing polymers in pipe lines is cost effective in some applications. When it was discovered that the winter cooling of crude oil in the Alaskan Pipeline had been underestimated, polymer additives offered an inexpensive alternative to increasing the number or power of the pumping stations. Long-chain molecules are presently used in waste-disposal plants, fire-fighting equipments and high-speed water-jet cutting. For surface ships or underwater vehicles practical considerations include cost of polymer, how, with what, and when to mix it, as well as how to eject it into the boundary layer, i.e., using pumps or pressurized tanks, through slots or porous surfaces, etc. A very important additional consideration is the portion of the payload that has to be displaced to make room for the additive. Oil companies appear to have concluded that the use of polymer for supertankers is just at the break-even point economically. For submarines and torpedoes, where space is a prime consideration, it seems that the volume occupied by the additive is a more important obstacle to its routine use.

Other substances that can be added to a clear fluid to reduce the skin-friction drag include surfactants (Savins, 1967; Patterson et al., 1969; Zakin et al., 1971; Aslanov et al., 1980), microbubbles (McCormick and Bhattacharya, 1973; Bogdevich and Malyuga, 1976; Bogdevich et al., 1977, Madavan et al., 1984; 1985; Merkle and Deutch, 1985; 1988), large length-to-diameter particles, e.g. fibers, (Hoyt, 1972; Lee et al., 1974; McComb and Chan, 1979; 1981), and spherical particles (Soo and Trezek, 1966; Pfeffer and Rosetti, 1971; Radi, 1974; Radin et al., 1975; Povkh et al., 1979). Lumley (1977; 1978) argues that the same mechanism discussed in conjunction with the polymer case governs the interaction of most of these substances with the turbulent boundary layer. The additives affect only the dissipative scales of the turbulence and thus increasing the scales of dissipation in the buffer layer. The energy containing eddies are also suppressed and, consequently, the momentum transport is reduced.

A surface-active substance, such as soap, is that when dissolved in water or an aqueous solution reduces its surface tension. The majority of surfactants appear to reduce the turbulent skin-friction drag only in the presence of electrolytes. These additives are stable against mechanical destruction, a certain advantage over polymer molecules which break down when the strain rate is sufficiently high. Although drag-reducing polymer flows undergo transition at a lower Reynolds number than Newtonian flows (Lumley and Kubo, 1984), some recent experiments indicate that surfactants are very effective in delaying laminar-turbulent transition   (A. J. Smits, private communication; Sabadell, 1988).

The idea of placing a thin layer of gas between the wall and its water boundary layer dates back to the last century. Substantial drag reductions are potentially possible due to the lower density of the gas. Unfortunately, the various instabilities associated with gas/liquid interface result in drag increase. Extremely small gas bubbles (microbubbles) injected through a porous wall or produced by electrolysis do not suffer from the stability problems of a gas film and result in skin-friction reduction as high as 80%. Maximum reduction is obtained when the gas volume fraction approaches the bubble packing limit. Skin friction is reduced because of the substantially lower density and also because of the usual increase in bulk viscosity due to particles (Batchelor and Green, 1972), which damps the small scale motions in the buffer layer. Legner (1984) presents a simple phenomenological model to predict microbubble drag-reduction in turbulent boundary layers. Application of the microbubbles technique for surface ships is quite feasible. Compressed atmospheric air is injected through a micro-porous skin over portions of the hull. This has the additional advantage of reducing fouling drag and fouling maintenance costs, since water is kept away from the hull surface. The situation is different for underwater vehicles. There, a source of air is not readily available and using electrolysis for bubble production will not yield net energy saving.

Spherical or large length-to-diameter particles can be used in both air and water flows. For spherical particles, drag reduction of up to 50% is feasible for certain parameter space, although drag increase is also possible. Heavy but small particles may reduce the drag by inducing a stable density stratification near the wall thus driving the turbulence towards relaminarization. Lighter particles that are large enough to interact with the smallest turbulence scales reduce the drag through the same mechanism as in the polymer case, i.e. suppressing the dissipative eddies and increasing the scale of dissipation.

McComb and Chan (1981) report up to 80% drag reduction using the naturally occurring macro-fiber chrysotile asbestos dispersed at a nominal concentration of 300 ppm by weight in a 0.5% aqueous solution of Aerosol OT. The individual fibrils of chrysotile asbestos have a mean diameter of 40 nm and a mean length-to-diameter ratio in an undegraded suspension in the range of 10^3-10^4. Like polymers, macro-fibers readily break down and have to be carefully dispersed into the solvent and delivered to the boundary layer. In principle, fibers can also be used in gases, although their drag-reduction potential has yet to be demonstrated.

5.3.6 METHODS INVOLVING GEOMETRY

Under this classification are some of the most recently researched techniques to reduce the turbulent skin-friction drag. These include large eddy breakup devices, riblets, compliant surfaces, wavy walls, and other surface modifications. With few exceptions, there is little theoretical basis for how these geometrical modifications affect the skin friction and most of the present knowledge comes from experiments. Needless to say that the lack of analytical framework makes optimization for a given flow condition as well as extension to other flow regimes very tedious tasks.

The large eddy breakup devices (LEBUs)[*] are designed to sever, alter or break up the large vortices that form the convoluted outer edge of a turbulent boundary layer. A typical arrangement consists of one or more splitter plates placed in tandem in the outer part of a turbulent boundary layer. It is of course very easy to reduce substantially the skin friction in a flat-plate boundary layer by placing an obstacle above the surface. What is difficult is to ensure that the device's own skin-friction and pressure drags do not exceed the saving. The original screen fence device of Yajnik and Acharya (1978) and the various sized honeycombs used by Hefner et al. (1980) did not yield a net drag reduction. In low-Reynolds-number experiments, very thin elements placed parallel to a flat plate have a device's total drag that is nearly equal to laminar skin friction. A net drag reduction of the order of 20% is feasible with two elements placed in tandem with a spacing of $0[10\delta]$ (Corke et al. 1980; 1981). These ribbons have typically a thickness and a chord of the order of 0.01δ and δ, respectively, and are placed at a distance from the wall of 0.8δ. Several experiments report a more modest drag reduction (Hefner et al., 1983) or even a drag increase, but it is believed that a slight angle of attack of the thin element can result in a laminar separation bubble and a consequent increase in the device pressure drag. Flat ribbons at small, positive angle of attack produce larger skin-friction reductions. This is consistent with the analytical result that a device producing positive lift away from the wall is more effective (Gebert, 1988). In any case, a net drag reduction should be achievable, at least for devices having chord Reynolds numbers $< 10^6$, if extra care is taken to polish and install the LEBU.

Net drag reduction has been documented even in the presence of favorable or adverse pressure gradient in the main flow (Bertelrud et al., 1982; Plesniak and Nagib, 1985). Additionally, Anders and Watson (1985) have demonstrated that a LEBU having an airfoil shape is nearly as effective in reducing the net drag as a flat ribbon. A wing is several orders of magnitude stiffer than a thin ribbon, a certain advantage for extending the technique to field conditions. At flight Reynolds and Mach numbers, the possibility of paying for transitional or turbulent skin-friction as well as wave drag on the device itself is real (Anders et al., 1984) and achieving net drag reduction in this circumstance is problematic at present. For an airplane, a LEBU is likely to take the form of a ring around the fuselage. Full-scale flight tests of an airfoil-shaped LEBU are planned in late 1988 on the fuselage of a Boeing 737 (Anders et al., 1988).

Two analytical attempts to explain the mechanisms involved with the large eddy breakup devices are noted in here. Basically, the LEBU acts as an airfoil on a gusty atmosphere and a vortex unwinding mechanism is activated. Dowling's (1985) inviscid model indicates that the vorticity shed from the LEBU's trailing edge as a result of an incident line vortex convected past the device tends to cancel the effect of the incoming vortex and to reduce the velocity fluctuations near the wall. Atassi and Gebert (1987) and Gebert (1988) model the incoming turbulent rotational flow as two- or three-dimensional harmonic disturbances. They use the rapid distortion approximation and unsteady aerodynamic theory to compute the fluctuating velocity downstream of thin-plate and airfoil-shaped devices. The fluctuating normal velocity component is most effectively suppressed for a range of frequencies that scales with the freestream velocity and the device chord. This important result determines the optimum size of a LEBU by

[*]These devices have also been called outer layer devices (OLDs), boundary layer alteration devices (BLADEs), manipulators, and ribbons.

selecting this frequency range to correspond to that of the large-scale eddies in a given turbulent boundary layer.

The second geometrical modification is the riblets. Small longitudinal striations in the surface, interacting favorably with the near-wall structures in a turbulent boundary layer, can produce a modest drag reduction in spite of the increase in surface area. The early work employed rectangular fins with height and spacing of $0[100\upsilon/u_\tau]$. The turbulent bursting rate was reduced by about 20% and a modest 4% net drag reduction was observed (Liu et al., 1966). In a later refinement of this technique, Walsh and his colleagues at NASA-Langley examined the drag characteristics of longitudinally ribbed surfaces having a wide variety of fin shapes that included rectangular grooves, V-grooves, razor blade grooves, semi-circular grooves, and alternating transverse curvature (Walsh and Weinstein, 1978; Walsh, 1980; 1982; 1983; Walsh and Lindemann, 1984). A net drag reduction of 8% is obtained with V-groove geometry with sharp peak and either sharp or rounded valley. Optimum height and spacing of the symmetric grooves are about $15\upsilon/u_\tau$. Although these dimensions would be extremely small for the typical Reynolds numbers encountered on an airplane or a submarine (peak-to-valley height ~ 35 μm), such riblets need not be machined on the surface. Thin, low specific gravity plastic films with the correct geometry on one side and an adhesive on the other side are presently manufactured by the 3M Corporation and existing vehicles could be readily retrofitted. In fact, these tapes were successfully tested at M = 0.7 on a T-33 airplane and on a Lear jet. The performance of the riblets in flight was similar to that observed in the laboratory (Anders et al., 1988). In water, riblets were employed on the rowing shell during the 1984 Summer Olympic by the United States rowing team. Similar riblets were also used on the submerged hull of the winner of the 1987 America's Cup yacht race, the Stars and Stripes, with apparent success.

Riblets are effective in the presence of moderate adverse and favorable pressure gradients. The loss of drag-reducing effectiveness as flow conditions vary off-design is gradual. Moreover, the percentage drag reduction slowly decreases to zero as the yaw angle between the flow and the grooves goes up to 30°. Surprisingly, drag does not increase at yaw angles >30°. Remaining practical problems include cost, weight penalty, particulate clogging, ultraviolet radiation, and film porosity and resistance to hydraulic fluids and fuel.

Curiously, fast sharks have a surface covering of dermal denticles with flow-aligned keels having the optimal riblet spacing as determined by Walsh (Bechert et al., 1985). As the shark grows, new keels are added onto the sides of the denticles without changing the keel-to-keel spacing. A clue to how riblets result in a net drag reduction despite the increase in wetted area is provided by the experiment of Hooshmand et al. (1983) who showed that, for an optimum V-grooved surface, C_f is reduced by 40% in the valleys and increased by 10% at the peaks. It appears that the riblets severely retard the flow in the valleys, which places a slip boundary condition upon the turbulence production process. Lumley and Kubo (1984) argue that the streamwise vortices in the near-wall region are forced to negotiate the sharp peaks of the riblets, causing increased losses. To stay in equilibrium, the eddies must increase their energy gain, and one way to do that is to grow larger (energy gain ~ $(size)^2$; loss ~ (size)). As was the case with polymers, the larger scales result in the secondary instability and the sharp change in profile slope occurring farther from the wall. The mean velocity is, thus, higher for the same friction velocity and the coefficient of skin friction is reduced.

The third geometrical modification to reduce the skin friction is fixed or moving wavy walls. Kendall (1970) observed that the integrated skin-friction in a turbulent boundary layer over rigid, sinusoidal waves can be as much as 25% below that over an equivalent flat surface. Kendall's waves were relatively shallow having a height-to-wavelength ratio of $h/\lambda = 0.028$, with $\lambda = 0[\delta]$. Sigal (1971) reported a reduction in integrated skin-friction of about 12% when using a 10% larger amplitude waves ($h/\lambda = 0.031$) than that used by Kendall. The transverse waves provide alternating regions of longitudinal concave and convex curvatures along with alternating adverse and favorable pressure gradients. The coincidence of convex curvature and favorable pressure gradient promote relaminarization (Section 5.4) as the wave crests are approached, and the near-wall momentum is reduced in regions having adverse pressure gradient. Accordingly, the integrated skin-friction drag is reduced. Unfortunately, viscous forces cause a downstream phase shift between the pressure distribution and the wave relative to the 180° out-of-phase relationship predicted from potential flow theory for a sinusoidal wave. This phase shift causes an additional pressure drag not present in the flat-surface case. The net result is that the total drag for the wavy surface increases despite the decrease in skin friction. Cary et al. (1980) argue that there may be two approaches to overcome this problem. First, since pressure drag varies as $(h/\lambda)^2$, further reduction of h/λ may yield a net drag reduction. Secondly, certain non-sinusoidal waves may have a less skewed pressure distribution relative to the waves and therefore less pressure drag, and more favorable pressure and curvature effects on the viscous drag. Cary et al.'s numerical results indicate the soundness of the first approach. Net drag reduction of about 13% is attained for the optimum h/λ of 0.005. This result seems to be insensitive to the ratio of wave height to the viscous scale at least for $hu_\tau/\upsilon < 33$. Experiments to test the second approach were only partially successful (Lin et al., 1983). Skewed waves with gradual, straight downstream-facing slope and steeper, sinusoidal upstream-facing surface have lower pressure drag than a symmetric sine-wave. However, the asymmetric surface is not as effective in reducing the viscous drag, with the result that the net drag reduction is only 1-2%, hardly worth the effort (Wilkinson et al., 1987; 1988).

Moving wavy walls are divided into two categories: driven, non-interactive walls and compliant surfaces. In the former technique, a flat or wavy wall is translated in the streamwise direction. The rectilinear wall motion essentially acts as a slip boundary condition and reduces the skin friction. Obviously the boundary layer could be completely eliminated when $U_{wall} = U_\infty$. A moving wavy wall can produce a thrust at high enough translational speed but causes an additional pressure drag at lower speed. In any case, this method of control is in general impractical and is used mainly to provide controlled experiments to determine what type of wave motion is required to achieve a given result, much the same as the laminar flow control case (Section 3.3.1).

Wall motion can also be generated by using a flexible coating with sufficiently low modulus of rigidity. The idea is very appealing if a particular kind of fluid/solid interaction yields a net drag reduction. Covering a vehicle with a compliant coating is relatively simple, does not require modification of existing design, does not require slots, ducts or internal equipment of any kind, and no energy expenditure is needed. The search for such a surface has been very elusive, however, despite reports of substantial drag reduction by some researchers. Irreproducibility seems to be an outstanding characteristic of the body of compliant coating experimental evidence (Bushnell et al., 1977; McMichael et al., 1980; Gad-el-Hak, 1986b; 1987b). The window of opportunity for a favorable coating response is rather narrow. For flow speeds below the transverse wave speed in the solid, no significant interaction between the fluid and the compliant surface takes place. Above the transverse wave speed, ample opportunity for interaction exists; however, hydroelastic instabilities appear and the resulting large-amplitude surface waves increase the drag because of the roughness-like effects (Gad-el-Hak et al., 1984a; Gad-el-Hak, 1986c). A second constrain concerns the density ratio of the fluid and the solid. The extremely low density of air makes finding a reasonable coating material highly unlikely, although in liquids the situation is different. Flexible surfaces that interact favorably with Tollmien-Schlichting waves in water boundary layers are readily available (Section 3.3.1). For turbulent boundary layers, the task is to find a coating that responds to the flow, particularly to its preburst wall pressure signature, with a surface motion that can alter, modify or interrupt the sequence of events leading to a burst. The analytical work of Purshouse (1977) indicates that an anisotropic coating would be more likely to achieve a net drag reduction in a turbulent boundary layer.

Other geometric modifications on the surface with a drag-reducing potential include fixed waves aligned with the flow (essentially large-scale riblets), micro air-bearings, compound or three-dimensional riblets, sieves, furry (wheatfield-type) surfaces, and sword-fish configuration. These techniques were reviewed by Bushnell (1983) and more recently by Wilkinson et al. (1987; 1988). The last of these methods is the most straightforward. As a boundary layer thickens, the Reynolds number increases and the coefficient of skin friction decreases, with the result that the local skin friction is higher in the forward part of a vehicle and quite low in the aft end. Consider, for example, a 50-meter-long aircraft with a unit Reynolds number of 10^7/m, and ignore pressure gradient and other effects. Near the nose, the skin-friction coefficient is about 0.003 (Figure 1). Near the tail, $Re \simeq 5 \times 10^8$ and $C_f \simeq 0.0018$, i.e. 40% lower . By substantially reducing the wetted area of the forebody, the sword-fish configuration, drag reduction may be attainable. This technique is particularly useful when combined with a convex aftbody (Section 5.5).

5.4 RELAMINARIZATION

Several articles define and explain the reversion of a turbulent flow to the laminar state, variously also known as retransition, inverse or reverse transition, or relaminarization (Preston, 1958; Patel and Head, 1968; Bradshaw, 1969). Narasimha and Sreenivasan (1979) provide a comprehensive review and analysis of the phenomenon. Instances of relaminarization may be found in a stably stratified atmosphere, spatially or temporally accelerated shear flows, coiled pipes, swirling flames, tip vortices behind finite lifting surfaces, far wakes of finite drag-producing bodies, air flow as it progresses from the trachea to the segmental bronchi in human lungs, and shear flows subjected to magnetic fields. Various authors use one or more of the following syndromes to diagnose the reversion of a turbulent flow: reduction in heat transfer or skin-friction coefficient, cessation of bursting events near the wall or entrainment of potential flow at the outer region of a boundary layer, reduction in high-frequency velocity fluctuations, breakdown of the log-law in the wall layer, spreading of intermittency to wall region, and overlapping of energy containing eddies and dissipating eddy scales. In the quasi-laminar state, the velocity fluctuations are not necessarily zero but rather their contribution to the dynamics of the mean flow becomes inconsequential. In other words, the turbulent fluctuations inherited from the previous history of the flow are no longer influencing the transport of mass, momentum or energy. Narasimha and Sreenivasan (1979) adopt a pragmatic definition of reversion. They maintain that a flow has relaminarized if its development can be understood without recourse to any model for turbulent shear flow.

The most obvious mechanism for the occurrence of relaminarization is dissipation. When the Reynolds number goes down in a turbulent flow, for example by enlarging a duct or by branching a channel flow, the viscous dissipation may exceed the production of turbulent energy and the flow may revert to a quasi-laminar state. This type of reversion tends to be rather slow. Turbulent energy can also be destroyed or absorbed by the work done against external forces such as buoyancy (e.g., in stably stratified fluids), centrifugal (e.g., in convex boundary layers), or Coriolis (e.g., in rotating channel flows). The governing parameter in these absorptive-type reversion is the Richardson number (Ri) and relaminarization proceeds rapidly once the critical value of Ri is exceeded. In both dissipative- and absorptive-type of reversion, the turbulence energy is decreased but more significantly the velocity components that generate the crucial Reynolds stresses are "decorrelated." This explains the strong effects on turbulence of, say, a very mild positive curvature (So and Mellor, 1973; Smits and Wood, 1985). In this case, the amplitude as well as the phase of the different components of the fluctuating motion are affected by the additional strain rate imposed upon the flow, particularly away from the wall.

The third mechanism to effect relaminarization is observed in highly accelerated flows. Narasimha and Sreenivasan (1973) argue that, in the outer layer of such flows, the pressure forces dominate over nearly frozen Reynolds stress field. The $\overline{uv}$ term remains at about the same level as the zero-pressure-gradient case over much of the boundary layer. Near the wall, however, dissipation dominates and the flow is stabilized by the acceleration. The turbulence-re-energizing bursts diminish in frequency and may stop altogether and a thin new laminar boundary layer grows from the wall within the old boundary layer. According to Morkovin (1988), the turbulence in the outer layer is "starved" and the inner laminar layer is "buffeted" by the decaying, wake-like turbulence of the outer region. The skin friction and the heat transfer rate in the relaminarized inner layer is less than those expected in a corresponding turbulent flow.

Narasimha and Sreenivasan (1979) asserts that a number of different reverting flows can be considered in light of the three archetypes summarized above. Typically, combination of mechanisms rather than a single one are operating. In internal flows, relaminarization can be accomplished by gentle wall injection (to avoid separation) or by wall heating (for gases). In both these cases, the flow accelerates thus activating the third mechanism above. Moreover, when a gas is heated its kinematic viscosity is increased resulting in a lower Reynolds number that may fall below the critical value, thus activating the first mechanism.

In external flows, equation (5.8) indicates that wall suction, surface heating in liquids or surface cooling in gases, may have the same effect on the near-wall region of a turbulent boundary layer as favorable pressure gradient. It appears that only the first of these stimuli has been experimentally demonstrated. Dutton (1960) observed that a suction coefficient $C_q \simeq 0.01$ is sufficient to relaminarize a boundary layer at a $Re = 0 [10^6]$. The initially turbulent flow approaches the laminar asymptotic state (Section 3.3.2) appropriate to the particular value of C_q used. Although the mean velocity gradient near the wall is increased, the Reynolds shear stress is reduced at a faster rate leading to a reduction in the turbulent energy production.

The question of immediate concern to this section is whether or not suction, or in fact any of the other reversion stimuli, is a viable drag reduction method for turbulent boundary layers. The suction rate necessary for relaminarization is too high to be applied profitably over an entire surface, although it may be feasible to apply massive suction through a spanwise slot (perhaps enough to ingest the entire mass flow in the boundary layer) followed by gentle suction (or any of the other stability modifiers discussed in Section 3.3) to prevent transition of the newly formed laminar sublayer. This issue will be discussed again in the next section.

Convex (or positive) curvature affects the boundary layer locally in much the same way as a large eddy breakup device. With a relatively large radius of curvature, of the order 10δ, positive Reynolds stress is produced in the outer flow and the wall shear is reduced. Downstream of the end of the curvature region, the flow relaxes very slowly to its unperturbed state. Wilkinson et al. (1987; 1988) suggest that a convex surface may be better suited for drag reduction than a LEBU device, particularly in supersonic flows where wave drag penalty on the LEBU blade is rather large. Obviously a vehicle surface cannot be all convex and short regions of concave curvature on the way to the convex portions are unavoidable. Bushnell (1983) speculates that the limited extent of the concave regions may not allow formation of any lasting alterations to the turbulent structures. Net reductions of drag of the order of 20% appear to be obtainable using surface curvature. Little research is available on the feasibility of using other relaminarization techniques to achieve drag reduction.

5.5   SYNERGISM

Ideally, one searches for combinations of drag reduction methods where the total favorable effect is greater than the sum. For example, Lee et al. (1974) have shown that the use of polymers and fibers together can produce much larger reduction in skin friction than either additive alone. A slightly different strategy, but perhaps no less beneficial, is to employ a second control method that reduces the drag penalty of a given technique while adding some saving of its own. An example of that is the use of fluid that is being withdrawn from a certain portion of a vehicle for injection into another location, thus avoiding the ram drag penalty associated with blowing freestream fluid. For an aircraft, tangential slot injection or normal distributed injection may be used to reduce the skin friction on portions of the fuselage as discussed in Section 5.3.4. Possible sources of low-loss air include LFC suction from the wings and empennage, relaminarization slots on the fuselage, and active or passive bleed air for separation control.

Several possible combinations of drag reduction techniques were briefly mentioned elsewhere in this article. The sword-fish configuration (small wetted area) may be used in the forward part of a vehicle where the local skin friction is high followed by the rest of the body where the boundary layer is thicker and the skin-friction coefficient is low. A nose spike followed by a short region of concave curvature then a longer convex portion may be used in supersonic flows. Massive suction through a spanwise slot may be used to ingest the entire mass flow in a turbulent boundary layer followed by any of the stability modifiers discussed in Section 3.3 to maintain the growing laminar boundary layer downstream.

The suction rate necessary for establishing an asymptotic turbulent boundary layer independent of streamwise coordinate ($d\delta_\theta / d x_1 = 0$) is much lower than the rate required for relaminarization, but still not low enough to yield net drag reduction. For $Re = 0\,[10^6]$, Favre et al. (1966), Rotta (1970) and Verollet et al. (1972), among others, report an asymptotic suction coefficient of $C_q \simeq 0.003$. From equation (5.8) rewritten for a zero-pressure-gradient boundary layer on a flat plate, the corresponding skin-friction coefficient is $C_f = 2\,C_q = 0.006$, indicating higher skin friction than if no suction was applied. The problem is that fluid withdrawn through the wall has to come from outside the boundary layer where the streamwise momentum per unit mass is at the relatively high level of U_∞, as was discussed in Section 3.3.2 in connection with using suction for transition delay. To achieve a net drag reduction with suction, the process must be further optimized. The results of Eléna (1975; 1984) and more recently of Antonia et al. (1988) indicate that suction causes an appreciable stabilization of the low-speed streaks in the near-wall region. The maximum turbulence level at $y^+ \simeq 13$ drops from 15 to 12% as C_q varies from 0 to 0.003. More dramatically, the tangential Reynolds stress near the wall drops by a factor of 2 for the same variation of C_q. The dissipation length scale near the wall increases by 40% and the integral length scale by 25% with the suction.

Gad-el-Hak and Blackwelder (1987c; 1989) suggest that one possible means of optimizing the suction rate is to be able to identify where a low-speed streak is presently located and apply a small amount of suction under it. Assuming that the production of turbulent energy is due to the instability of an inflection $U(y)$ velocity profile, one needs to remove only enough fluid so that the inflectional nature of the profile is alleviated. An alternative technique that could conceivably reduce the Reynolds stress is to inject fluid selectively under the high-speed regions. The immediate effect would be to decrease the viscous shear at the wall resulting in less drag. In addition, the velocity profiles in the spanwise direction, $U(z)$, would have a smaller shear, $\partial U/\partial z$, because the injection would create a more uniform flow. Since Swearingen and Blackwelder (1984) have found that inflectional $U(z)$ profiles occur as often as inflection points are observed in $U(y)$ profiles, injection under the high-speed regions would decrease this shear and hence the resulting instability. The combination of selective suction and injection is sketched in Figure 9. In Figure 9a, the vortices are sketched idealized by a periodic distribution in the spanwise direction. The instantaneous velocity profiles without transpiration at constant y and z locations are shown by the dashed lines in Figures 9b and 9c, respectively. Clearly, the $U(y_0,z)$ profile is inflectional, having two inflection points per wavelength. At z_1 and z_3, an inflectional $U(y)$ profile is also evident. The same profiles with suction at z_1 and z_3 and injection at z_2 are shown by the solid lines. In all cases, the shear associated with the inflection points has been reduced. Since the inflectional profiles are all inviscidly unstable with growth rates proportional to the shear, the resulting instabilities would be weakened by the suction/injection process.

The feasibility of the selective suction concept is indicated in Figure 10 taken from Gad-el-Hak and Blackwelder (1989). Low-speed streaks were artificially generated using the method of Gad-el-Hak and Hussain (1986), and a hot-film probe was used to record their near-wall signature (Figure 10a).

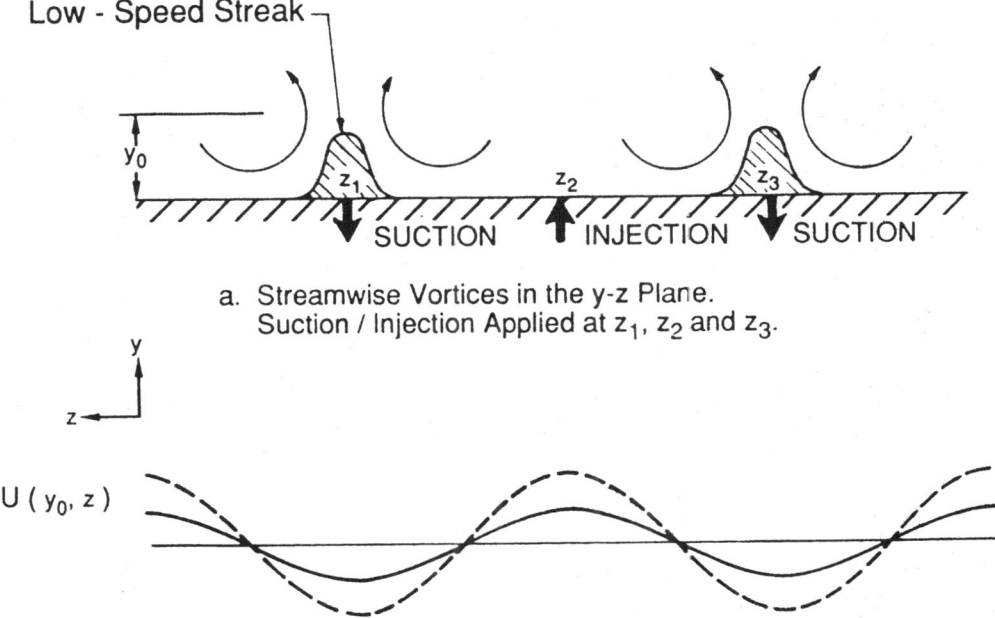

a. Streamwise Vortices in the y-z Plane.
Suction / Injection Applied at z_1, z_2 and z_3.

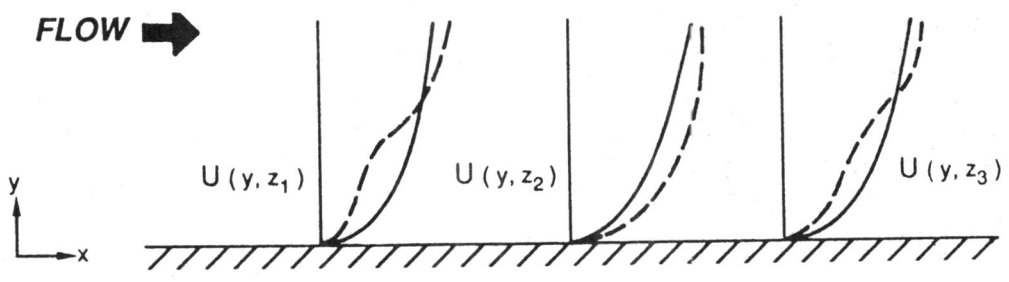

b. Resulting Spanwise Velocity Distribution at $y = y_0$.

c. Velocity Profiles Normal to the Surface.

Figure 9. Effects of Suction / Injection on Velocity Profiles. Dashed
Lines are Reference Profiles and Solid Lines are Profiles
with Transpiration. (from Gad-el-Hak and Blackwelder,
1987c).

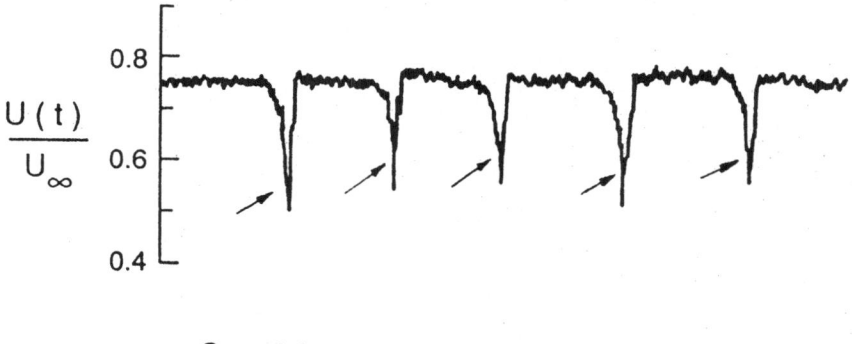

$$\frac{U(t)}{U_\infty}$$

a. $C_q = 0.0$.

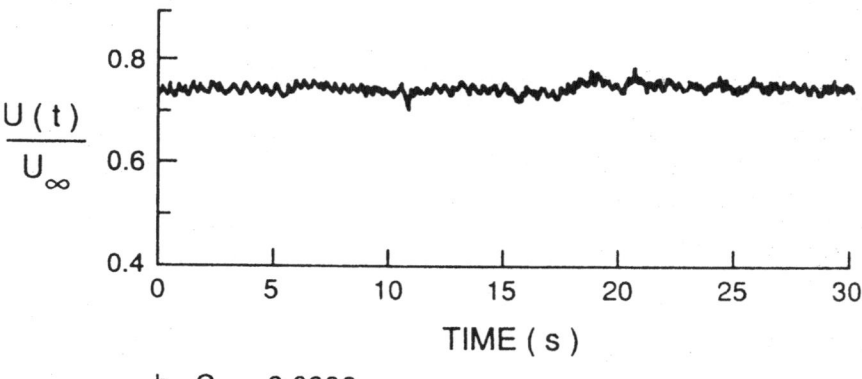

$$\frac{U(t)}{U_\infty}$$

TIME (s)

b. $C_q = 0.0006$.

Figure 10. Effects of Suction from a Streamwise Slot on Five
Artificially Induced Burst-Like Events in a Laminar
Boundry Layer. (from Gad-el-Hak and Blackwelder,
1988).

An equivalent suction coefficient of $C_q = 0.0006$ was sufficient to eliminate the artificial events and prevent bursting as shown in Figure 10b. This rate is five times smaller than the asymptotic suction coefficient for a corresponding turbulent boundary layer. If this result is sustained in a naturally developing turbulent boundary layer, a skin friction reduction of close to 60% would be attained. Gad-el-Hak and Blackwelder propose to combine suction with non-planar surface modifications. Minute longitudinal roughness elements if properly spaced in the spanwise direction greatly reduce the spatial randomness of the low-speed streaks (Johansen and Smith, 1986). By withdrawing the streaks forming near the peaks of the roughness elements, less suction should be required to achieve an asymptotic boundary layer. Recent experiments by Wilkinson and Lazos (1987) and Wilkinson (1988) combine suction/blowing with thin-element riblets. Although no net drag reduction is yet attained in these experiments, their results indicate some advantage of combining suction with riblets as proposed by Gad-el-Hak and Blackwelder (1987c; 1989).

A large eddy breakup device influencing the outer structures of a turbulent boundary layer may be combined with drag reduction methods that mainly influence the near-wall region. Walsh and Lindemann (1984) show that the performance of riblets in the presence of a LEBU is approximately additive, not quite a synergetic effect. Wilkinson et al. (1987; 1988) propose using a LEBU device to reduce the free mixing between tangentially injected fluid and the boundary layer flow, thus extending the low skin-friction region. Bushnell (1983) suggests that the suction mass flow rate required to achieve relaminarization of a turbulent boundary layer may be reduced by the use of a large eddy breakup device ahead of the suction region.

The LEBU's model of Gebert (1988) indicates that the fluctuating normal velocity component is most effectively suppressed for a particular incoming harmonic disturbance (Section 5.3.6). The modest drag reduction currently achieved using a LEBU device in a natural boundary layer may, therefore, be greatly enhanced by cyclically generating the large outer structures using the method developed by Gad-el-Hak and Blackwelder (1987b). In this case, the large eddies arrive at a given location at a precisely controlled period that matches the optimum frequency for the LEBU device. Alternatively, the Goodman's (1985) technique to generate closely spaced turbulent spots in the spanwise and streamwise directions may be used for tripping a laminar boundary layer. A LEBU device placed downstream will encounter more uniformly paced large eddies and its performance may be enhanced. Of course, in both schemes the penalty associated with artificially generating the large eddy structures must be considered.

Most of the combined drag reduction techniques discussed or proposed in this section are speculative. The reason being the scarcity of research in the area of synergism. Unguided by any theoretical framework and faced with the complexity of each of the methods used individually, one is tempted to stay away from the much more difficult situation when two or more techniques are combined. However, the potential for achieving a total effect larger than the sum should stimulate more research for combinations of method. In fact, in some situations one may not have much of a choice. Certain techniques, e.g. suction, may not yield net drag reduction and must, therefore, be combined with something else to achieve the desired goal. Other methods, e.g. riblets, yield modest drag reduction (less than 10%) that it may not be worth the effort implementing the individual technique.

6. TURBULENCE AUGMENTATION

For some applications, the efficient mixing or transport mechanisms of turbulence may be required. For example, a turbulent boundary layer is in general more resistant to separation than a laminar one; turbulence is used to homogenize fluid mixtures and to accelerate chemical reactions; and turbulent heat exchangers are much more efficient than laminar ones. As an additional example, earlier transition may be required for some low-speed wind or water tunnel testing to simulate high Reynolds number conditions.

Consider the three flow regimes discussed in Section 5.1. For a zero-pressure-gradient boundary layer, transition typically occurs at Reynolds number based on distance from leading edge of the order of 10^6. The critical Re below which perturbations of all wave numbers decay is about 6×10^4. To advance the transition Reynolds number, one may attempt to lower the critical Re, increase the growth rate of Tollmien-Schlichting waves, or introduce large disturbances that can cause bypass transition. The first two routes involve altering the shape of the velocity profile using wall motion (e.g., transition through deceleration, see Gad-el-Hak et al., 1984b), injection, adverse pressure gradient, or surface heating in gases or cooling in liquids. The third route is much simpler to implement though more difficult to analyze. Morkovin (1984) broadly classify the large disturbances that can cause bypass transition into steady or unsteady ones originating into the freestream or at the body surface. The most common example is single, multiple or distributed roughness elements placed on the wall. If the roughness characteristic-length is large enough, the disturbance introduced is nonlinear and bypass transition takes place. For a three-dimensional roughness element of height-to-width ratio of one, a transition Reynolds number $R_{\delta^*} \simeq 300$ (below the critical $R_{\delta^*} = 420$ predicted from the linear stability theory) is observed for a roughness Reynolds number based on its height and the velocity in the undisturbed boundary layer at the height of the element of about 10^3 (Tani, 1969). Transition occurs at $R_{\delta^*} \simeq 10^3$ for a roughness Reynolds number of about 600. For a smooth surface, transition typically takes place at $R_{\delta^*} \simeq 2.6 \times 10^3$.

Other large disturbances that could lead to early transition include high turbulence levels in the freestream, external acoustic excitations, particulate contamination, and surface vibration. These are often termed environmental tripping. Transition could also be effected by detecting naturally occurring T-S waves and artificially introducing in-phase waves. Transition could be considerably advanced, on demand, using this wave superposition principle.

Early transition could also be achieved by exploiting other routes to turbulence such as Taylor-Görtler or cross-flow vortices. For example, a very mild negative curvature of $(0.003/\delta^*)$ results in the generation of strong streamwise vortices. In this case, transition Reynolds number is lowered from $R_{\delta^*} \simeq 2600$ for the flat-plate case to $R_{\delta^*} \simeq 700$ for the curved surface (Tani, 1969). For high Mach number flows, the general decay in spatial amplification rate of T-S waves makes conventional tripping more difficult as the Mach number increases (Reshotko, 1976). For these flows, trips that generate oblique vorticity waves of appropriate wavelength may be most effective to advance the transition location.

The next issue is how to augment the turbulence for a shear flow that has already undergone transition, i.e. for Re = 10^6 and beyond. The obvious answer is to reverse the stimuli discussed in Section 5.4 in connection with relaminarization. Injection instead of suction in external flows, decelerating flows instead of accelerating ones, negative curvature (concave) instead of positive curvature (convex), and so on. These stimuli will lead to an increase in turbulence production and enhanced mass, momentum and heat transfers. Roughness will also enhance the turbulence, but its associated drag must be carefully considered. Other devices to enhance the turbulent mixing include vane-type vortex generators, which draw energy from the external flow, or wheeler-type or Keuthe-type generators, which are fully submerged with the boundary layer and presumably have less associated drag penalty (Rao and Kariya, 1988).

The size, performance and efficiency of many engineering systems are often limited by the ability to transfer heat to or from the system. Consider the supercomputer, where the speed of computing is limited by the distance through which the electrons must travel at nearly the speed of light. Electronic chips and other components must, therefore, be miniaturized and packaged tightly. Tremendous amounts of heat must be removed from a very modest surface area. The central processing unit of a present generation supercomputer is not much larger than, say, a refrigerator. In order to limit the temperature of the CRAY-2 to a tolerable range, its entire CPU is immersed in a dielectric liquid coolant (trade name FC-75). This is obviously an expensive, but perhaps unavoidable, approach to achieve the desired goal.

Convective heat transfer enhancement methods are reviewed by Bergles and Morton (1965), Bergles and Webb (1985), and Nakayama (1986). The simplest scheme involve the introduction of distributed roughness on the heat transfer surfaces. This would destabilize the wall region and intensify the mixing process leading to an increase in the convective heat transfer coefficient. Patera (1986) and Patera and Mikić (1986) introduce the concept of resonant heat transfer enhancement based on excitation of shear-layer instabilities in internal separated flows. In a two-dimensionally periodically grooved channel, Tollmien-Schlichting-like waves forced by the Kelvin-Helmholtz shear layer instabilities at the grooves' edge become unstable and take the form of self-sustained oscillations for Reynolds numbers greater than a critical value. For lower Re, oscillatory perturbation results in subcritical resonant excitation. For both laminar and turbulent flows, the resulting large-scale motions lead to significant lateral mixing and correspondingly dramatic enhancement of convective heat transfer coefficient.

7. NOISE CONTROL

Noise, or undesired sound, is generated from machines, vehicles moving in a fluid, or even natural phenomena. It is said that a single modern jet plane generates at takeoff more noise than the combined shouting power of the entire world's population! Noise control, a relatively young field of research, is required to maintain an acceptable (even pleasant) environment, to avoid detection of certain vehicles, and for the proper operation of sonars on underwater vehicles. Heckl (1988) broadly classify noise suppression methods as follows:

1 - Common sense application. For example, load or speed reduction, increase of
    mass, enclosures, or mufflers.

2 -   Add-on measures. These include multiple walls and floors, tuned absorbers and
vibration coatings, impedance mismatches in the propagation paths, and shift of
resonances away from exciting frequencies.

3 -   Changes at the source. For example, interruption of feedback loops or other
instabilities, reduction of roughness and free clearances, and reduction of time
derivatives.

4 -   Active methods of control such as sound reflection and absorption in wave guides,
or compensation of periodic excitation.

For the purpose of this article, flow-induced noise and its control are of primary interest. In 1952, Lighthill developed a very powerful theory that identified the sources of aerodynamic sound. Use the continuity equation (2.1) into the momentum equation and neglect body forces, (2.2) reads:

$$\frac{\partial (\rho U_i)}{\partial t} + \frac{\partial (\rho U_i U_j)}{\partial x_j} = \frac{\partial}{\partial x_j} \tau_{ij} \ . \tag{7.1}$$

By subtracting $\partial/\partial x_i$ of (7.1) from $\partial/\partial t$ of (2.1), an equation for the density results:

$$\frac{\partial^2 \rho}{\partial t^2} = \frac{\partial^2}{\partial x_i \partial x_j} (\rho U_i U_j - \tau_{ij}) \ . \tag{7.2}$$

Let $\rho = \rho_0 + \rho'$, where ρ' is the density perturbation due to sound, and subtract $c^2 \nabla^2 \rho'$ from both sides of (7.2), where c is the constant propagation speed of acoustic waves. The Lighthill's equation reads:

$$\frac{\partial^2 \rho'}{\partial t^2} - c^2 \nabla^2 \rho' = \frac{\partial^2 T_{ij}}{\partial x_i \partial x_j} \ , \tag{7.3}$$

where $T_{ij} \equiv \rho U_i U_j - \tau_{ij} - c^2 \rho' \delta_{ij}$, the Lighthill's stress tensor. The source field for the acoustic waves is a quadrupole distribution whose strength per unit volume is T_{ij}. Turbulence generates sound equivalently to a quadrupole source distribution. Organized structures in a turbulent flow play an important role in the source process and, therefore, controlling these eddies may be the key to noise suppression (Ffowcs Williams, 1977). At the low Mach number typically encountered in underwater applications ($O[0.01]$), unbounded turbulent flows have an extremely low acoustic efficiency according to Lighthill's (1952) theory. The large differences between the noise levels predicted by the quadrupole theory and those actually generated by a turbulent boundary layer are often explained via acoustic scattering processes (Crighton, 1984). Small rigid bodies, gas bubbles, or sharp edges can destroy the cancellation between opposing elements of the quadrupole and may lead to the more efficient radiation of a dipole or even a monopole. Thus, for noise suppression, these efficient sound generators should be avoided.

At present, sonars are placed at the nose of underwater vehicles where the boundary layer is laminar. Search rates of these listening devices is improved by placing additional hydrophones farther downstream. However, the turbulent-induced noise is sufficient to interfere with the proper operations of these devices, and must therefore be suppressed before the search rate could be improved. A second example comes from the aeronautics field. On commercial aircraft, the turbulent boundary layer outside the fuselage contributes significantly to cabin noise. Techniques to reduce flow-induced noise are therefore sought. Methods to delay transition (Section 3) or to relaminarize an already turbulent boundary layer (Section 5.4) should be useful to suppress the turbulent-induced noise in both examples above. It is feasible to delay transition on both vehicles, where the unit Re is about 10^7 /m, for the first few meters. Beyond that relaminarization or at least partial suppression of turbulence is the only option. As indicated in Section 5.4, achieving relaminarization over the entire length of an aircraft or a submarine may require large energy expenditure and is, therefore, not very practical. However, it may not be necessary to relaminarize an entire vehicle. Instead, selected portions of the boundary layer, where, for example, a hydrophone is to be placed, are treated.

A compliant coating that causes transition delay or reduction in turbulence skin friction will also lead to attenuation of sound radiated by the boundary layer. This is because the wall pressure fluctuations are dependent on turbulence levels, which in turn are related to the wall shear stress. The reverse is not necessarily true; i.e. a coating may attenuate the flow noise without affecting the hydrodynamic drag. In fact the technology exists today for manufacturing energy-absorbent compliant liners for sound absorption, vibration reduction, and noise shielding, while the search for a drag-reducing coating has thus far eluded researchers for about 30 years (Gad-el-Hak, 1986b; 1987b; Riley et al. 1988).

Flow noise is influenced by surface flexibility through two distinct mechanisms: either the surface acts as a sounding board excited by the turbulent pressure field, or the surface compliance induces a change in the turbulence structure and, hence, modifies the pressure fluctuations (Ffowcs Williams, 1965; Purshouse, 1976; Dowling, 1983; 1986). As mentioned earlier, flow noise is currently considered the limiting performance factor for sonar systems placed on surface ships, submarines, and towed arrays. Major advances in the reduction of "self-noise" have been achieved by exploiting the first mechanism above, and further reductions may be possible if the nature of the turbulent boundary layer and its wall pressure fluctuations can be altered (Von Winkle et al., 1982). Von Winkle (1961) and Barger and Von Winkle (1961) report pressure fluctuations measurements on a streamlined body of revolution free-falling in a water tank. Flush-mounted hydrophones fabricated from lead-zirconate titanite are used to measure the instantaneous pressure. Their experiment indicates a dramatic reduction in the pressure coefficient when the body is covered with a Krämer-type compliant coating. This result may, however, be due to transition delay caused by the flexible surface and not due to changes in the turbulence structure.

Active noise suppression systems are based on generating sound by auxiliary source with such an amplitude and phase that in the region of interest the sound wave interference from the original and auxiliary source results in considerable reduction of the noise levels. What makes these systems possible is, of course, the linearity of the governing equations.[*] The older systems with fixed gain in the feedback loop could not achieve high noise reduction. However, with the recent availability of adaptive filter systems, much more impressive suppression of noise is feasible (Tichy et al., 1984). These systems are capable of adjusting the feedback loop for the magnitude and phase relationship of the spectral components and can quickly compensate for the sound path changes. These active control devices seem to work best for low frequency sound, where passive silencers are relatively ineffective, and when the source is localized and accessible. Local control is obviously easier than global one. Moreover, effective control is achieved when the system response within the frequency band of interest is dominated by relatively few modes. Review papers on active control of noise are available by Warnaka (1980), Erickson et al. (1988), Van Laere and Sas (1988), and Warner et al. (1988). Anti-sound techniques seem to have gone from the laboratory to application in a remarkably short time.

8. CONCLUDING REMARKS

This article attempted the very ambitious task of presenting a unified view of the different control methods available or contemplated for external boundary-layer flows to achieve a variety of goals. These goals are not necessarily mutually exclusive and include transition delay, separation postponement, lift enhancement, drag reduction, turbulence augmentation, and noise control. In both laminar and turbulent, two-dimensional boundary layers, the effect of many of the control methods is explained in terms of the behavior of the spanwise vorticity flux at the wall. The fullness of the normal velocity profile is related to the direction of this flux, which in turn has a direct influence on the stability, separation, skin friction, and turbulence levels. The broad range of topics covered precludes in-depth review of a particular flow control technique and the reader is referred to selected original publications detailing these methods and referenced throughout this article.

Both the science and technology to maintain a laminar boundary layer to a Reynolds number of about 4×10^7 are well established, although some details remain to be worked out. The linear stability theory provides a solid analytical framework, at least for the important first stage of transition. Barring large disturbances in a conventional boundary-layer flow, the linear amplification of Tollmien Schlichting waves is the slowest of the successive multiple steps in the transition process. Stability modifiers inhibit this linear amplification and, therefore, determine the magnitude of the transition Reynolds number. Shaping to provide extended regions of favorable pressure gradient is the simplest method of control and is well suited for small underwater vehicles or for the wings of low- or moderate-speed aircraft. Flight tests have demonstrated the feasibility of using suction to maintain a laminar flow on a swept wing to $Re \simeq 4.7 \times 10^7$. The required suction rate is very modest and 20% net drag reduction is possible. Remaining problems are technological in nature and include maintainability and reliability of suction surfaces and further optimization of the suction rate and its distribution. Suction is less suited for underwater vehicles because of the abundance of particulate matters that can clog the suction surface as well as destabilize the boundary layer. For water applications, compliant coatings that increase the

---

[*]The pressure fluctuations associated with sound at the threshold of pain (140 dB) are only 0.1% of the atmospheric pressure.

transitional Reynolds number by a factor of 5-10 are available in the laboratory but performance in the field is still unknown. This technique is very appealing because of its simplicity and absence of energy requirement. Moderate surface heating also increases the transition Re by an order of magnitude, but a source of reject heat must be available to achieve net drag reduction. Additionally, a particle-defense mechanism is needed before the technique could successfully be used in the ocean. For futuristic aircraft using cryo-fuel, surface cooling may be a feasible method to delay transition.

The above stability modifiers change the shape of the velocity profile making it more full. Therefore, two-dimensional, steady separation can be postponed using the same techniques. Other separation control methods include passive ones, such as intentional tripping, fences or vortex generators, and active devices, such as tangential injection, acoustic excitations or oscillating surface flaps. A drag penalty is associated with both passive and active devices.

For time-dependent flows, upstream movement of the separation point results in a delay in boundary layer detachment, analogous to a wall moving in the same direction as the freestream. Three-dimensional boundary layers are in general more capable of overcoming an adverse pressure gradient without separation. In this case, the near-wall fluid is capable of moving in a direction in which the pressure gradient is more favorable, thus avoiding the adverse pressure in the direction of the main flow.

Techniques to reduce the pressure drag are more well established than turbulent skin-friction reduction techniques. Streamlining and other methods to postpone separation can eliminate most of the pressure drag. The wave and induced drag contributions to the pressure drag can also be reduced by geometric design. The skin friction constitutes about 50%, 90% and 100% of the total drag on commercial aircraft, underwater vehicles and pipelines, respectively. Most of the current research effort concerns reduction of skin-friction drag for turbulent boundary layers.

Three flow regimes are identified. First, for $Re < 10^6$, the flow is laminar and skin friction may be lowered by reducing the near-wall momentum. Adverse pressure gradient, blowing and surface heating/cooling could lower the skin friction, but increase the risk of transition and separation. Secondly, for $10^6 < Re < 4 \times 10^7$, active and passive methods to delay transition could be used, thus avoiding the much higher turbulent drag. Thirdly, at the Reynolds number encountered after the first few meters of a fuselage or a submarine, methods to reduce the large skin friction associated with turbulent flows are sought. These methods are classified in the following categories: Reduction of near-wall momentum; introduction of foreign substance; geometrical modification; relaminarization; and synergism.

The second category above leads to the most impressive results. Introduction of small concentration of polymers, surfactants, particles or fibers into a turbulent boundary layer lead to a reduction in the skin friction coefficient of as much as 80%. Among the practical considerations requiring further study are the cost of the additive, methods of delivering it to the boundary layer, potential for recovering and recycling, degradation, and the portion of the payload that has to be displaced to make room for the additive.

Recently introduced techniques mostly fall under the third category above and seem to offer more modest net drag reduction. These methods are, however, still in the research stage and include riblets (~ 8%), large eddy breakup devices (~ 20%), and convex surfaces (~ 20%). Potential improvement in these and other methods will perhaps involve combining more than one technique aiming at achieving a favorable effect that is greater than the sum. Due to its obvious difficulties, this area of research has not been very popular in the past but deserves future attention.

Along these lines, the selective suction technique, reviewed in Section 5.5, combines suction to achieve an asymptotic turbulent boundary layer and longitudinal riblets to fix the location of low-speed streaks. Although far from indicating net drag reduction, the available results are encouraging and further optimization is needed. Potentially the selective suction method is capable of skin friction reduction that approaches 60%.

Techniques to augment the turbulence for non-reacting mixing, combustion, heat transfer, and other applications were reviewed in Section 6. In the last section, methods to suppress noise, particularly flow-induced sound, were summarized. The most recent of these techniques exploits the linearity of the acoustic field. Active noise control using anti-sound is a new research area that, remarkably, seems to have gone into applications in just a few years.

ACKNOWLEDGMENT

The writing of this article, together with the research of the author for the past five years, has been supported by the National Science Foundation Grant ISI-8560825; the Air Force Office of Scientific Research Contract F49620-85-C-0131; the Office of Naval Research Contract N00014-81-C-0453; and the National Aeronautics and Space Administration-Langley Research Center Contracts NAS1-17951, NAS1-18292 and NAS1-18213. The author wishes to express his sincere appreciations to R.F. Blackwelder for numerous useful discussions and for commenting on the manuscript.

REFERENCES

Ackeret, J., Ras, M., and Pfenninger, W. (1941) "Verhinderung des Turbulentwerdens einer Grenzschicht durch Absaugung," Naturwissenschaften 29, pp. 622-623.

Ahuja, K.K., Whipkey, R.R., and Jones, G.S. (1983) "Control of Turbulent Boundary Layer Flows by Sound," AIAA Paper No. 83-0726.

Alvarez-Calderon, A. (1964) "Rotating Cylinder Flaps of V/STOL Aircraft," Aircraft Eng. 36, pp. 304-309.

Anders, J.B., Hefner, J.N., and Bushnell, D.M. (1984) "Performance of Large-Eddy Breakup Devices at Post-Transitional Reynolds Numbers," AIAA Paper No. 84-0345.

Anders, J.B., Walsh, M.J., and Bushnell, D.M. (1988) "The Fix for Tough Spots," Aerospace America 26, pp. 24-27.

Anders, J.B., and Watson, R.D. (1985) "Airfoil Large-Eddy Breakup Devices for Turbulent Drag Reduction," AIAA Paper No. 85-0520.

Antonia, R.A., Fulachier, L., Krishnamoorthy, L.V., Benabid, T., and Anselmet, F. (1988) "Influence of Wall Suction on the Organized Motion in a Turbulent Boundary Layer," J. Fluid Mech. 190, pp. 217-240.

Aslanov, P.V., Maksyutenko, S.N., Povkh, I.L., Simonenko, A.P., and Stupin, A.B. (1980) "Turbulent Flows of Solutions of Surface-Active Substances," Izvestiya Akademii Nauk SSSR, Mekhanika Zhidkosti i Gaza, No. 1, pp. 36-43.

Atassi, H.M., and Gebert, G.A. (1987) "Modification of Turbulent Boundary Layer Structure by Large-Eddy Breakup Devices," Proc. Int. Conf. on Turbulent Drag Reduction by Passive Means, Vol. 2, pp. 432-456, Royal Aeronautical Society, London, United Kingdom.

Bandyopadhyay, P.R. (1986) "Review-Mean Flow in Turbulent Boundary Layers Disturbed to Alter Skin Friction," J. Fluids Eng. 108, pp. 127-140.

Barger, J.E., and Von Winkle, W.A. (1961) "Evaluation of a Boundary Layer Stabilization Coating," J. Acoustical Soc. of America 33, p. 836.

Barker, S.J., and Gile, D. (1981) "Experiments on Heat-Stabilized Laminar Boundary Layers in Water," J. Fluid Mech. 104, pp. 139-158.

Batchelor, G.K. (1967) An Introduction to Fluid Dynamics, Cambridge University Press, London.

Batchelor, G.K., and Green, J.T. (1972) "The Determination of the Bulk Stress in a Suspension of Spherical Particles to Order c^2," J. Fluid Mech. 56, pp. 401-427.

Bechert, D.W., Hoppe, G., and Reif, W.-E. (1985) "On the Drag Reduction of the Shark Skin," AIAA Paper No. 85-0546.

Benjamin, T.B. (1960) "Effects of a Flexible Boundary on Hydrodynamic Stability," J. Fluid Mech. 9, pp. 513-532.

Bergles, E.A., and Morton, L.H. (1965) "Survey and Evaluation of Techniques to Augment Convective Heat Transfer," Dept. of Mech. Eng., Report No. EPL 5382-34, MIT, Cambridge, MA.

Bergles, E.A., and Webb, R.L. (1985) "A Guide to the Literature on Convective Heat Transfer Augmentation," Twenty-Third National Heat Transfer Conference: Advances in Enhanced Heat Transfer, Denver, CO.

Berman, N.S. (1978) "Drag Reduction by Polymers," Ann. Rev. Fluid Mech. 10, pp. 47-64.

Berman, N.S., and George, W.K. (1974) "Time Scale and Molecular Weight Distribution Contributions to Dilute Polymer Solution Fluid Mechanics," Proc. Heat Transfer Fluid Mech. Inst., eds. L.R. Davis and R.E. Wilson, pp. 348-364, Stanford Univ. Press, CA.

Bernard, J.J., and Siestrunck, R. (1958) "Échanges de Chaleur dans les Écoulements Présentant des Décollements," First Int. Cong. Aeronautical Sciences, Madrid, Spain.

Bertelrud, A., Truong, T.V., and Avellan, F. (1982) "Drag Reduction in Turbulent Boundary Layers Using Ribbons," AIAA Paper No. 82-1370.

Biringen, S. (1984) "Active Control of Transition by Periodic Suction-Blowing," Phys. Fluids 27, pp. 1345-1347.

Biringen, S., and Maestrello, L. (1984) "Development of Spot-Like Turbulence in Plane Channel Flow," Phys. Fluids 27, pp. 318-321.

Blackwelder, R.F. (1978) "The Bursting Process in Turbulent Boundary Layers," in Coherent Structures of Turbulent Boundary Layers, eds. C.R. Smith and D.E. Abbott, pp. 211-227, Lehigh University, Bethlehem, PA.

Blackwelder, R.F., and Haritonidis, J.H. (1983) "Scaling of the Bursting Frequency in Turbulent Boundary Layers," J. Fluid Mech. 132, pp. 87-103.

Bogdevich, V.G., Evseev, A.R., Malyuga, A.G., and Migirenko, G.S. (1977) "Gas-Saturation Effect on Near-Wall Turbulence Characteristics," Second Int. Conf. on Drag Reduction, Paper No. D2, BHRA Fluid Engineering, Cranfield, United Kingdom.

Bogdevich, V.G., and Malyuga, A.G. (1976) "The Distribution of Skin Friction in a Turbulent Boundary Layer of Water beyond the Location of Gas Injection," in Studies on the Boundary Layer Control (in Russian), eds. S.S. Kutateladze, and G.S. Migirenko, p. 62, Institute of Thermophysics, Novosibirsk, U.S.S.R.

Bradshaw, P. (1969) "A Note on Reverse Transition," J. Fluid Mech. 35, pp. 387-390.

Braslow, A.L., Burrows, D.L., Tetervin, N., and Visconti, F. (1951) "Experimental and Theoretical Studies of Area Suction for the Control of Laminar Boundary Layer," NACA Report No. 1025.

Bushnell, D.M. (1983) "Turbulent Drag Reduction for External Flows," AIAA Paper No. 83-0227.

Bushnell, D.M. (1989) "Applications and Suggested Directions of Transition Research," Fourth Symp. on Numerical and Physical Aspects of Aerodynamic Flows," Long Beach, CA, 16-19 January.

Bushnell, D.M., Hefner, J. N., and Ash, R.L. (1977) "Effect of Compliant Wall Motion on Turbulent Boundary Layers," Phys. Fluids 20, pp. S31-S48.

Bushnell, D.M., and Malik, M.R. (1988) "Compressibility Influences on Boundary-Layer Transition," Symp. on Physics of Compressible Turbulent Mixing, Princeton, NJ, 25-27 October.

Bushnell, D.M., Malik, M.R., and Harvey, W.D. (1988) "Transition Prediction in External Flows Via Linear Stability Theory, " Proc. IUTAM Symp. Transsonicum III, Gottingen, Germany, 24-27 May.

Bussmann, K., and Münz, H. (1942) "Die Stabilität der laminaren Reibungsschicht mit Absaugung," Jahrb. Dtsch. Luftfahrtforschung 1, pp. 36-39.

Cantwell, B.J. (1981) "Organized Motion in Turbulent Flow," Ann. Rev. Fluid Mech. 13, pp. 457-515.

Carpenter, P.W., and Garrad, A.D. (1985) "The Hydrodynamic Stability of Flow Over Kramer-Type Compliant Surfaces. Part 1. Tollmien-Schlichting Instabilities," J. Fluid Mech. 155, pp. 465-510.

Cary, A.M., Jr., Weinstein, L.M., and Bushnell, D.M. (1980) "Drag Reduction Characteristics of Small Amplitude Rigid Surface Waves," in Viscous Flow Drag Reduction, ed. G.R. Hough, pp. 144-167, AIAA Prog. in Astro. & Aero. 72, New York.

Chang, P.K. (1970) Separation of Flow, Pergamon Press, Oxford, England.

Chang, P.K. (1976) Control of Flow Separation, Hemisphere, Washington, D.C.

Charmichael, B.H. (1974) "Application of Sailplane and Low-Drag Underwater Vehicle Technology to the Long-Endurance Drone Problem," AIAA Paper No. 74-1036.

Chen, C.P., Goland, Y., and Reshotko (1979) "Generation Rate of Turbulent Patches in the Laminar Boundary Layer of a Submersible," in Viscous Flow Drag Reduction, ed. G.R. Hough, AIAA Progress in Astronautics & Aeronautics 72, pp. 73-89.

Corino, E.R., and Brodkey, R.S. (1969) "A Visual Investigation of the Wall Region in Turbulent Flow," J. Fluid Mech. 37, pp. 1-30.

Corke, T.C., Guezennec, Y., and Nagib, H.M. (1980) "Modification in Drag of Turbulent Boundary Layers Resulting from Manipulation of Large-Scale Structures," in Viscous Flow Drag Reduction, ed. G.R. Hough, pp. 128-143, AIAA Prog. in Astro. & Aero. 72, New York.

Corke, T.C., Nagib, H.M., and Guezennec, Y. (1981) "A New View on Origin, Role and Manipulation of Large Scales in Turbulent Boundary Layers," NASA Contractor Report No. 165861.

DiPrima, R.C., and Swinney, H.L. (1985) "Instabilities and Transition in Flow Between Concentric Rotating Cylinders, in Hydrodynamic Instabilities and the Transition to Turbulence, eds. H.L. Swinney and J.P. Gollub, Second Edition, pp. 139-180, Springer-Verlag, Berlin.

Dougherty, N.S., and Fisher, D.F. (1980) "Boundary Layer Transition on a 10-Degree Cone," AIAA Paper No. 80-0154.

Dowling, A.P. (1983) "Flow-Acoustic Interaction Near a Flexible Wall," J. Fluid Mech. 128, pp. 181-198.

Dowling, A.P. (1985) "The Effect of Large-Eddy Breakup Devices on Oncoming Vorticity," J. Fluid Mech. 160, pp. 447-463.

Dowling, A.P. (1986) "Mean Flow Effects on the Low-Wavenumber Pressure Spectrum on a Flexible Surface," J. Fluids Eng. 108, pp. 104-108.

Drazin, P., and Reid, W. (1981) Hydrodynamic Stability, Cambridge University Press, London.

Dutton, R.A. (1960) "The Effects of Distributed Suction on the Development of Turbulent Boundary Layers," ARC R&M No. 3155.

Eléna, M. (1975) "Etude des Champs Dynamiques et Thermiques d'un Ecoulement Turbulent en Conduit avec Aspiration à la Paroi," Thèse de Doctorat ès Sciences, Université d' Aix-Marseille, Marseille, France.

Eléna, M. (1984) "Suction Effects on Turbulence Statistics in a Heated Pipe Flow," Phys. Fluids 27, pp. 861-866.

Emmons, H.W. (1951) "The Laminar-Turbulent Transition in a Boundary Layer. Part I," J. Aero. Sci. 18, pp. 490-498.

Eriksson, L.J., Allie, M.C., Bremigan, C.D., and Gilbert, J.A. (1988) "Active Noise Control and Specificiations for Fan Noise Problems," Proc. Noise Control Design: Methods and Practice, ed. J.S. Bolton, pp. 273-278, Noise Control Foundation, Poughkeepsie, NY.

Falco, R.E. (1977) "Coherent Motions in the Outer Region of Turbulent Boundary Layers," Phys. Fluids 20, pp. S124-S132.

Falco, R.E. (1980) "The Production of Turbulence Near a Wall," AIAA Paper No. 80-1356.

Falco, R.E. (1983) "New Results, a Review and Synthesis of the Mechanism of Turbulence Production in Boundary Layers and its Modification," AIAA Paper No. 83-0377.

Favre, A., Dumas, R., Verollet, E., and Coantic, M. (1966) "Couche Limite Turbulente sur Paroi Poreuse avec Aspiration," J. Mecanique 5, pp. 3-28.

Ffowcs Williams, J.E. (1965) "Sound Radiation from Turbulent Boundary Layers Formed on Compliant Surfaces," J. Fluid Mech. 22, pp. 347-358.

Ffowcs Williams, J.E. (1977) "Aeroacoustics," Ann. Rev. Fluid Mech. 9, pp. 447-468.

Flatt, J. (1961) "The History of Boundary Layer Control Research in the United States of America," in Boundary Layer and Flow Control, Vol. 1, ed. G.V. Lachmann, pp. 122-143, Pergamon Press, New York.

Flettner, A. (1924) "Die Anwendung der Erkenntnisse der Aerodynamik zum Windantrieb von Schiffen," Jb. Schiffbautech. Ges. 25, pp. 222-251.

Francis, M.S., Keesee, J.E., Lang, J.D., Sparks, G.W., and Sisson, G.E. (1979) "Aerodynamic Characteristics of an Unsteady Separated Flow," AIAA J. 17, pp. 1332-1339.

Frick, C.W., and McCullough, C.B. (1942) "Tests of a Heated Low Drag Airfoil," NACA ARR, December.

Gadd, G.E. (1960) "Boundary Layer Separation in the Presence of Heat Transfer, NATO Advisory Group for Aerospace Research and Development, AGARD Report No. 280.

Gadd, G.E., Cope, W.F., and Attridge, J.L. (1958) "Heat-Transfer and Skin-Friction Measurements at a Mach Number of 2.44 for a Turbulent Boundary Layer on a Flat Surface and in Regions of Separated Flow, ARC R&M 3148.

Gad-el-Hak, M. (1986 a) "The Use of the Dye-Layer Technique for Unsteady Flow Visualization," J. Fluids Eng. 108, pp. 34-38.

Gad-el-Hak, M. (1986 b) "Boundary Layer Interactions With Compliant Coatings: An Overview," Appl. Mech. Rev. 39, pp. 511-523.

Gad-el-Hak, M. (1986 c) "The Response of Elastic and Viscoelastic Surfaces to a Turbulent Boundary Layer," J. Appl. Mech. 53, pp. 206-212.

Gad-el-Hak, M. (1987 a) "Unsteady Separation on Lifting Surfaces," Appl. Mech. Rev. 40, pp. 441-453.

Gad-el-Hak, M. (1987 b) "Compliant Coatings Research: A Guide to the Experimentalist," Fluids & Struct. 1, pp. 55-70.

Gad-el-Hak, M. (1988 a) "Review of Flow Visualization Techniques for Unsteady Flows," in Flow Visualization IV, ed. C. Véret, pp. 1-12, Hemisphere, Washington, D.C.

Gad-el-Hak, M. (1988 b) "Visualization Techniques for Unsteady Flows: An Overview," J. Fluids Eng. 110, pp. 231-243.

Gad-el-Hak, M., and Blackwelder, R.F. (1985) "The Discrete Vortices from a Delta Wing," AIAA J. 23, 961-962.

Gad-el-Hak, M., and Blackwelder, R.F. (1987 a) "Control of the Discrete Vortices from a Delta Wing," AIAA J. 25, pp. 1042-1049.

Gad-el-Hak, M., and Blackwelder, R.F. (1987 b) "Simulation of Large-Eddy Structures in a Turbulent Boundary Layer," AIAA J. 25, pp. 1207-1215.

Gad-el-Hak, M., and Blackwelder, R.F. (1987 c) "A Drag Reduction Method for Turbulent Boundary Layers," AIAA Paper No. 87-0358.

Gad-el-Hak, M., and Blackwelder, R.F. (1989) "Selective Suction for Controlling Bursting Events in a Boundary Layer," AIAA J. 27, No. 2.

Gad-el-Hak, M., Blackwelder, R.F., and Riley, J.J. (1981) "On the Growth of Turbulent Regions in Laminar Boundary Layers," J. Fluid Mech. 110, pp. 73-95.

Gad-el-Hak, M., Blackwelder, R.F., and Riley, J.J. (1984 a) "On the Interaction of Compliant Coatings With Boundary Layer Flows," J. Fluid Mech. 140, pp. 257-280.

Gad-el-Hak, M., Blackwelder, R.F., and Riley, J. J. (1985) "Visualization Techniques for Studying Transitional and Turbulent Flows," in Flow Visualization III, ed. W. J. Yang, pp. 568 575, Hemisphere, Washington, D.C.

Gad-el-Hak, M., Davis, S.H., McMurray, J.T., and Orszag, S.A. (1984 b) "On the Stability of the Decelerating Boundary Layer," J. Fluid Mech. 138, pp. 297-323.

Gad-el-Hak, M., and Ho, C.-M. (1985) "The Pitching Delta Wing," AIAA J. 23, pp. 1660-1665.

Gad-el-Hak, M., and Ho, C.-M. (1986 a) "Unsteady Flow Around An Ogive-Cylinder," J. Aircraft 23, pp. 520-528.

Gad-el-Hak, M., and Ho, C.-M. (1986 b) "Unsteady Vortical Flow Around Three-Dimensional Lifting Surfaces," AIAA J. 24, pp. 713-721.

Gad-el-Hak, M., and Hussain, A. K. M. F. (1986) "Coherent Structures in a Turbulent Boundary Layer. Part 1. Generation of 'Artificial' Bursts", Phys. Fluids 29, pp. 2124-2139.

Gampert, B., Homann, K., and Rieke, H.B. (1980) "The Drag Reduction in Laminar and Turbulent Boundary Layers by Prepared Surfaces with Reduced Momentum Transfer," Israel J. Technology 18, pp. 287-292.

Gebert, G.A. (1988) "Turbulent Boundary Layer Modification by Streamlined Devices," Ph.D. Thesis, University of Notre Dame, IN.

Gedney, C.J. (1983) "The Cancellation of a Sound-Excited Tollmien-Schlichting Wave with Plate Vibration," Phys. Fluids 26, pp. 1158-1160.

Goldstein, M.E. (1984) "Generation of Instability Waves in Flows Separating from Smooth Surfaces," J. Fluid Mech. 145, pp. 71-94.

Goodman, W.L. (1985) "Emmons Spot Forcing for Turbulent Drag Reduction," AIAA J. 23, PP. 155-157.

Granville, P.S. (1979) "Drag of Underwater Bodies," in Hydroballistics Design Handbook, Vol. 1, pp. 309-341, Naval Sea Systems Command, SEAHAC TR 79-1, Washington, D.C.

Heckl, M. (1988) "The Use of Mathematical Methods in Noise Control Design," Proc. Noise Control Design: Methods and Practice, ed. J.S. Bolton, pp. 27-38, Noise Control Foundation, Poughkeepsie, NY.

Hefner, J.N. (1988) "Dragging Down Fuel Costs," Aerospace America 26, pp. 14-16.

Hefner, J.N., Anders, J.B., and Bushnell, D.M. (1983) "Alteration of Outer Flow Structures for Turbulent Drag Reduction," AIAA Paper No. 83-0293.

Hefner, J.N., Weinstein, L.M., and Bushnell, D.M. (1980) "Large-Eddy Breakup Scheme for Turbulent Viscous Drag Reduction," in Viscous Flow Drag Reduction, ed. G.R. Hough, pp. 110-127, AIAA Prog. in Astro. & Aero. 72, New York.

Hendricks, E.W., and Ladd, D.M. (1983) "Effect of Surface Roughness on the Delayed Transition on 9:1 Heated Ellipsoid," AIAA J. 21, pp. 1406-1409.

Hinch, E.J. (1977) "Mechanical Models of Dilute Polymer Solutions in Strong Flows," Phys. Fluids 20, pp. S22-S30.

Hinze, J. O. (1975) Turbulence, Second Edition, McGraw-Hill, New York.

Holmes, B. J. (1988) "NLF Technology is Ready to Go," Aerospace America 26, pp. 16-20.

Holstein, H. (1940) "Messungen zur Laminarhaltung der Grenzschicht an einem Flügel," Lilienthal-Bericht S10, pp. 17-27.

Hooshmand, A., Youngs, R., Wallace, J.M., and Balint, J.-L. (1983) "An Experimental Study of Changes in the Structure of a Turbulent Boundary Layer Due to Surface Geometry Changes," AIAA Paper No. 83-0230.

Howard, F.G., and Goodman, W.L. (1985) "Axisymmetric Bluff-Body Drag Reduction Through Geometrical Modification," J. Aircraft 22, pp. 516-522.

Howard, F.G., and Goodman, W.L. (1987) "Drag Reduction on a Bluff Body at Yaw Angles to 30 Degrees," J. Spacecraft & Rockets 24, pp. 179-181.

Howard, F.G., Hefner, J.N., and Srokowski, A.J. (1975) "Multiple Slot Skin Friction Reduction," J. Aircraft 12, pp. 753-754.

Hoyt, J.W. (1972) "Turbulent Flow of Drag-Reducing Suspensions," Naval Undersea Center, Report No. TP 299, San Diego, CA.

Hoyt, J.W. (1979) "Polymer Drag Reduction - A Literature Review," Second Int. Conf. on Drag Reduction, Paper No. A1, BHRA Fluid Engineering, Cranfield, United Kingdom.

Hurley, D.G. (1961) "The Use of Boundary Layer Control to Establish Free Streamline Flows," in Boundary Layer and Flow Control, Vol. 1, ed. G.V. Lachmann, pp. 295-341, Pergamon Press, New York.

Iglisch, R. (1944) "Exakte Berechnung der laminaren Reibungsschicht an der längsangeströmten ebenen Platte mit homogener Absaugung," Schr. Dtsh. Akad. Luftfahrtforschung 8 B, pp. 1-51.

Illingworth, C.R. (1954) "The Effect of Heat Transfer on the Separation of a Compressible Laminar Boundary Layer," Quart. J. Mech. Appl. Math. 7, pp. 8-34.

Itoh, N. (1987) "Another Route to the Three-Dimensional Development of Tollmien-Schlichting Waves with Finite Amplitude," J. Fluid Mech. 181, pp. 1-16.

Jaffe, N.A., Okamura, T.T., and Smith, A. M. O. (1970) "Determination of Spatial Amplification Factors and Their Application to Predicting Transition," AIAA J. 8, pp. 301-308.

Johansen, J.B., and Smith, C.R. (1986) "The Effects of Cylindrical Surface Modifications on Turbulent Boundary Layers," AIAA J. 24, pp. 1081-1087.

Kachanov, Y.S., Koslov, V.V., and Levchenko, V. Ya. (1974) "Experimental Study of the Influence of Cooling on the Stability of Laminar Boundary Layers," Izvestia Sibirskogo Otdielenia Ak. Nauk SSSR, Seria Technicheskikh Nauk, Novosibirsk, No. 8-2, pp. 75-79.

Kannberg, L.D. (1988) "The Urgency Will Return,"Mechanical Engineering 110, p. 33.

Kaplan, R. E. (1964) "The Stability of Laminar Incompressible Boundary Layers in the Presence of Compliant Boundaries," Sc.D. Thesis, Massachusetts Institute of Technology, Cambridge, MA.

Kays, W.M., and Crawford, M.E. (1980) Convective Heat and Mass Transfer, McGraw-Hill, New York.

Kendall, J.M. (1970) "The Turbulent Boundary Layer Over a Wall with Progressive Surface Waves," J. Fluid Mech. 41, pp. 259-281.

Kim, H.T., Kline, S.J., and Reynolds, W.C. (1971) "The Production of Turbulence Near a Smooth Wall in a Turbulent Boundary Layer," J. Fluid Mech. 50, pp. 133-160.

Klebanoff, P.S., Schubauer, G.B., and Tidstrom, K.D. (1955) "Measurements of the Effect of Two-Dimensional and Three-Dimensional Roughness Elements on Boundary-Layer Transition," J. Aero. Sci. 22, pp. 803-804.

Klebanoff, P.S., Tidstrom, K.D., and Sargent, L.M. (1962) "The Three-Dimensional Nature of Boundary Layer Instability," J. Fluid Mech. 12, pp. 1-34.

Kline, S.J., Reynolds, W.C., Schraub, F.A., and Runstadler, P.W. (1967) "The Structure of Turbulent Boundary Layers," J. Fluid Mech. 30, pp. 741-773.

Koga, D.J., Reisenthel, P., and Nagib, H.M. (1984) "Control of Separated Flowfields Using Forced Unsteadiness," Illinois Institute of Technology, Fluids & Heat Transfer Report No. R84-1, Chicago, IL.

Kosecoff, M.A., Ko, D.R.S., and Merkle, C.L. (1976) "An Analytical Study of the Effect of Surface Roughness on the Stability of a Heated Water Boundary Layer," Dynamics Technology, Inc., Final Report PDT 76-131, Torrance, CA.

Krämer, M.O. (1960) Boundary Layer Stabilization by Distributing Damping," J. Am. Soc. Naval Engrs. 72, pp. 25-33.

Lachmann, G.V. (1961) Boundary Layer and Flow Control, Volumes 1 and 2, Pergamon Press, New York.

Ladd, D.M., and Hendricks, E.W. (1988) "Active Control of 2-D Instability Waves on an Axisymmetric Body," Exp. Fluids 6, pp. 69-70.

Landahl, M.T. (1962) "On the Stability of a Laminar Incompressible Boundary Layer Over a Flexible Surface," J. Fluid Mech. 13, pp. 609-632.

Landahl, M.T. (1973) "Drag Reduction by Polymer Addition," Proc. 13th IUTAM Congress, eds. E. Becker and G.K. Mikhailov, pp. 177-199, Springer, Berlin.

Landahl, M.T. (1977) "Dynamics of Boundary Layer Turbulence and the Mechanism of Drag Reduction," Phys. Fluids 20, pp. S55-S63.

Landau, L.D., and Lifshitz, E.M. (1963) Fluid Mechanics, translated from the Russian, Pergamon Press, Oxford.

Lange, R.H. (1954) "Present Status of Information Relative to the Prediction of Shock-Induced Boundary Layer Separation," NACA TN 3065.

Lankford, J. L. (1960) "Investigation of the Flow Over an Axisymmetric Compression Surface at High Mach Numbers, U.S. Naval Ordnance Laboratory, Report No. 6866.

Lankford, J.L. (1961) "The Effect of Heat Transfer on the Separation of Laminar Flow Over Axisymmetric Compression Surfaces; Preliminary Results at Mach Number 6.78," U.S. Naval Ordnance Laboratory, Report No. 7402.

Lauchle, G.C., and Gurney, G.B. (1984) "Laminar Boundary-Layer Transition on a Heated Underwater Body," J. Fluid Mech. 144, pp. 79-101.

Laufer, (1975) "New Trends in Experimental Turbulence Research," Ann. Rev. Fluid Mech. 7, pp. 307-326.

Lee, W.K., Vaseleski, R.C., and Metzner, A.B. (1974) "Turbulent Drag Reduction in Polymeric Solutions Containing Suspended Fibers," AIChE J. 20, pp. 128-133.

Lees, L. (1947) "The Stability of the Laminar Boundary Layer in a Compressible Fluid," NACA Report No. 876.

Legner, H.H. (1984) "A Simple Model for Gas Bubble Drag Reduction," Phys Fluids 27, pp. 2788-2790.

Libby, P.A. (1954) "Method for Calculation of Compressible Laminar Boundary Layer with Axial Pressure Gradient and Heat Transfer," NACA TN 3157.

Liebeck, R.H. (1978) "Design of Subsonic Airfoils for High Lift," J. Aircraft 15, pp. 547-561.

Liepmann, H.W., Brown, G.L., and Nosenchuck, D.M. (1982) "Control of Laminar Instability Waves Using a New Technique," J. Fluid Mech. 118, pp. 187-200.

Liepmann, H.W., and Fila, G.H. (1947) "Investigations of Effects of Surface Temperature and Single Roughness Elements on Boundary Layer Transition," NACA Report No. 890.

Liepmann, H.W., and Nosenchuck, D.M. (1983) "Active Control of Laminar-Turbulent Transition," J. Fluid Mech. 118, pp. 201-204.

Lighthill, M.J. (1952) "On Sound Generated Aerodynamically. I. General Theory," Proc. Roy. Soc. London A 211, pp. 564-587.

Lighthill, M. J. (1963) "Introduction - Boundary Layer Theory," in Laminar Boundary Layers, ed. L. Rosenhead, pp. 46-113, Clarendon Press, Oxford.

Lighthill, M.J. (1973) "On the Weis-Fogh Mechanism of Lift Generation," J. Fluid Mech. 60, pp. 1-17.

Lin, C.C. (1945) "On the Stability of Two-Dimensional Parallel Flows," Parts I, II and III, Q. Appl. Maths. 3, pp. 117-142, 218-234, 277-301.

Lin, J.C., Weinstein, L.M., Watson, R.D., and Balasubramanian, R. (1983) "Turbulent Drag Characteristic of Small Amplitude Rigid Surface Waves," AIAA Paper No. 83-0228.

Linke, W. (1942) "Über den Strömungswiderstand einer beheizten ebenen Platte," Luftfahrtforschung 19, pp. 157-160.

Liu, C.K., Kline, S.J., and Johnston, J.P. (1966) "Experimental Study of Turbulent Boundary Layer on Rough Walls," Dept. of Mechanical Engineering, Report No. MD-15, Stanford University, CA.

Lowell, R.L., and Reshotko, E. (1974) "Numerical Study of the Stability of a Heated Water Boundary Layer," Case Western University, Report No. FTAS/TR-73-93, Cleveland, OH.

Lumley, J.L. (1969) "Drag Reduction by Additives," Ann. Rev. Fluid Mech. 1, pp. 367-384.

Lumley, J.L. (1973) "Drag Reduction in Turbulent Flow by Polymer Additives," J. Polym. Sci.; Macromol. Rev. 7, pp. 263-290.

Lumley, J.L. (1977) "Drag Reduction in Two Phase and Polymer Flows," Phys. Fluids 20, pp. S64-S71.

Lumley, J.L. (1978) "Two-Phase and Non-Newtonian Flows," in Topics in Applied Physics 12, ed P. Bradshaw, Second Edition, pp. 289-324, Springer-Verlag, Berlin.

Lumley, J L. (1983) "Turbulence Modeling," J. Applied Mechanics 105, pp. 1097-1103.

Lumley, J.L. (1987) "Turbulence Modeling," Proc. 10th U.S. National Cong. of Applied Mechanics, ed. J.P. Lamb, pp. 33-39, ASME, New York.

Lumley, J.L., and Kubo, I. (1984) "Turbulent Drag Reduction by Polymer Additives: A Survey," Sibley School of Mech. & Aero. Eng., Report No. FDA-84-07, Cornell University, Ithaca, NY.

Luttges, M.W., Somps, C., Kliss, M., and Robinson, M. (1984) "Unsteady Separated Flows: Generation and Use by Insects," in Unsteady Separated Flows, eds. M.S. Francis and M.W. Luttges, U.S. Air Force Academy, Colorado Springs, CO.

Madavan, N.K., Deutsch, S., and Merkle, C.L. (1984) "Reduction of Turbulent Skin Friction by Microbubbles," Phys. Fluids 27, pp. 356-363.

Madavan, N.K., Deutsch, S., and Merkle, C.L. (1985) "Measurements of Local Skin Friction in a Microbubble-Modified Turbulent Boundary Layer," J. Fluid Mech. 156, pp. 237-256.

Maestrello, L., Badavi, F.F., and Noonan, K.W. (1988) "Control of the Boundary Layer Separation about an Airfoil by Active Surface Heating," AIAA Paper No. 88-3545-CP.

Malik, M.R., Weinstein, L.M., and Hussaini, M.Y. (1983) "Ion Wind Drag Reduction," AIAA Paper No. 83-0231.

Maltby, R.L. (1962) "Flow Visualization in Wind Tunnels Using Indicators," NATO Advisory Group for Aerospace Research and Development, AGARDograph No. 70.

Maskell, E.C. (1955) "Flow Separation in Three Dimensions," RAE Report Aero. 2565, Royal Aircraft Establishment, Farnborough, England.

Maxworthy, T. (1979) "Experiments on the Weis-Fogh Mechanism of Lift Generation by Insects in Hovering Flight. Part 1. Dynamics of the 'Fling'," J. Fluid Mech. 93, pp. 47-63.

Maxworthy, T. (1981) "The Fluid Dynamics of Insect Flight," Ann. Rev. Fluid Mech. 13, pp. 329-350.

McComb, W.D., and Chan, K.T.J. (1979) "Drag Reduction in Fibre Suspensions: Transitional Behavior due to Fibre Degradation," Nature 280, pp. 45-46.

McComb, W.D., and Chan, K.T.J. (1981) "Drag Reduction in Fibre Suspension," Nature 292, pp. 520-522.

McComb, W. D., and Rabie, L.H. (1979) "Development of Local Turbulent Drag Reduction Due to Nonuniform Polymer Concentration," Phys. Fluids 22, pp. 183-185.

McCormick, M.E., and Bhattacharyya, R. (1973) "Drag Reduction of a Submersible Hull by Electrolysis," Nav. Eng. J. 85, pp. 11-16.

McCroskey, W.J. (1977) "Some Current Research in Unsteady Fluid Dynamics," J. Fluids Eng. 99, pp. 8-39.

McCroskey, W.J. (1982) "Unsteady Airfoils," Ann. Rev. Fluid Mech. 14, pp. 285-311.

McMichael, J.M., Klebanoff, P.S., and Meese, N.E. (1980) "Experimental Investigation of Drag on a Compliant Surface," in Viscous Flow Drag Reduction, ed. G.R. Hough, pp. 410-438, AIAA Prog. in Astro. & Aero. 72, New York.

McMurray, J.T., Metcalfe, R.W., and Riley, J.J. (1983) "Direct Numerical Simulations of Active Stabilization of Boundary Layer Flows," Proc. Eighth Biennial Symp. on Turbulence, ed. J.L. Zakin & G.K. Patterson, Paper No. 36, Univ. Missouri, Rolla.

Mehta, R.D. (1985 a) "Aerodynamics of Sports Balls," Ann. Rev. Fluid Mech. 17, pp. 151-189.

Mehta, R.D. (1985 b) "Effect of a Longitudinal Vortex on a Separated Turbulent Boundary Layer," AIAA Paper No. 85-0530.

Merkle, C.L., and Deutsch, S. (1985) "Drag Reduction by Microbubbles: Current Research Status," AIAA Paper No. 85-0537.

Merkle, C.L., and Deutsch, S. (1988) "Microbubble Drag Reduction: A Survey," in Frontiers in Experimental Fluid Mechanics, ed. M. Gad-el-Hak, Springer-Verlag, New York.

Milling, R.W. (1981) "Tollmien-Schlichting Wave Cancellation," Phys. Fluids 24, pp. 979-981.

Milne-Thomson, L.M. (1968) Theoretical Hydrodynamics, Fifth Edition, Macmillan, London.

Moore, F.K. (1958) "On the Separation of the Unsteady Laminar Boundary Layer," in Boundary-Layer Research, ed. H.G. Görtler, pp. 296-310, Springer-Verlag, Berlin.

Morduchow, M., and Grape, R.G. (1955) "Separation, Stability, and Other Properties of Compressible Laminar Boundary Layer with Pressure Gradient and Heat Transfer," NACA TN 3296.

Morkovin, M.V. (1984) "Bypass Transition to Turbulence and Research Desiderata," in Transition in Turbines Symposium, NASA CP-2386.

Morkovin, M.V. (1988) "Recent Insights into Instability and Transition to Turbulence in Open-Flow Systems," AIAA Paper No. 88-3675.

Nagel, A.L., Alford, W.J., Jr., and Dugan, J.F. (1975) "Future Long-Range Transports- Prospects for Improved Fuel Efficiency," NASA Technical Memorandum No. X-72659.

Nakayama, W. (1986) Thermal Management of Electronic Equipment," App. Mech. Rev. 39, pp. 1847-1868.

Narasimha, R., and Ojha, S.K. (1967) "Effect of Longitudinal Surface Curvature on Boundary Layers," J. Fluid Mech. 29, pp. 187-199.

Narasimha, R., and Sreenivasan, K.R. (1973) "Relaminarization in Highly Accelerated Turbulent Boundary Layers," J. Fluid Mech. 61, pp. 417-447.

Narasimha, R., and Sreenivasan, K.R. (1979) Relaminarization of Fluid Flows," in Advances in Applied Mechanics, Vol. 19, ed. C.-S. Yih, pp. 221-309, Academic Press, New York.

Orszag, S.A. (1971) "Accurate Solution of the Orr-Sommerfeld Stability Equation," J. Fluid Mech. 50, pp. 689-703.

Panton, R. L. (1984) Incompressible Flow, Wiley-Interscience, New York.

Patel, V.C., and Head, M.R. (1968) "Reversion of Turbulent to Laminar Flow," J. Fluid Mech. 34, pp. 371-392.

Patera, A.T. (1986) "Spectral Element Simulation of Flow in Grooved Channels: Cooling Chips with Tollmien-Schlichting Waves," in Supercomputers and Fluid Dynamics, eds. K. Kuwahara, R. Mendez, and S.A. Orszag, pp. 41-51, Springer-Verlag, Berlin.

Patera, A.T., and Mikić B.B. (1986) "Exploiting Hydrodynamic Instabilities. Resonant Heat Transfer Enhancement," Int. J. Heat Mass Transfer 29, pp. 1127-1138.

Patterson, G.K., Zakin, J.L., and Rodriguez, J.M. (1969) "Drag Reduction. Polymer Solutions, Soap Solutions, and Solid Particle Suspensions in Pipe Flow," Indus. & Eng. Chem. 61, pp. 22-30.

Pfeffer, R., and Rosetti, S.J. (1971) "Experimental Determination of Pressure Drop and Flow Characteristics of Dilute Gas-Solid Suspensions," NASA Contractor Report No. 1894.

Pfenninger, W. (1946) "Untersuchungen über Reibungsverminderung an Tragflügeln, insbesondere mit Hilfe von Grenzschichtabsaugung," Reports of the Inst. of Aerodynamics, ETH Zürich, No. 13.

Phillips, O. (1979) The Last Chance Energy Book, Johns Hopkins Univ. Press, Baltimore.

Plesniak, M.W., and Nagib, H.M. (1985) "Net Drag Reduction in Turbulent Boundary Layers Resulting from Optimized Manipulation," AIAA Paper No. 85-0518.

Povkh, I.L., Bolonov, N.I., and Eidel'man, A.Ye. (1979) "The Average Velocity Profile and the Frictional Loss in Turbulent Flow of an Aqueous Suspension of Clay," Fluid Mech.-Soviet Research 8, pp. 118-124.

Prandtl, L. (1904) "Über Flüssigkeitsbewegung bei sehr kleiner Reibung," Proc. Third Int. Math. Congr, pp. 484-491, Heidelberg, Germany.

Prandtl, L. (1925) "Magnuseffeckt und Windkraftschiff," Naturwissenschaften 13, pp. 93 -108.

Prandtl, L. (1935) "The Mechanics of Viscous Fluids," in Aerodynamic Theory, Vol. III, ed. W. F. Durand, pp. 34-208, Springer, Berlin.

Preston, J.H. (1958) "The Minimum Reynolds Number for a Turbulent Boundary Layer and the Selection of a Transition Device," J. Fluid Mech. 3, pp. 373-384.

Pretsch, J. (1942) "Umschlagbeginn und Absaugung," Jahrb. Dtsch. Luftfahrtforschung 1, pp. 54-71.

Purshouse, M. (1976) "On the Damping of Unsteady Flow by Compliant Boundaries," J. Sound and Vibration 49, pp. 423-436.

Purshouse, M. (1977) "Interaction of Flow with Compliant Surfaces," Ph.D. Thesis, Cambridge University, London, United Kingdom.

Radin, I. (1974) "Solid-Fluid Drag Reduction," Ph.D. Thesis, University of Missouri, Rolla, MO.

Radin, I., Zakin, J.L., and Patterson, G.K. (1975) "Drag Reduction in Solid-Fluid Systems," AIChE J. 21, pp. 358-371.

Ragab, S.A., and Nayfeh, A.H. (1980) "A Comparison of the Second-Order Triple-Deck Theory and Interacting Boundary Layers for Incompressible Flows Past a Hump," AIAA Paper No. 80-0072.

Rao, D.M. and Kariya, T.T. (1988) "Boundary-Layer Submerged Vortex Generators for Separation Control - An Exploratory Study," AIAA Paper No. 88-3546-CP.

Ras, M., and Ackeret, J. (1941) "Über Verhinderung der Grenzschicht-Turbulenz durch Absaugung," Helv. Phys. Acta 14, p. 323.

Rayleigh, F. (1880) "On the Stability, or Instability, of Certain Fluid Motions," Proc. London Math. Soc. 11, pp. 57-70.

Reed, H.L., and Nayfeh, A.H. (1986) "Numerical-Perturbation Technique for Stability of Flat-Plate Boundary Layers with Suction," AIAA J. 24, pp. 208-214.

Reed, H.L., and Saric, W.S. (1987) "Stability and Transition of Three-Dimensional Flows," Proc. 10th U.S. National Cong. of App. Mech., ed. J.P. Lamb, pp. 457-468, ASME, New York.

Reed, H.L., and Saric, W.S. (1989) "Stability of Three-Dimensional Boundary Layers," Ann. Rev. Fluid Mech. 21, pp. 235-284.

Reischman, M.M., and Tiederman, W.G. (1975) "Laser-Doppler Anemometer Measurements in Drag-Reducing Channel Flows," J. Fluid Mech. 70, pp. 369-392.

Reshotko, E. (1976) "Boundary-Layer Stability and Transition," Ann. Rev. Fluid Mech. 8, pp. 311-349.

Reshotko, E. (1979) "Drag Reduction by Cooling in Hydrogen-Fueled Aircraft," J. Aircraft 16, pp. 584-590.

Reshotko, E. (1985) "Control of Boundary Layer Transition," AIAA Paper No. 85-0562.

Reshotko, E. (1987) "Stability and Transition, How Much Do We Know? "Proc. 10th U.S. National Cong. of App. Mech., ed. J.P. Lamb, pp. 421-434, ASME, New York.

Reynolds, G.A., and Saric, W.S. (1986) "Experiments on the Stability of the Flate-Plate Boundary Layer with Suction," AIAA J. 24, pp. 202-207.

Reynolds, W.C., and Carr, L.W. (1985) "Review of Unsteady, Driven, Separated Flows," AIAA Paper No. 85-0527.

Riley, J.J., and Gad-el-Hak, M. (1985) "The Dynamics of Turbulent Spots," in Frontiers in Fluid Mechanics, eds. S.H. Davis and J.L. Lumley, pp. 123-155, Springer-Verlag, Berlin.

Riley, J.J., Gad-el-Hak, M., and Metcalfe, R.W. (1988) "Compliant Coatings," Ann. Rev. Fluid Mech. 20, pp. 393-420.

Rosenhead, L. (1963) Laminar Boundary Layers, Clarendon Press, Oxford.

Rott, N. (1956) "Unsteady Viscous Flow in the Vicinity of a Stagnation Point," Q. Appl. Math. 13, pp. 444-451.

Rotta, J.C. (1970) "Control of Turbulent Boundary Layers by Uniform Injection and Suction of Fluid," Seventh Congress of the International Council of the Aeronautical Sciences," ICAS Paper No. 70-10, Rome, Italy.

Runyan, L.J., and Steers, L. L. (1980) "Boundary Layer Stability Analysis of a Natural Laminar Flow Glove on the F-111 TACT Airplane," in Viscous Flow Drag Reduction, ed. G.R. Hough, AIAA Progress in Astronautics & Aeronautics 72, pp. 17-32.

Ryskin, G. (1987) "Turbulent Drag Reduction by Polymers: A Quantitative Theory," Phys. Rev. Letters 59, pp. 2059-2062.

Sabadell, L.A. (1988) "Effects of a Drag Reducing Additive on Turbulent Boundary Layer Structure," M.Sc. Thesis, Princeton University, Princeton, NJ.

Saric, W.S., and Reed, H.L. (1986)" Effect of Suction and Weak Mass Injection on Boundary-Layer Transition," AIAA J. 24, pp. 383-389.

Savas, Ö, and Coles, D. (1985) "Coherence Measurements in Synthetic Turbulent Boundary Layers," J. Fluid Mech. 160, pp. 421-446.

Savins, J.G. (1967) "A Stress-Controlled Drag-Reduction Phenomenon," Rheologica Acta 6, pp. 323-330.

Schilz, W. (1965/66) "Experimentelle Untersuchungen zur Akustischen Beeinflussung der Strömungsgrenzschicht in Luft," Acustica 16, pp. 208-223.

Schlichting, H. (1959) "Einige neuere Ergebnisse über Grenzschichtbeein flussung," Proc. 1st Int. Congr. Aero. Sci., Madrid, Adv. in Aero. Sci. II, pp. 563-586, Pergamon Press, London.

Schlichting, H. (1979) Boundary-Layer Theory, Seventh Edition, McGraw-Hill, New York.

Schlichting, H., and Pechau, W. (1959) "Auftriebserhöhung von Tragflügeln durch kontinuierlich verteilte Absaugung," ZFW 7, pp. 113-119.

Schlichting, H., and Ulrich, A. (1940) "Zur Berechnung des Umschlages laminar-turbulent," Jahrb. Dtsch. Luftfahrtforschung 1, pp. 8-35.

Schubauer, G.B., and Skramstad, H.K. (1947) "Laminar Boundary-Layer Oscillations and Stability of Laminar Flow," J. Aero. Sci. 14, pp. 69-78.

Schubauer, G.B., and Spangenberg, W.G. (1960) "Forced Mixing in Boundary Layers," J. Fluid Mech. 8, pp. 10-32.

Scott, M.R., and Watts, H.A. (1977) "Computational Solution of Linear Two-Point Boundary Value Problems via Orthonormalization," J. Numerical Analysis 14, pp. 40-70.

Sears, W.R. (1956) "Some Recent Developments in Airfoil Theory," J. Aeronaut. Sci. 23, pp. 490-499.

Sigal, A. (1971) "An Experimental Investigation of the Turbulent Boundary Layer Over a Wavy Wall," Ph.D. Thesis, California Institute of Technology, Pasadena, CA.

Smith, A.M.O. (1957) "Transition, Pressure Gradient, and Stability Theory," Actes IX Congrès International de Mécanique Appliquée, Vol. 4, pp. 234-244, Université de Bruxelles, Belgique.

Smith, A.M.O., and Gamberoni, N. (1956) "Transition, Pressure Gradient and Stability Theory," Rep. ES-26388, Douglas Aircraft Company, El Segundo, CA.

Smits, A.J., and Wood, D.H. (1985) "The Response of Turbulent Boundary Layers to Sudden Perturbations," Ann. Rev. Fluid Mech. 17, pp. 321-358.

So, R.M.C., and Mellor, G.L. (1973) "Experiment on Convex Curvature Effects in Turbulent Boundary Layers," J. Fluid Mech. 60, pp. 43-62.

Soo, S.L., and Trezek, G.J. (1966) "Turbulent Pipe Flow of Magnesia Particles in Air," I&EC Fundamentals 5, pp. 388-392.

Squire, H.B. (1933) "On the Stability for Three-Dimensional Disturbances of Viscous Fluid Flow Between Parallel Walls," Proc. R. Soc. Lond. A 142, pp. 621-628.

Steinheil, E., Scherber, W., Seidl, M., and Rieger, H. (1977) "Investigations on the Interaction of Gases and Well-Defined Solid Surfaces with Respect to Possibilities for Reduction of Aerodynamic Friction and Aerothermal Heating," in Rarefied Gas Dynamics, ed. J.L. Potter, pp. 589-602, AIAA Progress in Aero. & Astro. 51, New York.

Stratford, B.S. (1959 a) "The Prediction of Separation of the Turbulent Boundary Layer," J. Fluid Mech. 5, pp. 1-16.

Stratford, B.S. (1959 b) "An Experimental Flow with Zero Skin Friction Throughout its Region of Pressure Rise," J. Fluid Mech. 5, pp. 17-35.

Strazisar, A.J., Reshotko, E., and Prahl, J.M. (1977) "Experimental Study of the Stability of Heated Laminar Boundary Layers in Water," J. Fluid Mech. 83, pp. 225-247.

Stuart, J.T. (1963) "Hydrodynamic Stability," in Laminar Boundary Layer Theory, ed. L. Rosenhead, pp. 492-579, Clarendon Press, Oxford.

Swearingen, J.D., and Blackwelder R.F. (1984) "Instantaneous Streamwise Velocity Gradients in the Wall Region," Bull. Am. Phys. Soc. 29, p. 1528.

Tani, I. (1969) "Boundary-Layer Transition," Ann. Rev. Fluid Mech. 1, pp. 169-196.

Tennekes, H., and Lumley, J.L. (1972) A First Course in Turbulence, MIT Press, Cambridge, MA.

Thomas, A.S.W. (1983) "The Control of Boundary-Layer Transition Using a Wave Superposition Principle," J. Fluid Mech. 137, pp. 233-250.

Tichy, J., Warnaka, G.E., and Poole, L.A. (1984) "A Study of Active Control of Noise in Ducts," J. Vibration, Acoustics, Stress, and Reliability in Design 106, pp. 399-404.

Tiederman, W.G., Luchik, T.S., and Bogard, D.G. (1985) "Wall-Layer Structure and Drag Reduction," J. Fluid Mech. 156, pp. 419-437.

Tobak, M., and Peake, D.J. (1982) "Topology of Three-Dimensional Separated Flows," Ann. Rev. Fluid Mech. 14, pp. 61-85.

Tollmien, W. (1935) "Ein allgemeines Kriterium der Instabilität laminarer Geschwindigkeitsverteilungen," Nachr. Wiss. Fachgruppe, Göttingen, Math. Phys. Kl. 1, pp. 79-114.

Toms, B.A. (1948) "Some Observations on the Flow of Linear Polymer Solutions Through Straight Tubes at Large Reynolds Numbers," Proc. 1st Int. Congr. Rheol. 2, pp. 135-141, North-Holland, Amsterdam.

Truckenbrodt, E. (1956) "Ein einfaches Näherungsverfahren zum Berechnen der laminaren Reibungsschicht mit Absaugung," Forschg. Ing.-Wes. 22, pp. 147-157.

Ulrich, A. (1944) "Theoretische Untersuchungen über die Widerstandsersparnis durch Laminarhaltung mit Absaugung," Schriften Dtsch. Akad.Luftfahrtforschung 8 B, p. 53.

Van Ingen, J.L. (1956) "A Suggested Semiempirical Method for the Calculation of the Boundary-Layer Transition Region," Rep. V.T.H.74, Dept. of Aero. & Eng., Institute of Technology, Delft, Holland.

Van Laere, L., and Sas, P. (1988) "Principles and Applications of Active Noise Cancellation," Proc. Noise Control Design: Methods and Practice, ed. J.S. Bolton, pp. 279-284, Noise Control Foundation, Poughkeepsie, NY.

Verollet, E., Fulachier, L., Dumas, R., and Favre, A. (1972) "Turbulent Boundary Layer with Suction and Heating to the Wall," in Heat and Mass Transfer in Boundary Layers, Vol. 1, eds. N. Afgan, Z. Zaric and P. Anastasijevec, pp. 157-168, Pergamon Presss, Oxford.

Viets, H. (1980) "Coherent Structures in Time Dependent Shear Flows," in Turbulent Boundary Layers, AGARD/NATO CPP-271, Paper No. 5, Nevilly Sur Seine, France.

Viets, H., Ball, M., and Bougine, D. (1981) "Performance of Forced Unsteady Diffusers," AIAA Paper No. 81-0154.

Viets, H., Palmer, G.M., and Bethke, R.J. (1984) "Potential Applications of Forced Unsteady Flows," in Unsteady Separated Flows, eds. M.S. Francis and M.W. Luttges, pp. 21-27, U.S. Air Force Academy, Colorado Springs, CO.

Virk, P.S. (1975) "Drag Reduction Fundamentals," AIChE J. 21, pp. 625-656.

Virk, P.S., Merrill, E.W., Mickley, H.S., Smith, K.A., and Mollo-Christensen, E.L. (1967) "The Toms Phenomenon: Turbulent Pipe Flow of Dilute Polymer Solutions," J. Fluid Mech. 30, pp. 305-328.

Von Winkle, W.A. (1961) "An Evaluation of a Boundary Layer Stabilization Coating," Naval Underwater Systems Center, Tech. Memo. No. 922-111-61, New London, CT.

Wagner, R.D., and Fischer, M.C. (1984) "Fresh Attack on Laminar Flow," Aerospace America 22, pp. 72-76.

Wagner, R.D., Maddalon, D.V., and Fischer, M.C. (1984) "Technology Development for Laminar Boundary Control on Subsonic Transport Aircraft," AGARD CP-365, Paper No. 16.

Walsh, M.J. (1980) "Drag Characteristics of V-Groove and Transverse Curvature Riblets," in Viscous Flow Drag Reduction, ed. G.R. Hough, pp. 168-184, AIAA Prog. in Astro. & Aero. 72, New York.

Walsh, M.J. (1982) "Turbulent Boundary Layer Drag Reduction Using Riblets," AIAA Paper No. 82-0169.

Walsh, M.J. (1983) "Riblets as a Viscous Drag Reduction Technique," AIAA J. 21, pp. 485-486.

Walsh, M.J., and Lindemann, A.M. (1984) "Optimization and Application of Riblets for Turbulent Drag Reduction," AIAA Paper No. 84-0347.

Walsh, M.J., and Weinstein, M. (1978) "Drag and Heat Transfer on surfaces with Small Longitudinal Fins," AIAA Paper No. 78-1161.

Warnaka, G.E. (1982) "Active Attenuation of Noise: The State of the Art," Noise Control Eng. 18, pp. 100-110.

Warner, J.V., Waters, D.E., and Bernhard, R.J. (1988) "Adaptive Active Noise Control in Three Dimensional Enclosures," Proc. Noise Control Design: Methods and Practice, ed. J.S. Bolton, pp. 285-290, Noise Control Foundation, Poughkeepsie, NY.

Wazzan, A.R., Okamura, T.T., and Smith, A.M.O. (1968) "Stability of Water Flow Over Heated and Cooled Flat Plates," J. Heat Transfer 90, pp. 109-114.

Wazzan, A.R., Okamura, T.T., and Smith, A.M.O. (1970) "The Stability and Transition of Heated and Cooled Incompressible Boundary Layers," Proc. 4th Int. Heat Transfer Conf., Vol. 2, ed. U. Grigull and E. Hahne, FC 1.4, Elsevier, New York.

Weis-Fogh, T. (1973) "Quick Estimates of Flight Fitness in Hovering Animals, Including Novel Mechanisms for Lift Production," J. Exp. Biol. 59, pp. 169-230.

Wells, C.S., Jr., and Spangler, J.G. (1967) "Injection of a Drag-Reducing Fluid into Turbulent Pipe Flow of a Newtonian Fluid," Phys. Fluids 10, pp. 1890-1894.

Whitcomb, R.T. (1956) "A Study of the Zero-Lift Drag-Rise Characteristics of Wing-Body Combinations Near the Speed of Sound," NACA Report No. 1273.

White, A., and Hemmings, J.A.G. (1976) Drag Reduction by Additives: Review and Bibliography, BHRA Fluid Engineering, Cranfield, United Kingdom.

Whites, R.C., Sudderth, R.W., and Wheldon, W.G. (1966) "Laminar Flow Control on the X-21," Astro. & Aero. 4, pp. 38-43.

Wilkinson, S.P. (1988) "Direct Drag Measurements on Thin-Element Riblets with Suction and Blowing," AIAA Paper No. 88-3670-CP.

Wilkinson, S.P., Anders, J.B., Lazos, B.S., and Bushnell, D.M. (1987) "Turbulent Drag Reduction Research at NASA Langley-Progress and Plans," Proc. Int. Conf. on Turbulent Drag Reduction by Passive Means, Vol. 1, pp. 1-32, Royal Aeronautical Society, London, United Kingdom.

Wilkinson, S.P., Anders, J.B., Lazos, B.S., and Bushnell, D.M. (1988) "Turbulent Drag Reduction Research at NASA Langley: Progress and Plans," Int. J. Heat and Fluid Flow 9, pp. 266-277.

Wilkinson, S.P., and Lazos, B.S. (1987) "Direct Drag and Hot-Wire Measurements on Thin-Element Riblet Arrays," in Turbulence Management and Relaminarization, eds. H.W. Liepmann and R. Narasimha, pp. 121-131, Springer-Verlag, New York.

Williams, J.C., III (1977) "Incompressible Boundary-Layer Separation," Ann. Rev. Fluid Mech. 9, pp. 113-144.

Williams, T. I. (1987) The History of Invention, Facts on File Publications, New York.

Willis, G.J.K. (1986) "Hydrodynamic Stability of Boundary Layers Over Compliant Surfaces," Ph.D. Thesis, University of Exeter, United Kingdom.

Willmarth, W.W. (1975) "Structure of Turbulence in Boundary Layers," Adv. Applied Mech., Volume 15, ed., C.-S. Yih, pp. 159-254, Academic Press, New York.

Wood, C.J. (1961) "A Study of Hypersonic Separated Flow," Ph.D. Thesis, University of London, United Kingdom.

Wooldridge, C.E., and Muzzy, R.J. (1966) "Boundary-Layer Turbulence Measurements with Mass Addition and Combustion," AIAA J. 4, pp. 2009-2016.

Wu, J., and Tulin, M.P. (1972) "Drag Reduction by Ejecting Additive Solutions into a Pure-Water Boundary Layer," ASME: J. Basic Eng. 94, pp. 749-756.

Wuest, W. (1961) "Survey of Calculation Methods of Laminar Boundary Layers With Suction in Incompressible Flow," in Boundary Layer and Flow Control, Vol. 2, ed. G.V. Lachmann, pp. 771-800, Pergamon Press, New York.

Wygnanski, I., Sokolov, M., and Friedman, D. (1976) "On a Turbulent 'Spot' in a Laminar Boundary Layer," J.Fluid Mech. 78, pp. 785-819.

Yajnik, K.S., and Acharya, M. (1978) "Non-Equilibrium Effects in a Turbulent Boundary Layer Due to the Destruction of Large Eddies," in Structure and Mechanisms of Turbulence, Vol. 1, ed. H. Fiedler, pp. 249-260, Springer-Verlag, New York.

Yeo, K.S., and Dowling, A.P. (1987) "The Stability of Inviscid Flows Over Passive Compliant Walls," J. Fluid Mech. 183, pp. 265-292.

Young, A.D. (1953) "Boundary Layers," in Modern Developments in Fluid Dynamics: High Speed Flow, Vol. 1, ed. L. Howarth, pp. 375-475, Clarendon Press, Oxford.

Zakin, J.L., Poreh, M., Brosh, A., and Warshavsky, M. (1971) "Exploratory Study of Friction Reduction in Slurry Flows," Chem. Eng. Prog. Symp. Seri., No. 67, Vol. 111, pp. 85-89, AIChE, New York.

Zilberman, M., Wygnanski, I., and Kaplan, R. (1977) "Transitional Boundary Layer Spot in a Fully Turbulent Environment," Phys. Fluids 20, pp. S258-S271.

MICROBUBBLE DRAG REDUCTION

C. L. Merkle and S. Deutsch
Department of Mechanical Engineering and
The Applied Research Laboratory
The Pennsylvania State University
University Park, PA 16802

I. INTRODUCTION

Over the past twenty-five years there has been extensive and sustained research aimed both at determining techniques for reducing skin friction drag and at explaining how successful techniques work. Certainly, one reason for pursuing this research is because of our need to drive systems faster and farther for the same power, but this is only part of the motivation. An important additional rationale for the work is bound up with our continuing fascination with boundary layers -- particularly turbulent boundary layers -- with what makes them work the way they do and how we might intervene and change them.

Research on skin friction drag reduction has two branches. The first, laminar flow control, is concerned with the delay of transition. This is usually accomplished by making the boundary layer velocity profile more stable near the wall by "filling it out". Suction, body shaping, and heating in water or cooling in air have all been employed and have shown great success in the laboratory (Hefner et al., 1977). The implementation of these successes in practical application has proven to be more difficult.

The second branch of research has been concerned with reducing the skin friction drag of the turbulent boundary layer. Potential payoffs are almost as spectacular as for laminar flow control -- but appear to be more realizable for actual systems. We have known for some time that the reduction of skin friction drag in a turbulent boundary layer is possible. Polymer, fibre and particle solutions are some of the best known drag reducers. Several good reviews of these and other techniques are available (Bushnell, 1983).

In the current paper we review one of the most promising areas of drag reduction research, that of microbubble injection. Although research into drag reduction by gas injection into a liquid turbulent boundary layer has been done for about twenty years, there exists no detailed review. In this first review, then, we shall describe and discuss the early results, both in the U.S. and in the Soviet Union. As most of the work done in the last ten years has been done in our laboratory at Penn State, we shall consider that work in detail. We also describe the results of the few computational and theoretical studies that have been attempted. Finally, we summarize what we do and do not know, and how we might go about finding out more.

II. <u>EARLY RESEARCH</u> (Historical Perspective)

A. <u>Work of McCormick and Bhattacharya</u>

McCormick and Bhattacharya (1973) reported the first microbubble experiments. We do not know the immediate impetus for these experiments, but the idea of reducing drag by placing a thin layer of air between a ship and its water boundary layer was patented in the 1800's (see Crewe and Eggington, 1960). It may also have been possible to postulate the potential for drag reduction from experiments in particulate flows (slurries in pipes) as much of that work seems to have occurred at about the same time.

McCormick and Bhattacharya (1973) used an axisymmetric body of 915 mm length and 127 mm maximum diameter for their experiment. Total drag was measured as the body was towed in a tank at speeds ranging from 0.32 to 2.6 m/s. These correspond to x Reynolds numbers of 0.3 and 2.4 million, respectively. Hydrogen bubbles were generated in the boundary layer by passing a current through copper wire wrapped around the hull. The number of coils, their location and the surface area over which they extended were all varied. On the basis of the reported electrical power utilization, we estimate that the volume flow rate of gas varied from 0.14 to 0.4 x 10^{-3} m^3/s. To place this gas flow rate in perspective, we estimate that the volumetric flow of water in the boundary layer was about 1.3 and 7.9 x 10^{-3} m^3/s for the two Reynolds number cases. (These estimates are based on assuming a one-seventh power law flat plate turbulent boundary layer over one-half the length of the body and should be considered as order of magnitude estimates.) Consequently, their peak volume fraction of bubbles in the boundary layer was about 30% at the lower speed but only 5% at the higher speed.

These initial experiments showed that microbubbles could reduce total drag, and that the drag reduction increased with increasing gas generation rate and decreasing speed. Drag reductions as high as 30% were observed. We present in Figure 1 some typical data (McCormick and Bhattacharya, 1973). Although the lack of detailed flowfield measurements and the presence of separation and transition make these data difficult to interpret, they represent the first successful total drag data reported and have served as the impetus for much of the later work.

More recent measurements of total drag reduction have been reported by Meng (1987) and Baba (1986). Meng measured the total drag on an axisymmetric body in a tow tank and observed drag reductions that were generally consistent with the Penn State skin friction measurements discussed below. Baba measured the total drag on a towed flat plate and also found drag reductions that correlated well with our results.

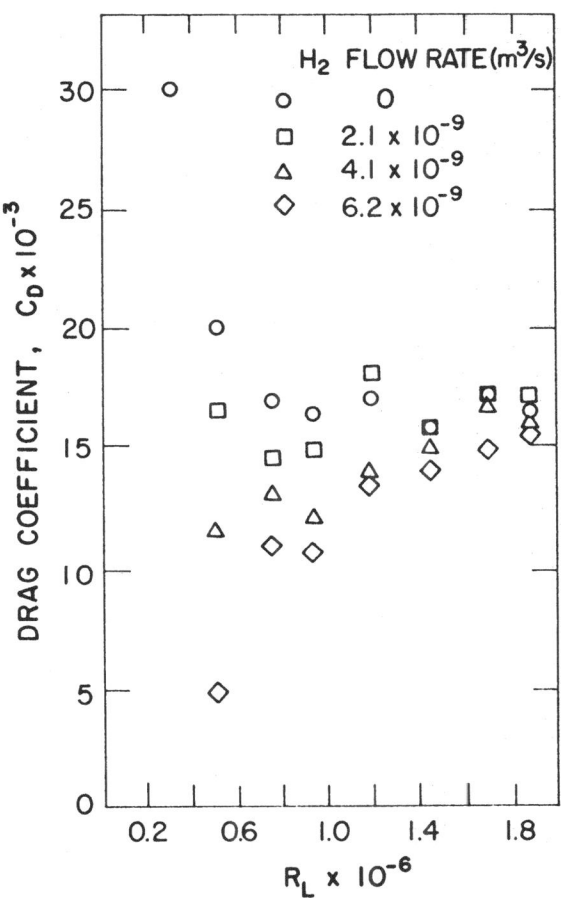

Fig. 1   Total drag coefficient as a function of Reynolds number and hydrogen generation rate. Measurements by McCormick and Bhattacharya, 1973.

B. Soviet Work

1. Experimental Facilities and Procedures

Although microbubble drag reduction was not pursued in the U.S. for about ten years after the McCormick and Bhattacharya experiments, three separate experimental studies were reported in a series of papers from the Soviet Union (Migirenko and Evseev, 1974; Bogdevich and Evseev, 1976; Bogdevich and Malyuga, 1976; and Dubnischev, Evseev, Sobolev, and Utkin, 1975). In these experiments, microbubbles were produced by injecting air through a porous surface into a zero pressure gradient turbulent boundary layer whose detailed characteristics were not reported. The Soviet data include local skin friction measurements from floating element balances, surface mounted hot film probes and near-wall velocity profiles, as well as some concentration measurements and some surface pressure fluctuations on the wall. The measurements show skin friction reduction as high as 80%. To distinguish the three experiments, we classify them by facility as experiments A, B, and C. The characteristics of each of these experimental facilities are summarized in Table I, along with those of our experiments which are discussed later.

TABLE I

COMPARISON OF MICROBUBBLE EXPERIMENTS

Experi-ment	Facility	Experimental Configuration	Porous Section Size	Freestream Velocity	Reynolds Number	Measurement Location[1]	Instrumentation and Measurements	Refs.
A	Pipe Flow Facility 30x40 mm^2	Porous Section in Lower Wall	250x24 mm^2	8.3 m/s	2.2 x 10^5 Re$_{DH}$	x = 50 mm	1)Bubble concentration by Laser Transmission 2)LDA Profile Measurements 3)Local τ_w by Flush-mounted hot films.	Migirenko & Evseev (1974); Dubnischev, et al., (1975)
B	Water Tunnel	925x244 mm^2 Flat Plate Measurements on Top Surface Porous Sections on Top and Bottom Surface:	3 Porous Sections 285x244 mm^2 (325x244 including spacers)	4.36 m/s 6.55 8.7 10.9	1.5-3.7x 10^6	x = 50 mm	1)Bubble concentration by capacitance probe 2)Tunnel pressure gradient 3)Local τ_w by floating element balance 4)Wall pressure fluc-	Bogdevich & Evseev (1976); Dubnischev, et al., (1975)
C	Water Tunnel 180x50 mm^2	Porous Section on Lower Wall	180x140 mm^2	2 m/s 4 m/s 6 m/s		x = 65 mm 115 215 310 435 695	1)Local τ_w by floating element balance (26 mm dia.) 2)Bubble concentration by capacitance wire 3)Various Porous materials	Bogdevich & Malyuga (1976)
D	Rectangular Water Tunnel	Porous Section on Tunnel Wall. Multiple gravitational orientation.	165x89	4.5-17 m/s	2.75 - 10.4x10^6	x = 65 mm 115 166 210 230	1)Integrated τ_w by floating element balance 250x102 2)Local τ_w by Flush-mounted hot films	Madavan, et al. (1984a), (1985); Madavan (1985)
E	12" Round Water Tunnel	4" Axisymmetric Body	360x12 mm	4.5-17 m/s	2.75 - 10.4x10^6		1)Floating element drag balance on cylindrical section	Deutsch & Castano (1986)
F	12" Round Water Tunnel	950x300 mm^2 flat plate	50x89 mm^2	4.5-17 m/s	2.75 - 10.4x10^6		1)Local τ_w by flush mounted hot films 2)Bubble size by top & side photography 3)Bubble cloud location by laser trans-mission	Pal, et al. (1988)

[1]Distance from trailing edge of porous material.

Experiment A was conducted in a fully developed pipe flow facility (Migirenko and Evseev, 1974; Dubnischev, et al., 1975) of rectangular cross-section (30 x 40 mm). The porous section was 254 x 24 mm and was placed on the lower wall of the pipe. A single water velocity of 8.3 m/s, which corresponds to a Reynolds number based on hydraulic diameter of 2.2 x 10^5, was used. Wall shear stress was measured by a flush-mounted hot film probe placed 50 mm downstream of the trailing edge of the porous section. Estimates of the bubble concentration profiles were made from laser transmission measurements, and near-wall segments of the velocity profiles were measured using LDV.

Experiment B was conducted on a 955 x 244 mm flat plate in a water tunnel (Bogdevich and Evseev, 1976; Dubnischev et al., 1975). Porous sections were placed on both the top and bottom surfaces of the plate, but measurements were reported on only the top surface. The tunnel pressure gradient was assumed constant on the basis of measurements from three pressure taps on the tunnel wall. A single floating element force balance was mounted approximately 50 mm downstream of the trailing edge of the porous section and was used to measure local shear stress. The size of this balance is not given. Bubble concentration profiles across the boundary layer were measured by means of capacitance probes, and turbulent pressure fluctuations on the wall were measured both in the presence and the absence of microbubbles. Measurements at free-stream speeds of 4.36, 6.55, 8.7 and 10.7 m/s were reported. These correspond to Reynolds numbers (based on x) at the measurement station of from 1.5 to 3.7 million. The Reynolds number at the leading edge of the porous surface ranged from 0.3 to 0.8 million, which suggests that the boundary layer over all or part of the porous surface may have been laminar at the lower speeds. The initial increase in drag at low airflow rates observed in these experiments was likely the result of the bubbles tripping this laminar boundary layer.

In Experiment C, the lower wall of a rectangular water tunnel was used for microbubble measurements (Bogdevich and Malyuga, 1976). Here, velocities of 2, 4 and 6 m/s were used. The porous section was 180 x 140 mm. Skin friction was measured at six locations downstream of the porous section. Distances from the trailing edge are noted in Table I. Floating element balances were used for the C_f measurements. Bubble concentration profiles were again measured by capacitance probes and the effect of various pore sizes and spacings was measured.

2. Synopsis and Assessment of Soviet Experimental Results

Perhaps the most significant capsule statement that can be made concerning the Soviet measurements is that all experiments showed substantial drag reduction. This is readily seen from the representative skin friction data in Fig. 2. In this Figure, the ratio of skin friction with bubbles to that without bubbles, C_f/C_{fo}, is plotted against the volumetric fraction of air in the boundary layer, C_v for Experiments A, B and C. This non-dimensional parameter is defined as $C_v = Q_A/(Q_A + Q_W)$, where Q_A and Q_W are the volumetric flow rates of air and water in the boundary layer. (The boundary layer characteristics in the Soviet experiments were estimated from available information.) The data in Fig. 2 show that local reductions exceeding 80% are reported in Experiments B and C, while a maximum reduction of 40% is reported at the single speed in Experiment A. For the most part, the Soviet data show that skin friction reduction increases with gas flow rate although a saturation effect at high injection rates is observed in most cases.

A very important point to note in regards to the skin friction ratio, C_f/C_{fo}, is that the density used in non-dimensionalizing both C_f and C_{fo} is always taken as the density of water. This point has seemed to cause some

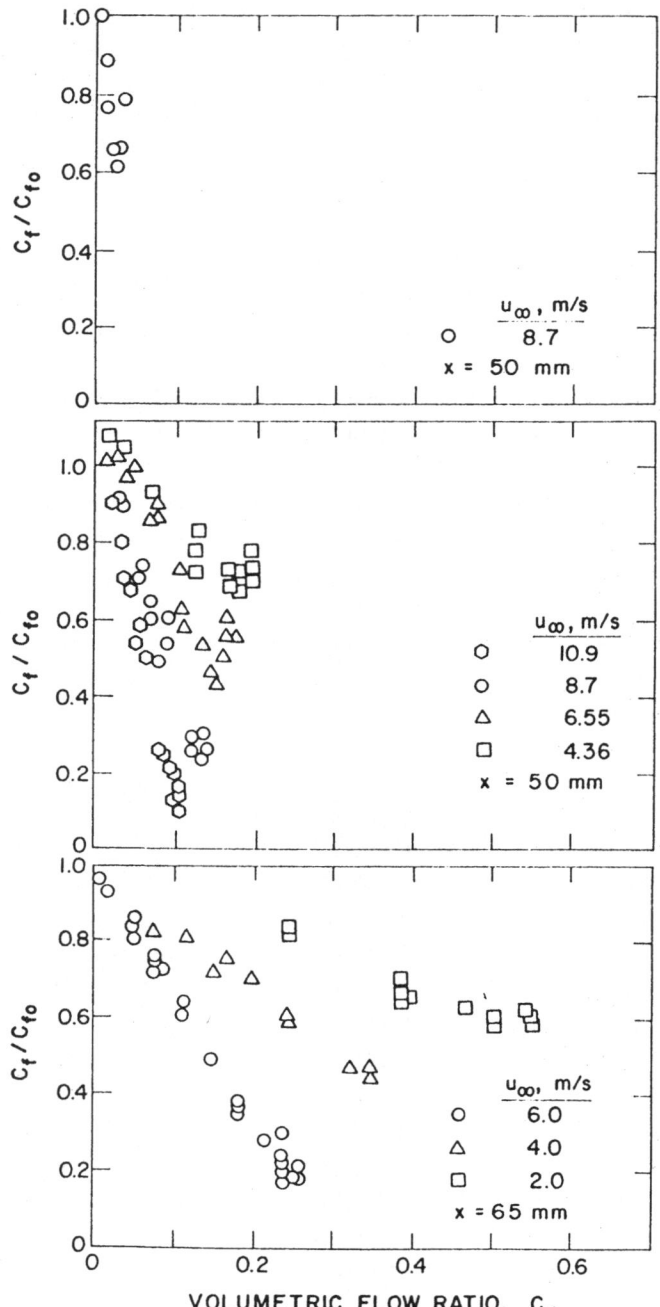

Fig. 2   Local skin friction drag reduction as a function of volumetric flow
ratio in Soviet experiments. Top, Experiment A; Middle, Experiment B;
Bottom, Experiment C. Curves are plotted to allow comparison of airflow
rates in all three experiments.

confusion from time to time, and a moment's perspective is appropriate to clarify this issue. The misunderstanding arises because the density of the air-water mixture varies strongly from the wall to the freestream, and there is a natural inclination to ask, which is the most proper value of ρ to use in defining C_f? There are two reasons that make this point moot. First, the air-water density is generally unknown at <u>all</u> points in the boundary layer so that even if it were agreed that the density at the wall, or at a y/δ of 0.2, for example, were the proper density to use, its value is unknown and so one would not be able to proceed. Second, although C_f itself serves as an important measure of boundary layer characteristics, in comparing C_f to C_{fo}, we are not interested in comparing boundary layers, we are interested in comparing the force on the wall. By using the same density to define both C_f and C_{fo}, the ratio C_f/C_{fo} becomes equal to τ_w/τ_{wo} (the ratio of shear stresses) or D/D_0 (the ratio of drags). The choice of a ratio of skin friction coefficients rather than a ratio of forces is made solely in observance of the general practice of preferring non-dimensional quantities over dimensional ones. Consequently, the skin friction ratio which is used here, in the rest of the paper, and in nearly all the references is based on the same density (that of water) for both C_f and C_{fo}.

In evaluating the results of the Soviet experiments, we should bear in mind that all the data reported is for the orientation in which the plate is underneath the boundary layer. For these "plate-on-bottom" results, buoyant effects would tend to remove the bubbles from the boundary layer. This is particularly important as most of the Soviet experiments were conducted at rather low tunnel velocities. For example, Experiments B and C (Fig. 2) show that drag reduction increases as speed increases (a notion on everyone's wish list). We shall find that this is contrary to the Penn State flat plate results. We strongly suspect that these low speed, plate-on-bottom results were dominated by buoyancy. At higher speeds the bubbles stayed in the boundary layer longer (shorter convective times and smaller bubbles) and thereby produced more drag reduction.

A final point concerning the Soviet C_f/C_{fo} results has to do with the magnitudes of the airflow needed to obtain a given level of C_f reduction. Although they presented most of their original data in terms of the airflow parameter $C_Q = Q_A/SU_\infty$ (where S is the area of the porous section and U_∞ is the freestream velocity), this parameter does not collapse the data from any of the individual experimental configurations, let alone the entire data set. One reason for this is that, as is shown later, the area of the porous material is of little significance in the drag reduction process. Of the several alternative parameters we have tried, the volumetric air-water flow ratio C_V, which we have used here, appears to be best. Using this normalization the airflow rates needed to achieve a given C_f/C_{fo} reduction in the Soviet Experiment C are about the same as in our experiments, while airflow requirements in Experiment B are about a factor of two smaller. (When compared in terms of C_Q, our experiments require about five times more airflow than do Experiments B and C.) Even when compared in terms of C_V, Experiment A remains an anomaly: The airflow rates for Experiment A are an order of magnitude smaller than either of those in Experiments B or C, or those of our experiments. Although we cannot explain this order of magnitude change, we do note that Experiment A is the only pipe flow experiment reported.

These large variations in airflow requirements were initially very perplexing, and although the choice of the volumetric fraction of air as the independent variable brings the results into much better congruence, airflow requirements still differ by a factor of two. As we shall show later, the magnitude of the C_f reduction is also quite sensitive to the location of the bubbles in the boundary layer; consequently, strict airflow congruence on the basis of volumetric concentrations is not to be expected.

A second parameter measured by the Soviets was the concentration profiles in the boundary layer. Some typical concentration profiles are reproduced in Figure 3. Two things are of particular interest. The first is the magnitude of the maximum volumetric bubble concentration, which can evidently be as high as .6 or .7. The second is the apparent return to zero of the concentration profiles as the wall is approached. Peak values of the concentration profiles occur quite close to the wall ($y/\delta = 0.1$ or 0.2) particularly at the highest velocity.

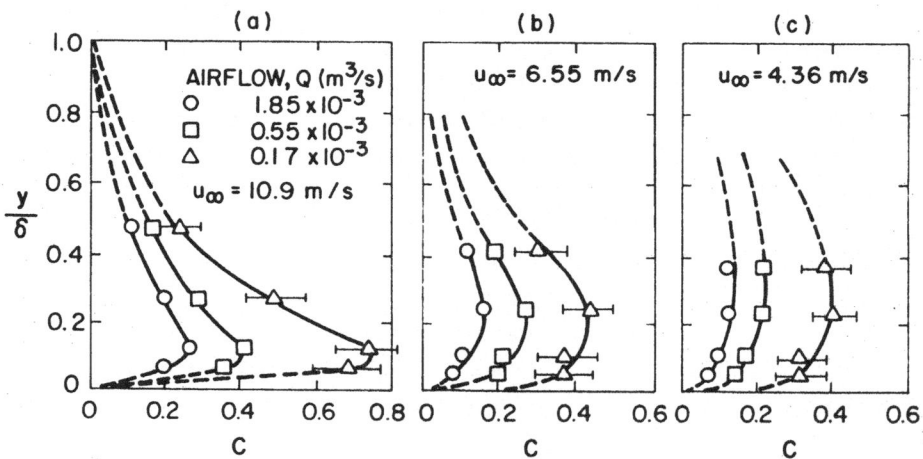

Fig. 3 Microbubble concentration distributions measured by Migirenko and Evseev (1974) with a conductivity probe at three freestreams velocities and airflow rates. C represents volume fraction of air.

Although Migirenko and Evseev (1974) attributed the lack of bubbles near the walls to viscous lift on the bubbles, our later observations suggest that the primary reason is because of buoyancy. The presence of buoyant effects also explains why the concentration peaks occur closer to the wall at the higher speeds. Finally, it is worth noting that suitable integrations of these concentration profiles only give the quoted gas flow rate back to within a factor of two or three. This illustrates the difficulty involved in making these measurements. It is probable that no more than the general trend of the data can be considered reliable.

A third important parameter, the persistence of the drag reduction, was measured in Experiment C and is reproduced in Fig. 4. Substantial drag reduction is maintained for some fifteen boundary layer thicknesses. Again, this data is for plate-on-bottom and very low speeds and we suspect that strong buoyant effects are accounting for the rapid return to zero reduction.

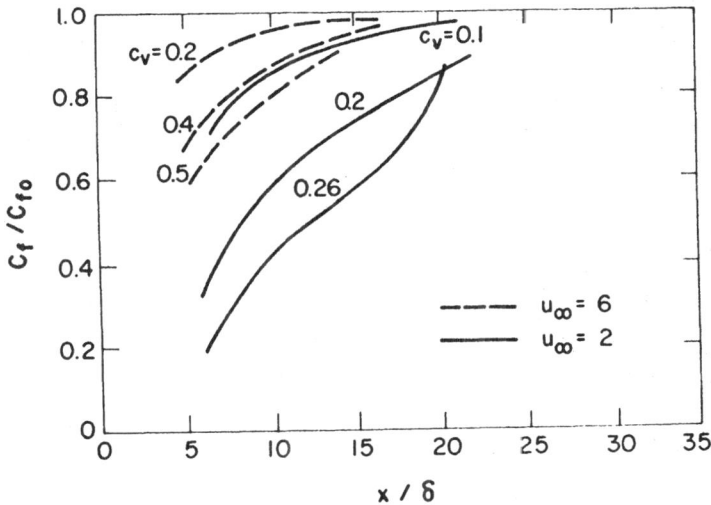

Fig. 4   Persistence of skin friction reduction as a function of distance downstream normalized by boundary layer thickness in Experiment C.

Other details of the Soviet experiments include some very near-wall velocity profiles and some wall spectral data from pressure and flush-mounted hot film probes. All the reported spectral data show a loss of high frequency fluctuations as gas is injected. The near-wall velocity profiles in the presence of bubbles are the only ones of their kind that have been reported. These data, which were taken in the fully developed pipe flow facility of Experiment A, are replotted as Fig. 5. There seems to be a monotonic decrease of the velocity gradient at the wall as the gas flow is increased. The Soviets report that this

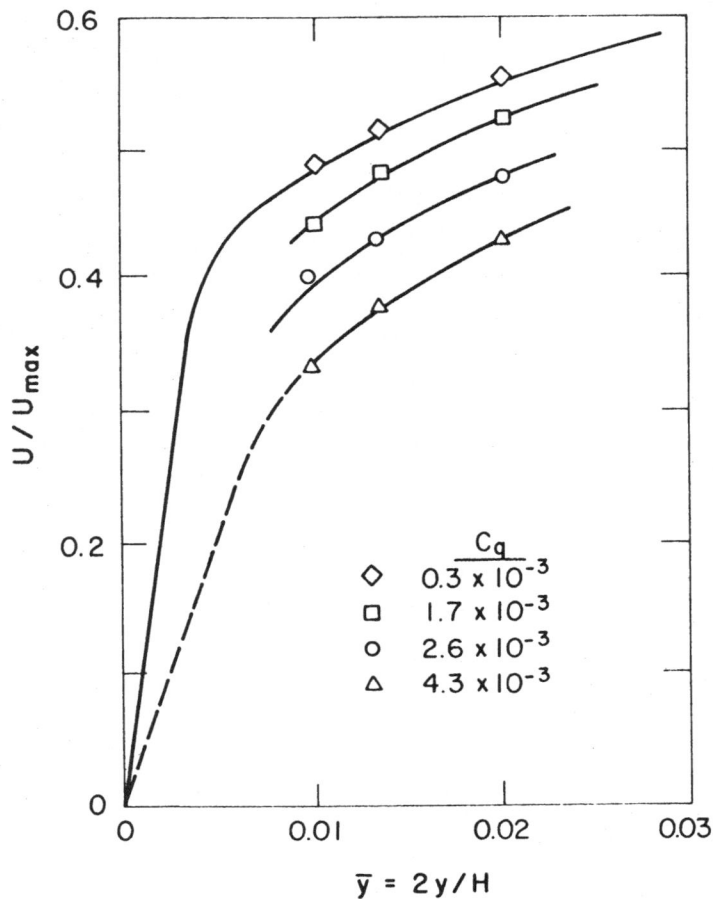

Fig. 5   LDA velocity profiles measured by Dubnishchev et al. (1975) with
microbubbles in the boundary layer. $C_q = Q/U_\infty S$. Channel height, H =
30 mm.

reduction in wall gradient correlates well with the measured drag reduction.
Both the spectra and the velocity profiles will be of some importance to our
discussion of the theoretical attempts made on the microbubble problem to date.

     Finally, we should note that the Soviet papers contain contradictory
statements as to the significance of the size and type of the porous material.
On the one hand, Migirenko and Evseev (1974) and Bogdevich and Evseev (1976) give
detailed characteristics of the painstaking methods used to manufacture their
porous material, and Migirenko and Evseev (1974) describe a hypothesis that drag
reduction cannot be obtained above a critical bubble size of 50 µm.   In
distinction to this, Bogdevich and Evseev (1976) present data which show pore
size has little effect.   The overall impression, however, is that the porous

material is extremely critical to drag reduction, and that the resulting bubbles must be (and are) very small. Although no supporting data is presented, these considerations had some effect on the nature of the next set of experiments which were done in our laboratory.

III. PENN STATE WORK

A. Description of Experiments

Three different experimental configurations have been used for microbubble testing at Penn State. The first and simplest (Madavan, Deutsch, Merkle, 1984a) used the constant-pressure tunnel wall boundary layer in the two-dimensional (rectangular) test section of the 12-inch water tunnel. The second (Deutsch and Castano, 1986) was an axisymmetric body, which mounted in the circular section of the 12-inch water tunnel, while the third (Pal, Merkle and Deutsch, 1988) was a flat plate mounted along the diameter of the circular section. This third configuration was fabricated to allow improved optical access as compared to the first installation. Characteristics of each of these configurations are reported in Table I as Experiments D, E and F. Detailed descriptions of the models follow a description of the tunnel and its two test sections.

All the experiments we shall report were done in the 12-inch water tunnel at The Applied Research Laboratory of The Pennsylvania State University. The rectangular test section has dimensions of 508 mm x 114 mm x 762 mm long, while the interchangeable circular section is 305 mm x 762 mm long. For either configuration, test section velocity can be varied continuously to nearly 20 m/sec while the pressure can be varied between 0.2 and 4 atmospheres. Two of the features of the tunnel were of particular use: first, the tunnel test section could be rotated to study buoyant effects; and second, a bypass system which continuously interchanged about 15% of the tunnel volume flow rate at maximum tunnel velocity could be used to keep the freestream relatively gas free. Tunnel freestream turbulence intensity was measured to be 0.3% to 0.5% over the velocity range of interest.

1. Tunnel Wall Boundary Layer (Experiment D)

This experimental configuration consisted of an injector and instrumentation mounted on one wall of the rectangular test section. Centered spanwise on the test plate was a 102 mm x 178 mm section for gas injection, followed by a 102 mm x 254 mm force balance (Fig. 6). For the most part, a sheet of porous sintered stainless steel filter material manufactured by Mott Metallurgical was used for the gas injection. Many of the early results were reported for a material with a 0.5 micron filter rating. This filter has a nominal pore size of 5 μm. Larger filter sizes as well as other types and sizes of injection sections were also employed. The force measurements on this plate were supplemented by

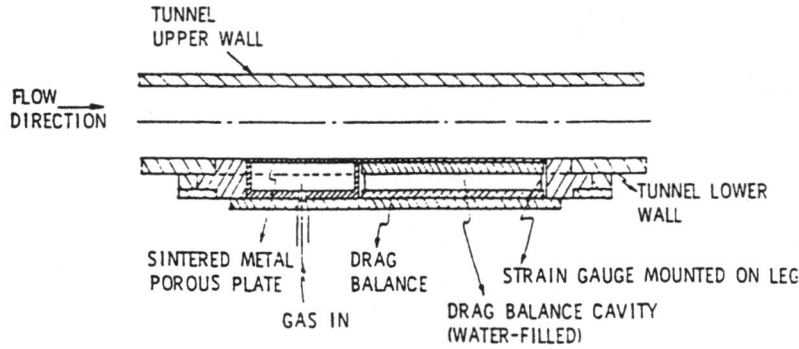

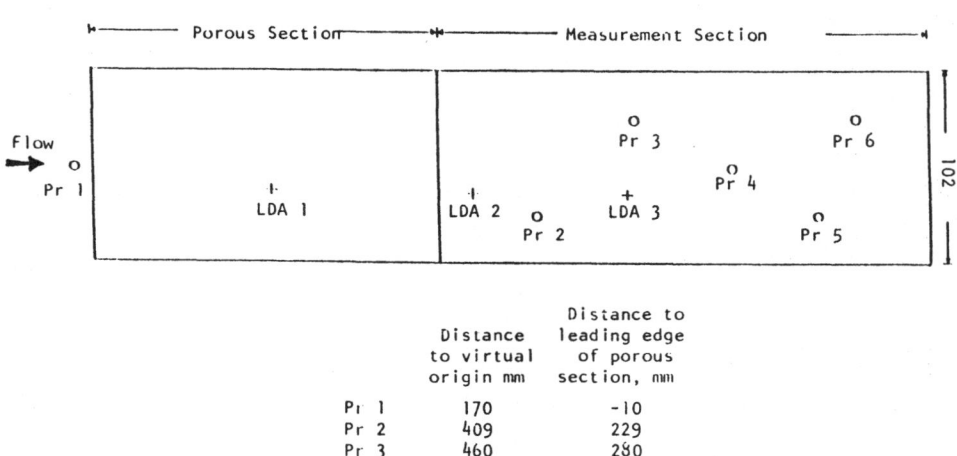

	Distance to virtual origin mm	Distance to leading edge of porous section, mm
Pr 1	170	-10
Pr 2	409	229
Pr 3	460	280
Pr 4	511	331
Pr 5	555	375
Pr 6	575	395
LDA 1	272	92
LDA 2	374	194
LDA 3	454	274

Fig. 6   Schematic diagram of tunnel wall boundary layer, Experiment E, showing drag balance section, injection section and measurement locations.

flush-mounted hot film measurements and the characteristics of the undisturbed boundary layer were documented by LDA surveys. The pressure distribution along the surface was determined by a series of 20 pressure taps mounted on the opposite wall.

2.   Axisymmetric Body (Experiment E)

Figure 7 shows the 89 mm diameter, 632 mm long axisymmetric model. An isolated cylindrical section, which was 273 mm long and 196 mm from the nose of the body, was instrumented with a stiff shear member and used to measure

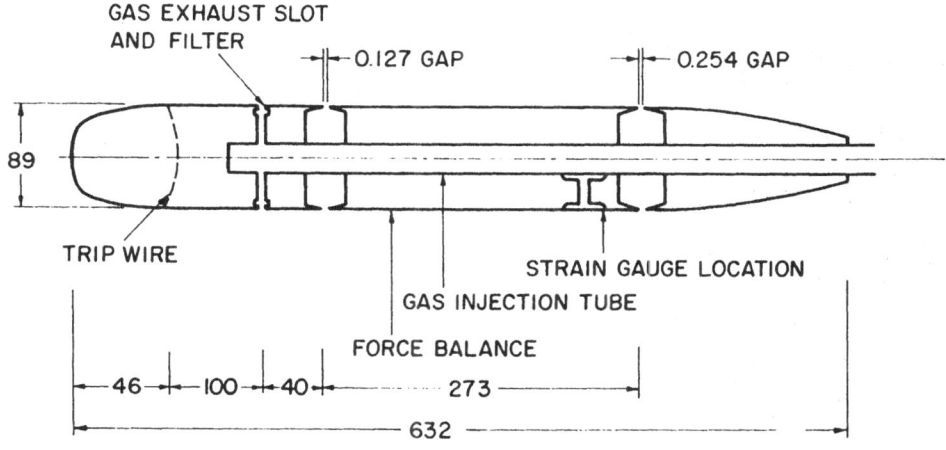

Fig. 7   Schematic of axisymmetric body, Experiment E. All dimensions are in mm.

integrated skin friction. Gas was injected some 40 mm upstream of the leading edge of the force balance through a 6.35 mm long cylindrical sintered porous plastic section, with a nominal pore size of 5 microns. Visual inspection of the gas cloud showed no evidence of bubble cloud asymmetry at the injection site. Pressure gradient measurements, which were made at 6 axial stations for each of 4 circumferential locations on the tunnel wall, showed no evidence of flow asymmetry either with or without gas injection.

The injection section length on the axisymmetric body (6.35 mm) was much shorter than either the one used in the tunnel wall experiment (178 mm) or the one used in the flat plate experiment (51 mm). Also, while the flat plate flows were naturally tripped boundary layers, the axisymmetric model flow was tripped by a steel wire, 0.35 mm in diameter, which was mounted 46 mm from the nose. In this regard, it is useful to point out that the unit Reynolds numbers in these three experiments (ranging from 5 to 20 million per meter) were considerably higher than those typically used in laboratory experiments. Calculated three-dimensional trip sizes for any of these boundary layers are typically a few micrometers, rivaling the surface roughness of a "smooth" surface. Thus, even well-polished surfaces effectively act as distributed roughness trips to these thin boundary layers. Because of the lack of understanding of the transition mechanisms in nosetip boundary layers, the wire trip was used to ensure transition in the strongly favorable pressure gradient boundary layer on the axisymmetric body.

3. Flat Plate (Experiment F)

The flat plate was 1.1 m in length, was centered in the circular test section and spanned the width of that section. Gas was injected through a 76 mm x 51 mm porous section flush-mounted on one surface of the plate. The leading edge of the porous section was 250 mm downstream of the leading edge of the plate (see Fig. 8). The porous filter used throughout these experiments was again a 3 mm thick sheet of porous stainless steel, except here a nominal filter size of 40 microns was used for most of the experiments. This larger pore size was chosen so that the pressure drop across the filter remained at about the same manageable level as that used in Experiment D.

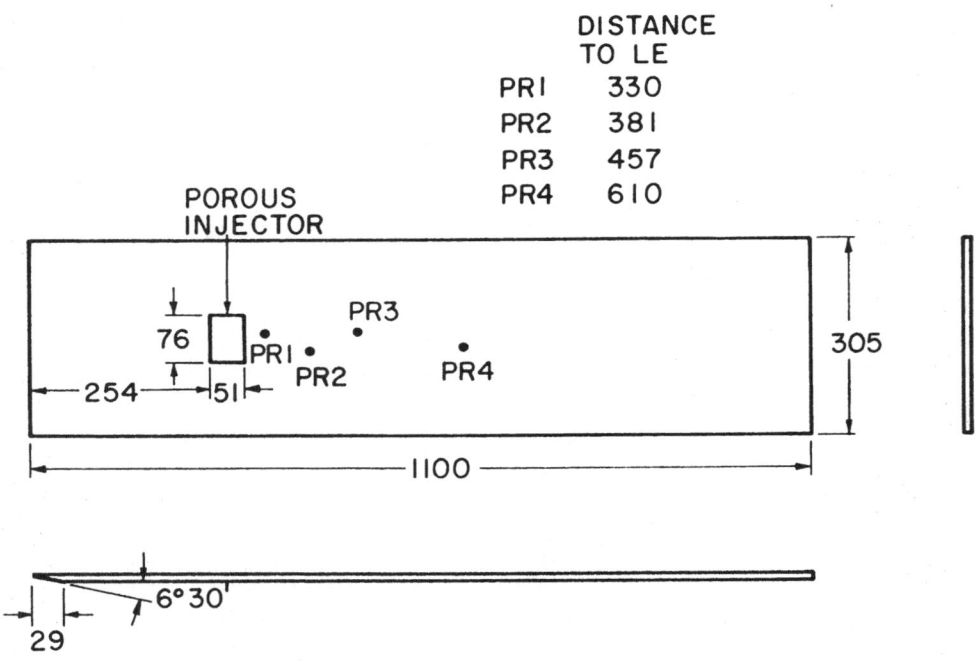

Fig. 8   Schematic of flat plate, Experiment F. All dimensions are in mm.

Excellent optical access to this plate was available from both sides and from the direction normal to the injection surface. This allowed us to employ optical techniques for the determination of bubble sizes and trajectories. Access for LDA was also good, and we hoped to exploit the bubble/no bubble interface to look at velocity profiles, but thus far this has proven to be difficult. Skin friction reduction on the plate was measured by means of flush-mounted hot film probes. The verification of zero pressure gradient was made by 42 pressure taps on the plate surface.

B. Documentation of the Undisturbed Flowfield

We have documented the undisturbed boundary layer for all three experimental configurations by using a combination of pressure measurements and LDV profiles. In addition to providing a carefully documented description of the boundary layers in which we were working, these profiles also serve to provide in situ calibration of our flush-mounted hot film probes. For all three configurations, we found the boundary layers to be fully developed zero pressure gradient turbulent boundary layers in which skin friction characteristics could be predicted by standard correlations (see, for example, White, 1974). Table II summarizes the three sets of undisturbed boundary layer measurements. For the tunnel wall boundary layer, the virtual origin was extrapolated from plots of displacement and momentum thickness to the 5/4 power versus distance. For the axisymmetric body, the origin of the boundary layer was assumed to be at the trip location while for the flat plate the origin was taken as the leading edge. Comparison of the data from the flat plate of Experiments D or F with the axisymmetric data of Experiment E at equivalent distances from the origin show that they have very similar outer and inner scales.

TABLE II
COMPARISON OF LDA DATA

Tunnel Wall Boundary Layer (Experiment E)

Location*	u_∞ (m/sec)	R_θ	$u+$(m/s)	y at $y^+ = 5 (\mu m)$
272	4.68	3372	0.195	25.6
	10.90	7358	0.423	11.8
374	4.75	4215	0.190	26.3
	10.30	8862	0.392	12.7
454	4.78	5127	0.188	26.6
	10.47	10627	0.392	12.7

*From virtual origin

Axisymmetric Body (Experiment F)

Location**	u_∞ (m/sec)	R_θ	u^+(m/s)	y at $y^+ = 5 (\mu m)$
155	5.26	2635	0.21	23.8
	10.71	5062	0.405	12.3
	16.94	8837	0.607	8.24
277	5.25	3491	0.21	23.8
	10.77	6764	0.406	12.3
	16.86	10144	0.617	8.10
403	5.30	3775	0.210	23.8
	10.90	8191	0.413	12.11
	16.88	14197	0.594	8.42

**From trip wire

As intimated above, a major difference between the boundary layers produced in this high speed facility and those in many other laboratory experiments is the manner in which high x-Reynolds numbers are achieved. In the present case, high Reynolds numbers are generated by high velocity flows over relatively short distances, whereas in most laboratory studies high Reynolds numbers are reached by using long boundary layer runs at low speeds. The present approach produces high unit Reynolds number boundary layers with skin friction velocity (u_τ's) representative of those observed on real field vehicles. This implies that the viscous lengths are very small (as they are in high speed vehicle applications) and translates directly into difficulties in making measurements in the very-near-wall (small y^+) region. The Re_θ's created in the present experiment are, however, typical of those observed in other laboratory boundary layers and are smaller than those generally encountered in vehicle applications.

C. Integrated Skin Friction Measurements

1. Techniques

Integrated skin friction was measured on both the tunnel wall boundary layer of Experiment D and the axisymmetric body of Experiment E by force balances. Both balances used strain-gauged members to infer drag. The balances were calibrated using a pulley and weights arrangement and excellent linearity was observed up to the maximum shear stress (maximum freestream velocity). Both balances were required to cover a large speed range, and in practical terms this resulted in some data scatter at the lower speeds where the absolute forces (particularly in the presence of C_f reduction) were between one and ten percent of full scale. For the tunnel wall installation, the drag member was in the bending mode, while for the axisymmetric body it was in shear. The latter design is advantageous because of the reduced travel afforded by its relatively greater stiffness.

2. Tunnel Wall Boundary Layer Results (Experiment D)

Some typical integrated skin friction data taken with the force balance of the tunnel wall configuration are shown in Fig. 9. These data are for the plate-on-top orientation. In a global sense, these integrated skin friction data agree very well with the data presented by the Soviets. They show that microbubble injection does reduce skin friction drag, that the amount of reduction increases with increasing gas flow, and that the maximum reduction is near 80%. In contrast to the Soviet data, however, we find that these plate-on-top data show that the amount of drag reduction decreases as the freestream speed increases. When the airflow rate is normalized by the freestream velocity and an arbitrary area (here chosen as the wetted area of the porous section) to form the non-dimensional parameter C_Q, the plate-on-top data collapse nearly to a single line as shown in Fig. 10.

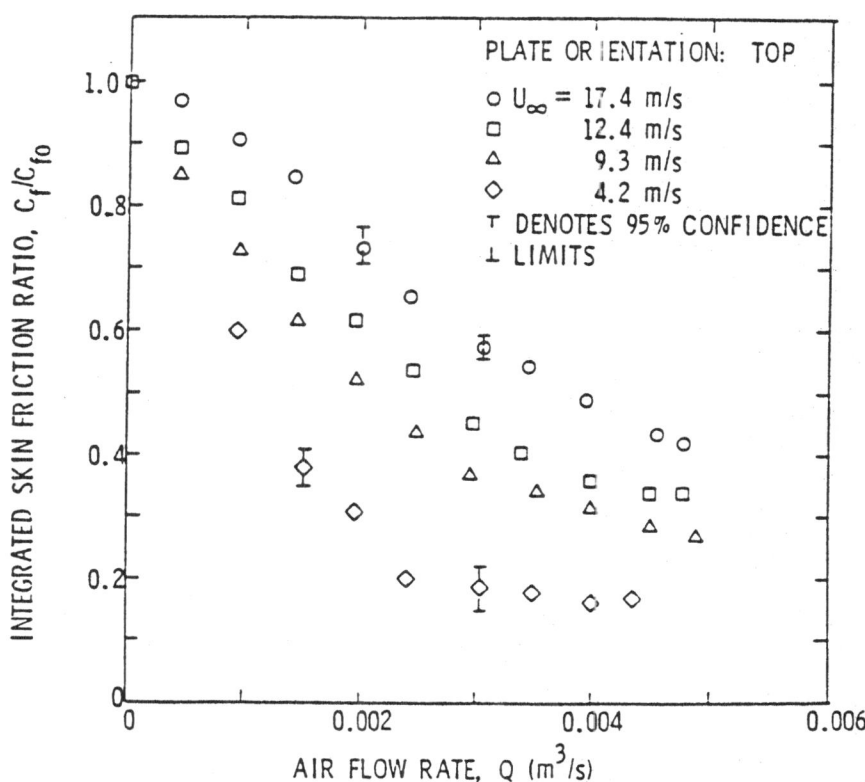

Fig. 9   Representative results showing effect of skin friction reduction with
          airflow at various freestream velocities; Experiment D. Integrated C_f
          measurements. Plate-on-top orientation.

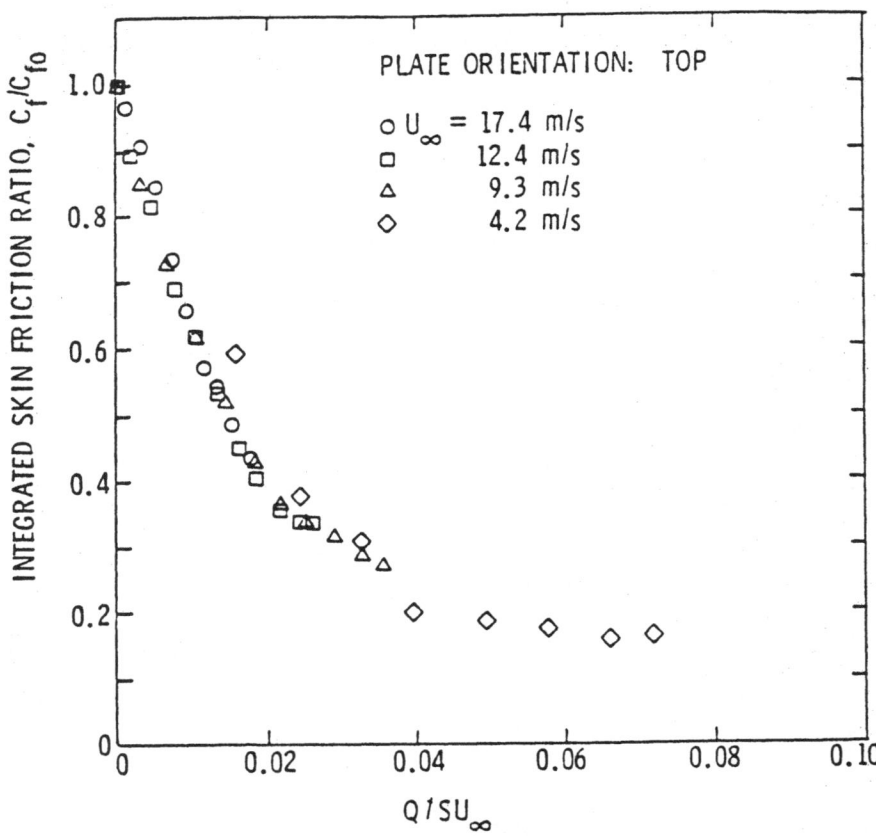

Fig. 10 Integrated skin friction results for flat plate Experiment D showing
data collapse achieved by normalizing airflow rate by freestream
velocity. Plate-on-top orientation.

This collapse indicates that for these data the airflow requirement for a
given amount of drag reduction scales directly with freestream velocity, i.e.,
that it is the volumetric fraction of air to water that is important. In view of
later results which show (not surprisingly) that the location of the bubbles in
the boundary layer is also of importance, these data suggest that the
concentration profiles for the various freestream speeds are in some sense
similar. As we show later for the plate-on-bottom orientation, this collapse
does not always work -- an indication that the concentration profiles are in
general different at different speeds.

As a demonstration that the wetted area of the porous surface is not a
proper area with which to non-dimensionalize the airflow rate, we present data in

Fig. 11 for injection through a porous section half the size of that used for the data in Fig. 10. (The downstream half of the porous section was used, see Fig. 6.) If these data for injection through the half section are normalized by the smaller injection area, they fall far away from the collapsed data line on Fig. 10. By contrast, if they are normalized by the same area as the data on Fig. 10 (i.e., by the area of the full porous section) they again collapse on the data line as Fig. 11 shows. From this we conclude that the drag reduction is not strongly dependent on the injection area. A more appropriate area to normalize Q/u by is the cross-sectional area of the boundary, δb, where b is the spanwise width of the injection section and δ is the boundary layer height. Detailed comparisons (Madavan, Deutsch and Merkle, 1984b), however, show that the ratio of volumetric flow rates, C_V, is more effective at normalizing data than is $Q/u\delta b$.

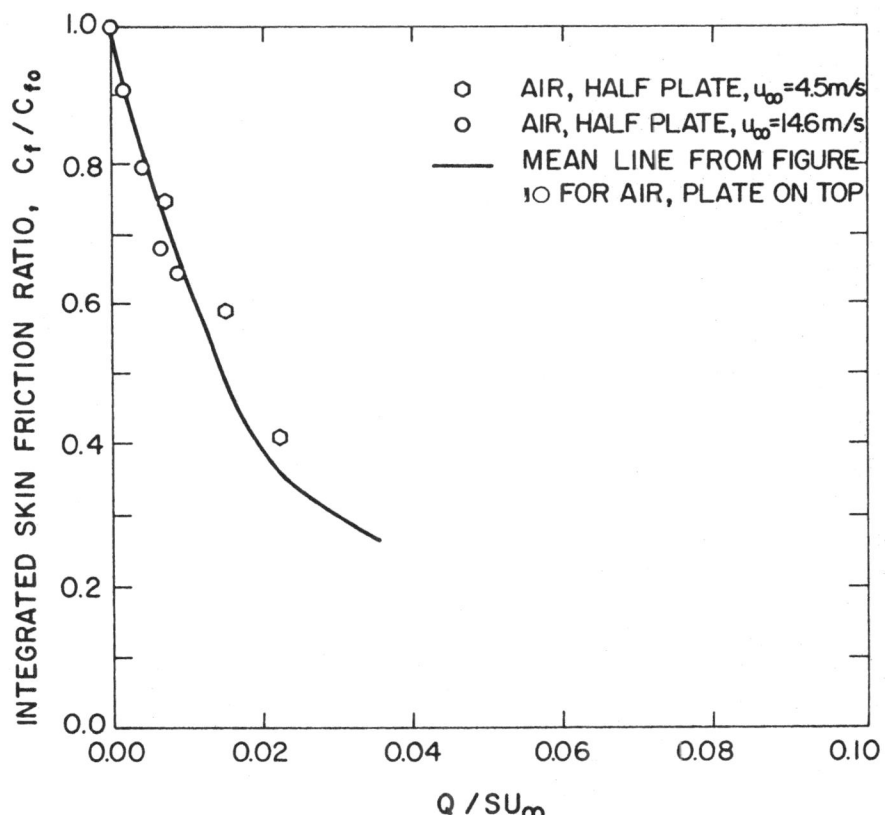

Fig. 11 Comparison of skin friction reduction obtained with half-length injector (89 mm) with that obtained with full length injector (178 mm); Experiment D. Note S in abscissa is based on full length (178 x 102) in both cases. Plate-on-top.

Before leaving this issue, it is useful to point out that no matter how the volumetric airflow rate (Q) is scaled, it is clear that the width of the injection section (the spanwise extent of the treated boundary layer) must always be a scaling parameter. If the spanwise width is doubled, the volumetric air requirement will likewise be doubled.

A comparison of some plate-on-bottom and plate-on-top data for three tunnel speeds is given on Fig. 12. Here, the C_f/C_{fo} measurements are plotted against the airflow, Q, in dimensional form. At the lower speeds, the skin friction reduction for the plate-on-bottom configuration is considerably less than that for the plate-on-top orientation. As the speed is increased, this deficiency fades and becomes nearly unnoticeable at the highest speed. Because the only difference between these two sets of measurements is the gravitational orientation, it is clear that the effectiveness of the bubbles is reduced for the plate-on-bottom data because buoyancy is causing the bubbles to drift out of the boundary layer. (This drift due to buoyancy is verified later by direct optical observations.) Again, this is evidence that the total volume of air in the boundary layer is by itself not sufficient to define the bubbles' effectiveness. Something must also be known of the bubble cloud location.

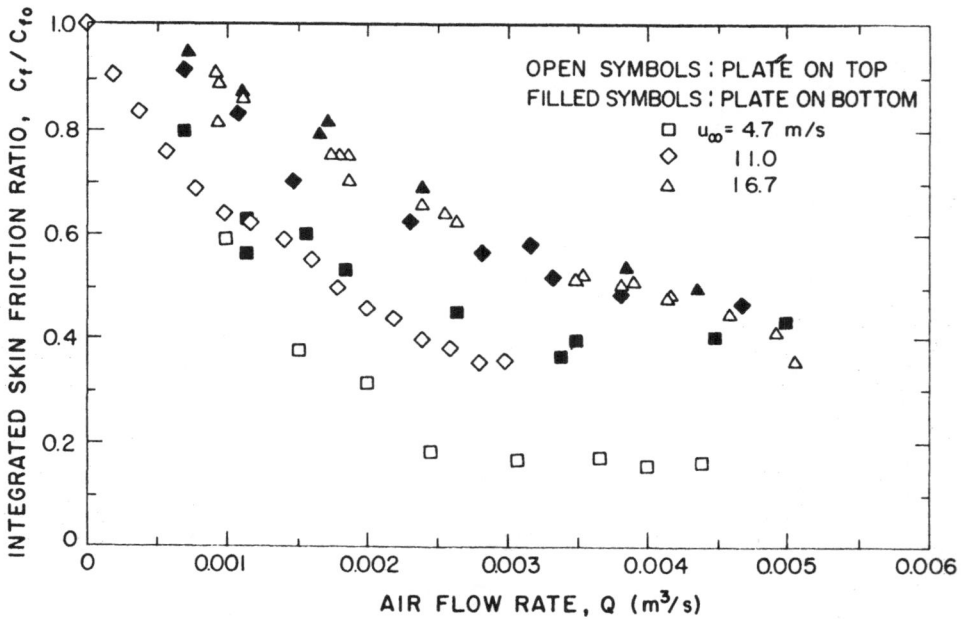

Fig. 12 The effect of buoyancy on the integrated skin friction reduction. Comparison of data taken with the plate above and below the boundary layer. Experiment D; 0.5 μm porous plate.

A final observation from Experiment D concerns the sensitivity of the drag reduction to the method of injection. The Soviets implied that appropriate construction of the porous section was crucial to the success of the microbubble experiment, and for that reason we did extensive measurements with porous materials of various sizes, tediously prepared etched stacked plates to simulate the Soviet material, and finally with injection through a one-eighth inch slot. At no time did we notice a first order effect on drag reduction, although some second order effects were observed.

An example of the extent to which the airflow was sensitive to pore size is shown in Fig. 13. The previously noted collapse of C_f/C_{fo} data when plotted against C_Q is shown in Fig. 13a for a 0.5 μm filter size. When re-run with a 100 μm filter size the data no longer collapse to a single line, as Fig. 13b shows, but retain a dependency on velocity. Comparisons of Figs. 13a and b show that the pore size has no effect at low speeds, but that injection through 0.5 μm material is somewhat more effective at higher speeds. The only apparent reason for this difference is that a change in pore size induces a change in bubble size. In a later section, we show in fact that bubble size increases slowly with pore size. This would suggest that the bubbles from the 100 μm section are somewhat larger than those from the 0.5 μm section. Such a change in bubble size would cause a redistribution of bubbles in the boundary layer which could explain the differences between Fig. 13a and 13b.

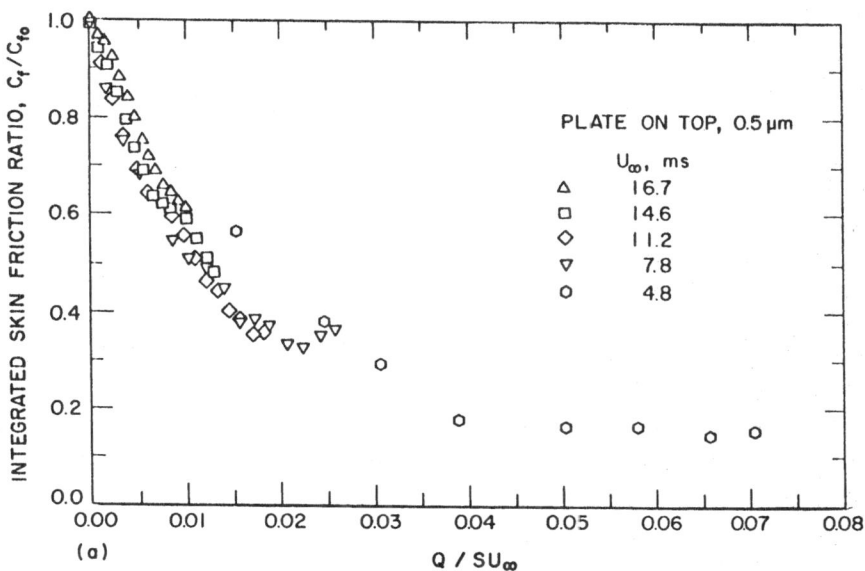

Fig 13   Effect of pore size on integrated skin friction reduction.   (a) 0.5 μm filter size;

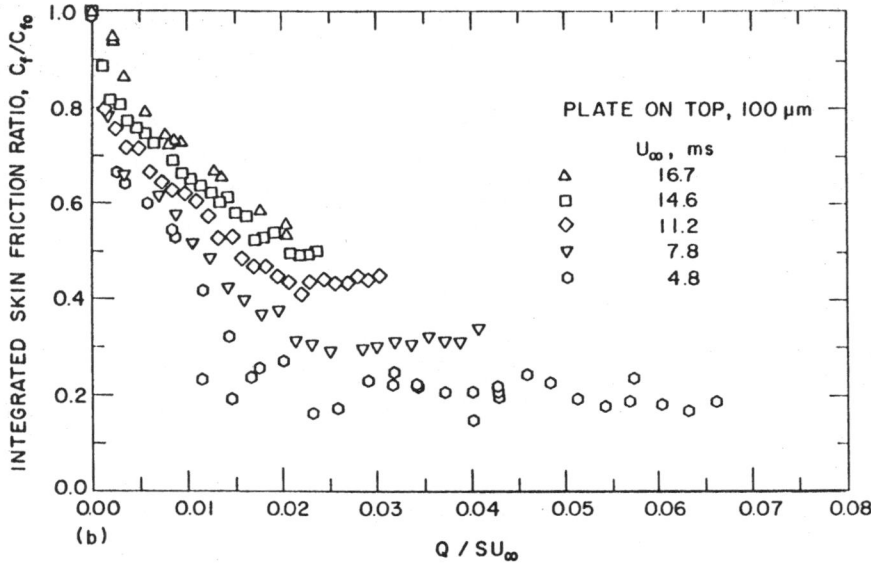

(b) 100 μm filter size. Data taken from Experiment D using
    sintered stainless steel porous injector.

Although such higher order effects could be pursued in practical
applications, one would be likely to choose the material, assuming sufficient
strength, that provides the smallest pressure drop. For the experimental results
reported here, we have, in general, selected material porosities which give large
enough pressure drops to provide reliable control over the flow rate.

3. Axisymmetric Body Results (Experiment F)

Because the axisymmetric body was tested in the same tunnel as the tunnel
wall experiment (D), the unit Reynolds numbers and arclength distances on the two
configurations are close enough that drag reduction comparisons may be made
directly one against the other. Direct comparison does, however, require an
adjustment for the "spanwise" dimension of the injector because it is clear that
airflow requirements do scale as the width of the treated area. For the tunnel
wall boundary layer, the injector width was 102 mm (Fig. 6) while for the
axisymmetric body the injector "width" corresponded to the body perimeter, 89 π
mm (Fig. 7). This corresponds to a span-wise ratio of about 2.74 between the two
bodies.

Comparisons of the skin friction reductions for the tunnel wall boundary
layer and the axisymmetric body are given on Fig. 14 as a function of Q/b (where
b is the injector width). For the tunnel wall boundary layer, the drag reduction
at the lowest speed (4.7 m/s) is quite large for either gravitational
orientation, while for the axisymmetric body the bubbles are almost ineffective

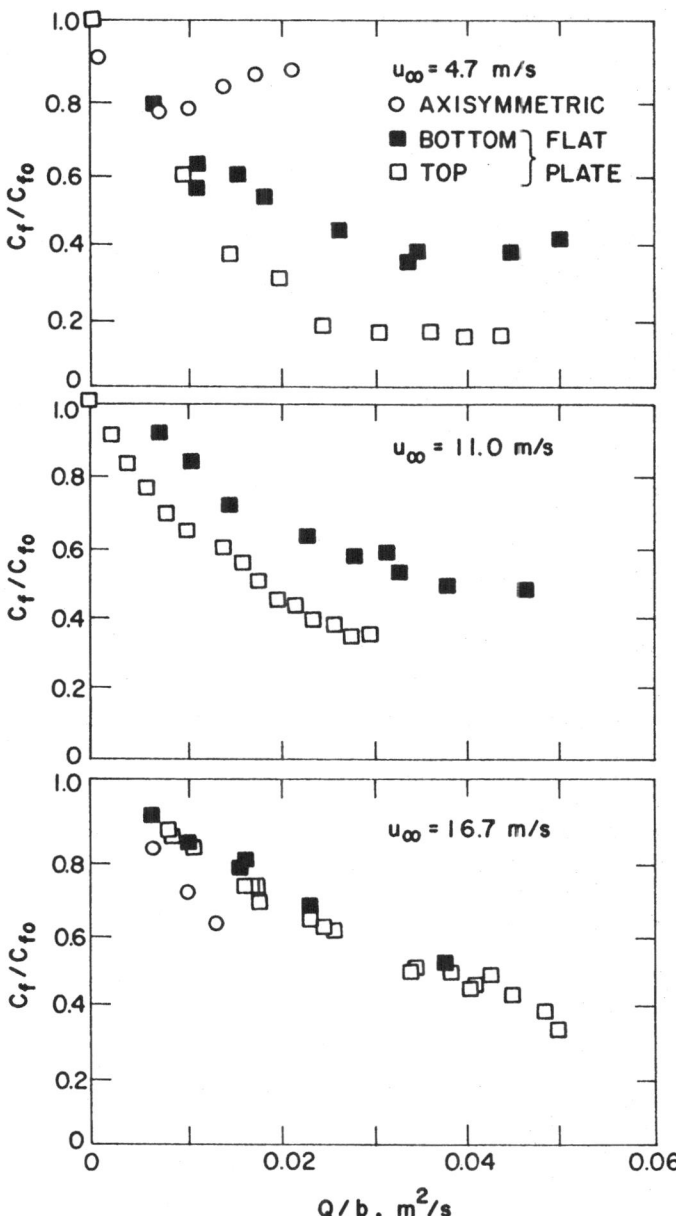

Fig. 14 Comparison of skin friction reduction as a function of airflow per unit span on flat plate and axisymmetric body. Experiments D and E. Integrated skin friction data.

at this speed. The ineffectiveness of the bubbles on the axisymmetric body is the result of a tendency for the bubble cloud to bifurcate and roll up on the horizontal sides of the axisymmetric body. A similar phenomenon can also be seen on a flat surface when it is rotated to a vertical orientation.

At 16.7 m/s, the data suggest that a given amount of air is slightly more effective for the axisymmetric body than it is for either gravitational orientation of the flat plate. Comparison of Figs. 14a, b and c shows that drag reduction on the axisymmetric body increases with velocity. This increased effectiveness if reminiscent of that reported by the Soviets. In both cases, the change stems from the decreased influence of buoyancy at higher speeds, but the details of the physical mechanisms are different. In the Soviet work, the influence of buoyancy decreases gradually as the speed increases, while on the axisymmetric body a critical speed and airflow rate exists below which a bifurcation appears.

4. Comparison of Air and Helium Injection

An assessment of the relative advantage of different injection gases is also of interest. To obtain a handle on the sensitivity of the C_f reduction to the type of gas injected, we have tested helium and air injection back-to-back in both the tunnel wall boundary layer (Experiment D) and the axisymmetric body (Experiment E). The results, which are summarized on Figs. 15 and 16 provide a contrasting but incomplete picture. The integrated drag measurements, Fig. 15, show that helium and air provide similar levels of drag reduction. The axisymmetric data, Fig. 16, show that helium is somewhat more effective than air at the higher speeds. Perhaps one explanation for this is that all flat plate data with helium are for the plate-on-top orientation. Additional flat plate tests with other gravitational orientations might help to resolve these discrepancies.

D. Local Skin Friction Measurements

1. Techniques

To date, the only information that has been obtained about the local skin friction or the turbulence characteristics of the microbubble boundary layer has been obtained from measurements at the wall. The harsh conditions present (optically opaque, high dynamic pressure, thin boundary layers, small turbulence scales, and strong two phase character) are likely to sustain this condition for the near future. For this reason, we have relied on flush-mounted hot film probes for which the shear stress may be related to the voltage output of a constant temperature anemometer (Bellhouse and Schultz, 1966; Brown, 1967; Ramaprian and Tu, 1983). Calibration is accomplished in place using the characteristics of classical turbulent boundary layers.

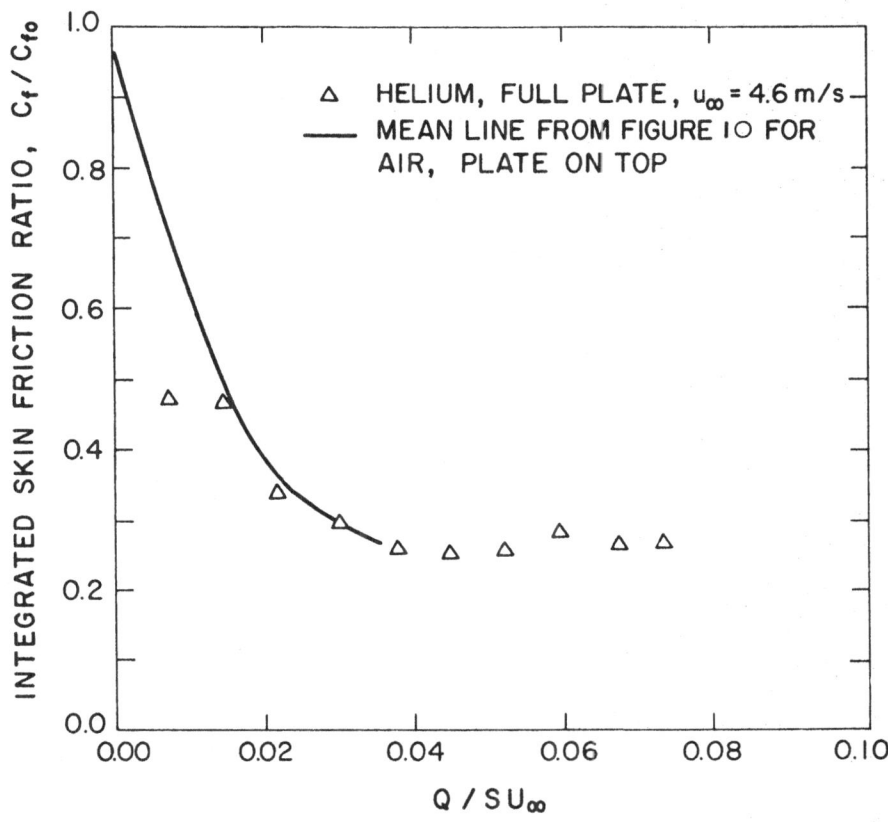

Fig. 15 Integrated drag measurements for helium injection in Experiment D.
Plate-on-top. Solid line shows results for air taken from Fig. 10.

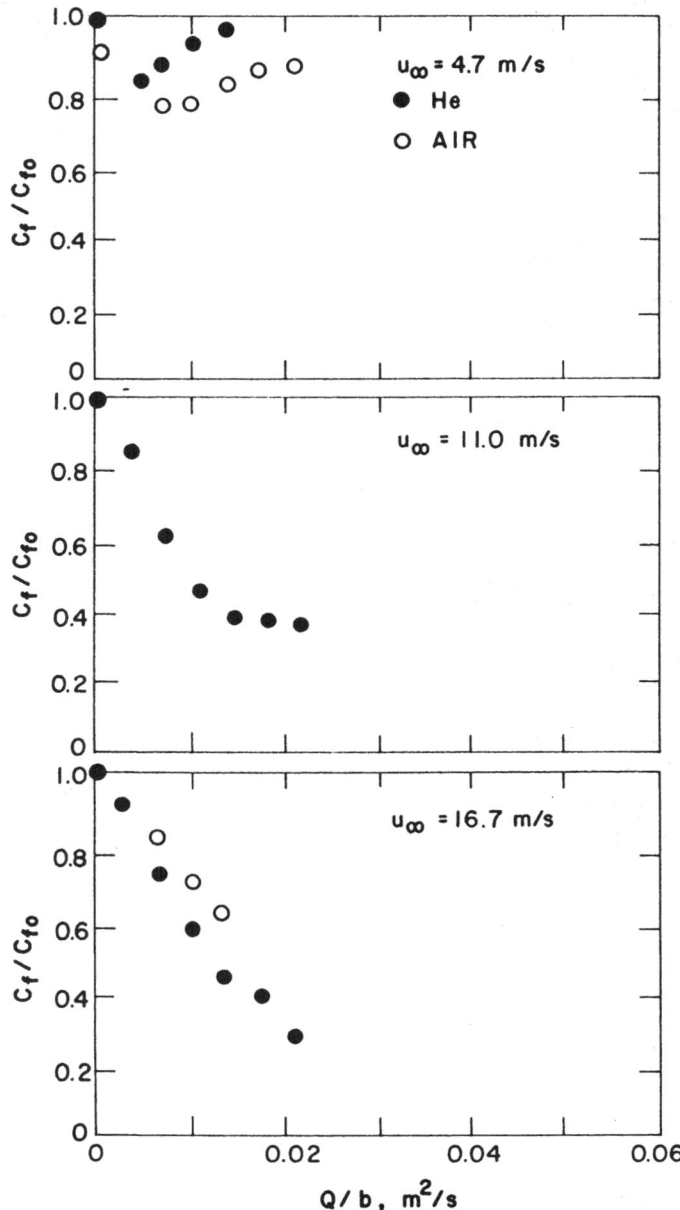

Fig. 16 Comparison of effects of helium and air injection for axisymmetric body;
Experiment E. Integrated C_f results.

Hot film calibrations for microbubble boundary layers are performed in a bubble-free boundary layer in which the skin friction is varied by changing the freestream velocity. Their application in microbubble boundary layers, therefore, warrants some discussion. We note first that it is reasonable to be more comfortable operating the probes in a plate-on-bottom configuration as, in that case, both the Soviet concentration measurements and our own results (see Section E) indicate a measurable bubble-free region exists very close to the wall. More care is needed for the plate-on-top data, and here two experimental results have proven useful. First, we have observed from probes immersed in the boundary layer that the impact of a bubble on a hot film probe is signaled by a large negative spike; second, the exposure of a hot film probe to air results in a voltage which is less than the "zero velocity" voltage of the probe. We have evaluated our raw hot film data in terms of these observations and have found no evidence of either of these forms of contamination for the plate-on-top data, except under the simultaneous conditions of our lowest speeds and our highest gas flow rates. These data have been thrown out. The plate-on-bottom data are trouble-free in all cases.

One additional reasonable question is whether the proximity of the bubbles, short of impingement, might change the heat transfer characteristics of the film, (and so the calibration), by changing the local fluid properties. Direct demonstration that this was not a problem was difficult, so we devised a comparison of the hot films and the force balance. For this comparison, we mounted a series of hot film probes on the surface of the drag balance of the tunnel wall test configuration. In this way, we were able to compare directly the integrated skin friction measurements from the force balance with an integration of the local skin friction measurements from the probes. The locations of the probes is shown in Fig. 6. We present the comparisons, which are quite good for all airflow rates and both plate orientations, in Fig. 17. These results verify that the hot film probes give C_f reductions that are both qualitatively and quantitatively similar to those inferred from the hot films. Hence, we conclude that the pure water calibration of the hot films is acceptable.

2. Local Skin Friction Measurements in the Tunnel Wall Boundary Layer

Our local skin friction measurements obtained with hot film probes have, in general, been in qualitative agreement with the integrated C_f measurements. The peak reductions in the local shear stresses are as much as 90% and the dependence upon airflow rate and velocity are similar to that observed with the integrated C_f measurements. In general, the reduction in C_f is largest immediately downstream of the injection section and eventually relaxes back toward zero with

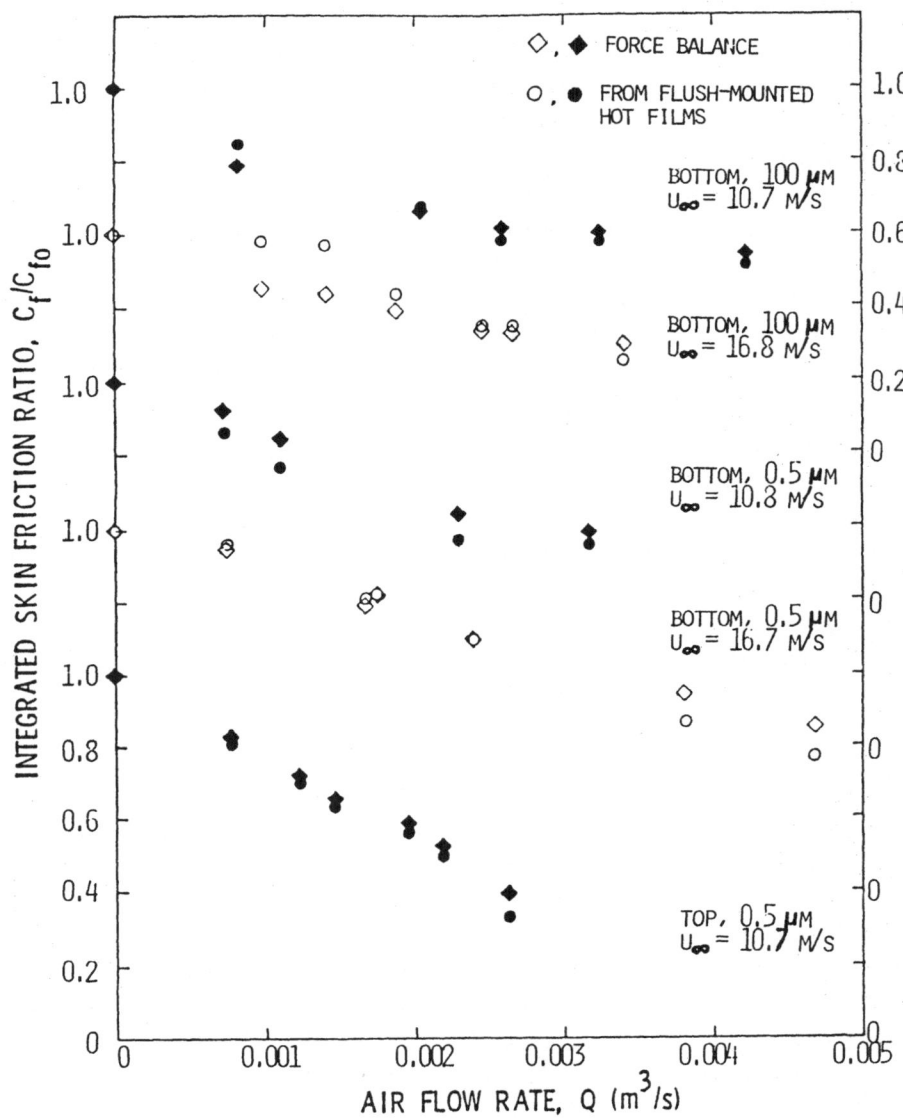

Fig. 17 Comparison of integrated skin friction measurements taken from numerical integration of local skin friction measurements with direct force-balance measurements; Experiment D. (Hot film mounted on force balance surface.)

increasing distance from the injector. Perhaps the best way to summarize the results is to display the drag reduction at a constant volume fraction of air against downstream distance. This we do for both the plate-on-top and plate-on-bottom orientations in Figs. 18 and 19. Both figures show that the rate of relaxation is somewhat slower for the plate-on-top data than for the plate-on-bottom results. These differences are relatively insignificant at the highest test speed of 16.8 m/sec (Fig. 18), but are quite large at the slower speed of 10.7 m/s (Fig. 19). These variations again show the influence of buoyancy at the lower speeds, as well as the (near) absence of it at higher speeds.

The data at 16.8 m/s (Fig. 18) show effective persistence of the reduction for more than 30 boundary layer thicknesses for both plate orientations. At 10.7 m/s (Fig. 19), the differences between the two gravitational orientations is larger as noted above. Obviously, buoyancy is more important at this speed. This is not only because the smaller convective speed gives the buoyant force more time to remove the bubbles from the boundary layer, but also because the bubble sizes are apt to be bigger at the higher gas flow rates and lower speeds. The results of both Silberman (1957) and Hughes, Reischman and Holzmann (1979) suggest that the bubble size increases as the square root of the ratio of air flow to mean speed. Our measurements of bubble size (Section E) also indicate an increase in bubble size with an increase in airflow rate and a decrease with freestream velocity.

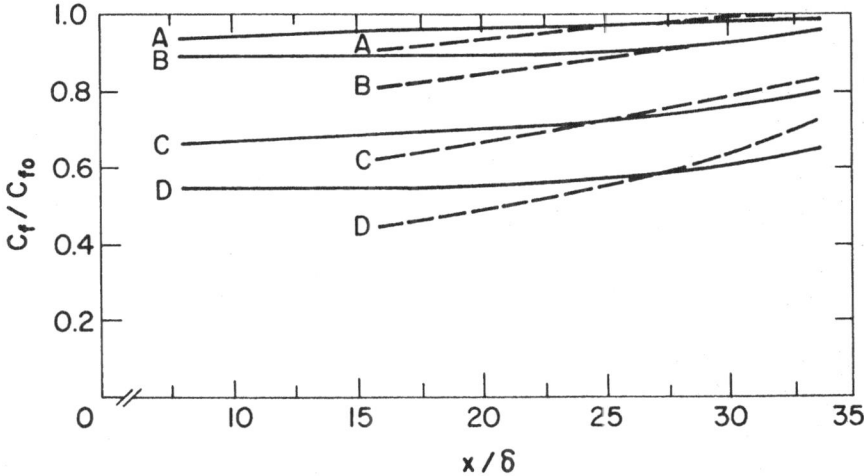

Fig. 18   The downstream persistence of the skin friction reduction. Comparison of data taken with the plate above and below the boundary layer. U_∞ = 16.8 m/s. The solid line represents the plate-on-top, the dashed line the plate-on-bottom. $Q_a/(Q_a + Q_w)$ is 0.13 for A, 0.18 for B, 0.27 for C and 0.34 for D.

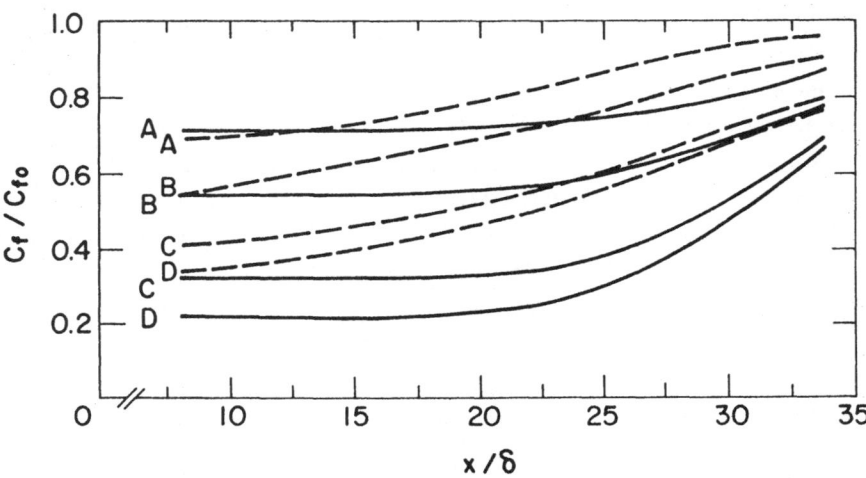

Fig. 19   The downstream persistence of the skin friction reduction. Comparison
of data taken with the plate above and below the boundary layer. U_∞ =
10.7 m/s. The solid line represents the plate-on-top, the dashed line
the plate-on-bottom. $Q_a/(Q_a + Q_w)$ is 0.17 for A, 0.24 for B, 0.34 for C
and 0.39 for D.

We should note that although the data presented here were taken from the
tunnel wall boundary layer experiment, the companion data taken in the flat plate
experiment (Pal, Merkle, Deutsch, 1988) generally support these results.

   3. Statistical Information

   Statistical information gleaned from the hot film probes has been most
interesting. A typical trace of the linearized hot film signal is shown as a
function of time for various airflow rates (and drag reduction levels) in Fig.
20.

   In the Figure, the location of τ_{ref} marks the average shear stress level for
this test condition without bubble injection. The drop in average shear stress
with the addition of bubbles is quite apparent. Also apparent is the loss of
high frequency content of the signal as the bubble concentration is increased.
Note the increasingly one-sided character of the signal at the higher airflow
rates. Plots of the higher moments of the data (given later) verify this
observed trend. Finally, we note that even for the largest drag reduction case
(62%), the signal has not become intermittently laminar, but retains a
turbulent-like character.

   The frequency spectra of the shear stress fluctuations in the presence of
bubbles may be most appropriately compared by normalizing by inner variables
based upon the local (reduced) shear stress. One such plot is shown in Fig. 21.

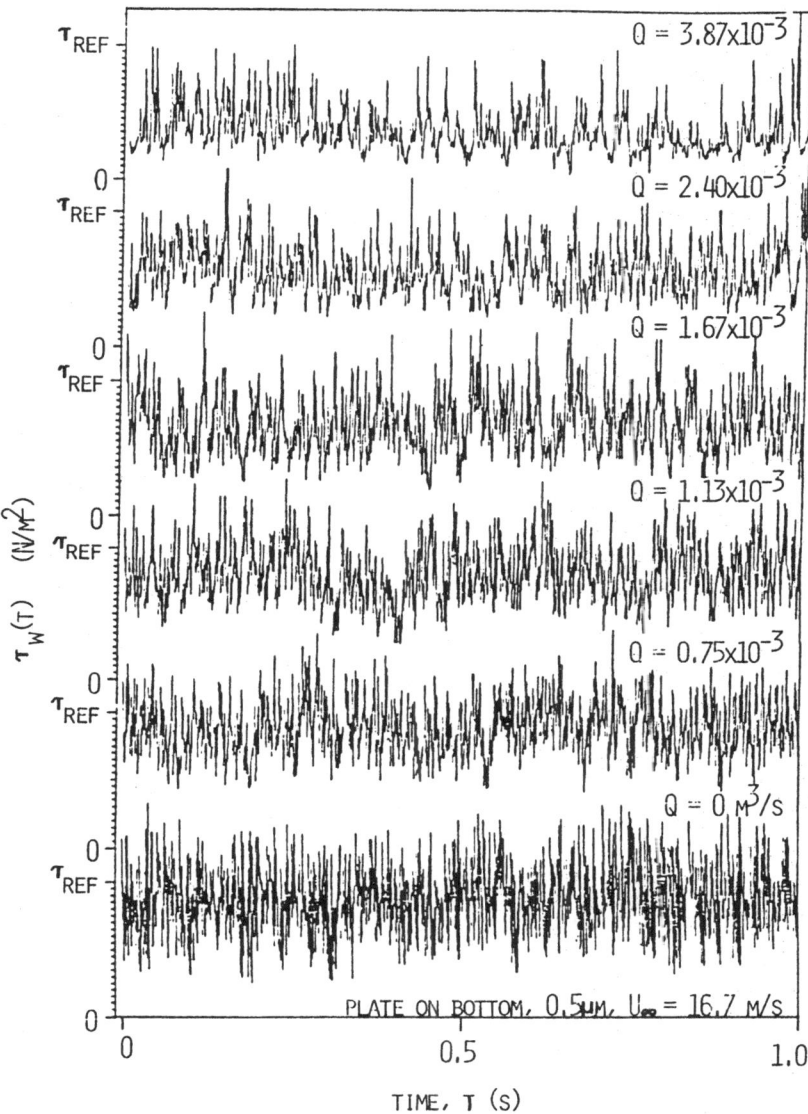

Fig. 20 Linearized hot film signals as a function of time for various airflow
rates. Experiment D. Freestream velocity 16.7 m/s; plate-on-bottom.

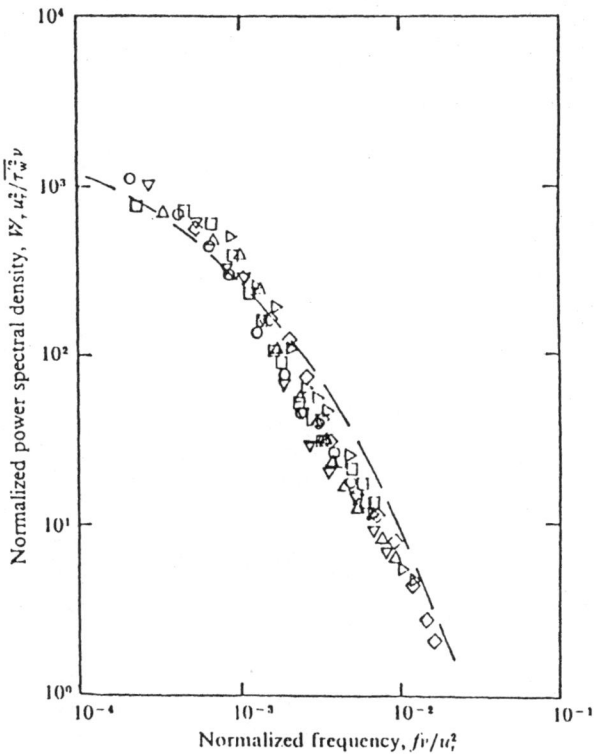

Fig. 21   Normalized shear stress spectra for microbubble boundary layers.
$\square$, U_∞ = 4.6 m/s; $\triangle$, 7.6 m/s; $\bigcirc$, 10.7 m/s; ∇, 13.8 m/s; $\diamondsuit$, 16.8 m/s.

With an inner variable scaling one would expect a good collapse of the data in the high frequency range if the microbubble boundary layer retains the characteristics of a classical turbulent boundary layer. This appears to be the case. In particular, note that there is no general trend with airflow rate.

Fortuna and Hanratty (1972) have measured shear stress spectra for polymer drag reduction. A comparison of the polymer and the microbubble results show very strong similarities. The collapse of the spectra with an appropriate inner scaling suggests that the turbulent dynamics remain more or less the same.

The similarity of the polymer and microbubble results, as well as the one-sidedness of the microbubble time trace, led us to perform back-to-back microbubble/polymer experiments on the flat plate. Details may be found in Pal et al. (1988). We used Separan (AP 30) in a solution of 500 ppm. Both the Separan solution and the microbubbles were injected at the same location. We found remarkable similarities in the higher order statistics between the two solutions at comparable values of the drag reduction. In Figs. 22 and 23, we show the skewness (representing the one-sidedness of the signal) and the kurtosis. The data show a strong similarity which suggests that polymers and microbubbles

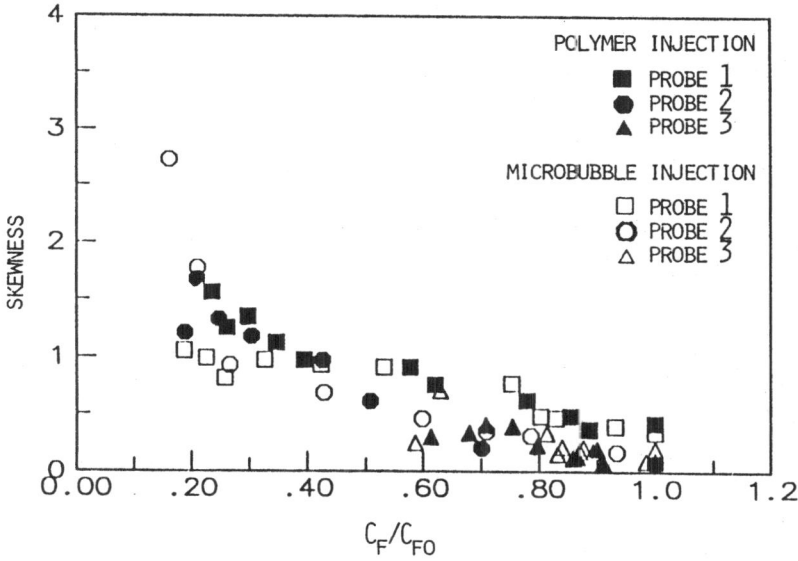

Fig. 22 Skewness of shear stress signals versus C_f/C_{fo}.

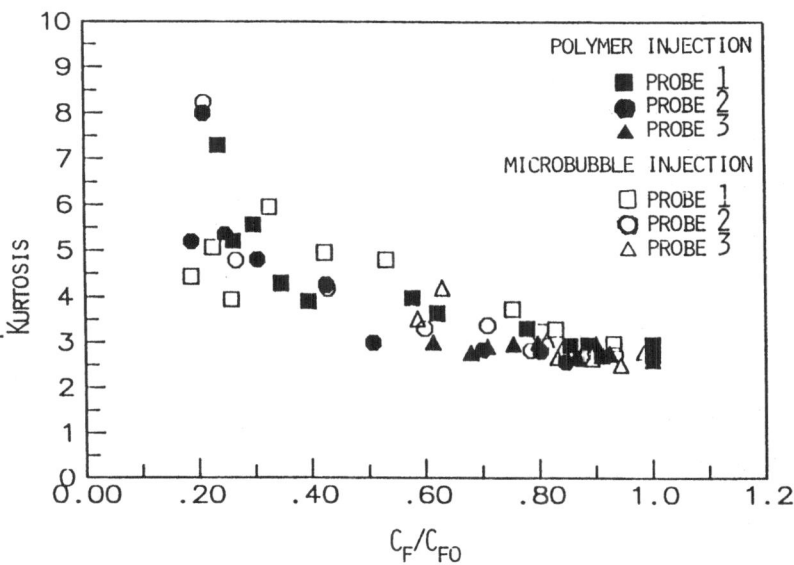

Fig. 23 Kurtosis of shear stress signals versus C_f/C_{fo}.

have a similar effect on the turbulence and that in some sense the mechanisms may be similar. Also note that both sets of data depart from their near Gaussian values for drag reductions greater than about 40%. This suggests that simple scaling arguments are probably no longer appropriate for large values of the drag reduction. As we have shown that the flow is not relaminarizing, this implies that for large amounts of skin friction reduction significant changes are occurring in the dynamics of the turbulence.

E. Bubble Sizes and Trajectories

1. Techniques

Using the flat plate of Experiment F, we have made estimates of bubble sizes and documented the trajectories of the bubble clouds. Bubble sizes were measured from still photographs of the bubbly flow. Bubble trajectories were deduced by traversing a laser beam through the bubble cloud and picking up the intensity of the light that passed through. Bubble cloud location measurements were made at the three locations on the plate shown in Fig. 24; bubble size measurements were made at the first and last of these positions. At these same three locations, we have documented the flow and the shear stress by mounting hot film probes there, and by making LDV profile measurements. Measurements were made at the freestream speeds of 4.6 and 10.7 m/s.

2. Bubble Size Measurements

As noted earlier, the gas flow ratios encountered at substantial drag reduction levels are as high as 50%. It is probably fair to say that at these concentrations, coupled with the very thin bubble clouds, the steep concentration gradients, and the high velocities present in a microbubble boundary layer, we stand a fairly good chance of never being able to make an exhaustive survey of the bubble characteristics. The bubble clouds are completely opaque, and photographs show a bubble smear rather than identifiable bubbles. Still, a knowledge of the bubble size is a most crucial element in understanding how the bubbles affect a turbulent boundary layer.

Because of the difficulty of making detailed observations of the bubbles at drag reducing conditions, we have made bubble size measurements at very low airflow rates from top view photographs like that shown in Fig. 25. At these airflows, the bubbles are individually distinguishable, even though they provide little or no drag reduction. We note, however, that photos taken at higher airflow rates where significant drag reduction is present, even though indistinct, present the impression that the bubble sizes are not much different than those in the low airflow photos. In addition, simple dimensional estimates of the bubble injection process or of the effect of turbulence on bubble size show that it is difficult to change the bubble size dramatically.

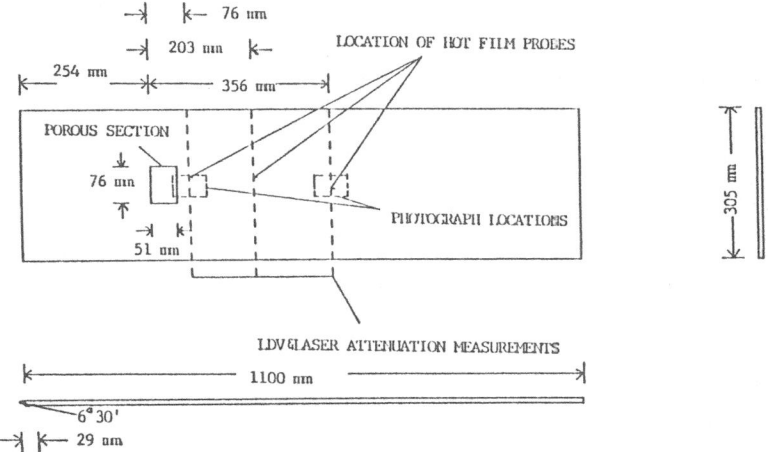

Fig. 24 Schematic of the flat plate showing location and size of porous section
and instrumentation locations for LDV measurements, flush-mounted
hot-film probes, and top and side view photographs.

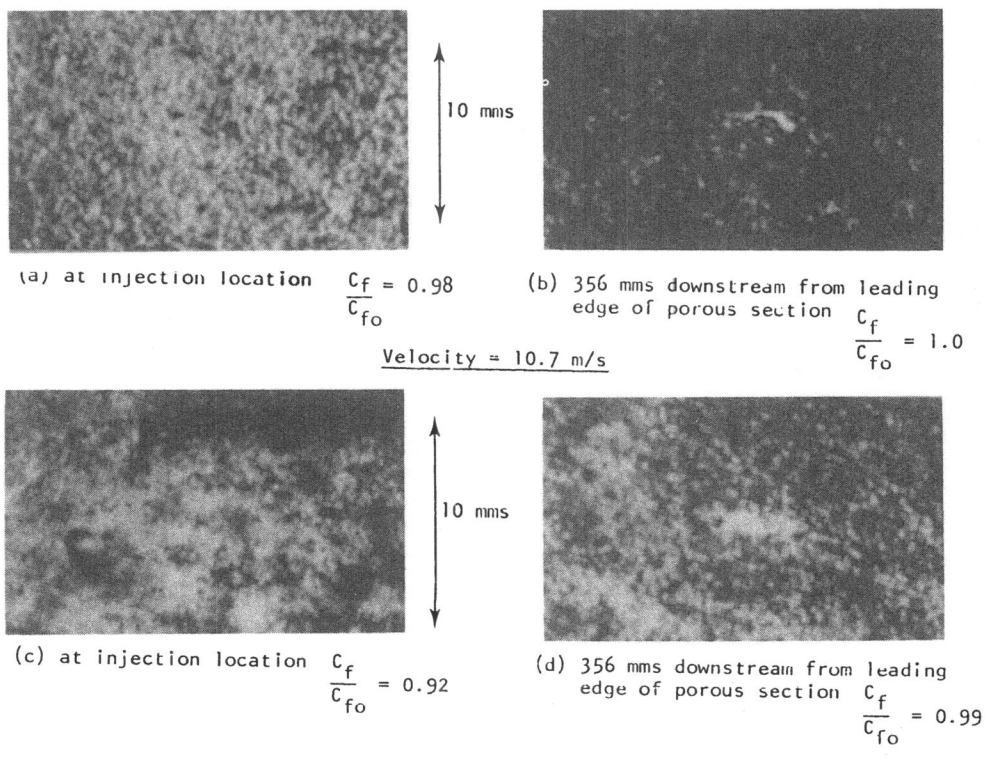

(a) at injection location $\dfrac{C_f}{C_{fo}} = 0.98$

(b) 356 mms downstream from leading
edge of porous section $\dfrac{C_f}{C_{fo}} = 1.0$

Velocity = 10.7 m/s

(c) at injection location $\dfrac{C_f}{C_{fo}} = 0.92$

(d) 356 mms downstream from leading
edge of porous section $\dfrac{C_f}{C_{fo}} = 0.99$

Fig. 25 Top view comparison of airflow II between a freestream velocity of 4.6
m/s (a) and (b) and 10.7 m/s (c) and (d). Plate-on-bottom orientation.

As an example, Hinze (1955) has shown through a balance of pressure and capillary forces that the maximum bubble size that is stable against breakup is:

$$d_1 \sim (\frac{\sigma}{\rho})^{3/5} \, \varepsilon^{-2/5} \tag{1}$$

where σ is the surface tension, ρ the liquid density, and ε the dissipation. Following Tennekes and Lumley (1968), we express the dissipation in the boundary layer as,

$$\varepsilon \sim \frac{u*^3}{\theta} \tag{2}$$

and using the definition of the friction velocity, Eqn. 1 becomes,

$$d_1 \sim \tau_w^{-3/5} \tag{3}$$

A parallel analysis for bubble coalescence based on similar physical reasoning reported by Thomas (1981) gives the minimum bubble size stable to coalescence as,

$$d_2 \sim \tau_w^{-3/8} \tag{4}$$

From these equations we estimate that for a factor of two decrease in the wall shear stress, the breakup size, d_1, would increase by about 50% while the coalescence size, d_2, would increase by about 30%. Within the spirit of these estimates, then, the low airflow rate data are of value.

A synopsis of the average bubble size measurements deduced from these photographs is shown in Fig. 26. In general, the averages were computed from samples of 50 to 100 bubbles. Bubble sizes are seen to lie between 150 and 1200 microns. (If we assume a proportionality constant of one in Eqns. 1 and 2 above, we predict bubble sizes from 500 to 1100 microns for the low speed case and from 50 to 500 microns for the high speed case.) Note that the smallest bubble sizes are still an order of magnitude larger than the sublayer thickness (about 10 microns here) while the largest are still an order of magnitude smaller than the boundary layer thickness (about 10 mm here). This perhaps explains the difficulty of changing bubble sizes by an order of magnitude (see Eqns. 3 and 4) and may help explain the relative insensitivity of the drag reduction to injection method which is discussed above. One might expect then that an order of magnitude change in bubble size would be required to dramatically influence the drag-reduction/volumetric-flowrate relation.

Two additional conclusions can be drawn from the bubble size data. First, the relatively rapid growth of bubble size with distance from the injector, even

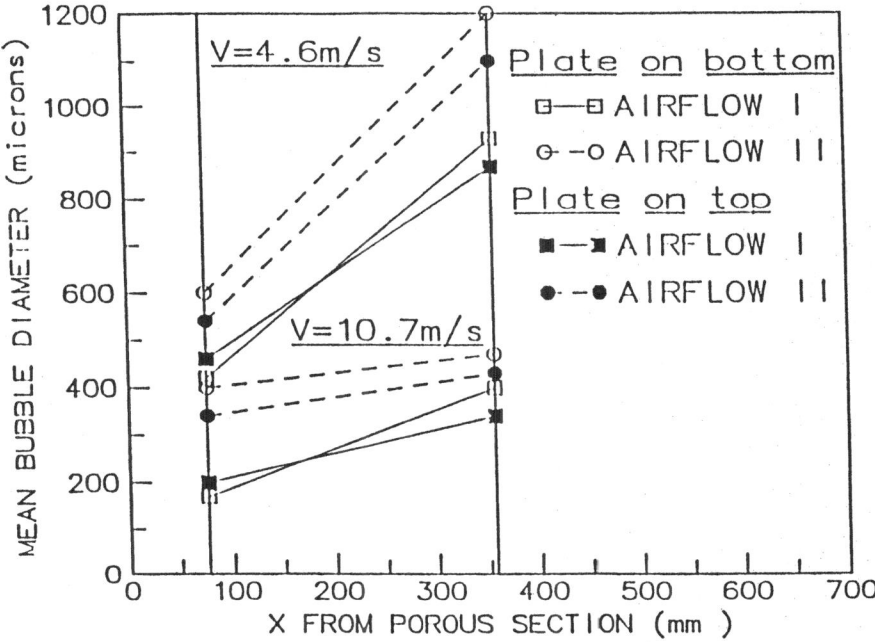

Fig. 26   Bubble size measurements taken from photographic surveys. Airflow rates given in Table III.

at low bubble densities, would seem to suggest that interbubble interactions (in particular collisions that result in coalescence) are quite significant; and second, the data show that the differences in bubble size between the plate-on-top and plate-on-bottom configurations is within the scatter of the data.

3. Bubble Cloud Trajectories

To obtain information about bubble cloud trajectories, a 1 mm diameter beam from a low power He-Cd laser was passed through the cloud and the attenuated signal received by a photomultiplier on the opposite side of the tunnel. To map the location of the cloud, the laser beam-photomultiplier set-up was traversed from the plate surface to the edge of the bubble cloud in step sizes as small as 0.0254 mm. Because of the difficulties cited above, quantitative calibration of this experiment is seemingly impossible, but it is possible to deduce important information from it. In particular, there is a surprising amount of information available from the case in which the bubbles are not present in the near wall region.

We present percent laser attenuation for both gravitational orientations at
a freestream speed of 4.6 m/s in Fig. 27. The figure has six embedded plots:
the bottom three represent laser attenuation traverses for the plate-on-bottom
configuration; the top three represent the plate-on-top data. The abscissae of
the figure represent the distance along the flat plate measured from the
injection section, while the ordinate is the distance normal to the wall. Note
that the two vertical lines at each station marked 0.0 and 1.0 represent the
photomultiplier output for the beam transmitting through clear water and for the
beam turned off, respectively. Airflow rates and their corresponding drag
reduction are given in Table III.

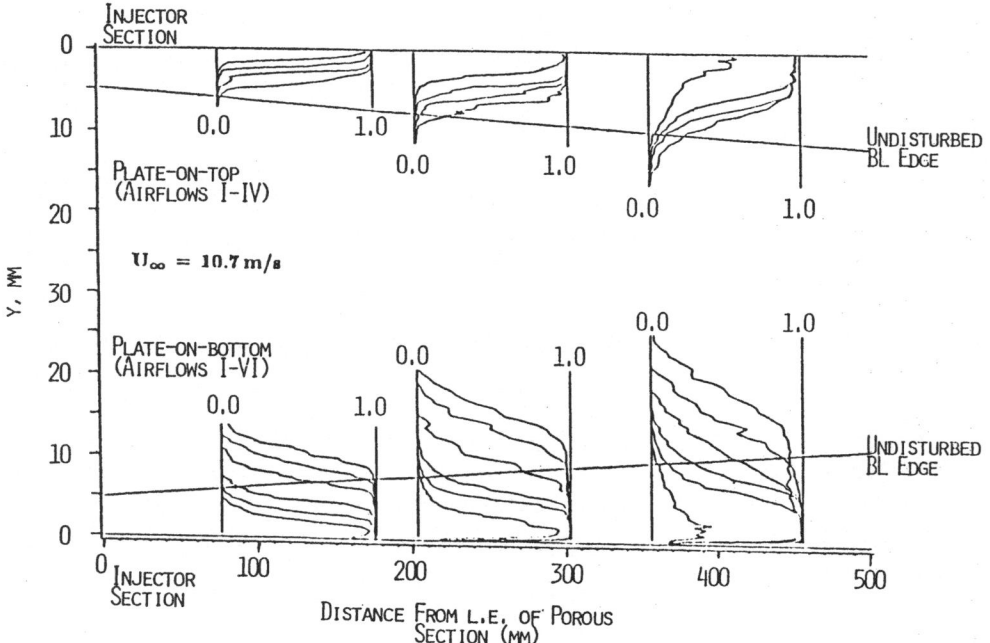

Fig. 27   Bubble cloud locations based upon laser attenuation measurements.
          Airflow rates and corresponding drag reductions are given in Table III.

Apparent from the plate-on-bottom results in Fig. 27 is the growth of a
bubble-free region with distance downstream from the injector. The existence of
this bubble-free region -- which agrees with the Soviet plate-on-bottom
observations -- is not apparent in the plate-on-top data. We cannot determine
from the data whether or not this means that such a region does not exist for the
plate-on-top data or that we cannot resolve it. Clearly though, the difference
between the top and bottom data show that the major cause of this region is

TABLE III
AIRFLOW RATE CLASSIFICATIONS AND
CORRESPONDING SKIN FRICTION RESULTS
Experiment F

AIRFLOW RATE	DIMENSIONAL AIRFLOW (x10^{-3} m^3/s)	VOLUMETRIC AIRFLOW RATIO $\frac{Q_A}{Q_A + Q_W}$ Vel = 4.6m/s	VOLUMETRIC AIRFLOW RATIO Vel = 10.7m/s	C_f/C_{fo} Freestream Velocity=4.6 m/s Top*/Bottom* x=76mms	Top*/Bottom* x=203mms	Top*/Bottom* x=356 mms	Freestream Velocity=10.7m/s Top*/Bottom* x=76mms	Top*/Bottom* x=203mms	Top*/Bottom* x=356mms
I	0.05	0.03	0.01	0.98 / 0.99	0.98 / 0.98	0.99 / 1.0	0.98 / 0.98	0.98 / 0.99	0.99 / 1.0
II	0.14	0.07	0.04	0.95 / 0.98	0.95 / 0.92	0.98 / 1.0	0.92 / 0.92	0.95 / 0.98	0.98 / 0.99
III	0.28	0.13	0.07	0.42 / 0.93	0.40 / 0.82	0.75 / 0.96	0.85 / 0.90	0.92 / 0.93	0.97 / 0.98
IV	0.46	0.19	0.11	0.05 / 0.51	0.04 / 0.92	0.08 / 0.91	0.58 / 0.78	0.86 / 0.90	0.92 / 0.94
V	1.13	0.37	0.23	/ 0.26	/ 0.29	/ 0.98	/ 0.12	/ 0.56	/ 0.82
VI	1.96	0.51	0.35	/ 0.18	/ 0.22	/ 1.01	/ 0.22	/ 0.11	/ 0.60

*Top and bottom signify plate-on-top and plate-on-bottom.

buoyancy. Note further that even for the plate-on-top data, in which buoyancy inhibits motion away from the wall, the outer edge of the bubble cloud is getting farther from the wall with distance downstream. Thus, we see that turbulent diffusion of bubbles is significant and also contributes to a loss of effectiveness with increasing distance from the injector.

Comparison with data taken at the higher velocity of 10.8 m/s (which is not presented) show that the bubble cloud is thinner. This occurs both because the shorter convection times allow less time for buoyant and turbulent diffusion, and also because although the volumetric flow of air may be the same for the two speeds, the relative concentration of bubbles is smaller at the higher velocity. We note that comparisons at constant volumetric flow ratio (we have chosen our airflow rates so that this is easily done -- see Table III), make the bubble profiles more nearly similar. This would seem to explain why we have found the volumetric flow ratio to be the best of the normalizations.

We might check the scaling of the bubble profiles with inner variables by considering a plot of the bubble cloud location as a function of y^+, which we present in Fig. 28. Although the effect of the bubbles is to reduce τ_W and, hence, to increase the inner scale, the outer edge of the bubbles still moves farther out from the plate as more air is added. In the figure we present a case in which the inner edge of the bubble cloud has migrated out past a y^+ of 200. For this data set, as for all data in which this is the case, there is no drag reduction. We conclude then that drag reduction is only possible when bubbles are present in the very near wall region.

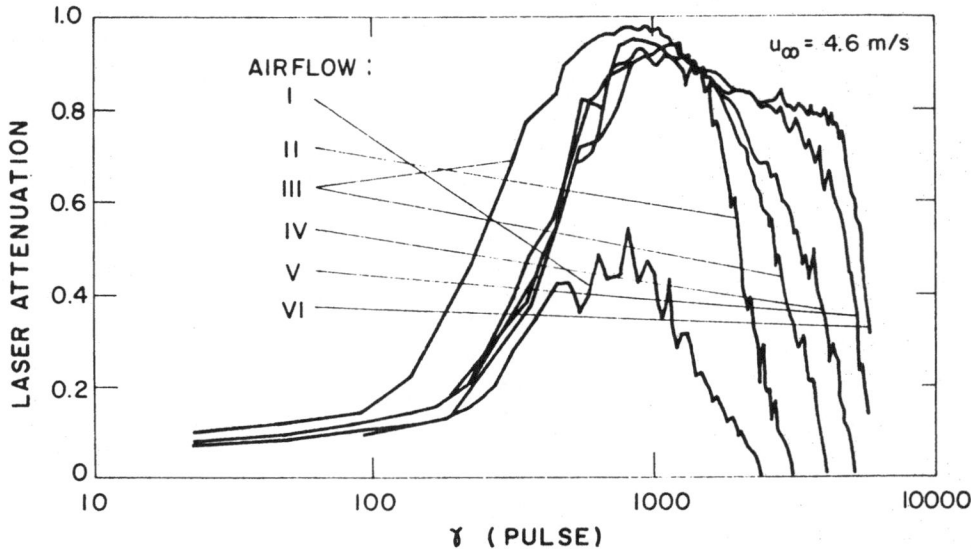

Fig. 28 Laser attenuation measurements of bubble cloud location. $u_\infty = 4.7$ m/s. Experiment F. Airflow rates and corresponding C_f reduction given in Table III.

F. Theoretical and Computational Studies

1. Some Simple Ideas

In Madavan, Deutsch and Merkle (1984a), we compared the mechanism for microbubble drag reduction to that for polymer reduction. We argued, following Lumley (1973, 1977), that the drag reduction proceeded through an increased viscosity in the buffer region which destroyed the turbulent production scales there, effectively increasing the sublayer thickness. Scaling arguments show that doubling the sublayer thickness will approximately halve the drag. For either mechanism to work in this way, it must be ineffective in the sublayer. It was not our intention in that article to do more than suggest a way of looking at the mechanism -- from, say, the order of a mixing length level of complexity. The mechanism whereby each of these drag reduction techniques affects the flow dynamically is fascinating but not well understood.

On the basis of our subsequent measurements, we feel that the analogy between the drag reduction mechanisms in polymer solutions and microbubbles is useful, and is possibly representative of an entire class of drag reducing agents that affect the near wall region. Our one glimpse into the dynamics of these drag reducers supports this comparison, but suggests that the characteristics of the turbulence may deviate from their familiar equilibrium turbulent boundary layer character when the drag reduction exceeds 40%.

2. A Computational Study

On the basis of the physical reasoning given above, we attempted a simple computational study (Madavan, Merkle and Deutsch, 1985). A homogeneous flow model in which the presence of the bubbles was assumed to modify the fluid properties near the wall through an increase in kinetic viscosity (decreased density, increase absolute viscosity) was postulated. Based on the available concentration measurements, profiles which peaked in the near wall region but went to zero at the walls, were specified. The presence of bubbles modified the turbulence in the near wall region by decreasing the turbulence Reynolds number. The turbulence model and, hence, the dynamics of the turbulence was left unchanged.

The model was inserted into a well tested compressible boundary layer code. Reductions of the order of 50% were calculated. The predictions, in agreement with the bubble trajectory data, highlighted the importance of the location of the bubble cloud with reference to the wall. The bubbles were most effective when near the buffer region.

Marie (1987) recently used quite similar reasoning to produce a simple analytical model for the phenomena.

3. Theoretical Studies

Legner (1984) performed the first theoretical study of the microbubble phenomenon. Again, a simple mixing length model was used for the turbulence. The manner in which the density and absolute viscosity was taken to vary with concentration was similar to that done in the computational work. Legner assumes that the bubble concentration profile is constant through the near wall region (in fact to the wall), and that the addition of bubbles does not change the velocity gradient at the wall.

Comparison of Legner's model with the Soviet data shows good agreement. Several of the assumptions made, however, such as assuming that the turbulent eddy viscosity is much larger than the molecular viscosity, although valid in the near wall should not be made through the sublayer and at the wall. One potential improvement in the model would be to embed it in a multi-layer model of the turbulent boundary layer in which the sublayer could be treated separately. The assumptions on conditions at the wall could then be relaxed.

Legner's model includes the first rational concept for the modification of the turbulence in a microbubble boundary layer. He assumes the presence of the bubbles and their compressibility leads to a bulk viscosity effect which damps the turbulence. Legner presents theoretical justification for supposing that the ratio of mixing lengths with and without bubbles is

$$\frac{q\Lambda}{q_0\Lambda_0} \simeq (1-\phi) \tag{5}$$

where ϕ is the bubble concentration and q and Λ represent the turbulent energy and length scale, respectively. The subscript zero refers to the case without bubbles. The incorporation of this effect into computational studies would be appropriate.

More recently, Meng and Uhlman (1986) have suggested that the energy required for bubble splitting could serve as a significant energy sink for turbulence. Estimates of bubble splitting in a microbubble boundary layer at concentrations of interest for drag reduction verify that the magnitude is sufficient to have a significant impact on turbulence levels. This mechanism would again be expected to be ineffective in the sublayer and to peak in the buffer and log-law regions. Further study of this mechanism would also appear promising.

It is probably fair to say that modeling of the microbubble phenomenon has not kept pace with the experiments. Two phase models, in which the bubbles are treated separately and their trajectories calculated and coupled to the turbulence, although of great difficulty, would be valuable.

IV. SUMMARY

The injection of gas into a liquid turbulent boundary layer to form bubbles reduces skin friction drag locally by as much as 80%. In general, the phenomenon is most effective at speeds where convective times are short compared to either buoyant or turbulent diffusion times. For very low speed conditions and for particular geometric configurations, buoyancy becomes so dominant that microbubble injection is ineffective. As the free stream speed is increased above these conditions, a regime is encountered in which the bubble effectiveness increases with speed. This characteristic, however, is limited to the buoyancy dominated regime. At higher speeds the bubbles can become more or less effective as speed is increased, depending upon local conditions, and no general statement can be made.

The reason for the diversity of bubble effectiveness with changes in speed is the complexity of the flowfields of interest. The degree to which the bubbles alter the boundary layer characteristics depends upon both their concentration and their location in the boundary layer. In general, the amount of C_f reduction increases with airflow rate, but to be effective the bubbles must find their way into the near wall region of the boundary layer. Here, they can interact with the turbulent flow in the buffer layer to cause a change of order unity in the skin friction. In this respect their effect seems to be closely related to that achieved by polymer additives. Microbubbles, like polymer solutions, appear to destroy the energy producing fluctuations near the buffer region. The resulting growth of the sublayer thickness is a manifestation of the drag reduction. Both polymer solutions and microbubbles appear to have very strong effects on the dynamics of turbulence for drag reductions greater than about 40%.

The size of the bubbles is clearly a parameter of importance. Their diameters have much to do with their trajectories and, hence, affect both their concentration and their location in the boundary layer. In addition, the effects of buoyancy vary strongly with bubble diameter. Finally, a detailed understanding of the mechanisms by which they interact with the turbulence and the mean flow requires at least an order of magnitude knowledge of bubble size.

The very first estimates of bubble size have only recently appeared. These measurements show that the bubbles generated by injecting through a porous material are of the same order of magnitude as those generated by injection from a single pore. In addition, the two appear to exhibit the same sorts of trends. Specifically, the bubble size decreases when free stream speed is increased and increases when airflow rate is increased, but appears to show little dependence on the injection procedure. The bubble sizes in a microbubble cloud can be set by any of three competing mechanisms: the initial formation at the wall; bubble splitting and break-up by turbulence action; and bubble coalescence upon collision. It is of note that the bubbles are quite close together and do exhibit decided growth in size as they are swept downstream in the boundary layer.

Perhaps the most significant characteristic of the measured bubble sizes is their diameter in comparison to the boundary layer scales. The measurements show that, during their period of activity, the bubbles appear to range between an order of magnitude larger than the sublayer thickness and an order of magnitude smaller than the boundary layer thickness. Thus, fairly substantial changes in bubble size would be needed to alter the manner in which the bubbles interact with the boundary layer. Evidence to date suggests that such dramatic changes in bubble size may be quite difficult to accomplish. Nevertheless, possibilities for controlling bubble sizes and trajectories and ultimately for optimizing gas flow requirements remain high priority items in microbubble research. Improved understanding of bubble dynamics is also crucial for predicting effects of scale-up to practical vehicle sizes.

V. REFERENCES

BABA, E. (1986) Discussion in 16th Symposium on Naval Hydrodynamics, pp. 214-215.

BELLHOUSE, B. J. & SCHULTZ, D. L. (1966) "Determination of Mean and Dynamic Skin Friction, Separation and Transition in Low-Speed Flow with a Thin-Film Heated Element", J. Fluid Mech. 24, p. 379.

BOGDEVICH, V. G. & EVSEEV, A. R. (1976) "Effect of Gas Saturation on Wall Turbulence", in Investigations of Boundary Layer Control (in Russian)(eds. S. S. Kutateladze & G. S. Migirenko), p. 49. Thermophysics Institute Publishing House.

BOGDEVICH, V. G. & MALYUGA, A. G. (1976) "The Distribution of Skin Friction in a Turbulent Boundary Layer of Water Beyond the Location of Gas Injection", in Investigations of Boundary Layer Control (in Russian)(ed. S. S. Kutateladze & G. S. Migirenko), p. 62. Thermophysics Institute Publishing House.

BROWN, G. L. (1967) "Theory and Application of Heated Films for Skin Friction Measurements", Proc. 1967 Heat Transfer and Fluid Mech. Inst., p. 361. Stanford University Press.

BUSHNELL, D. M. (1983) "Turbulent Drag Reduction for External Flows", AIAA Paper 83-0227.

CREWE, P. R. & EGGINGTON, W. J. (1960) Trans. Roy. Inst. Nav. Arch. 102, 315.

DEUTSCH, S. & CASTANO, J. (1986) "Microbubble Skin Friction Reduction on an Axisymmetric Body", Physics of Fluids, 29, 11, pp. 3590-3597.

DUBNISCHEV, Y., EVSEEV, A. R., SOBOLEV, V. S. & UTKIN, E. N. (1975) "Study of Gas-Saturated Turbulent Streams Using a Laser-Doppler Velocimeter", J. Appl. Mech. Tech. Phys. 16, 1, p. 114. Translated from Zhur. Prikl. Mech. Tekh. Fiz., No. 1, p. 147.

FORTUNA, G. & HANRATTY, T. J. (1972) "The Influence of Drag-Reducing Polymers on Turbulence in the Viscous Sublayer", J. Fluid Mech. 53, p. 575.

HEFNER, J. N., BUSHNELL, D. M., WHITCOMB, R. T., CARY JR., A. M. & ASH, R. L. (1977) "Concepts for Aircraft Drag Reduction", AGARD/VKI Special Course on Concepts for Drag Reduction, AGARD-R-654.

HOUGH, G. R. (1980) "Viscous Flow Drag Reduction", Prog. Astro. & Aero., p. 72.

HUGHES, N. H., REISCHMANN, M. M. & HOLZMANN, J. M. (1979) "Digital Image Analysis of Two-Phase Flow Data", 6th Bienn. Symp. on Turbulence, University of Missouri, Rolla.

LEGNER, H. H. (1984) "A Simple Model for Gas Bubble Drag Reduction", Phys. of Fluids, 27, pp. 2788-2790.

LUMLEY, J. L. (1977) "Drag Reduction in Two-Phase and Polymer Flows", Phys. of Fluids, 20, 10, Pt. II, pp. 564-571.

LUMLEY, J. L. (1973) J. Polymer Sci., Macromol. Rev. 7, p. 263.

MADAVAN, N. K. (1984) "The Effects of Microbubbles on Turbulent Boundary Layer Skin Friction", Ph.D. Thesis, The Pennsylvania State University.

MADAVAN, N. K., DEUTSCH, S. & MERKLE, C. L. (1984a) "Reduction of Turbulent Skin Friction by Microbubbles", Phys. Fluids 27, p. 356.

MADAVAN, N. K., DEUTSCH, S. & MERKLE, C. L. (1984a) "The Effect of Porous Material on Microbubble Skin Friction Reduction", AIAA Paper 84-0348.

MADAVAN, N. K., DEUTSCH, S. & MERKLE, C. L. (1984b) "Reduction of Turbulent Skin Friction by Microbubbles", Phys. of Fluids, 27, 2, pp. 356-363.

MADAVAN, N. K., DEUTSCH, S. & MERKLE, C. L. (1985) "Measurements of Local Skin Friction in a Microbubble Modified Turbulent Boundary Layer", J. Fluid Mech. 156, p. 237-256.

MADAVAN, N. K., MERKLE, C. L. & DEUTSCH, S. (1985) "Numerical Investigations into the Mechanisms of Microbubble Drag Reduction", J. Fluids Engr., 107, pp. 370-377.

MARIE, J. L. (1987) "A Simple Analytical Formulation for Microbubble Drag Reduction", J. Physico-Chemical Hydrodynamics, pp. 213-220.

MCCORMICK, M. E. & BHATTACHARYA, R. (1973) "Drag Reduction of a Submersible Hull by Electrolysis", Nav. Eng. Jnl. 85, p. 11-16.

MENG, J. C. S. & UHLMAN, J. S. (1986) "Theoretical Analysis and Numerical Simulation of Bubble Formation Dynamics in a Turbulent Boundary Layer", Gould Ocean Systems Division Report, OSD-771-HYDRO-CR-86-04, Newport, R. I.

MENG, J. C. S. (1987) Personal communication.

MIGIRENKO, G. S. & EVSEEV, A. R. (1974) "Turbulent Boundary Layer with Gas Saturation", in Problems of Thermophysics and Physical Hydrodynamics (in Russian), Novosibirsk, Nauka.

PAL, S., MERKLE, C. L. & DEUTSCH, S. (1988) "Bubble Characteristics and Trajectories in a Microbubble Boundary Layer", Physics of Fluids, 31, 4, pp. 744-751.

PURTELL, L. P., KLEBANOFF, P. S. & BUCKLEY, F. T. (1981) "Turbulent Boundary Layer at Low Reynolds Number", Phys. Fluids 24, p. 802.

RAMAPRIAN, B. R. & TU, S. W. (1983) "Calibration of a Heat Flux Gage for Skin Friction Measurement", Trans. ASME I: J. Fluids Eng. 104, p. 455.

SILBERMAN, E. (1957) "Production of Bubbles by the Disintegration of Gas Jets in Liquid", in Proc. 5th Midwestern Conf. on Fluid Mech., University of Michigan, p. 263.

TIEDERMAN, W. G., LUCHIK, T. S. & BOGARD, D. G. (1985) "Wall-Layer Structure and Drag Reduction", J. Fluid Mech. 156, pp. 419-437.

WHITE, F. M. (1974) Viscous Fluid Flow, McGraw-Hill Publishing Co.

VI. ACKNOWLEDGEMENT

This work was sponsored by the Office of Naval Research under Contract No. N00014-81-K-0431.

Unsteady Pulsing of Cylinder Wakes

D. R. Williams & C. W. Amato

Fluid Dynamics Research Center
Illinois Institute of Technology
Chicago IL 60616

An overview of recent progress in modelling the behavior of low Reynolds number cylinder wakes is presented. The discussion examines ways in which unsteady forcing has been used to test hypotheses and to control the behavior of the wake. In addition to the review a new method of reducing the wake momentum defect using pulsating jets is demonstrated for flow around a circular cylinder at a Reynolds number of 370. The line of pulsating jets is embedded in the trailing generator of the cylinder. There is no net mass added by the pulsating jets on an average over the cycle, but there is net momentum addition to the flow by the second-order streaming effect. The jets are most effective in modifying the wake when pulsating at twice the Karman shedding frequency. The streaming flow generated by the pulsation suppresses the Karman vortex street and reduces the momentum defect.

1. Introduction

Substantial progress has been made in the last five years toward understanding and controlling the wakes behind bluff bodies. Much of the progress is a result of laboratory and numerical experiments in which a disturbance is deliberately superposed on the natural wake. The disturbance may be caused by a periodic oscillation of the body or the flow. The disturbance interacts in some way with the natural wake resulting in a "forced" wake. Another type of disturbance is a transient change in conditions, such as an impulsive or step change in velocity. The response of the flow to the transient is observed to obtain the desired flow state. Modification of a flow by changing its geometry, such as with splitter plates, additional wires or other appendages that alter the shape of the body are also effective in controlling flows. These techniques may also be considered as a form of forcing, especially when feedback from the downstream body affects the upstream flow. However, they have only been briefly discussed in this paper. One can read about these methods in the review by Zdravkovich(1981).

The most common methods of disturbing the wake are by an external sound source or by oscillating the cylinder. Forcing by sound disturbs the entire flow field. Forcing by oscillation of the cylinder is a more localized disturbance, but even small displacement amplitudes may change the near field by the nonlinear streaming effect. A new method which is very localized and only disturbs the base region of the bluff body is forcing by pulsating jets. This technique will be described in detail later. When any of the forcing methods are applied to the cylinder wake, one of three possibilities can occur depending on the disturbance frequency and amplitude. If the forcing level is very weak, then the natural wake oscillations dominate and the forcing has little or no effect. When the amplitude and frequency are comparable in magnitude to the natural oscillations, interaction between the two oscillators occurs. This situation is the most common and produces the most interesting results in terms of being able to test hypotheses or modify the wake. At large amplitudes and for very high or low frequencies, the forcing flow field may not couple with the wake oscillation. Steady base bleed is an example of the limit of low frequency forcing with large amplitude.

Generally, the experimentalist has one of two purposes in mind when forcing the flow. Either the flow is to be modified to obtain a more desirable altered flow state, or the response of the flow to the forcing is studied in order to determine the fundamental principles governing the wake behavior. This paper will examine some recent experiments in both categories, which have made important contributions to our understanding of the flow around bluff bodies. In addition to a brief survey of the literature, an experiment will be discussed which demonstrates the ability of a new type of forcing technique to modify the wake.

1.A The cylinder wake in its natural state

If a cylinder is placed in a flow and the velocity of the flow is slowly increased, then the wake can be seen to pass through a number of successive states. Morkovin (1964) has categorized six different regimes on the basis of Reynolds number, $Re = U_o d / v$ where U_o is the freestream veloctiy, d is the cylinder diameter and v is the kinematic viscosity.

$3-5 < Re < 30-40$	Twin Vortex Stage (steady)
$30-40 < Re < 80-90$	Incipient Karman Range
$80-90 < Re < 150-300$	Pure Karman Range
$150-300 < Re < (1.0-1.3) \times 10^5$	Subcritical
$(1.0-1.3) \times 10^5 < Re < 3.5 \times 10^6$	Critical and Postcritical
$3.5 \times 10^6 < Re$	Transcritical

The flow around the cylinder is tremendously rich in fluid dynamic phenomena that can be seen proceding through the different stages. In the Incipient Karman Range, vortices form in the far wake region as a result of a laminar wake instability (Kovasznay, 1949). The vortex formation, which occurs directly behind the body in the Pure Karman Range is a well-organized, self-excited global instability. The attention of this paper is centered on recent work aimed at clarifying the nature of the instability which develops into the Karman vortex street, i.e., the change in the flow as the Reynolds number is increased across the Incipient Karman Range to the Pure Karman Range.

Prior to the onset of vortex shedding, the wake is in a steady state with a pair of counter-rotating vortices in the base region of the cylinder. The near-wake region extends from the separating boundary layers with the corresponding low-pressure base region, to several diameters downstream where the pressure returns to its ambient value. Eventually, the wake reaches a state of uniform pressure in the far-wake region.

When the critical Reynolds number (approximately 46) is exceeded, an instability develops which leads to unsteady vortex shedding. The wake usually sheds vortices in a staggered pattern, similar to the model of von Karman (1911). In the instantaneous view of the near-wake region, the counter-rotating vortices are not present. However, if the time-averaged field is measured, then the recirculation region is seen.

Under some circumstances, which are not well understood, the varicose mode of instability is observed (Gerrard 1978). This seems to occur around Reynolds numbers of 500 or higher. In the varicose mode, the vortices form side by side in an symmetric pattern about the centerline of the wake. Morkovin (1987) pointed out a common misconception that the finite amplitude Karman vortex street is an antisymmetric configuration of vortices. Actually, in terms of the streamfunction the varicose mode is antisymmetric about the wake centerline, because of the change in sign of vorticity across the streamwise axis, as shown in Figure 1a. A true sinuous mode would be an arrangement of symmetric vortices configured as shown in Figure 1b. The superposition of a true antisymmetric vortex street (varicose) and true symmetric vortex street (sinuous) street with twice the wavelength of the varicose mode produces the finite strength Karman street of Figure 1c.

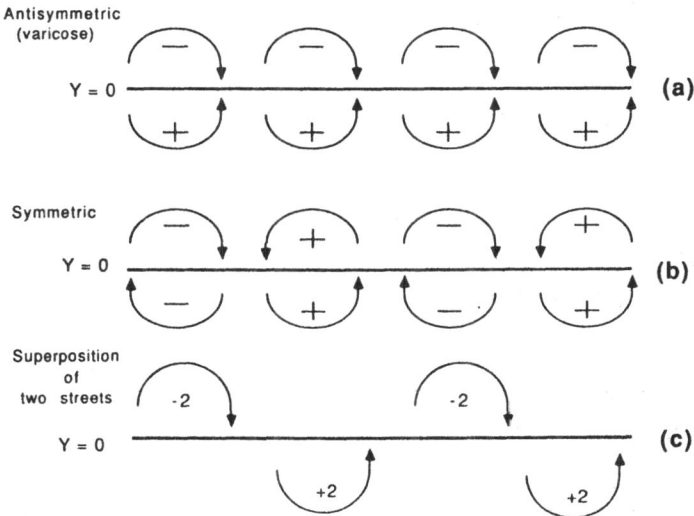

Figure 1. Symmetries of vortex streets (after Morkovin(1987))

On the basis of this observation, Morkovin suggested that *symmetric* forcing at twice the Karman shedding frequency could be an effective frequency for modifying the structure of the wake. This will be shown to be correct for the forcing with unsteady jets described later.

Data presented by Nishioka & Sato (1978) and Unal & Rockwell (1988) have shown that the velocity fluctuation levels are quite large in the near wake. The disturbance amplitude close to the body is approximately 2 percent of U_0 when the Reynolds number is near 120, and it reaches a maximum of 35 percent near three diameters downstream. The amplitude of the velocity and pressure fluctuations are large enough for nonlinear effects to be important throughout the near wake region. The large amplitude of the velocity fluctuations is one clue that feedback to the boundary layer separation points on the body are an important part of the vortex formation process. Furthermore, this supports the idea that the wake is in a self-excited or resonant state of oscillation.

If one asks an engineer to predict the shedding frequency or the steady and unsteady forces acting on a body with as simple a shape as a circular cylinder, then he or she would most likely go to a handbook to find an empirical coefficient obtained from experiment. The coefficient and consequently the engineer's prediction are not likely to be more precise than two significant figures. The inaccuracy in the coefficients obtained from experiments can be attributed to the effects of blockage, aspect ratio, background turbulence, etc.

An alternative for the engineer would be to conduct a numerical simulation. The computer may be used to simulate a specific flow case provided the Reynolds number is not too large. But even then, trends with parameter variations cannot be established without repeating the lengthy computations many times. Some examples of simulations of the two-dimensional wake using the Navier-Stokes equations were conducted by Son & Hanratty (1969), Thoman & Szewczyk (1969), Jordan & Fromm (1972), Fornberg (1982) and more recently by Braza, Chassaing & Ha Minh (1986). Three-dimensional effects begin to become important near Reynolds number 150. Only now are the codes which simulate the complete Navier-Stokes equations beginning to be able to deal with three-dimensional disturbances in the near wake of a cylinder. If the Reynolds number is very large, then a turbulence model would have to be employed in the simulation in order to account for the effect of Reynolds stresses. The turbulence models are an approximation to the Navier-Stokes equation, and do not allow one to compute detailed information about the flow. The model would presumably have been justified by comparison with experiment, and could not be considered to be any more accurate.

Therefore, fluid mechanics is in the interesting position of knowing the equations that govern the majority of commonly occurring flow fields, namely, the Navier-Stokes equations, but not being able to fully use them. This predicament has fostered a great deal of research attempting to "understand the physics of the problem". What is really meant by this phrase is that partial models simpler than the complete Navier-Stokes equations are being sought, which are capable of describing specific aspects of flow behavior. Of course, any partial model will be restricted to a limited range of application. One example of a simpler model that provides insight into the "physics" is von Karman's (1911) vortex street (see Lamb 1945). This simple model idealizes the experimental observations of the wake at low Reynolds number. It predicts one stable configuration of an infinite row of inviscid point vortices. However, the model of von Karman does not tell us how the vortices are formed, or how the vortex street is related to the geometry of the body, or what the shedding frequency of the vortices from a specific type of body will be. As Roshko (1954) discussed, the connection between the vortex street and the body will have to be established through the vortex formation mechanism in the near wake. Clearly, another model is needed for this region.

1.B Use of forcing to clarify fundamental issues

Forcing with transient and periodic disturbances has been used to test theories that characterize the near wake region and vortex formation mechanism. In this section the progress in linear, weakly nonlinear and nonlinear models for the cylinder wake at low Reynolds numbers will be examined. Evidence that the near wake resonance is a result of a global instability will be presented. The ability of the Landau equation to model the growth and nonlinear saturation of the disturbances in the near wake is shown. Finally, some recent results are discussed which show that modern dynamical systems theory is capable of predicting some characteristics of the response of the wake to external excitation.

The evidence that the cylinder wake is a self-sustained oscillator comes from both numerical and laboratory experiments in which transient disturbances were introduced and the response of the wake was studied. Braza, et al. (1986) performed numerical experiments on a two-dimensional wake at supercritical Reynolds numbers 100, 200 and 1000. Because of the near perfect symmetry in numerical simulations, the wake was initially in a stable state with no oscillations. After the mean flow was established an artificial perturbation was introduced, which corresponded to a quick transient rotation of the cylinder about its axis. Periodic waves then began to amplify in the wake, and after a few cycles the wake reached a state of saturated finite amplitude oscillation. As Braza points out, the initial perturbation was only the catalyst for the change from steady to periodic flow.

The energy to sustain the oscillations in Braza's numerical simulation could not have been provided by the initial perturbation. The vorticity of the perturbation was swept far downstream during the time it took for the saturated state to be reached. The feedback from the oscillations in the near wake to the shear layers separating from the body resulted in a strong, single frequency spectrum. Therefore, the wake must be a self-excited oscillator. As long as any external disturbances are not strong enough to interrupt the feedback mechanism, then the wake will continue to oscillate at its natural frequency. The type of instability which leads to this self-excited state is referred to as *global instability*. A more detailed discussion of the vorticity dynamics related to global instability is given by Morkovin (1988).

In contrast to the global instability found in the wake behind a cylinder, the flow behind a streamlined body behaves quite differently. For example, Stuber and Gharib (1988) have studied the wake formed by a symmetric airfoil NACA 63A008. The spectrum of the velocity fluctuations in the naturally occurring wake shows that a relatively broad band of frequencies are amplified. They used small heating strips on the surface of the airfoil to introduce low amplitude disturbances to the boundary layer. Provided the disturbance frequency was in the correct range for amplification, the wake would form a vortex street with the same frequency as the forcing. With the gentle forcing the spectrum was no longer broad band, but showed discrete peaks associated with the forcing frequency. In this type of flow, the frequency of oscillation in the wake will vary continuously as the input frequency is changed. When the forcing is turned off, the wake returns to its natural state with a broad band spectrum. The wake behind streamlined bodies acts like a filter and amplifier of external disturbances, rather than a self-excited oscillator. This type of flow is said to be convectively unstable.

The concepts of global, absolute and convective instability have gone a long way toward explaining the fundamental differences between instabilities found in a wide variety of flows. Huerre and Monkewitz (1985) and Huerre (1987) used the absolute and convective instability concepts to examine the differences between spatial and temporal linear instabilities in shear layers†. Basically, one considers the linear stability of a velocity profile to an *impulsive and localized* disturbance. Parallel flow is also assumed. The impulsive disturbance develops into a wave packet which spreads due to dispersion of the individual waves. Each travelling wave may amplify in space or in time, or both. If the wave packet convects downstream (group velocity is positive), so that the disturbance amplitude at the input location eventually goes to zero and the flow returns to its initial undisturbed state, then the instability is the convective type. In this case, spatial stability theory is appropriate for modelling linear disturbances in a shear layer. On the other hand, if the disturbance has zero group velocity and the envelope of the wavepacket spreads both upstream and downstream, then the amplitude of the fluctuations at the initial input location of the disturbance grow exponentially, and the instability is the absolute type. The Blasius boundary layer is an example of convective instability. It will amplify external disturbances in the streamwise direction, but once the disturbance excitation is removed, the flow returns to its initial undisturbed state (Gaster 1981). In contrast, Taylor-Couette flow between two rotating cylinders is absolutely unstable when the critical Taylor number is exceeded. Any small disturbance will lead to exponential amplification of a specific wave. Eventually, nonlinear effects cause the disturbance to saturate and form the counter-rotating cells of vortices. Reviews of these concepts can be found in Bechert (1985) and Huerre (1987).

There is strong evidence that the concepts of convective and absolute instability for linear disturbances may be useful to explain the frequency selection mechanism in the flow behind bluff bodies, even though the near wake is a highly nonlinear and nonparallel flow. In particular, the reason why the bluff-body wake behaves like a self-sustained oscillator and the streamlined body wake does not have this resonance may be explained. Working toward the goal of explaining the frequency selection mechanism, Koch (1985), Monkewitz & Nguyen (1987) and Triantafyllou, et al. (1986) studied the stability characteristics of velocity profiles representative of those found in the near wake region behind bluff bodies. Their results showed that these profiles can be absolutely unstable to sinuous disturbances when the velocity defect is sufficiently large. If a wake has a region of absolute instability near the body, then at some point farther downstream the wake profile instability changes to convective type, because the velocity defect will decrease in the streamwise direction. Initially, it was hoped that some characteristic of the region of absolute instability could be found which could be used to explain the frequency selection mechanism behind bluff bodies. The nonlinear effects

---

† Originally, the theory of absolute and convective instability was developed for the field of plasma physics by Briggs (1964) and Bers (1975).

associated with the disturbance field were not expected to alter the frequency of oscillation very far from that predicted by the linear theory. However, since the actual flow is nonparallel, one has a variety of different wake profiles and conditions to choose from in order to predict the frequency of oscillation. Monkewitz and Nguyen (1987) examined several different criteria that had been proposed for the resonance, although not all of the problems were resolved.

More recently, Chomaz, Huerre & Redekopp (1987) modelled nonparallel effects with the Ginzburg-Landau equation. They have shown that a finite region of absolute instability in the streamwise direction is required in order to have resonance between an upstream and downstream travelling wave, i.e., global instability. Extending this result to the near wake of a cylinder indicates that it is not sufficient to have absolute instability at just one location in the wake. Instead, a finite region of absolute instability is necessary before a global instability is established.

With the findings of Chomaz, et al., Monkewitz (1988) further examined the stability characteristics of velocity profiles similar to those found in the near wake of cylinders. He determined the following sequence of instabilites occurred as Reynolds number is increased. At Re = 5 the wake profiles became convectively unstable. The first appearance of local absolute instability occurred near Re = 25, which was much lower than the experimentally determined value for the occurence of Karman vortex shedding, namely, Re = 47. Assuming that the onset of vortex shedding at the critical Reynolds number is the indicator of the global instability, then the results of Monkewitz are consistent with the model of Chomaz, et al. It is not yet clear how the required critical region of absolute instability is related to a specific geometry. Even if the critical Reynolds number is known, it is not apparent how the exact frequency of oscillation is to be determined. Nevertheless, Monkewitz has shown that the range of possible frequencies can be predicted.

Experiments also support the notion that global instability is not present below Re = 47, although the lower Reynolds number limit below which the wake is completely stable to all infinitesimal disturbances has not yet been well established. Taneda (1963) using visualization of the far wake behind an oscillating cylinder found a lower Reynolds number limit of approximately Re = 5, which agrees with the calculations of Monkewitz. Unfortunately, it is not clear if the disturbance input by the oscillating cylinder was small enough to be considered linear. Nishioka & Sato (1978) demonstrated linear behavior of the disturbance input, but found no disturbance amplification below Reynolds number 20. Similarly, Sreenivasan, et al. found a lower limit of Re = 27 for amplification of externally excited disturbances. However, these experiments were not specifically searching for the lower limit of amplification, and it is not clear if they covered a wide enough range of disturbance frequencies to be conclusive. The issue of the lower limit of convective instability is important for verification of the theory, and should be resolved with a suitably designed experiment.

1.C Nonlinear disturbance behavior near the critical Reynolds number

The nonlinear behavior of the near wake has been explored in laboratory experiments conducted by Sreenivasan, Strykowski and Olinger (1987). Somewhat analogous to the numerical experiments of Braza, they studied the transient growth of small amplitude waves to the nonlinear saturated state. In the laboratory it is not possible to achieve the metastable state at a supercritical Reynolds number, because background noise is always present and always perturbing the wake. Instead, they used a modified wind tunnel to produce a sudden step change in Reynolds number. By jumping from a subcritical (stable state) to a supercritical Reynolds number (unstable state), they could observe the transient formation of the Karman vortex street as the critical Reynolds number was exceeded. The temporal oscillations in the wake amplified exponentially over several cycles before nonlinear effects limited the growth to the saturation amplitude. This amounts to experimental verification of the global instability of the cylinder wake, similar to the numerical experiments of Braza, et al. Disturbance decay rates were obtained by oscillating the cylinder at subcritical Reynolds numbers. Once the oscillation was stopped, the rate at which the flow returned to a steady state was measured.

The experiments of Sreenivasan, et al. have gone beyond demonstrating that the wake at supercritical Reynolds numbers is a nonlinear saturated oscillator. In particular, they have shown the descriptive ability of an equation far simpler than the Navier Stokes equations. In the region of Reynolds numbers close to critical, the disturbance has characteristics of a Hopf bifurcation (see Drazin & Reid 1981). Furthermore, they showed that locally the Landau equation gives an accurate description of the temporal behavior of the growth of disturbances in the wake. The Landau equation describes a Hopf bifurcation, so it should also be descriptive of the near wake of the cylinder. Similar conclusions were obtained in an independent study of the cylinder wake by Mathis, Provensal & Boyer (1987). It is also apparant that the wake has characteristics universal to nonlinear oscillators. Identification of the universal properties will make prediction of the disturbance behavior possible for a larger class of flows around bluff bodies. For instance, the variation in saturation amplitude with Reynolds number, and the variation of frequency with disturbance amplitude are predictable assuming the Landau coefficients for the flow are known. Of course, the Landau equation is only a partial theory and does not answer all the questions about the wake.

1.D Response of the wake to external excitation

The evidence is strong that the Karman vortex street is a self-excited oscillator resulting from the saturated state of a global instability. Since it is self excited, we expect that the wake will not be very receptive to external disturbances. Then we may ask how the resonance responds to finite amplitude excitation, and how control techniques can be used to modify and control the wake? The answers to these questions have both fundamental and practical consequences.

Koopman (1967) studied cylinders that were forced to oscillate perpendicular to the flow direction the flow, and showed that the vortex shedding frequency. The wake response frequency would lock on to the driving frequency and could change by 20 percent from the natural value, if the excitation amplitude of the cylinder was large enough. A great deal of effort has been expended on the study of vortex-induced vibrations for the practical importance to the design of structures. A comprehensive review of the flow around oscillating bluff bodies was given by Bearman (1984). In another review, Griffin (1984) discussed experiments by Overvik(1984) in which a loosely supported cylinder coupled with the oscillations of its own wake. As the cylinder oscillated, it developed drag and fluctuating lift coefficients which were 250 percent larger than the undisturbed values!

On a more fundamental level, the saturated nonlinear state of the wake has been shown to be the result of a global instability. As previously discussed, the global instability has a well-defined spectrum and is not very receptive to low-amplitude external disturbances. A second oscillator can be added to the system by sound excitation or some other technique. The purpose of the second oscillator may be as a means of controlling the wake. We want to be able to predict how the wake will respond to the additional oscillator.

The low Reynolds number wake forcing experiments by Olinger and Sreenivasan (1988) have shown that the interaction of the natural wake oscillation and the periodic disturbance introduced by oscillating the cylinder behave according to modern dynamical systems theory. Depending on the excitation frequency and its amplitude, the response frequency may lock-in to any rational multiple of the excitation frequency. Excitation amplitude vs. frequency plots of the lock-in regions form patterns called Arnol'd tongues. A boundary (critical line) which marks the onset of chaotic behavior can be plotted on the same diagram. The regions of lock-in along the critical line can be replotted on a graph of response frequency vs. excitation frequency. The resulting graph forms a self-similar devil's staircase, which has the universal property of a fractal dimension of 0.87. Furthermore, when excitation occurs at the critical golden mean point, the power spectrum that is obtained is predictable by the sine circle map. Their demonstration that certain "universal" properties apply to the wake means that the interaction of the two oscillators is predictable. One of the exciting aspects of this work, is that the Navier-Stokes equations are not required to predict the behavior of this low Reynolds number system. A simpler model, i.e., the sine circle map can be used instead.

In recent years there has been a resurgence of interest in unsteady and passive techniques to control the forces acting on both bluff and streamlined bodies. In particular, control techniques which involve unsteady motion have often been found to be effective for modifying flow fields. For example, unsteady pitching motion of an airfoil can generate higher lift coefficients than the same airfoil at any steady angle of attack. The benefits of flow control to the areas of hydro- and aerodynamics can appear as noise reduction, enhanced lift and maneuverability, and reduced friction and form drag around all types of bluff bodies. The ability to control the system of vortices shed from the forebodies of aircraft is of immense importance to the stability and maneuverability of the aircraft. In addition to the practical applications, it is very often possible to obtain a deeper understanding of the fundamental fluid dynamic mechanisms at work when a control technique is applied.

One measure of the success of a control method is the ability to change the drag coefficient. The drag on bluff bodies is a combination of viscous friction and the pressure distribution, or so-called form drag, around the body. At Reynolds numbers of practical importance, the majority of the drag is due to the form drag. Thom (1933, see Goldstein 1965) examined the relative contributions of skin friction and form drag to the total drag of a circular cylinder, and found that the friction decreases in importance relative to the form drag as the Reynolds number increases. At a Reynolds number of 1000, the skin friction is only 13 percent of the total drag compared to 87 percent associated with form drag. Thus, the most reasonable approach for drag reduction is to reduce the form drag by modifying the pressure distribution around the body. Since most of the form drag is a result of the negative base pressure region, any increase in the base pressure would be an indication of reduced drag. A few examples of experiments which have used three fundamentally different flow control techniques are described below. The methods are boundary-layer control, splitter plates and steady base bleed. Although the techniques are different, each results in a net drag reduction by modifying the separated flow region behind the body.

A classical experiment which demonstrates drag reduction by modifying the flow field around a bluff body was conducted by Prandtl (1914, see Schlichting 1979) with a sphere. Prandtl placed a wire ring on the surface of the sphere ahead of its equator to trip the normally laminar boundary layer into a turbulent state. The turbulent boundary layer remained attached to the surface of the sphere for a greater distance than the laminar boundary layer. As a result, the area of separated flow and form drag were reduced. Besides showing the importance of the boundary layer state to the overall flow pattern, Prandtl's experiment illustrated the connection between the boundary-layer separation and the base pressure.

Another example of the effects of a modified flow pattern is the experiment conducted by Roshko (1954). By placing a splitter plate in the separated flow region behind a circular cylinder, he was able to disrupt the vortex formation process. In this way he established the importance of the interaction between the two separating shear layers in the near wake to the vortex formation process. He also suggested the possible application of splitter plates as practical devices to suppress unsteady oscillations that result from the formed vortices.

In addition to suppressing the formation of large-scale vortices, it is also known that splitter plates reduce the drag on bluff bodies. The drag reduction is reflected by an increase in the base pressure coefficient, has been investigated in some detail by Bearman (1967) using airfoil-shaped models with blunt ends. By measuring the velocity fluctuation amplitudes he determined that splitter plates force the vortices to form farther from the base of the body. He also discovered that the base pressure decreased with the inverse of the vortex formation length, which implied that the vortex formation process and drag are related.

A different approach to controlling the flow field and disturbing the vortex formation process is the base-bleed method. The base-bleed technique uses steady ejection of fluid from a slot in the trailing edge of the body to modify the flow in the separated region. The method has been used on airfoil sections with blunt trailing edges by Wood (1964, 1967), Bearman (1967), and Cimbala and Park (1987). The reduction in drag due to the fluid ejection is primarily a result of changes in the base pressure region, rather than by thrust from the ejected fluid. Determining how the base bleed actually results in a net drag reduction is a key element in determining the more fundamental issue of the relationship between the wake structure and the base pressure. Monkewitz and Nguyen suggested that the Wood's results support the idea that the Karman vortex street is a result of an absolute (global) instability. With sufficient base bleed, the wake is no longer globally unstable, so conditions for resonance are no longer present. Therefore, the Karman vortex street cannot form.

Wood (1964) believed that the effect of steady base bleed was to delay the formation of the large-scale vortices by interfering with the interaction of the shear layers. Bearman (1967) found two different regimes of wake behavior with varying bleed rates. At low bleed rates the base pressure increases with the shedding frequency, while at high bleed rates the base pressure decreases with increasing frequency. He also found that base bleed increased the vortex formation length in a manner similar to the splitter plate experiments. Once again he showed that the base pressure was dependent on the vortex formation length. Also in 1967 Wood used quantitative flow visualization to estimate the strength of the vortices in the wake which was modified by base bleed. He determined that the strength of the vortices decreased as the formation length increased, which supported the trends observed by Bearman. In addition, Wood suggested that the increase in formation length was necessary for the wake to entrain the additional fluid supplied by the base bleed.

More recently Cimbala and Park (1987) investigated the character of the turbulence in the far wake behind an airfoil-shaped body with blunt end. Sufficient base bleed was used to obtain zero net momentum defect, hence zero net drag. They found a decrease in the width of the mean wake profile and an increase in the rate of turbulence decay. They also demonstrated that a net thrust with a jet-like behavior could be obtained with high rates of base bleed. The turbulence decay was fastest when the momentum defect was zero.

2. A new method of localized forcing

When an unsteady disturbance is used to control an unsteady flow field, then it may modify the original flow by

a.   superposition of the mean component of the control flow, or

b.   by the direct interaction between the unsteady forcing field and the original unsteady field.

Cancellation of Tollmien-Schlichting waves in a boundary layer with periodic heating strips is one example of the latter method of control with an unsteady flow. The periodic disturbances created by the heating strips are out of phase with the Tollmien-Schlichting waves, and cancel them. But the heating also contributes a mean heat flux to the flow by rectification of the heating signal. Presumably, the mean flow is not modified significantly, and the control effect is by the interaction of the disturbances. The current hope is that the direct interaction of the unsteady field will require less energy input for a specific amount of control than by direct mean flow modification.

The purpose of the remainder of this paper is to demonstrate an entirely new technique for suppressing the vortex formation and reducing the form drag on bluff bodies. The new method involves *unsteady base bleed* with *zero net mass addition*. Similar to the steady base bleed experiments, fluid is ejected from the trailing edge of the body; however, in the unsteady base bleed case the same amount of fluid is drawn back into the body during the low-pressure side of the cycle as is ejected during the high-pressure side of the cycle. In contrast to the steady base bleed case, one does not need a supply of bleed fluid with the unsteady pulsing jet control method. Therefore, it seems that the new flow control technique would be easier to apply to practical situations. Furthermore, the controlling jets may be turned on and off and their amplitude and frequency may be varied as desired. In the experiments described below, the effects of forcing frequency and forcing amplitude on the development of the wake have been explored using flow visualization and quantitative measurements with a scanning laser anemometer.

3. Experimental Procedure

The experiments were conducted in a small water channel whose test section was 10.2 cm in width, 15 cm in height and 200 cm in length. The channel was completely enclosed except for a slot cut into the top to allow the cylinder to be moved to different streamwise locations. The freestream velocity, U_o, for the tests was 3.83 cm/s with a spatial variation of 0.6% U_o across the central 9 cm of the channel. The 1.0% U_o noise level of the measurement system prohibited an accurate measurement of the freestream turbulence, which we believe had a lower value. The water temperature was 29°C with a corresponding kinematic viscosity $v = 0.818 \times 10^{-6}$ m²/sec.

The cylinder used in the experiments was a hollow plexiglas tube with a diameter, d = 7.94 mm. Circular end plates were placed 90 mm apart in order to minimize the three-dimensional effects from the ends of the cylinder. The cylinder was mounted vertically in the channel and the top of the cylinder extended above the free surface. For all of the results presented the Reynolds number based on diameter was Re = 370. The unforced shedding frequency of the cylinder was measured at f_K = 0.95 Hz. The cylinder was mounted on a traversing system on top of the water channel. The water channel and laser anemometer were stationary, and the cylinder was moved to different locations in order to change the streamwise measurement position. In this experiment x was the streamwise direction, y was perpendicular to the cylinder axis and z was along the cylinder axis. The experiment cannot be considered to be particularly "clean" since the aspect ratio of the cylinder is only 11.3 and the blockage of the cross-sectional area of the water channel was approximately 8 percent. Nevertheless, we believe the conclusions drawn from the results are not significantly affected by these limitations of the experiment.

The wake was forced by a column of pulsating jets positioned along the trailing "edge" of the cylinder as shown in Figure 2. The jets were formed by drilling a line of small holes along the axis of the cylinder. The holes were 0.79 mm in diamter with 1.06 mm spacing between centers. The top of the hollow cylinder was connected to Tygon tubing which was closed at one end. The Tygon tubing was squeezed by a rotating cam connected to a variable-speed Bodine motor. The periodic squeezing created a sinusoidal pressure fluctuation to drive the jets. Since the tubing was closed at one end, there was zero net mass addition over the forcing cycle. A bypass line and valve were included to allow adjustments to the pressure fluctuation amplitude. The pressure fluctuations were measured with a Validyne DP-103 pressure transducer. The tubing

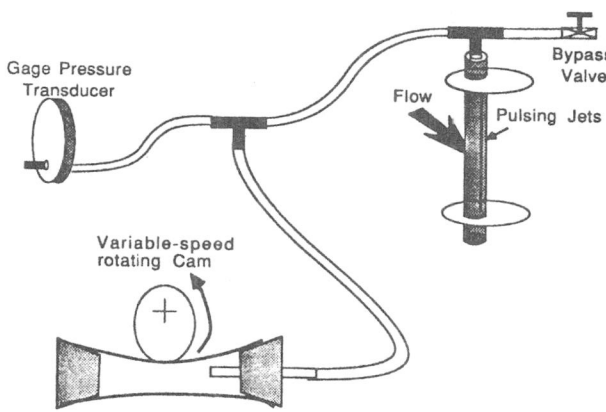

Figure 2. Diagram of pulsating jet system. The jets are formed by a line of small holes drilled into the trailing edge of the cylinder.

connecting the pressure transducer to the cylinder was 6mm in diameter. Tests measuring the change in height of a water column were used to determine dynamic response of the transducer. There was no detectable attenuation of the signal over the frequency range (0.1 - 3Hz) of the experiment. Measurements along the axis of the cylinder showed a variation of approximately 10% U_o in the velocity fluctuation amplitude, so the pulsating jets were considered to be reasonably two-dimensional.

Measurements of the streamwise component of velocity were made with the scanning laser anemometer shown in Figure 3. The design is an extension of the type developed by Chehroudi and Simpson (1984). In this design three optical scanners were used which allowed scanning along the optical axis and perpendicular to the plane shown in the Figure 3. For this experiment the scanning motion was along the optical axis rather than perpendicular to it. The scanning rate was at 10 Hz, which amounts to 20 scans/sec across the flow. The scan velocity was kept constant across the 9 cm scan length. Frequency shifting at 40 MHz with a Bragg cell and then down mixing to 50 kHz, which was twice the maximum Doppler frequency, allowed measurements to be made in the reversed flow region and reduced the angular biasing effect. The LDA system was a forward-scatter type, which used a TSI model 1990-C processor. The seeding particles were 2μm latex spheres, which were added to water that had been filtered to a nominal level of 0.5μm. Seeding particles were added to give an average sampling rate of 2600 samples/sec while scanning. The sampling rate with a scanning system should exceed a minimum sampling rate given by

minimum sampling rate = (2 x scanlength x scanrate/spatial resolution), in order to have an average of one measurement per length of the probe volume per scan. The spatial resolution was approximately 0.8mm which gave a minimum sampling rate of 2250 samples/sec; therefore, the average data rate was sufficient for good spatial resolution. Only time-averaged results have been presented, although the instantaneous data has been stored for further processing. Residence-time weighting was used to reduce the biasing due to velocity fluctuations. Scanner positioning, data acquisition, real-time display and processing of the data were done with a Masscomp 5500 series computer. More detailed information about the scanning laser anemometer system can be found in Williams and Economou (1987) and Economou (1986).

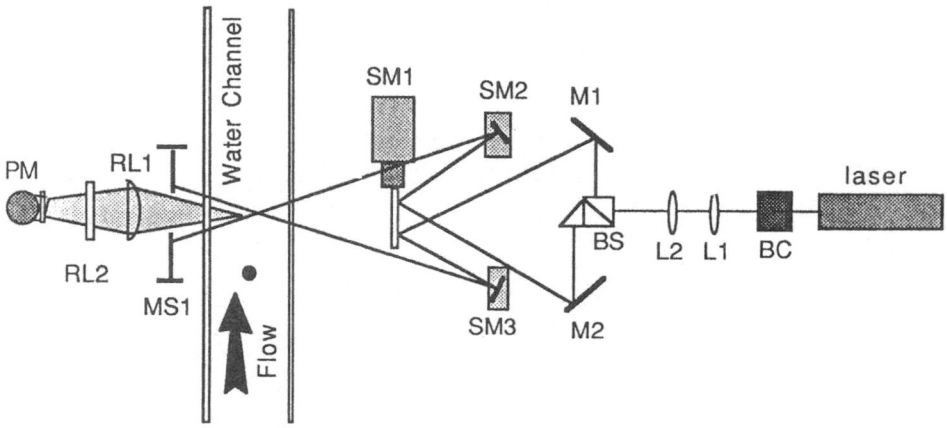

Figure 3. Schematic of the scanning laser anemometer system. BC-Bragg cell; L1 & L2 - beam focusing lenses; BS - beam splitter and prism; M1 & M2 - mirrors; SM1, SM2 & SM3 - optical scanners; RL1 & RL2 - cylindrical lenses; PM - photomultiplier tube.

4. Results

The frequency and amplitude of the pulsating jets were found to be the most significant forcing parameters in the control of the wake defect. The frequency could be easily measured, but the forcing amplitude was more difficult to define. The pulsation frequency, f_p was normalized by the Karman vortex shedding frequency, f_K, to define the frequency parameter, $F = f_p / f_K$. We chose to base the forcing level on the peak-to-peak pressure fluctuation amplitude, p', in the interior of the cylinder. However, the displacement of the air/water interface inside the cylinder is also a useful measure of the forcing level. It was found that the air/water interface displacement varies with both frequency and pressure amplitude. If the pressure fluctuation amplitude is kept constant while the fequency is increased, then the displacement amplitude decreases. A calibration of this effect has been plotted in Figure 4. The velocity at the exit of the jet can be estimated from the oscillation frequency and the displacement of the air/water interface. For a frequency of 1.8 Hz and a displacement of 10 cm, the root mean square velocity at the jet exit is V_{rms} = 12.5 cm/s. It has not yet been determined if the pressure fluctuation level or the displacement level is more important as a scaling parameter.

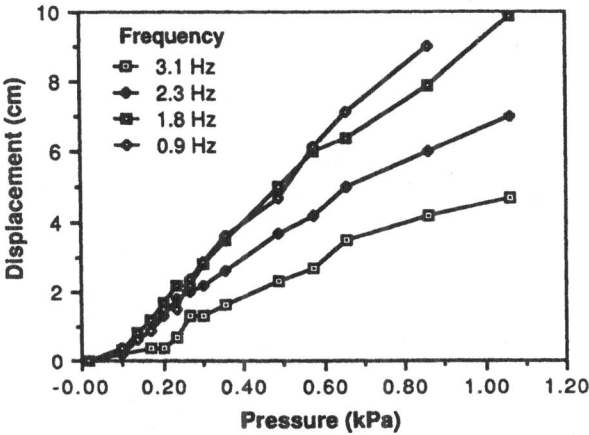

Figure 4. Calibration of the displacement of the water column inside the cylinder during the forcing cycle as the pressure amplitude is varied.

Flow visualization with hydrogen bubbles of the unforced wake are shown in Figures 5a, b and c. The large-scale Karman vortices are reasonably two-dimensional with smaller spanwise waves developing as originally described by Hama (1957). The pulsating jets were set at a frequency of 1.8 Hz (F = 1.9) and an amplitude of 0.52 kPa in Figure 5c. It is apparent that the finer scale of turbulence generated by the pulsating jets is less organized than the three-dimensional waves that occur in the unforced wake.

In order to visualize the effect of the pulsating amplitude and frequency on the overall wake behavior, a variety of different conditions were visualized with hydrogen bubbles. The results are displayed in matrix form in Figure 6. Each column of photographs corresponds to a fixed pressure amplitude and each row corresponds to a specific pulsing frequency. When F = 1.0, there is little effect on the wake development. Increasing the frequency to almost twice the natural wake shedding frequency, F = 1.9, had a profound effect on the development of the wake. The wake seems to be most sensitive to forcing at twice the Karman shedding frequency, which confirms the speculation by Morkovin (1987). At the amplitude of 0.52 kPa (V_{rms} = 6.2 cm/s) the Karman vortex street is replaced by a narrow wake without the strong organized vortices. For the same pressure amplitude at other frequencies, the discrete vortices appear, but in general they are not as well organized as in the unforced case. Although it is not shown in Figure 6, the vortex street could be completely suppressed at F > 1.9

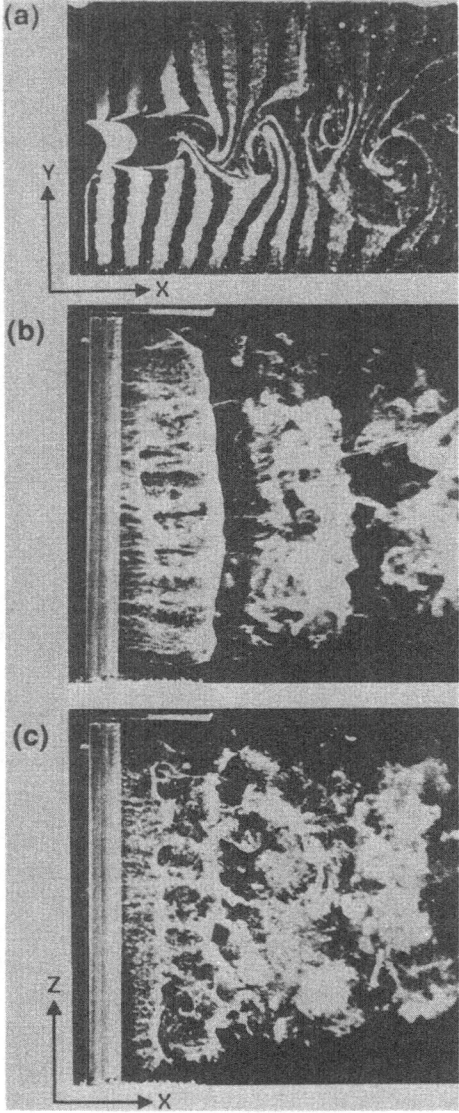

Figure 5. Hydrogen bubble flow visualization. a) View along the axis of the cylinder without forcing. b) View perpendicular to the cylinder axis without forcing. c) Flow field with pulsating jets activated.

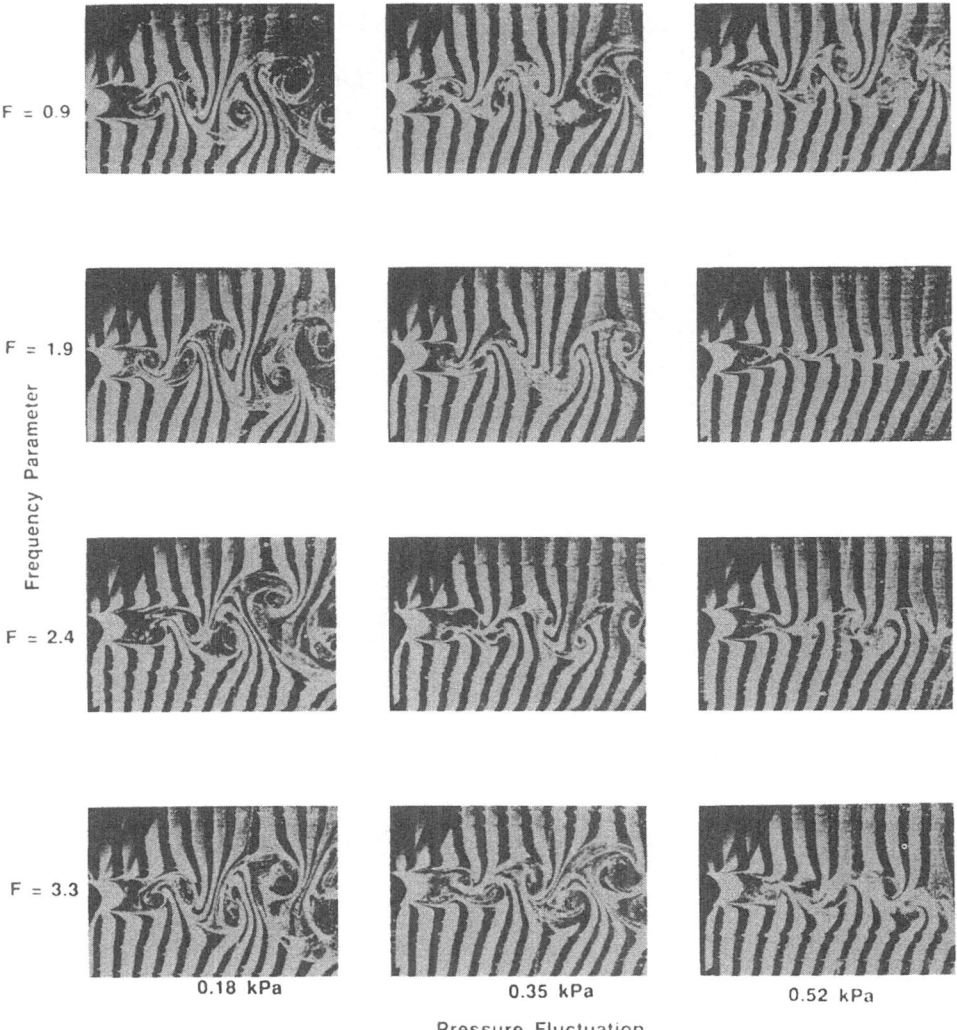

F = 0.9

Frequency Parameter

F = 1.9

F = 2.4

F = 3.3

0.18 kPa   0.35 kPa   0.52 kPa

Pressure Fluctuation

Figure 6. Hydrogen bubble flow visualization showing the effect of the forcing at different amplitudes and frequencies.

when the amplitude was increased. It is possible that the displacement amplitude may be the appropriate scaling parameter, but this has not yet been thoroughly investigated. It should also be noted that when the vortices do form, there does not appear to be any significant increase in the formation length of the vortices as the forcing amplitude is increased. Here formation length is defined as the first location downstream of the cylinder where hydrogen bubbles are convected across the axis of the wake.

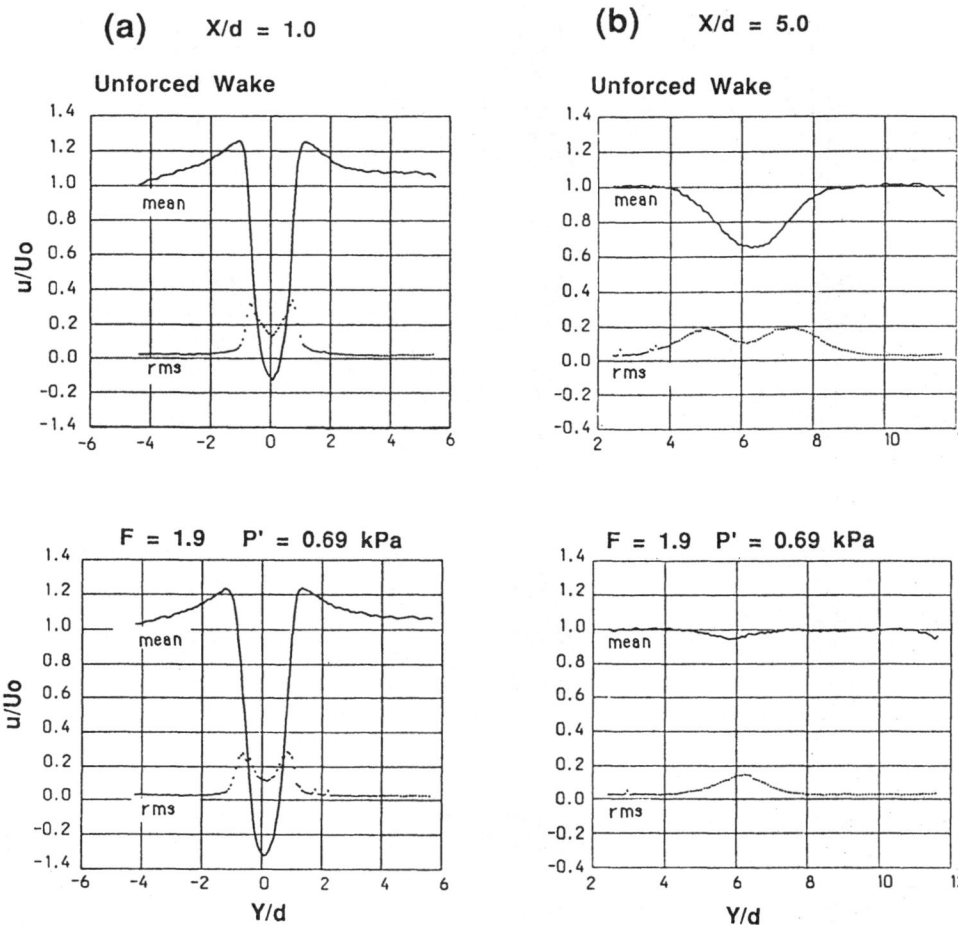

Figure 7. Comparison of mean and r.m.s. velocity distributions with and without jets activated. a) Near the base of the cylinder, x/d = 1.0, the effect of pulsing is to increase the magnitude of the reversed flow. b) At x/d = 5.0, the mean velocity defect is significantly reduced by the pulsating jets and the r.m.s. fluctuation level is reduced.

Quantitative data were obtained with the scanning laser anemometer described earlier. Mean and r.m.s. (root-mean-square) velocity distributions in the near wake at x/d = 1.0 and farther downstream at x/d = 5.0 are shown in Figures 7a and 7b, respectively. In the near wake region of the cylinder the mean flow has a reversed flow region which increases from U/U_o = -0.10 to U/U_o = -0.30 when the jets are activated at F = 1.9 and the amplitude is 0.52 kPa. This effect can be seen in Figure 7a. Farther downstream the effect of the pulsation is to increase the centerline velocity, U_{cl}, and reduce the wake velocity defect as shown in Figure 7b. The r.m.s. velocity distribution is bimodal in the unforced wake, due to the organized vortex street. However, when the jets are activated, only a single peak is found on the centerline which is weaker than the two peaks in the unforced case. This is an indication that the Karman vortex street has been suppressed by the pulsating jets.

To get a better indication of the effect of forcing on the wake the centerline velocity normalized by the freestream velocity, U_{cl}/U_o, has been plotted against x/d in Figure 8. The trailing edge of the cylinder is at x/d = 0.5, where the boundary condition must be U_{cl}/U_o = 0, since there is no net mass addition by the pulsating jets. The unforced wake shows a trend to approach the U_{cl}=0 condition. The strongest reversed flow occurs at x/d = 1.5. The cases with the pulsed jets show an even stronger reversed flow closer to the base of the cylinder, which must be associated with the modified base pressure.

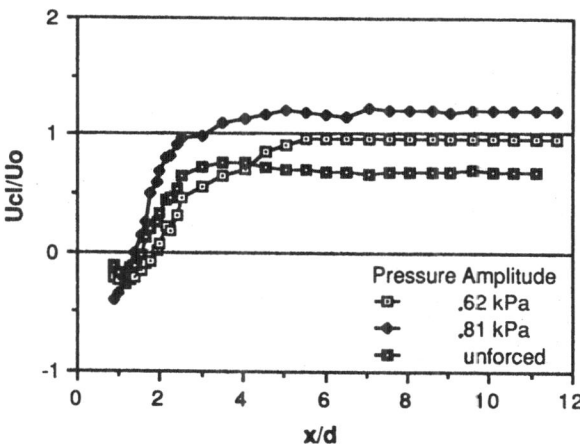

Figure 8. Streamwise variation in centerline velocity with F = 1.9 and different amplitudes.

The centerline velocity increases rapidly for the first three diameters downstream of the cylinder, then it levels off. In the forced wake cases the stronger level of forcing causes the centerline velocity to approach the freestream value more rapidly. An overshoot in the velocity is present with forcing amplitudes in excess of 0.71 kPa. The overshoot is a result of fluid entrainment by the pulsing jets. It will be shown later that the overshoot also corresponds to an addition of momentum by the pulsating jets, which is in excess of the loss of momentum by the wake.

The streamwise variation in the r.m.s. maxima for the pulsed and unforced cases are shown in Figure 9. When forcing is applied the peak values of flucutuation are on the centerline of the wake. In the natural wake the r.m.s. profiles are bimodal with peak amplitudes off center, near the location of maximum mean velocity gradient. The forced wake has maxima which occur on the centerline. After $x/d = 2.0$, the decay rate is faster for the forced wake. Unlike the splitter plate and steady base bleed experiments, the maxima in the r.m.s. do not show any increase in the vortex formation length as the forcing level is increased. Even though the r.m.s. profiles are not the same shape, and the comparison may not be strictly a fair one, the fact that the turbulence intensity in the forced cases decays faster than the unforced cases is an interesting effect. A similar phenomenon was observed by Cimbala (1987).

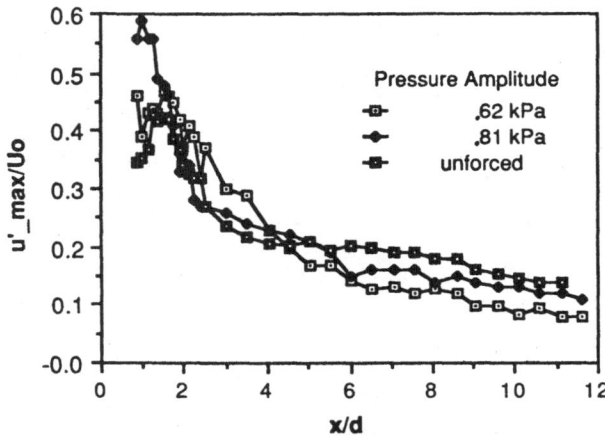

Figure 9. Streamwise variation in r.m.s. maxima at variable y-locations.

The momentum defect, θ, is defined for steady flow as

$$\theta = \int U(y)/U_o \, [1-U(y)/U_o] \, dy.$$

The momentum defect can be related to the drag coefficient by $C_d=2\theta/d$ provided θ is measured in the region of zero pressure gradient. From the integral form of the momentum equation it is easy to see that a variation in θ corresponds to a region of pressure gradient. The momentum defect variation with streamwise distance is shown in Figure 10. The near wake region $x/d < 2.5$ has the strongest variation in θ which indicates a region of strong adverse pressure gradient. For $x/d > 2.5$ the value of θ gradually approaches a region of uniform pressure and the value representative of the drag on the cylinder. It appears that the flow in the larger amplitude case of 0.81 kPa is experiencing a favorable pressure gradient, since the value of θ is continually decreasing. The reason for this trend is not yet known, but it may only require a longer distance to reach equillibrium. In the unforced case, θ approaches 0.48 cm which corresponds to $C_d=1.2$. This is about 8 percent lower than the values reported in the literature (see Schlichting 1979).

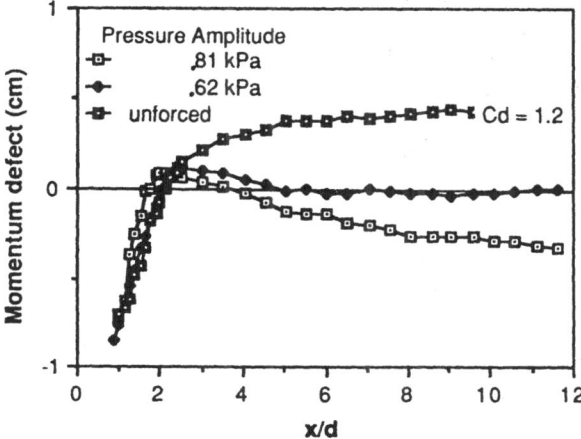

Figure 10. Streamwise variation in the momentum defect with F = 1.9 and different amplitudes of pulsation.

The sensitivity of the momentum defect to the pulsing amplitude is shown in Figure 11. The frequency parameter was F = 1.9 and the momentum defect was measured at x/d = 5.0. Clearly, the wake defect decreases as the level of pulsing amplitude increases. At pressure fluctuation amplitudes greater than 0.71 kPa, the wake begins to show an overshoot in the mean centerline velocity and the momentum defect becomes negative. If the flow is indeed in a region of zero pressure gradient, then the jets have created a net thrust on the cylinder.

There appear to be three separate regions that can be identified. The first region is the vortex dominant region, which is found at very low forcing levels from 0 - 200 Pa. Remembering that the wake is a self-excited oscillator makes it clear that a critical amplitude of the disturbance will have to be superposed on the wake before any effect can be detected.

The second region is the interactive region. The region is narrow, occurring near 200 Pa forcing level. When the control disturbance is roughly the same amplitude as the wake instability, then interaction between the two oscillators may occur. The very low value in the momentum thickness at this location was due to the presence of the varicose mode. When this mode was present the "dead region" of fluid behind the cylinder became extremely large compared to the Karman shedding mode. Monkewitz and Nguyen showed that the varicose mode of instability was convectively unstable. Thus, global instability was not possible for this mode. Evidently the symmetric forcing caused it to appear.

Increasing the forcing level above 200 Pa leads again to periodic shedding and to a region of linear decay in the momentum defect. In this region the wake defect behaves as if the momentum flux from the pulsing jets were linearly superposed on the flow. Therefore, we designated this region as the jets dominant region.

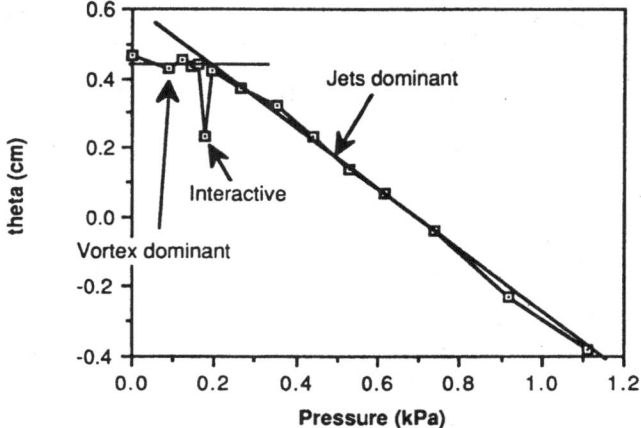

Figure 11. Variation in momentum defect with pressure fluctuation level.

5. Discussion

Clearly the pulsating jets have a large influence on the development of the wake, even though there is no net mass addition to the flow. If the forcing level is large enough, then the Karman vortex street is not formed. In early experiments we believed that the reduction in drag was a result of the disrupted vortex street. Although this is probably true, it cannot account for the overshoot in centerline velocity and the negative momentum defect generated by the jets. The ideal case of potential flow would only lead to $\theta = 0$ without an overshoot in velocity.

To get a better understanding of the effect of the jets, the water channel was turned off and the jets were activated in the initially quiescent flow. The frequency of pulsation was 1.8 Hz (F = 1.9) and the amplitude was 0.52 kPa. After a short time the jets established a recirculating region downstream of the cylinder. Figure 12 shows a velocity profile of the flow induced by the jets at x/d = 2.1. An estimate of the momentum flux associated with the jet gives $C_T = 0.22$, where C_T is the thrust normalized by the dynamic pressure, $\frac{1}{2}\rho U_\infty^2$. The corresponding wake flow with the jets activated showed a reduction of C_d from 1.1 to 0.25 at the same forcing conditions. Thus, no more than 26% of the drag reduction can be attributed to the momentum flux from the jet. The remainder must be associated with the modification in base pressure.

In Figure 13 the variation of maximum induced velocity with streamwise distance is shown. It should be emphasized that the induced flow pattern was generated in a quiescent channel without the background flow present in the wake experiments. The jet-like flow of fluid downstream from the cylinder is created solely by entrainment, since the mean velocity at the jet exit must be zero. The point is that a net momentum flux has been generated by the pulsing jets even though there has not been any net mass addition.

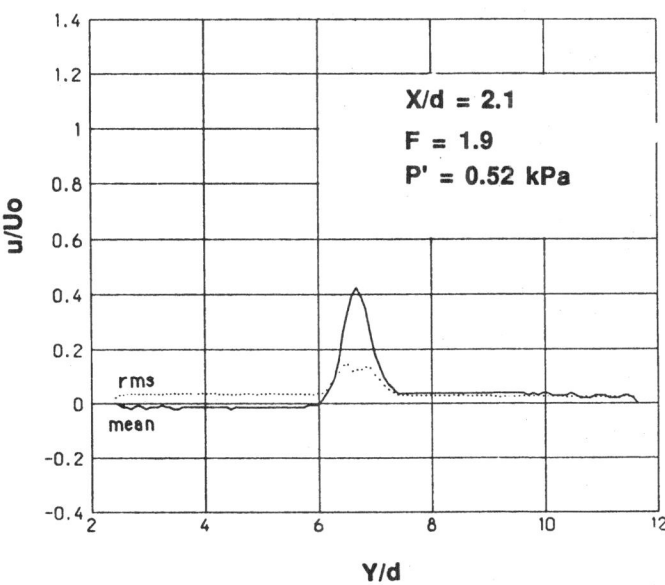

Figure 12. Mean and root mean square velocity distributions with F = 1.9, amplitude of 0.62 kPa in a quiescent water channel. Data has been normalized by 3.83 cm/s for comparison with other data.

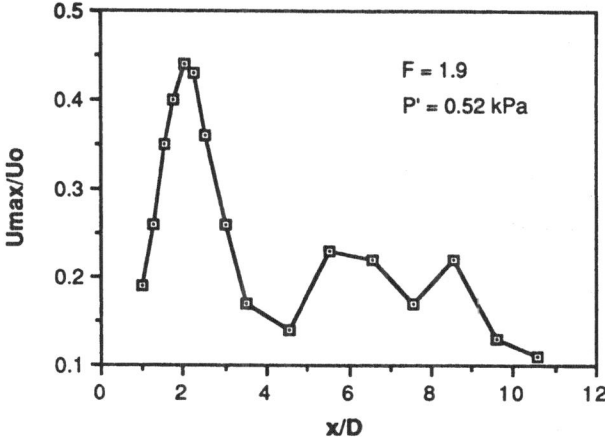

Figure 13. Streamwise variation in mean velocity maxima induced by the pulsating jets. F = 1.9 and amplitude is 0.62 kPa.

The mechanism for the addition of momentum to the flow by the pulsating jets appears to be analogous to acoustic streaming. The streaming effect was first used by Rayleigh (1945) to explain the accumulation of dust by sound waves in a Kundt tube. Later the term was applied to the phenomenon of a recirculating mean flow which formed around an oscillating cylinder in a quiescent channel (see Schlichting 1979). Although the term "streaming" often implies the presence of a viscous boundary layer, for this paper streaming refers to the second-order mean flow generated by a finite amplitude harmonic oscillation through the nonlinear terms in the momentum equation. In the jets-dominant range, the pulsing jets have large enough amplitude for a mean flow to be generated downstream of the jet exit. Since on average, no mass is added to the flow, the addition of momentum to the mean flow must be a result of the streaming effect.

In addition to changes in the mean flow by streaming, the flow near the exit of the pulsating jets experiences a lower pressure. The jet flow is directed downstream during the outflow portion of the cycle. During the suction side of the cycle, there is a sink-like behavior around the base of the cylinder, and the fluid is drawn back into the cylinder, primarily from the sides of the cylinder. The lack of flow field symmetry during the pressure cycle results in a net streamwise flux of momentum, and a rectified pressure signal at the exit of the jet. Thus, in the far wake the effect of the pulsation is to add a net momentum flux to the flow as if a constant base bleed were applied. However, in the near-wake region the rectification of the pressure reduces the value of the mean pressure. This effect is completely different from the steady base bleed case.

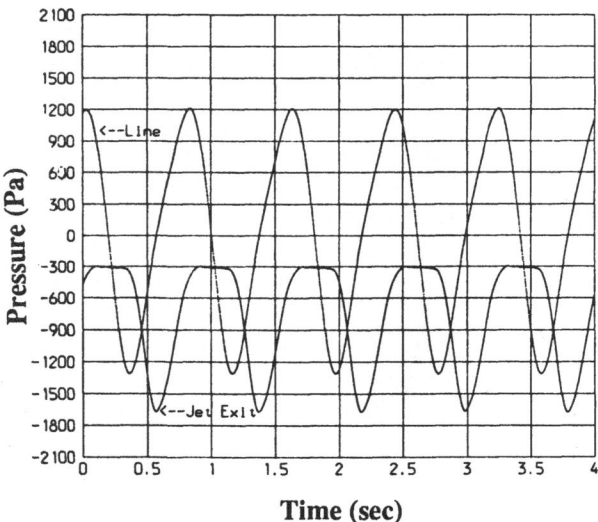

Figure 14. Interior and exit pressure variation for a single pulsating jet

To demonstrate the pressure rectification effect, a single pulsating jet with a diameter of 0.32 cm was connected to the fluid pulsation device shown in Figure 2. One pressure transducer was connected to the line between the cam and motor. A second transducer measured the pressure at the exit of the jet. The results are shown in Figure 14. From the time the line pressure is at its maximum value until it reaches its minimum value, the fluid is leaving the jet and the exit pressure is constant. Inflow corresponds to the line pressure going from minimum to maximum values. During these phases the exit pressure drops below the ambient pressure. The net result is a half-wave rectified exit pressure. Therefore, the mean pressure in the exit of the jet is lower than ambient. The reduction in exit pressure tends to shrink the recirculating region behind the cylinder. Even though the base pressure may be lower than in the natural case, the drag is still reduced because the surface area over which the base pressure acts has been reduced even more.

6. Conclusions

The effectiveness of the pulsating jets as a means of controlling the development of the wake behind a circular cylinder has been demonstrated. Flow visualization shows that the organization of vorticity shed from the body can be strongly influenced by unsteady jets. With sufficiently high forcing levels and frequencies, the formation of the Karman vortex street is suppressed. It is believed that the unsteady pulsing jets interfere with the mechanism of resonance, preventing a global instability from being established.

The momentum defect and corresponding drag coefficient can be reduced and even made negative. The mechanism for the control appears to be the second-order effect of streaming. The addition of momentum to the flow was shown by measurements in an initially quiescent water channel in which the pulsating jets were activated. The jets created a mean flow in the streamwise direction by entrainment, even though there is no net mass addition by the jets. The interaction between the streaming motion and the wake of the cylinder is believed to account for the reduction in the wake momentum defect.

The many helpful discussions and the encouragement from Professor M. Morkovin are greatly appreciated. The experiments were conducted with the support of the Air Force Office of Scientific Research contract #F49620-86-C-0133

REFERENCES

Bearman, P. W. 1967 The effect of base bleed on the flow behind a two-dimensional model with a blunt trailing edge. *Aero. Quart.* **18**, pp. 207-224.

Bearman, P. W. 1984 Vortex shedding from oscillating bluff bodies. *Ann. Rev. Fluid Mech.* **16** pp. 195 - 222.

Bechert, D. W. 1985 Excitation of instability waves. IUTAM Symp. *Aero and Hydro-Acoustics* Lyon.

Bers, A. 1975 Linear waves and instabilities. in *Physique des Plasmas* (eds C. DeWitt and J. Peyraud) pp. 117-213.

Braza, M. Chassaing, P. & Ha Minh, H. 1986 Numerical study and physical analysis of the pressure and velocity fields in the near wake of a circular cylinder. *J. Fluid Mech.* **165** pp. 79 - 130.

Briggs, R. J. 1964 Electron Stream Interaction with Plasmas. Research Monograph No. 29, Cambridge, Mass. MIT Press.

Chehroudi, B. and Simpson, R. L. 1984 A rapidly scanning laser Doppler anemometer. *J. Phys. E.* **17,** p. 131.

Chomaz, J. M., Huerre, P. & Redekopp, L. G. 1987 Local and global bifurcations in spatially developing flows. *Phys. Rev. Lett.* **60** pp. 25 - 31.

Cimbala, J. M. and Park, W. J. 1987 Mean velocity profiles and flow visualizations of a two-dimensional momentumless wake. *Bulletin Am. Phys. Soc.,* November. p 2044.

Drazin, P. & Reid, W. 1981 *Hydrodynamic Stability* Cambridge University Press.

Economou, M. 1986 Design and Performance of a Scanning Laser Doppler Velocimeter. M.S. Thesis, Illinois Institute of Technology.

Fornberg, B. 1980 A numerical study of steady viscous flow past a circular cylinder. *J. Fluid Mech.* **98** (4), pp. 819 - 855.

Gaster, M. 1981 On Transition to Turbulence in Boundary Layers in *Transition and Turbulence,* (ed. by R.E. Meyer) Academic Press.

Gerrard, J. H. 1978 The wakes of cylindrical bodies at low Reynolds number. *Phil. Trans. Roy. Soc.* London **288 A** pp. 351 - 382.

Goldstein, S. 1965 *Modern Developments in Fluid Dynamics* Fluid Motion Panel of the A.R.C. (S. Goldstein, ed.) Vol II. Dover Publications, p. 425.

Griffin, O. M. 1984 Vibrations and flow-induced forces caused by vortex shedding. In *Symposium on flow induced vibrations* ASME Winter Annual Meeting, New Orleans.

Hama, F. R. 1957 Three-dimensional vortex pattern behind a circular cylinder. *J. Aero. Sci.,* **24,** pp. 156-157.

Huerre, P. 1987 Spatio-temporal instabilities in closed and open flows. *Instabilities and Nonequilibrium Structures.* E. Tirapequi and D. Villarroel, eds., Reidel Publ. Co., pp. 141 - 177.

Huerre, P. & Monkewitz, P. A. 1985 Absolute and convective instabilities in free shear layers. *J. Fluid Mech.* **159** pp. 151 - 168.

Jordan, S. K. & Fromm, J. E. 1972 Oscillatory drag, lift and torque on a circular cylinder in a uniform flow. *Phys. Fluids* **15** pp. 371 - 376.

Koch, W. 1985 Local instability characteristics and frequency determination of self-excited wake flows. *J. Sound and Vibration.* **99** (1), pp. 53 - 83.

Koopman, G. H. 1967 The vortex wakes of vibrating cylinders at low Reynolds numbers. *J. Fluid Mech.* **28** pp. 501 - 512.

Kovasznay, L. S. G. 1949 Hot-wire investigation of the wake behind cylinders at low Reynolds numbers. *Proc. Roy. Soc.* **A 198** pp. 174 - 190.

Lamb, H. 1945 *Hydrodynamics.* Sixth edition, Dover Publications. pp. 174 - 190.

Mathis, C., Provansal, M. & Boyer, L. 1984 The Benard - Von Karman instability: an experimental study near the threshold. *J. Physique Lett.* **45** pp. 483 - 491.

Monkewitz, P. A. & Nguyen, L. N. 1987 Absolute instability in the near wake of two-dimensional bluff bodies. *J. Fluids and Struct.* **1** pp. 165 - 184.

Monkewitz, P. A. 1988 The absolute and convective nature of instability in two-dimensional wakes at low Reynolds numbers. *Phys. Fluids* **25** pp. 1137 - 1143.

Morkovin, M. 1964 Flow around circular cylinders - a kaleidoscope of challenging fluid phenomenon. *ASME Symposium on Fully Separated Flows,* pp. 102 - 118.

Morkovin, M. 1987 On Symmetry Properties of Free and Forced Vortex Streets. Bulletin of Am. Phys. Soc. **32** (10) Nov. p2059.

Morkovin, M. 1988 Recent insights into instability and transition to turbulence in open-flow systems. AIAA paper 88-3675. Presented at 1st Nat. Fluid Dyn. Cong., Cincinnati.

Nishioka, M. & Sato, H. 1978 Mechanism of determination of shedding frequency of vortices behind a cylinder at low Reynolds numbers. *J. Fluid Mech.* **89** pp. 59 - 82.

Olinger, D. J. & Sreenivasan, K. R. 1988 Nonlinear dynamics of the wake of an oscillating cylinder. *Phys. Rev. Lett.* **60** pp. 797 - 800.

Overvik, T. 1982 Hydroelastic Motion of Multiple Risers in a Steady Current. Ph.D. Thesis, Norwegian Institute of Technology.

Prandtl, L. 1914 Der Luftwiderstand von Kugeln. *Nachr. Ges. Wiss.* Goettingen, Math. Phys. Klasse. pp. 177 - 190.

Rayleigh, L. 1945 *Theory of Sound.* Dover Publications, p. 333.

Roshko, A. 1954 On the development of turbulent wakes from vortex streets. NACA Rep. 1191.

Schlichting, H. 1979 *Boundary-Layer Theory.* Fourth edition, McGraw-Hill Publish. Co., p. 41.

Son, J. S. & Hanratty, T. J. 1969 Numerical solution for the flow around a cylinder of Reynolds numbers 40, 200 and 500. *J. Fluid Mech.* **35** , p 651.

Stuber, K. & Gharib, M. 1988 Experiments on the forced wake of an airfoil - Transition from order to chaos. Proc. 1st Nat. Fluid Dynamics Cong., **2** Cincinnati, pp. 723 - 730.

Thom, A. 1933 The flow past circular cylinders at low speeds. *Proc. Roy. Soc. A* **141** ,p 651.

Thoman, D. C. and Szewczyk, A. A. 1969 Time dependent viscous flow over a circular cylinder. *Phys. Fluids Supp. II* pp. s.76 - s.86.

Triantafyllou, G. S., Triantafyllou, M. S. & Chryssostomidis, C. 1986 On the formation of vortex streets behind stationary cy linders. *J. Fluid Mech.* **170** pp. 461 - 477.

Unal, M. F. and Rockwell, D. 1988 On vortex formation from a cylinder. Part 1. The initial instability. *J. Fluid Mech.* **190** pp. 491-512.

von Karman, T. 1911 Ueber den Mechanismus des Widerstandes, den ein bewegter Koerper in einer Fluessigkeit erfaehrt. Goettinger Nach. Math. Phys. Kl. pp. 509 - 517.

Williams, D. R. and Economou, M. 1987 Scanning laser anemometer measurements of a forced cylinder wake. *Phys. Fluids* **30,** pp. 2283-2285.

Wood, C. J. 1964 The effect of base bleed on a periodic wake. *J. Roy. Aero. Soc.* **68** p. 477.

Wood, C. J. 1967 Visualization of an incompressible wake with base bleed. *J. Fluid Mech.* **29,** pp. 259-272.

Zdravkovich, M. M. 1981 Review and classification of various aerodynamic and hydrodynamic means for suppressing vortex shedding, *J. Indus. Aero. and Wind Engrg.* **7,** pp. 145 - 189.

VORTEX DYNAMICS OF DELTA WINGS

Mario Lee and Chih-Ming Ho

Department of Aerospace Engineering

University of Southern California

Los Angeles, California 90089-1191

ABSTRACT

The typical angle of attack for maximum lift of a delta wing is about 35°, which is much higher than for a two-dimensional airfoil. The delta wing is, therefore, suitable for highly maneuverable aircraft. In this paper, experimental results for delta wings is reviewed. The review is made from the perspective of fundamental fluid dynamic mechanisms. In particular, the balance between vorticity generation on the surface and freestream convection of it is used to understand how different parameters affect the leading edge vortices which dominate the aerodynamics of a delta wing at high angles of attack.

I. INTRODUCTION

The flow over delta wings has been the subject of extensive research in the pursuit of high lift at large angles of attack. For example, a two dimensional NACA 0012 airfoil would attain maximum lift at about 15° angle of attack. At higher angles, the flow on the suction side of the airfoil does not follow the surface. Vortices are formed and shed downstream. The separated flow results in a loss of suction; the airfoil stalls. In contrast, a 70° swept delta wing can continue to increase its lift until the angle of attack reaches approximately 40°. This is due to the presence of two <u>stationary</u> separation vortices. The potential flow has a convex curvature near the leading edge that produces a suction which contributes to increase C_L even after the flow has <u>separated</u> at the leading edge. As the angle of attack increases, the vortex breaks down with the vortex burst propagating from the trailing edge to the apex. With increasing angle of attack. The $3D$ equivalent to $2D$ stall occurs when the flow over the suction side separates globally.

The leading edge region of the delta wing is the principal source of vorticity. The vorticity diffuses from the surface and is convected downstream with the free stream velocity. The strength and other properties of the leading edge vortices are determined by the balance between diffusion and convection at different operating conditions. Therefore, we can understand delta wing aerodynamics if the fundamental vorticity balance is clarified. This is the basic approach chosen for the review of the existing experimental work on this topic. In the present paper, a brief summary of fundamental concepts is first presented. Next, the different parameters affecting steady and unsteady flow over delta wings will be discussed. Finally, the results of different schemes employed to control the wing will be examined.

II. VORTICES AND AERODYNAMIC FORCES

2.1 General Features of Leading Edge Vortices

The two spiralling vortices on a delta wing are formed from the separated shear layers which originate from the leading edge (Fig. 1). On a wing with rounded leading edge, Earnshaw and Lawford (1964) found that these vortices did not appear until the angle of attack was more than 5°. For sharp leading wing, the separation vortices start at smaller angle of attack (Ericsson and Reding 1977). The size of the L. E. vortices is in the order of the half span. Small vortices which develop from the Kelvin-Hemholtz instability waves of the separating shear layer are superimposed on the large vortices (Gad-el-Hak and Blackwelder 1987). These small vortices (Fig. 2, Payne et al. 1988) are scaled with the thickness of the shear layer which is usually one order of magnitude smaller than the span. Under the L. E. vortices and slightly outboard from the cores, there is a pair of secondary vortices (Figs. 1-a, 3-a), induced by the primary vortices.

Lift on the wing is maintained by the spiralling vortices, which produce suction footprints on top of the wing. The pressure measurements by Fink and Taylor (1967)

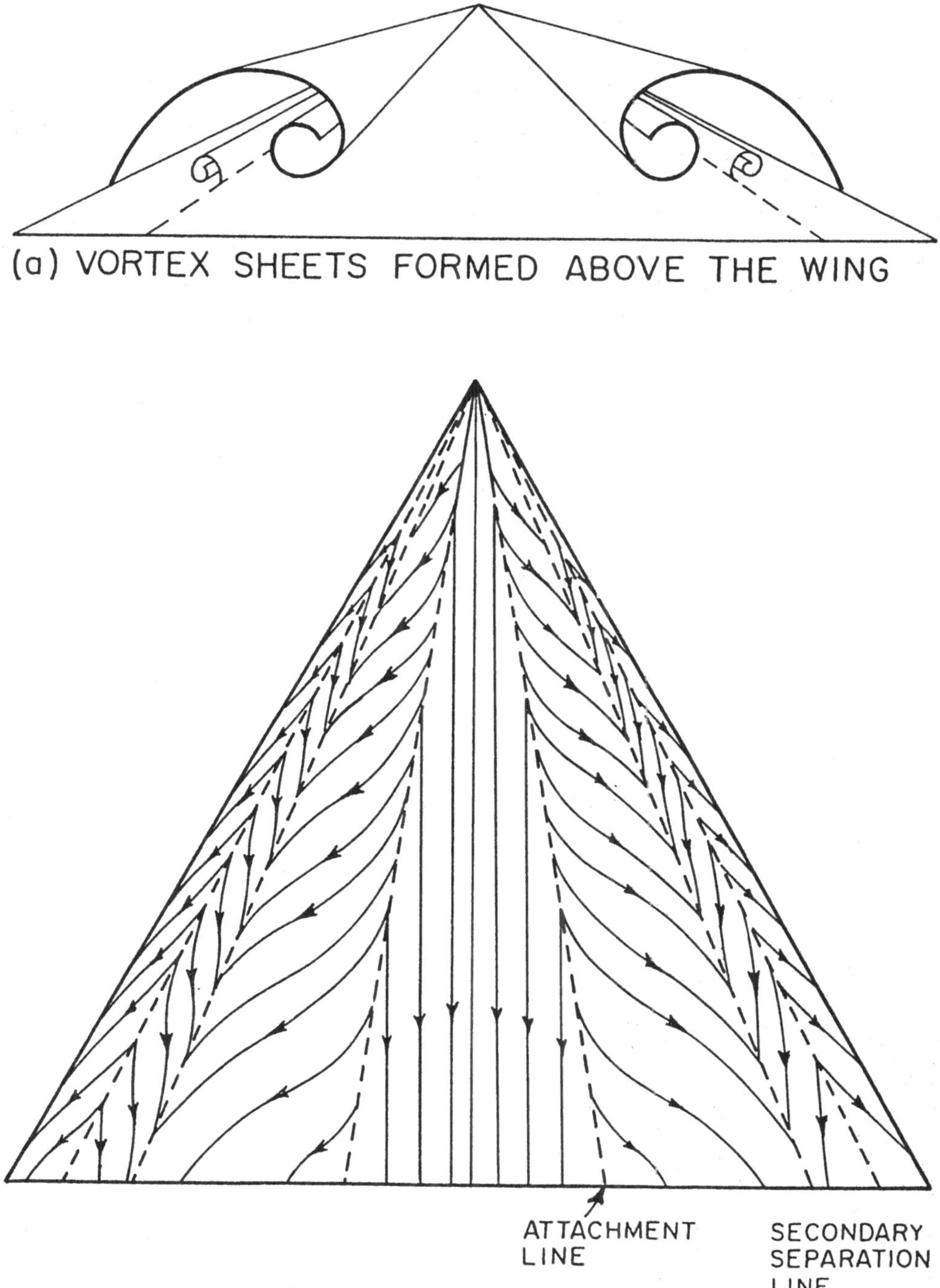

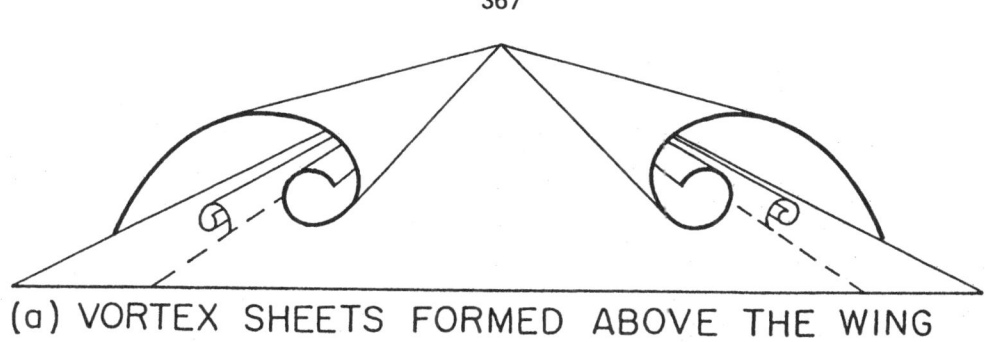

(a) VORTEX SHEETS FORMED ABOVE THE WING

ATTACHMENT
LINE

SECONDARY
SEPARATION
LINE

(b) SURFACE FLOW PATTERN

Fig. 1: [a] Leading edge vortices and [b] Surface flow pattern, Earnshaw and Lawford

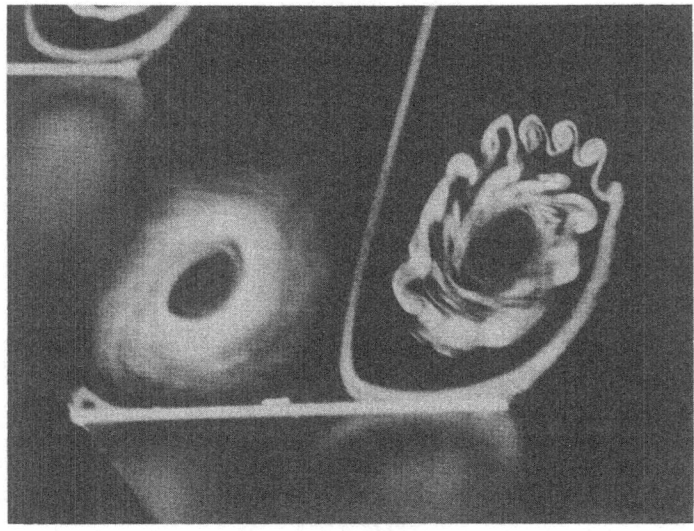

Fig. 2: Secondary vortices superimposed on L. E. separation vortices, Payne [1987]

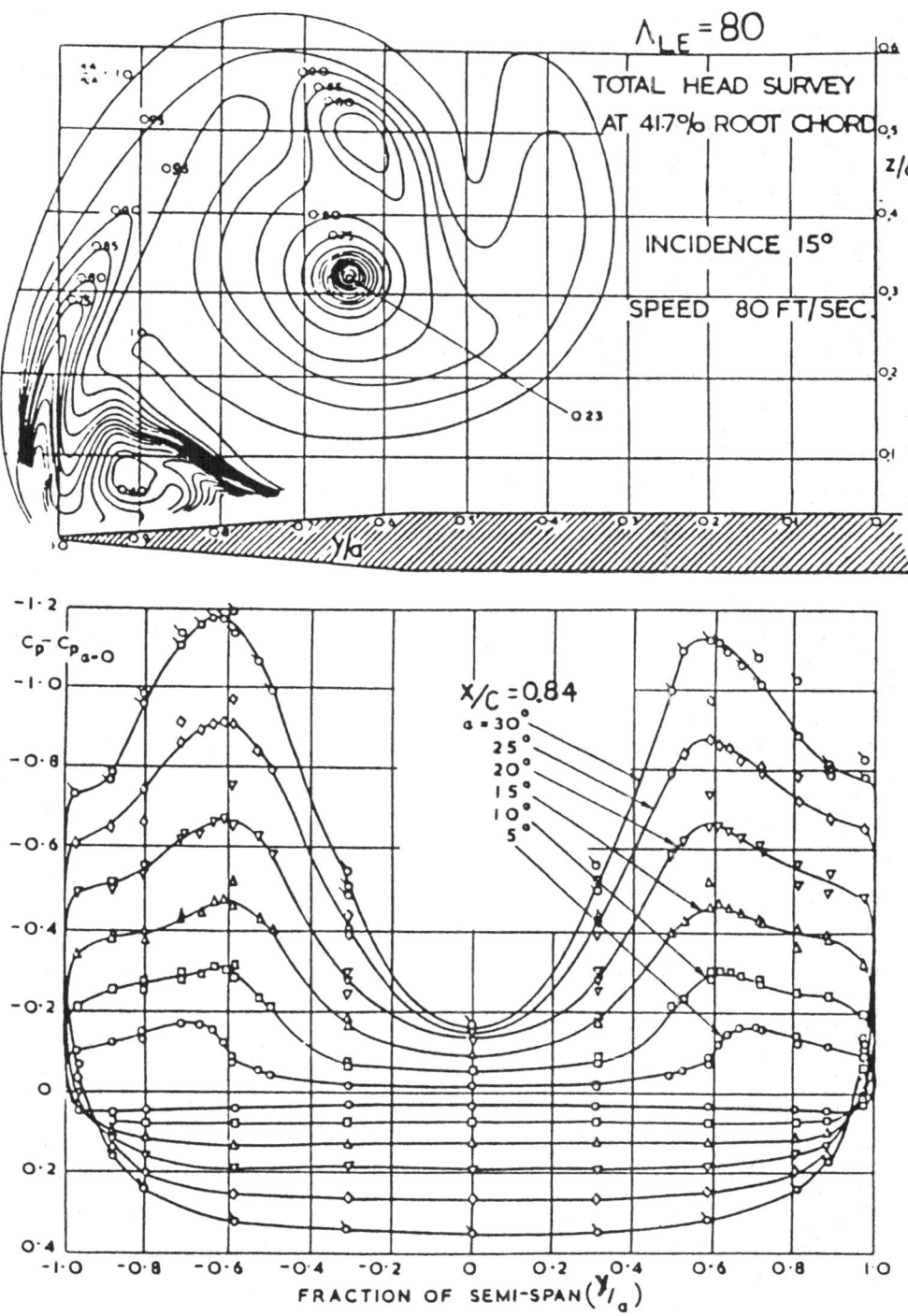

Fig. 3: [a] Total head measurement of the L. E. vortices [b] Surface pressure measurements, Fink and Taylor [1967]

in Fig. 3 show that the location of maximum suction is very close to the center of the primary vortex.

At large angles of attack, the L.E. vortices will suddenly expand in size. This is coupled with a sharp increase of the dynamic pressure and a decrease in axial velocity. This phenomenon is called vortex burst or breakdown. Flow visualizations obtained by Lambourne and Bryer (1961, Fig. 4) show that the transformation can take on a spiral (the top one in Fig. 4) or a bubble form (the bottom one in Fig. 4). The measurements made by Hummel (1965) in Fig. 5 reveal that the vorticity is spread out over a larger region after the breakdown. The vortex breakdown is associated with the loss of lift.

2.2 Balance of Vorticity

The importance of the vorticity balance concept was emphasized by Reynolds and Carr (1985) in their review of separation driven flows. The underlying assumption is that two and three dimensional separated flows are dominated by large concentrated vortices. At the fluid boundary, vorticity is generated on the surface and diffuses into the boundary layer. In an attached flow, the vorticity is removed continuously by free stream convection. When the flow separates, a local shear layer forms (Didden and Ho 1986). The vorticity in the shear layer is lumped into large vortical structures which are convected downstream with the velocity of the freestream. The balance between surface vorticity flux and vorticity transport in the freestream dictates the vorticity accumulation or depletion above solid surfaces. Reynolds and Carr (1985) further shown that the surface vorticity flux is governed by the local pressure gradient, transpiration through the surface, and motion of the surface itself. On a delta wing, the amount of vorticity shed into the vortex depends on the condition of the boundary layer at the leading edge prior to separation.

Next, let us look at the effects of vorticity convection. Suction is generated on the top surface of delta wings by the swirling motion of the vortex core. The swirl in the core depends on the amount of bounded vorticity fed through the shear layer. A stationary leading edge vortex is achieved only when the convection of vorticity along the core axis balances the vorticity generation from the boundary layer of the leading edge. On a

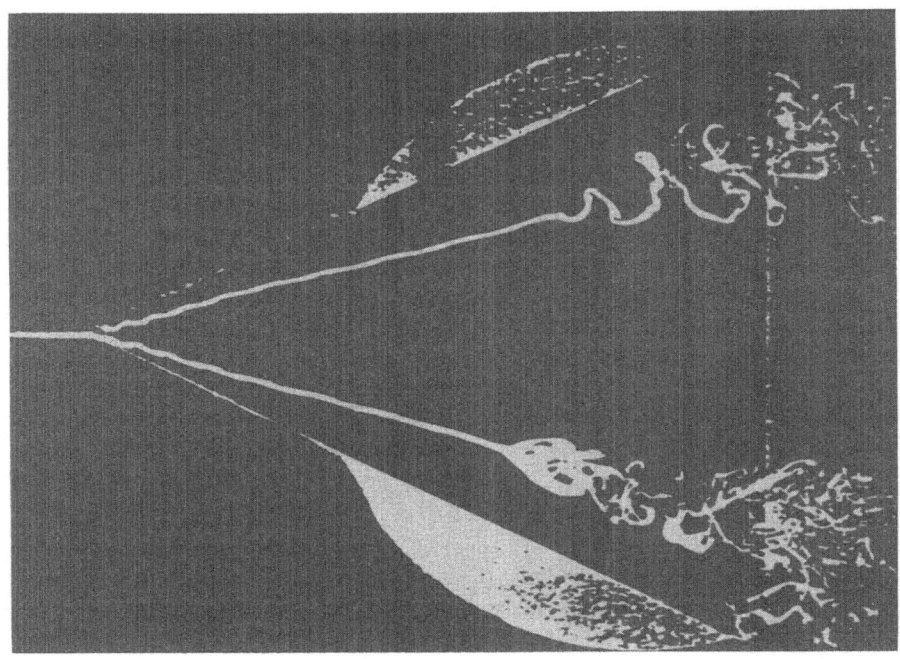

Fig. 4: Patterns of vortex breakdown, Lambourne and Bryer [1961]

372

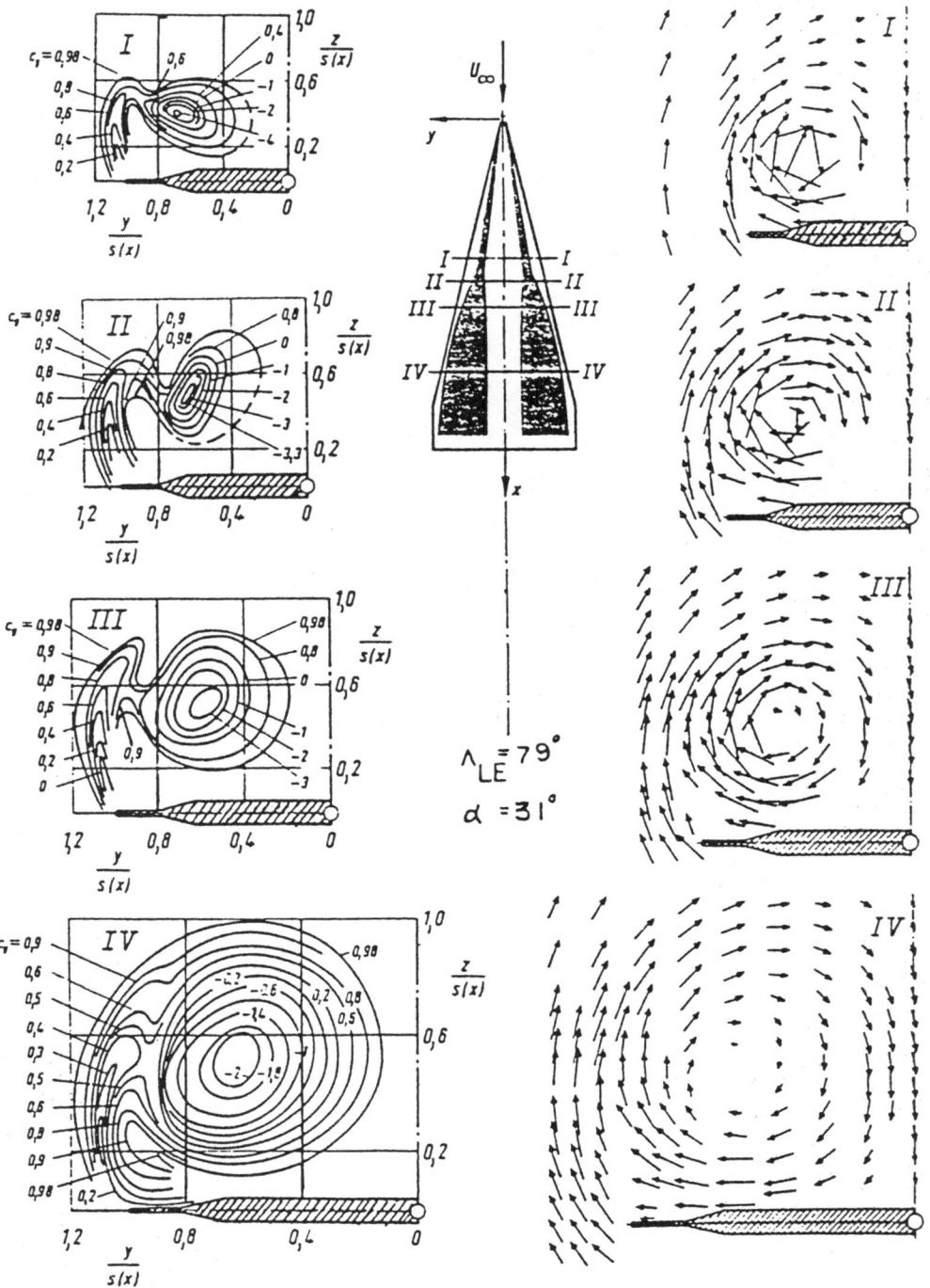

Fig. 5: Velocity distributions before and after vortex breakdown, Hummel [1965]

two dimensional airfoil, separation from the leading edge does not lead to a stationary vortex, because there is no intrinsic convection to remove vorticity from the core. The unorganized vorticity is simply convected away by the freestream. Rossow (1978) tried to stabilize the vortex on a two dimensional wing by providing artificial suction from the side and using a vertical fence to shield the vortex from the freestream. He found that the vortex was still delicate to maintain stationary, and an inclination of 10° to 20° from the leading edge was observed. This shows that axial convection along the core is essential in maintaining a stationary vortex. The magnitude of axial convection along the vortex core is determined by the component of the freestream in the direction parallel to the leading edge. For a delta wing, the component of the freestream along the leading edge is a function of the angle of attack, α, angle of sideslip, β, and sweepback angle, Λ_{LE}, as shown in Fig. 6. At the trailing edge, an adverse pressure gradient, prescribed by the Kutta condition, forces the flow to slow down, thereby reducing axial convection in the vortex core. This causes an imbalance of the vorticity budget, resulting in vortex burst. Based on the above analysis, the swirl angle, ψ, between the perpendicular and axial velocity components along the leading edge would reflect the vorticity balance necessary for a stationary vortex. Lambourne and Bryer (1961) suggested that ψ is important in determining the location of vortex breakdown over a delta wing. The evidence supporting this view will be provided in section III. A more detailed review of the theories relating swirl angle at the edge of the viscous core and breakdown has been given by Leibovich (1984).

2.3 Lift Prediction

Before we look at the different parameters affecting delta wings in steady flow, we will look at the leading edge suction analogy offered by Polhamus (1971). The leading edge suction analogy predicts the total lift on a delta wing by separating the normal force into potential and vortex components. The potential lift term is based on the lifting-surface theory, taking into account the Kutta condition at the trailing edge. The vortex lift term is modeled by the suction force generated by the equivalent attached flow around the edge. The only condition necessary for the analogy to hold is that the separation

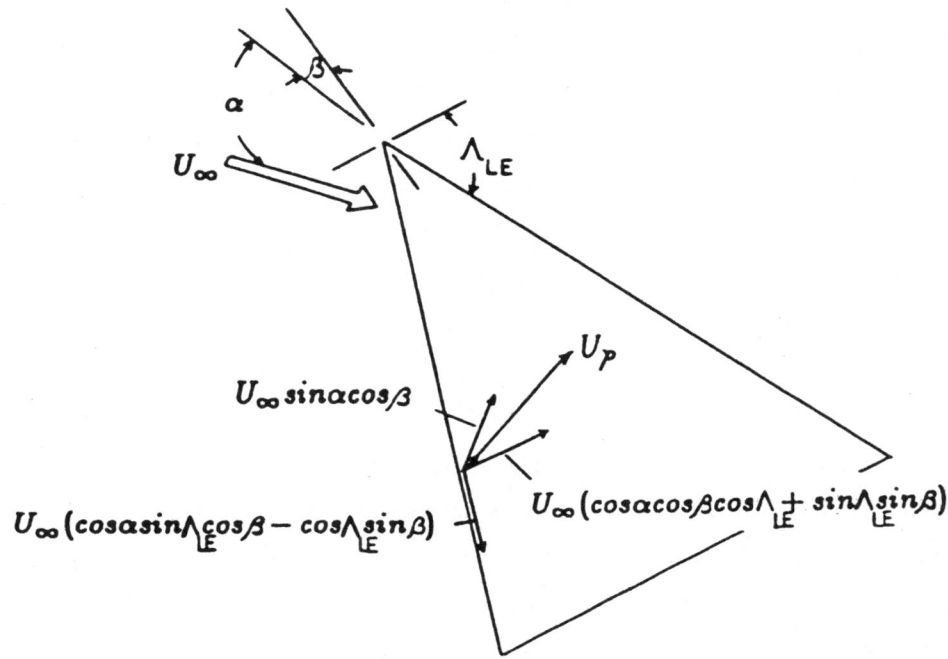

Fig. 6: Geometric relation of velocity and delta wing

must reattach on the upper surface. No detailed knowledge of the shape, strength and position of the vortex is required. It is interesting to note that the vortex lift term is only sensitive to angles of attack. The sweepback angle dependence comes solely from the potential lift term (Fig. 7). According to this theory, the airfoil depends more and more on the vortex for lift as the sweepback angle increases. Polhamus' theory (1971) accurately predicts the lift of delta wings until vortex breakdown occurs over the wing surface. Fig. 8 shows the limiting effect of vortex bursting on the lift at a given angle of attack, compiled by Campbell (1976). A review of various lift prediction methods was provided by Parker (1976).

III. STEADY LIFT Of DELTA WINGS

Parameters which affect steady flow over a delta wing include angle of attack, sweepback angle, angle of sideslip, leading edge profile, trailing edge geometry, Reynolds number, Mach number, and freestream turbulence. With the exception of angle of attack, trailing edge geometry, and Mach number, the parameters affect the lift only through the vortex term according to Polhamus' theory. Since only limited data is available on drag, our discussion will deal mostly with lift and vortex breakdown.

3.1 Effects of Angles of Attack, Sweepback, and Sideslip

The affects on lift of angle of attack, α, and sweepback angle, Λ_{LE}, on lift are shown in Fig. 9 (Earnshaw and Lawford 1964). The delta wings with $\Lambda_{LE} = 65°$ and $70°$ produced the best performance in terms of maximum lift. A comparison between the measurements and the predictions by the leading edge suction analogy is shown in Fig. 10. The breakdowns of the vortices at the trailing edge and the apex are also indicated in the figure. The C_L curves show that full vortex-induced lift is achieved up until vortex breakdown occurred at the trailing edge. The breakdown occurs at higher angles of attack for larger sweepback angle. Therefore, the prediction can better match the measurements for larger sweepback angles. For $\Lambda_{LE} > 75°$, vortex asymmetry occurs before breakdown and full vortex lift cannot be achieved. A semi-empirical fit of the lift

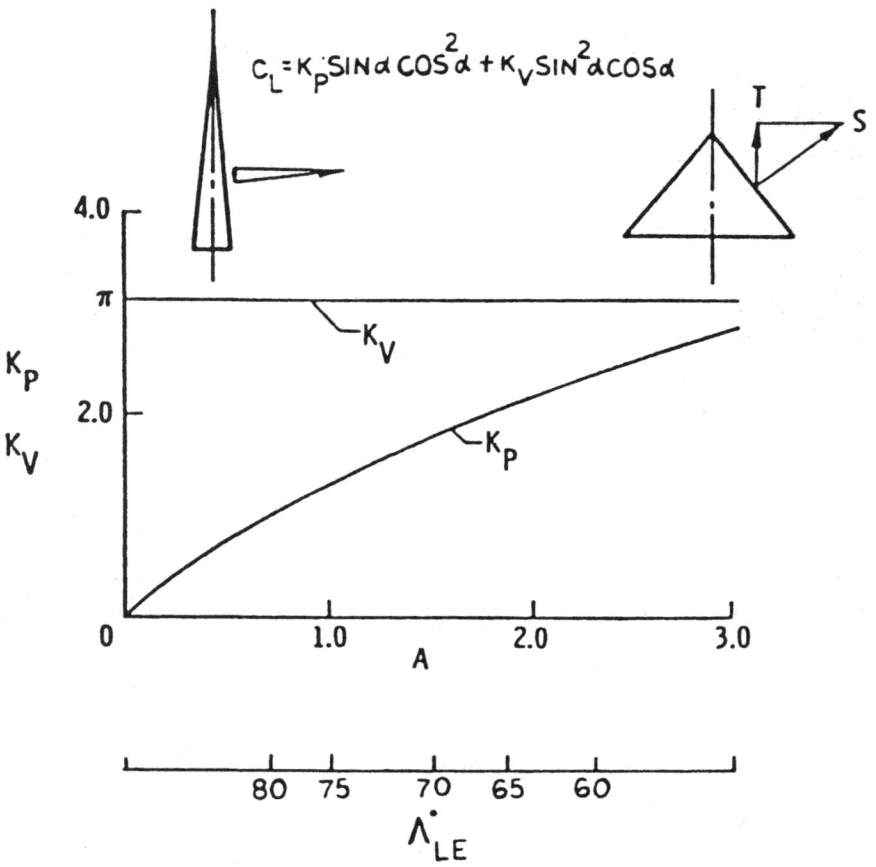

Fig. 7: Vortex and potential lift, Polhamus [1971]

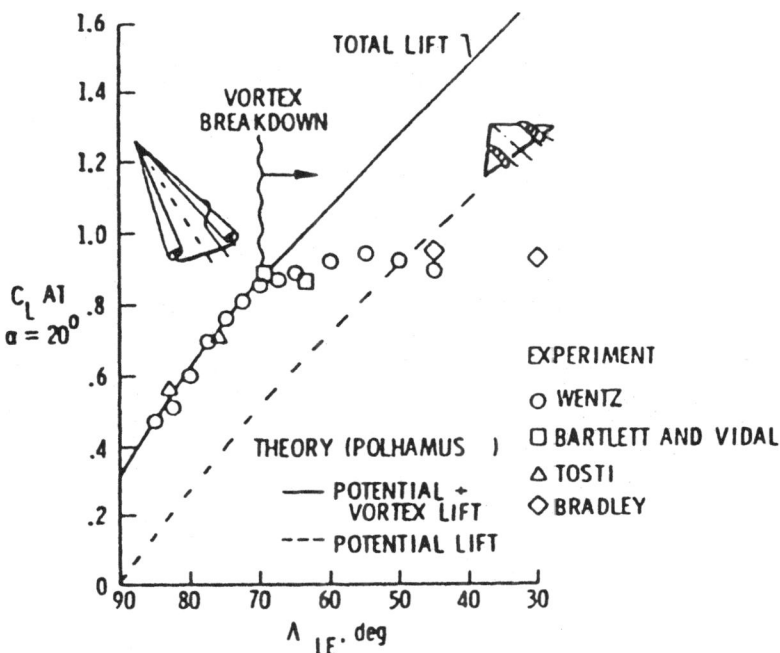

Lift capability of delta wings at $\alpha = 20°$.

Fig. 8: Vortex breakdown and the leading edge suction analogy, Campbell [1976]

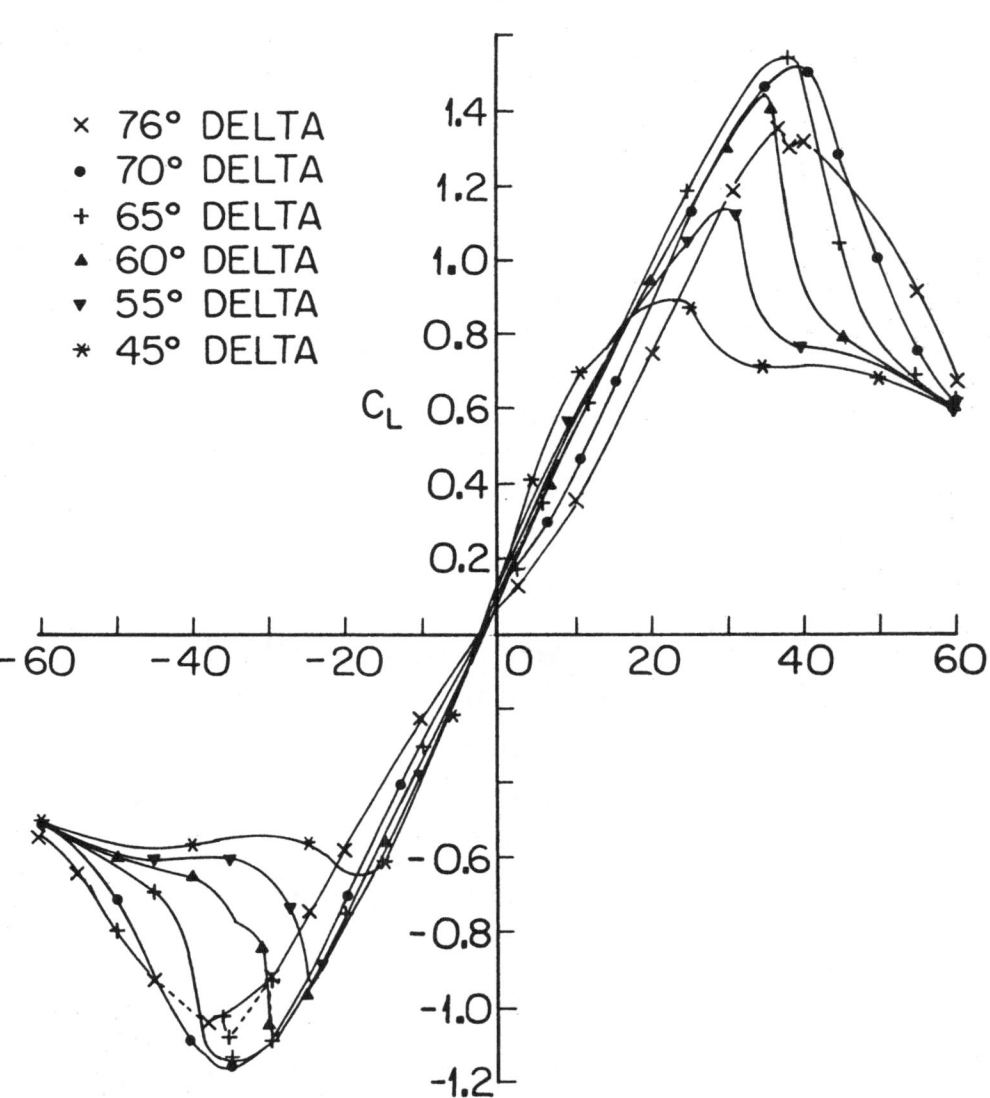

Fig. 9: Lift coefficient vs angle of attack, Earnshaw and Lawford [1964]

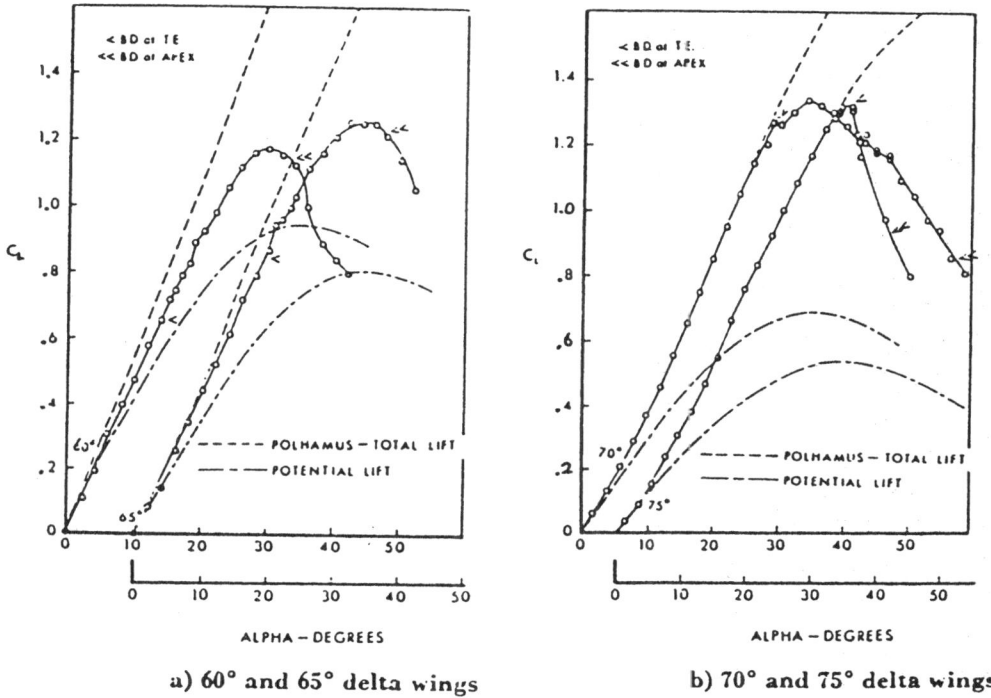

a) 60° and 65° delta wings   b) 70° and 75° delta wings

Fig.10: Comparison between experiments and Polhamus' prediction, Wentz and Kohlman [1971]

curve for different Λ($\theta_{LE} = \pi/2 - \Lambda_{LE}$) is provided by Ericsson and Reding (1977) in Fig. 11 based on Polhamus' theory.

The effects of yaw angle, β, on the position of vortex breakdown is shown in Fig. 12 from the data of McKernan and Nelson (1983). The vortex would burst closer to the apex on the windward side and farther away on the leeward side. The pressure measurements of Hummel (1965) in Fig. 13 shows the corresponding shift. However, the normal force measured by Harvey (1958) in Fig. 14 shows only a weak dependence on yaw. Ericsson and Reding (1977) suggested that the effect of β can be included in an effective Λ_{LE} for small angle of attack. Hence, the normal force does not change much with β.

The variation of the vortex bursting position described above can be traced to an effective change in the balance between vorticity surface flux and convection. Fig. 15 shows a summary of leeside vortex breakdown locations versus swirl angle, ψ, for different angles of attack, sweepback and yaw angles. The collapse of the data suggests that all three parameters affect the stability of the vortex through the same mechanism. An increase in the angle of attack, decrease in sweepback angle or an increase in yaw would increase the ratio between circumferential to axial flow (Fig. 6). Sforza et al. (1978) measured the flow over a $\Lambda_{LE} = 75°$ delta wing and showed that the swirl velocity at the core increases with angle of attack (Fig. 16). At the same time a stronger adverse pressure gradient at the trailing edge forces the axial flow to slow down faster. The vortex eventually becomes more unstable and bursts closer to the apex.

Payne (1987) surveyed the measurements made of the swirl angle at the edge of the viscous core upstream of vortex breakdown in vortex tube experiments. He reported that the angle varied from 38° to 55° for a spiral breakdown. Payne's own measurements over a delta wing ($\Lambda_{LE} = 85°$) was 44° at $\alpha = 40°$ while Hummel (1965) measured 53°($\Lambda_{LE} = 79°$) at $\alpha = 31°$. These values are consistent with the present data which varies from 32° to 66° even though the angles are deduced based on geometry at the leading edge.

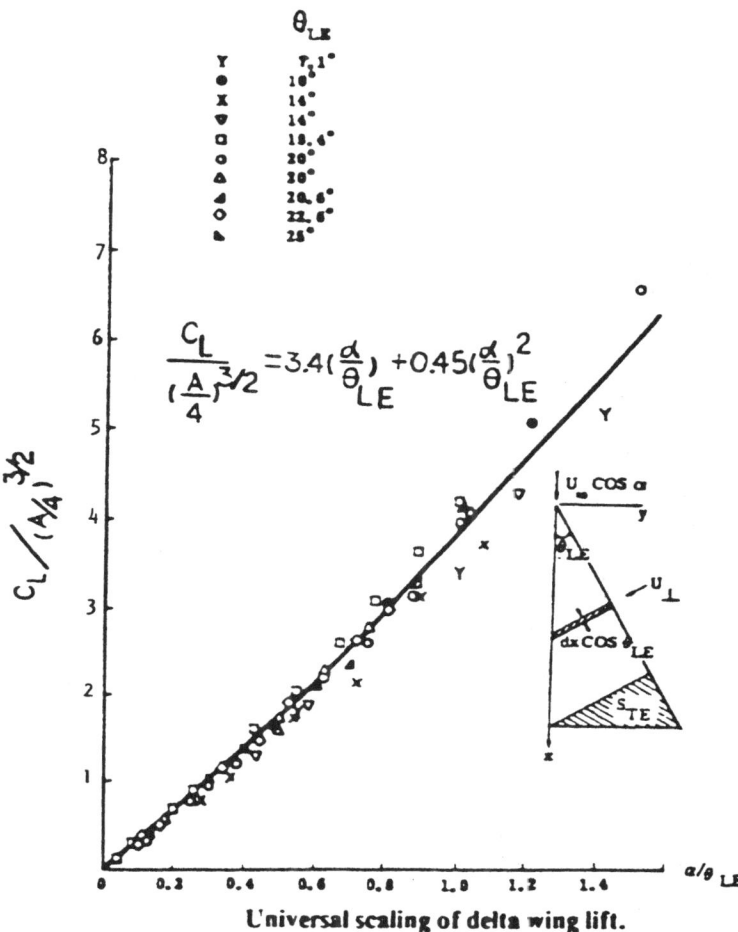

Fig.11: A scaling law of the delta wing lift, Ericsson and Reding [1977]

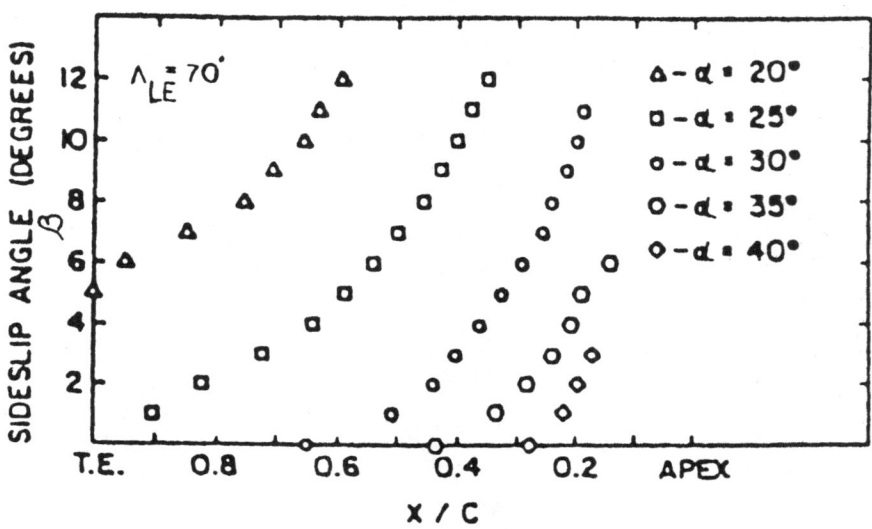

Fig.12: Effect of yaw on the vortex breakdown position, McKernan and Nelson [1983]

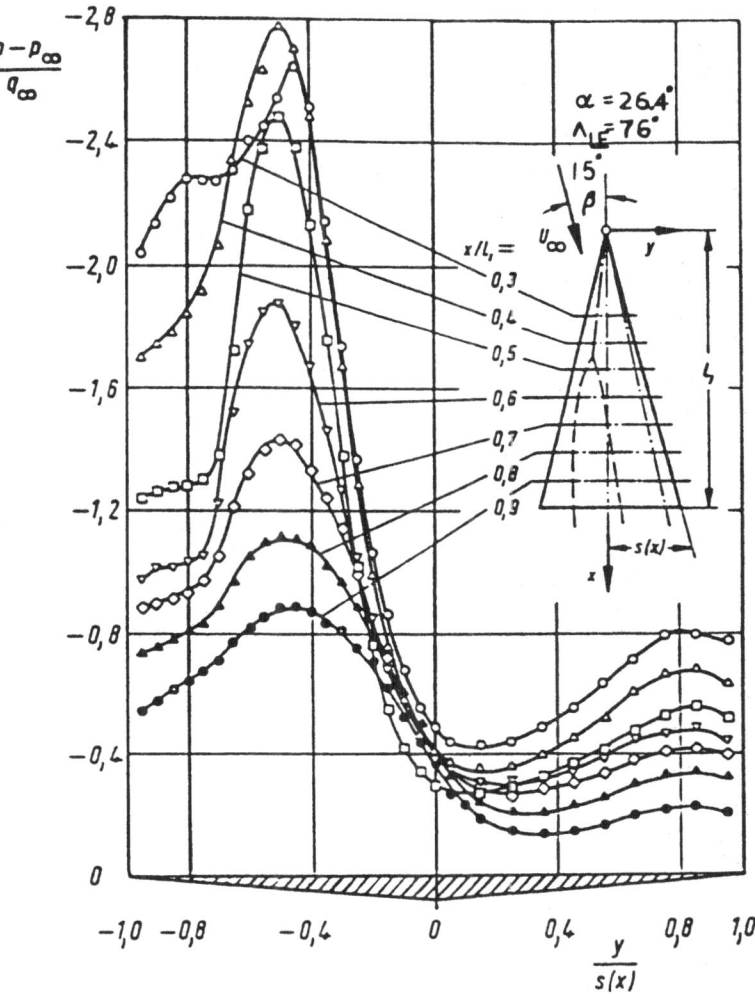

Fig.13: Surface pressure measurement of a yawed delta wing, Hummel [1965]

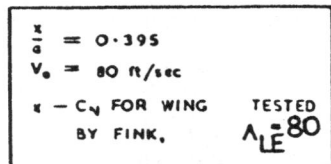

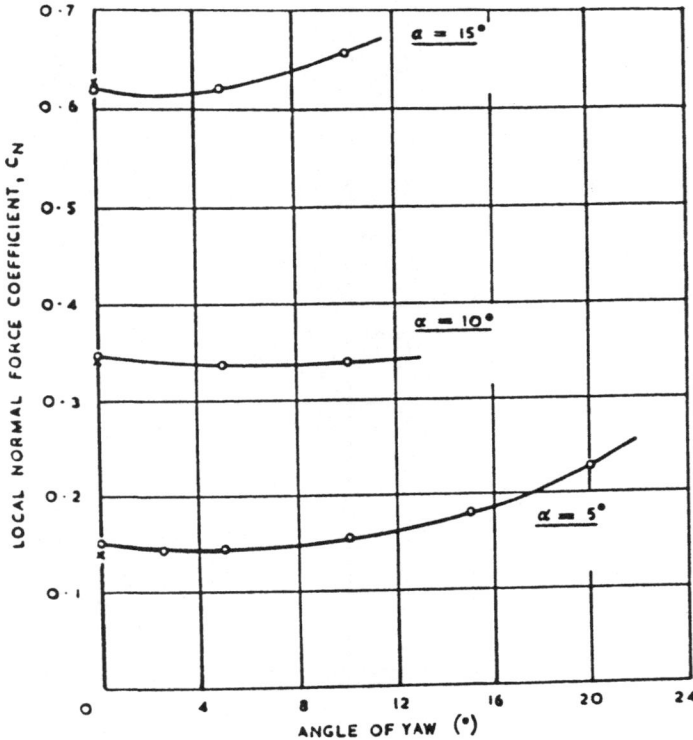

The normal force acting on the wing.

Fig.14: Normal force of a yawed delta wing, Harvey [1958]

Vortex Breakdown

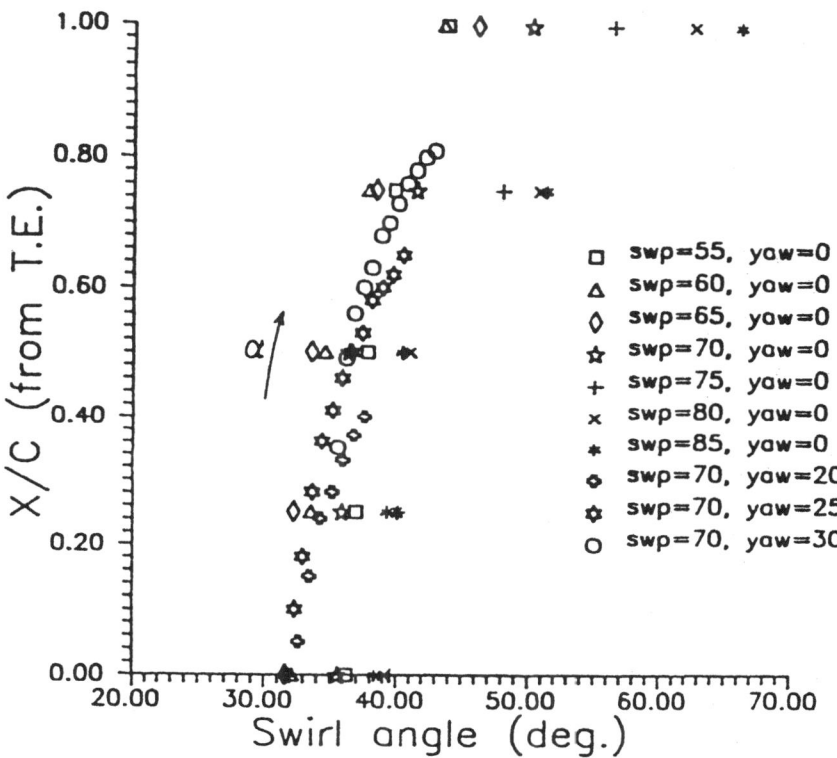

Fig.15: A scaling law of the vortex breakdown positions

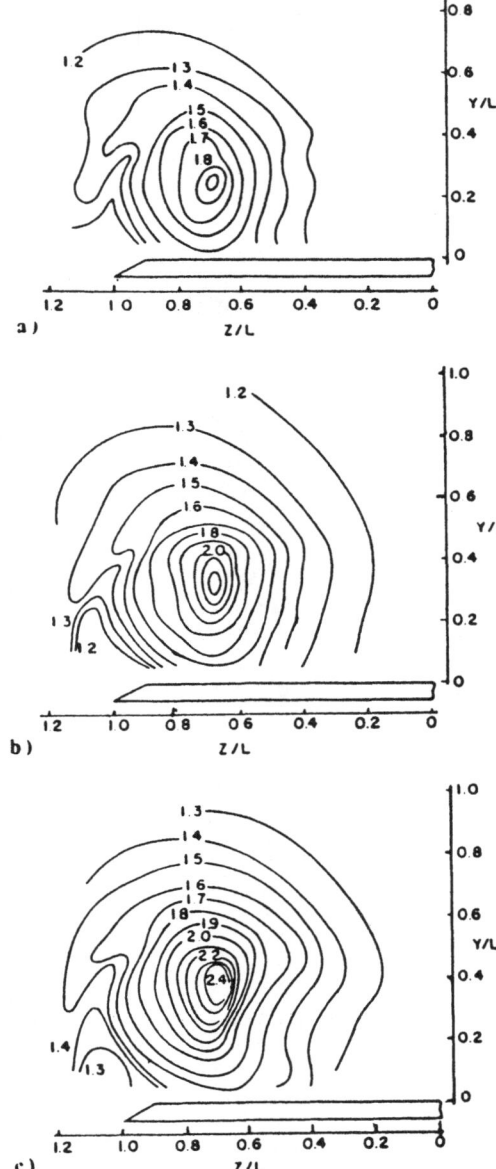

Contours of constant velocity magnitude over 75-deg sweepback plane delta wing. Numbers denote values of V/V_∞. The angle of attack for this is a) 20 deg, b) 25 deg, c) 29 deg.

Fig.16: Swirl velocity contours Sforza, et al. [1978]

3.2 Leading Edge Profile

Bartlett and Vidal (1955) studied the effects of leading-edge sharpness on three wings with different symmetric cross sections and thickness-to-cord ratios. They found that the wing with beveled edges produced the maximum slope for C_L vs α as shown in Fig. 17. The flow separated right at the sharp edge and produced a high suction. On the wings with elliptic or round edges, separation and vortex formation were delayed and the lift was lower. Wentz (1972) looked at the effects of spanwise camber and measured a lower lift coefficient for his conical and apex cambered wings at small to moderate angles of attack (Fig. 18). However, maximum lift was increased somewhat. The pressure measurements in Fig. 19 show that the vortex is more sensitive to camber at the apex than anywhere else on the leading edge. This delay in the roll-up of the vortex was due to the smaller adverse pressure gradient at the apex while separation was postponed further inboard onto the convex surface. Other experiments by Lamar (1977) with a linearly twisted bat wing (Fig. 20) and the inverted wing experiments of Earnshaw and Lawford (1964) show that a gain in lift was achieved when a convex surface was placed on the suction side of the airfoil. This gain in lift is due to a longitudinal camber effect as indicated by the non-zero lift generated at a zero angle of attack. Earlier, Lambourne and Bryer (1961) demonstrated through flow visualizations that longitudinal camber with a convex upper surface can delay vortex breakdown.

3.3 Trailing Edge Geometry

For a given sweepback angle, Polhamus' theory (1971) predicted a higher lift for an arrow wing compared with a diamond wing through the trailing edge boundary condition and aspect ratio effects for both vortex and potential lift terms. Experimental data by Wentz and Kohlman (1971) in Fig. 21 substantiated these results although the difference was not as large as predicted. The theory over predicts the arrow wing and under predicts the diamond wing. Ericsson and Reding (1977) corrected the difference by using the argument of equivalent delta wings. Wentz and Kohlman also found that the trailing edge apparently has no effect on the vortex breakdown location. This is reasonable since the trailing edge does not affect the vorticity balance which is dictated by the swirl

Leading Edge Shape

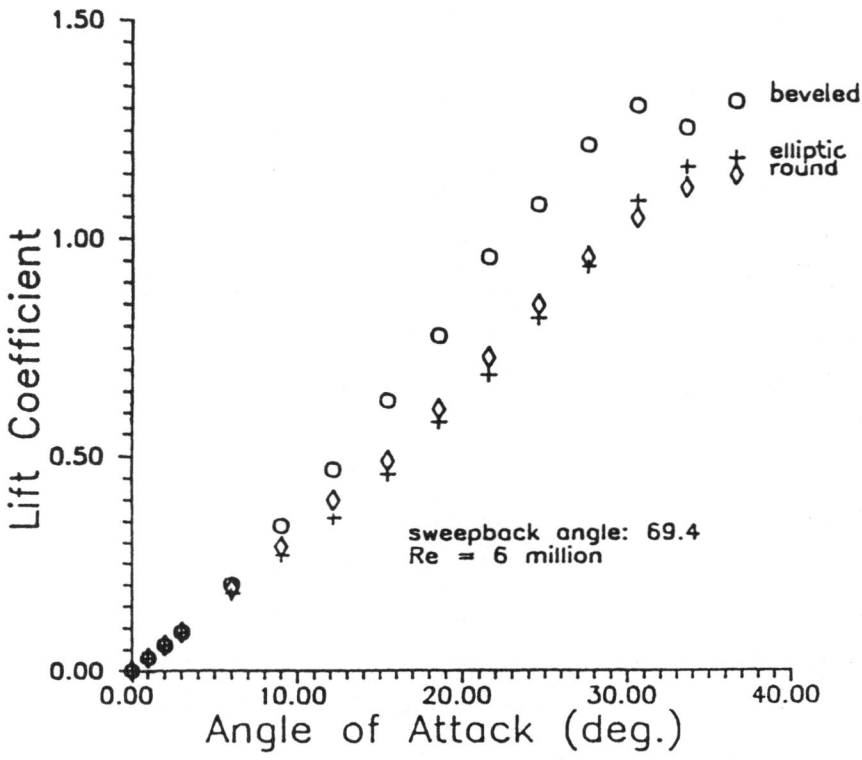

Fig.17: Effect of leading edge profile on the lift coefficient, Barlet and Vidal [1955]

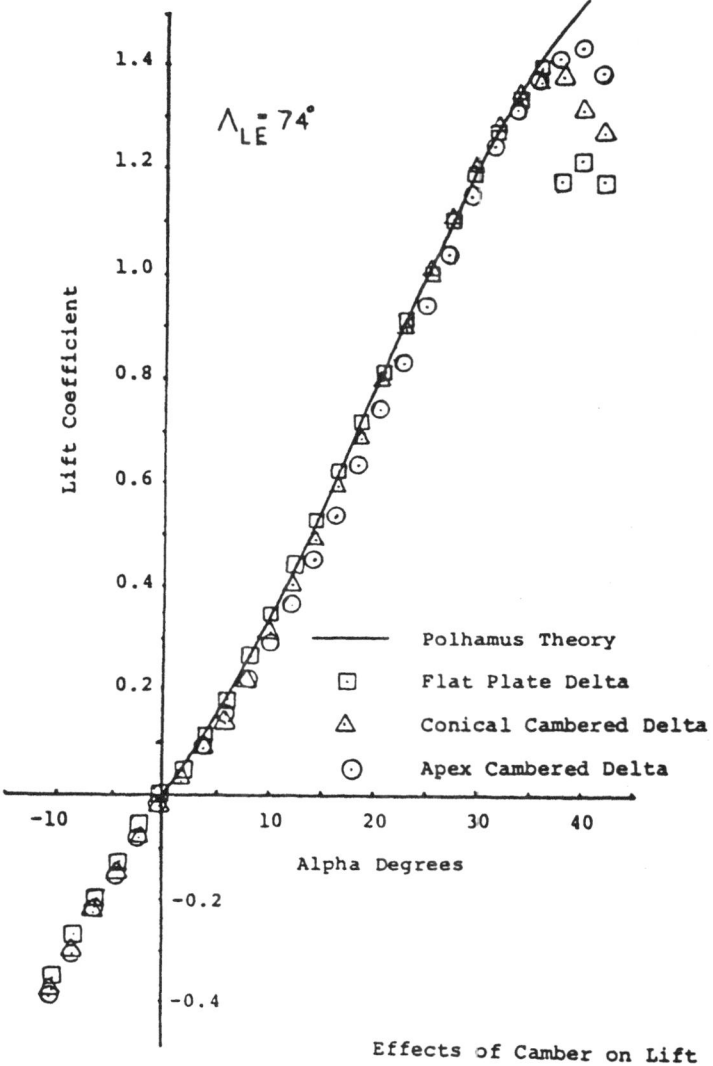

Fig.18: Effects of Camber on Lift, Wentz [1972]

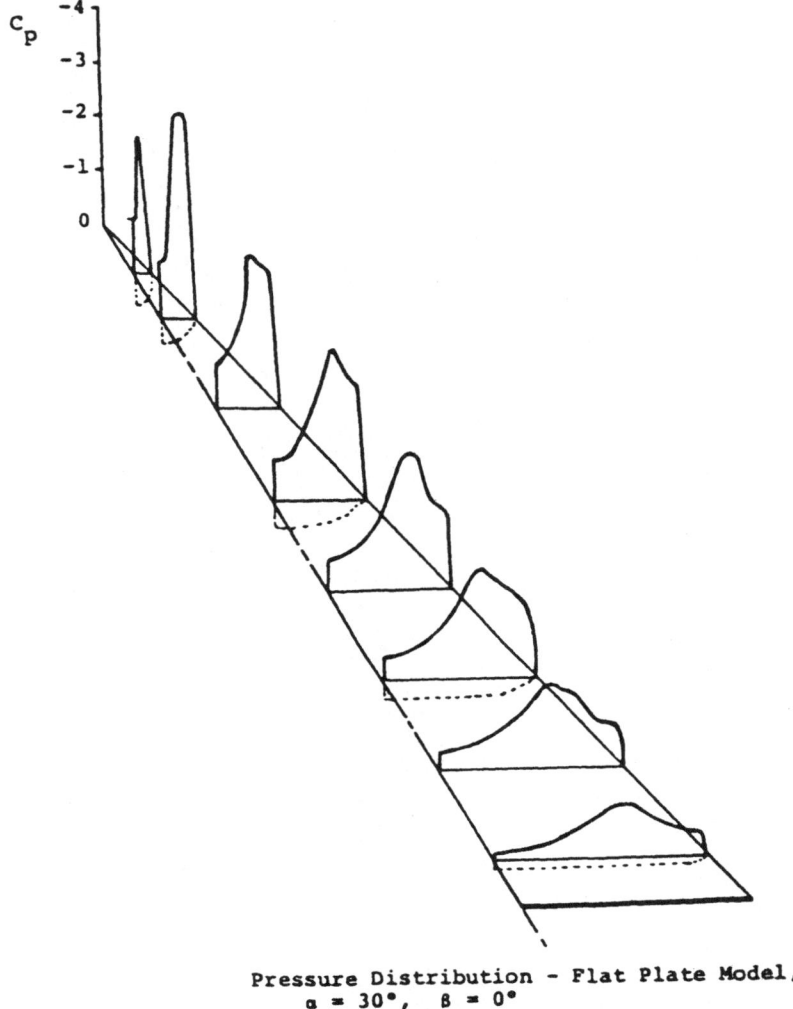

Pressure Distribution – Flat Plate Model,
$\alpha = 30°$, $\beta = 0°$

(A)

Fig.19: Effects of the camber on the pressure distributions, Wentz [1972]

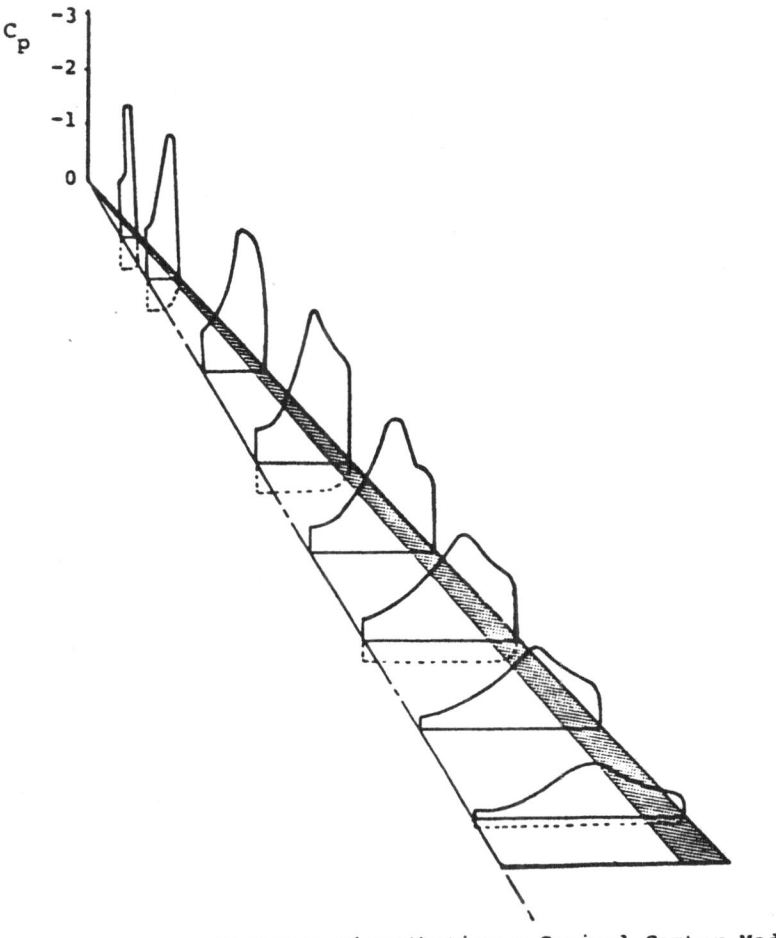

Pressure Distribution – Conical Camber Model,
$\alpha = 30°$, $\beta = 0°$

(B)

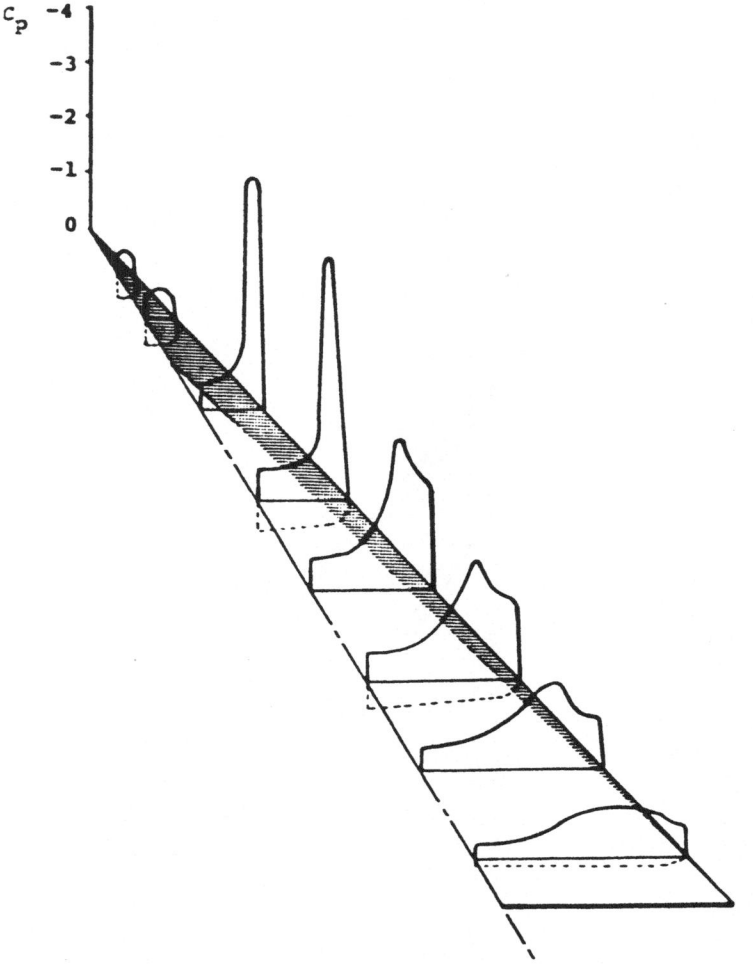

Pressure Distribution - Apex Camber Model,
$\alpha = 30°$, $\beta = 0°$

(C)

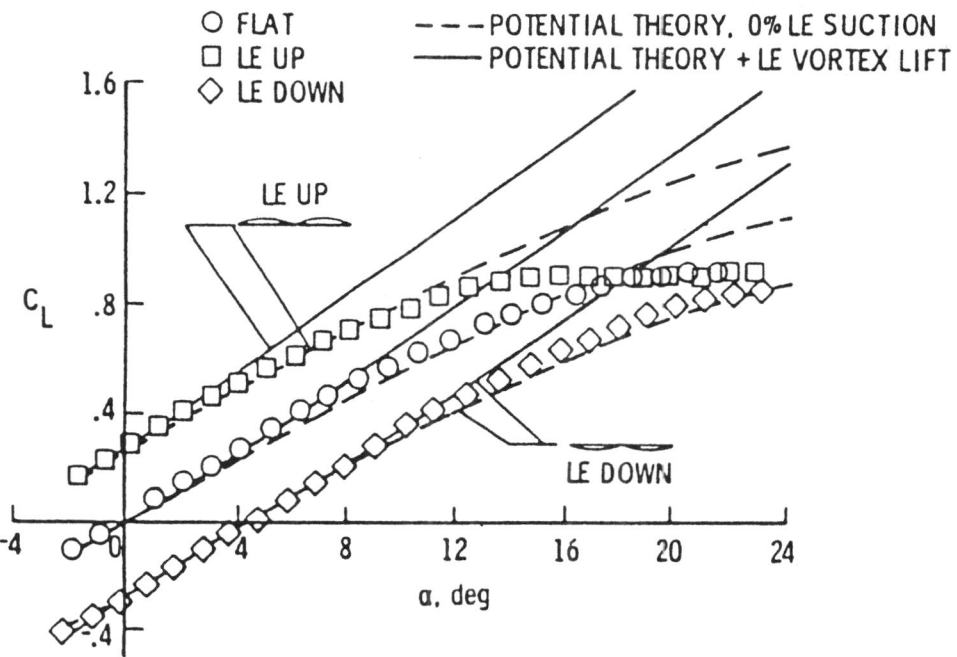

$M_\infty \approx 0$

Lift characteristics for 45-deg delta wing, flat and linearly twisted ("bat wing"), $M_\infty \approx 0$.

Fig.20: Lift coefficient of linearly twisted wings, Lamar [1977]

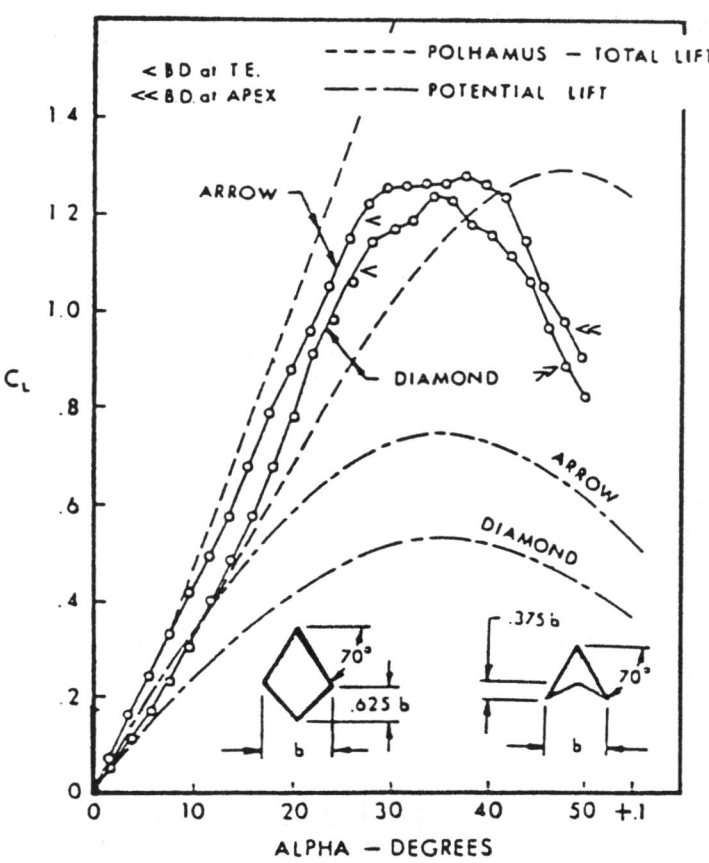

Fig.21: Trailing edge geometry and its effects on lift, Wentz and Kolman [1971]

angle between the perpendicular and parallel velocity along the vortex. However, trailing edge flaps deflected upward and downward can advance or retard vortex breakdown by modifying the adverse pressure gradient which affects the axial vorticity convection.

3.4 Reynolds Number

Elle (1961) studied the location of a vortex over sharp edged delta wings in both water and air and concluded that the flow is insensitive to Reynolds number. Fig. 22 shows a compilation of data by Erickson (1982) taken from water and wind tunnels and in flight. These results confirmed previous experiments which showed that vortex location and breakdown is governed by an inviscid mechanism. Lift data (Fig. 23) taken by Lee, Shih and Ho (1987) further show that Reynolds number insensitivity extends to aerodynamic forces acting on the flow. They also show that the vortex breakdown location is not a function of the Reynolds number for sharp edged wings. However, Erickson (1982) suggested that the flow is insensitive to Reynolds number for sharp edged wings only because the separation is fixed along the edge. He reasoned that wings with round leading edges and flaps would still be sensitive to boundary layer laminar/turbulent transition effects. In addition, the strength and location of the secondary separation on the surface induced by the primary vortex would also be affected since this phenomenon is also viscous in nature. His arguments were substantiated by Lee (1955) who showed variations in the secondary separation line on the surface through oil-film visualizations from $Re = 5 \times 10^5$ to 2×10^6.

The size of the L. E. vortex is independent of Reynolds number. As the freestream velocity increases, the viscous core decreases in size due to a thinning of the boundary layer at the leading edge. This effect by itself, apparently is not strong enough to affect the vorticity balance which is governed by the swirl angle and pressure gradient. A more in-depth discussion of Reynolds number sensitivity over delta wings is given by Payne (1987). In vortex tube experiments (Escudier and Zehnder 1982), the Reynolds number was a dominating parameter. It should be noted that there are fundamental differences between vortices generated in a tube and over a delta wing. As pointed out by Leibovich (1984) vorticity is shed into the center of the tube by a vane type generator producing a

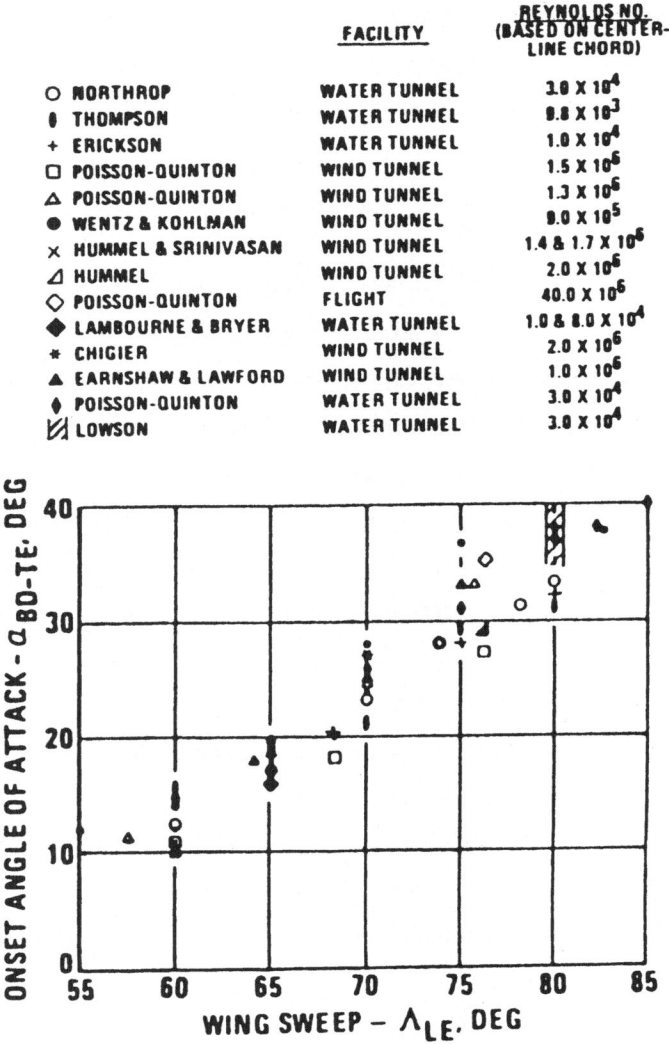

	FACILITY	REYNOLDS NO. (BASED ON CENTER-LINE CHORD)
O NORTHROP	WATER TUNNEL	1.0×10^4
● THOMPSON	WATER TUNNEL	8.8×10^3
+ ERICKSON	WATER TUNNEL	1.0×10^4
□ POISSON-QUINTON	WIND TUNNEL	1.5×10^5
△ POISSON-QUINTON	WIND TUNNEL	1.3×10^6
● WENTZ & KOHLMAN	WIND TUNNEL	9.0×10^5
× HUMMEL & SRINIVASAN	WIND TUNNEL	$1.4 \ \& \ 1.7 \times 10^6$
◿ HUMMEL	WIND TUNNEL	2.0×10^6
◇ POISSON-QUINTON	FLIGHT	40.0×10^6
◆ LAMBOURNE & BRYER	WATER TUNNEL	$1.0 \ \& \ 8.0 \times 10^4$
✳ CHIGIER	WIND TUNNEL	2.0×10^6
▲ EARNSHAW & LAWFORD	WIND TUNNEL	1.0×10^6
◆ POISSON-QUINTON	WATER TUNNEL	3.0×10^4
▨ LOWSON	WATER TUNNEL	3.0×10^4

(A)

Fig.22: Effects of Reynolds number on the vortex breakdown and vortex positions, Erickson [1982]

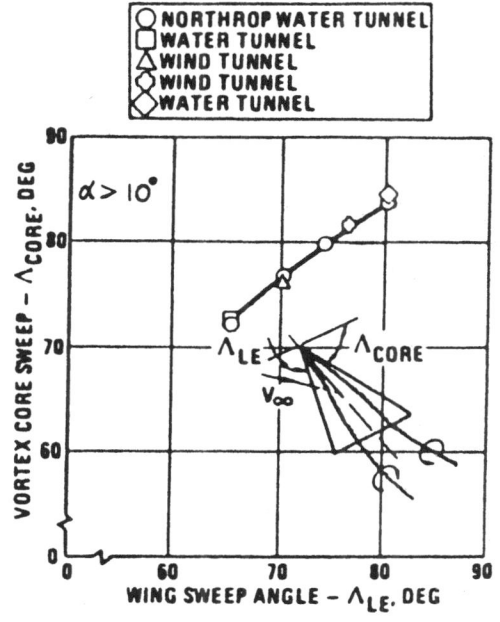

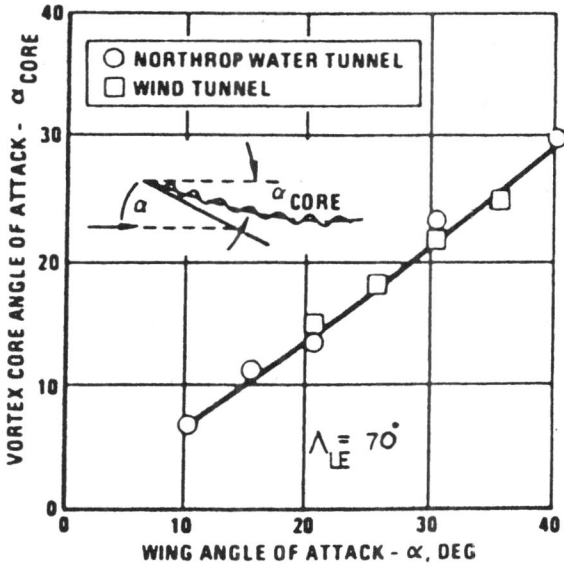

(B)

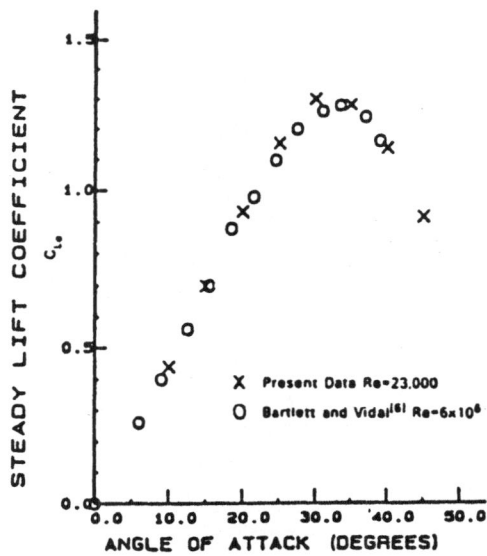

Fig.23: Effects of Reynolds number on lift coefficient, Lee, Shih and Ho [1987]

spiral vortex with constant vorticity. On a delta wing vorticity is constantly fed into the core from the leading edge resulting in an almost linear increase of vorticity along the vortex. Wedmeyer (1982) found that velocity profiles measured in vane type generators did not compare well with those measured over delta wings.

3.5 Mach Number

In the supersonic regime, the separated leading edge vortex is replaced by a serious of attached shock waves and Prandtl-Meyer type expansions depending on Mach number, angle of attack and wing geometry. Stanbrook and Squire (1964) showed that the important parameters in supersonic flow is the angle of attack and Mach number perpendicular to the leading edge. They found that the flow can take on three different forms as illustrated by Squire (1976) in Fig. 24. Later on, Wood and Miller (1985) further classified the flow into six regimes depending on the existence and nature of shock induced separation on the top surface. It is interesting to note that leading edge separation reappears at a large enough angles of attack regardless of Mach number.

Polhamus (1971) modified his theory and predicted a drop in lift with Mach number because the separation line on the pressure surface would gradually move outboard toward the leading edge resulting in a weaker vortex (Fig. 25). The vortex would totally vanish when the stagnation line reached the leading edge where no flow reversal occurred. This corresponds to the situation when the Mach cone coincides with the leading edge at the Stanbrook-Squire Boundary between regions A and B in Fig. 24. Therefore, this effect would be more severe for delta wings with smaller sweepback angles. In addition, Squire, Jones and Stanbrook (1961) pointed out that the lift would be further lowered at large angles of attack and high Mach numbers when the low pressure region in the vortex core reaches the vacuum limit. A comparison between the theory and experimental data is given by Polhamus (1971) in Fig. 26.

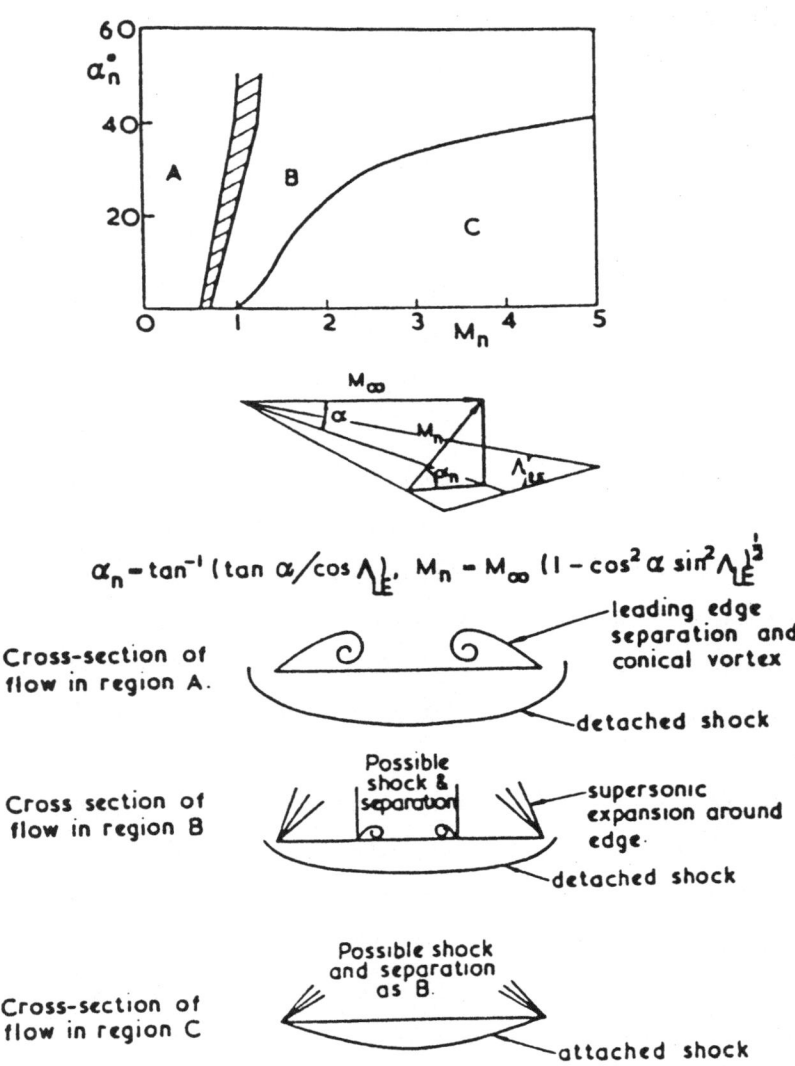

$$\alpha_n = \tan^{-1} \left(\tan \alpha / \cos \Lambda_{LE} \right), \quad M_n = M_\infty \left(1 - \cos^2 \alpha \, \sin^2 \Lambda_{LE} \right)^{\frac{1}{2}}$$

Cross-section of flow in region A.

leading edge separation and conical vortex

detached shock

Cross section of flow in region B

Possible shock & separation

supersonic expansion around edge.

detached shock

Cross-section of flow in region C

Possible shock and separation as B.

attached shock

Fig.24: Flow field of a supersonic delta wing, Squire [1976]

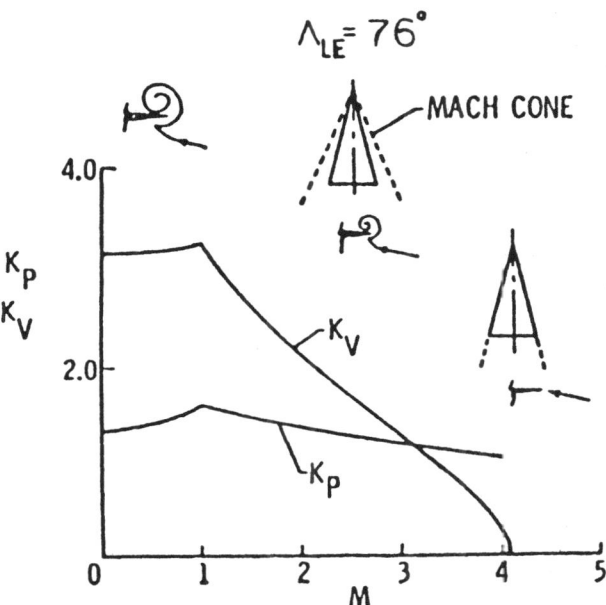

Fig.25: Leading edge suction analogy for a supersonic delta wing, Polhamus [1971]

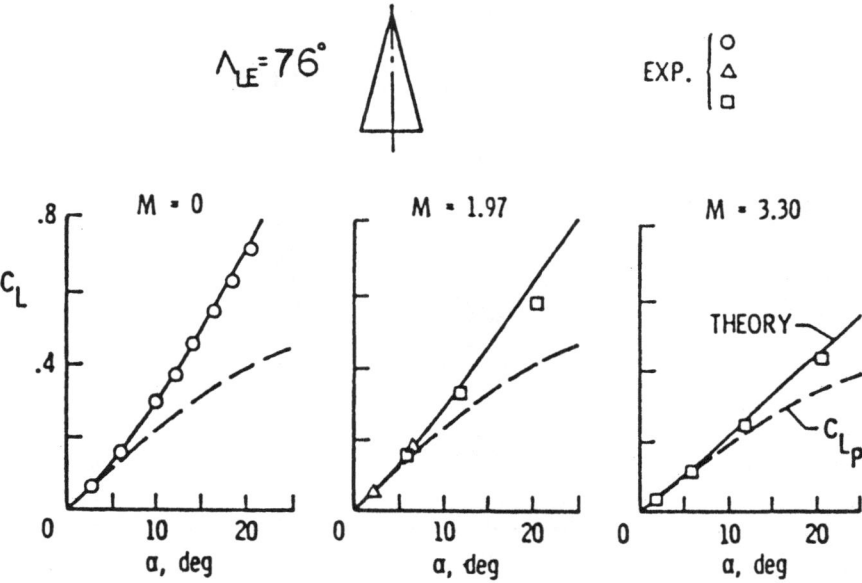

Fig.26: Comparison of theory an experiment of supersonic delta wings, Polhamus [1971]

3.6 Freestream Disturbance

Very little work is reported in the literature concerning freestream effects on the vortex. Lambourne and Bryer (1961) found no effect on the vortex when a spoiler was placed either at or in front of the apex. However, Lee, Shih and Ho(1987) found that the breakdown location became very unstable when freestream turbulence was increased from 0.5% to 1.5%. It appears that more work is required in this area.

IV. UNSTEADY DELTA WING FLOW

Experimental investigation of unsteady delta wing aerodynamics has been limited due to the difficulty in producing well controlled time-varying free streams in the laboratory. The unsteady freestream needs to be produced by accelerating or decelerating the whole fluid mass in the test section. The interaction between the control device and non-linear characteristics of the pump or the blower is non-trivial. In a vertical water channel, Shih, Lee and Ho (1987) was able to achieve various velocity waveforms by operating it in a constant head mode. In this case, the non-linear feature of the pump is isolated. Most other experiments subject the wing to some form of periodic motion in a steady freestream. Pitching involves varying the angle of attack by pivoting the wing about a certain chordwise location. This motion produces a continuous change not only in the angle of attack, but also in the effective free-stream velocity approaching the flow. Other modes are the plunging and heaving motion, where the airfoil is in up-down or forward-back movements. Both modes can produce stepwise or continuous change of the effective angle of attack. Wing rock, another unsteady phenomenon which involves back and forth rolling about the centerline axis, has been observed in real flight. This and other unsteady effects can drastically influence the performance of delta wings.

The response time of the unsteady wing depends on the mechanism governing the stability of the leading edge vortex. The vorticity balance concept dictates two time scales. First, changes in the vorticity generation along the leading edge is transmitted to the core in one local turn-over time. This time is maximum at the trailing edge and decreases towards the apex. Therefore, the flow around the apex is more sensitive to disturbance than any other place along the leading edge. Second, upstream disturbances is convected throughout the vortex in a time, C/U, which scales with the streamwise velocity along the vortex. Since the latter time is always longer, streamwise convection becomes the limiting factor for the vortex to respond to any imposed disturbance. In the following sections, we will look at how these time scales play a role in different types of unsteady motion.

4.1 Pitching, Plunging and Heaving

In a flow visualization experiment with a pitching delta wing, the two leading edge vortices were observed (Gad-el-Hak and Ho 1985, Gad-el-Hak 1987)to roll up at the trailing edge tip and migrated toward the apex. An organized wake behind the wing was formed when the reduced oscillating period, UT/C, equalled 1. In a series of plunging experiments, Lambourne, Bryer and Maybrey (1969) also found that the vortex required a time period $UT/C = 1$ to reach its new equilibrium position. Similar results by Maltby et al. (1963) were reported for a heaving delta wing. These experiments suggest that the slow convection along the vortex limits the response time of the flow such that all periodic disturbances should scale with the characteristic time of C/U as discussed above. In a related experiment, Patel (1980) subjected the delta wing to vertical gusts generated by a movable section of his wind tunnel. Fig. 27 shows the measured lift amplitude and phase as a function of the oscillating periods which were longer than the convection limit. The independence of these parameters on the oscillating frequency suggests that the vortex delay effect was not important in this frequency range and that the potential flow dominated. The constant phase lag which appeared in the data was probably due to the difference between the convection speed of the vertical disturbance and the freestream velocity as mentioned by the author.

4.2 Unsteady Freestream

Freymuth (1987) used titanium tetrachloride to visualize the response of a delta wing in a flow starting from zero speed. The sequence of pictures showed the formation of

405

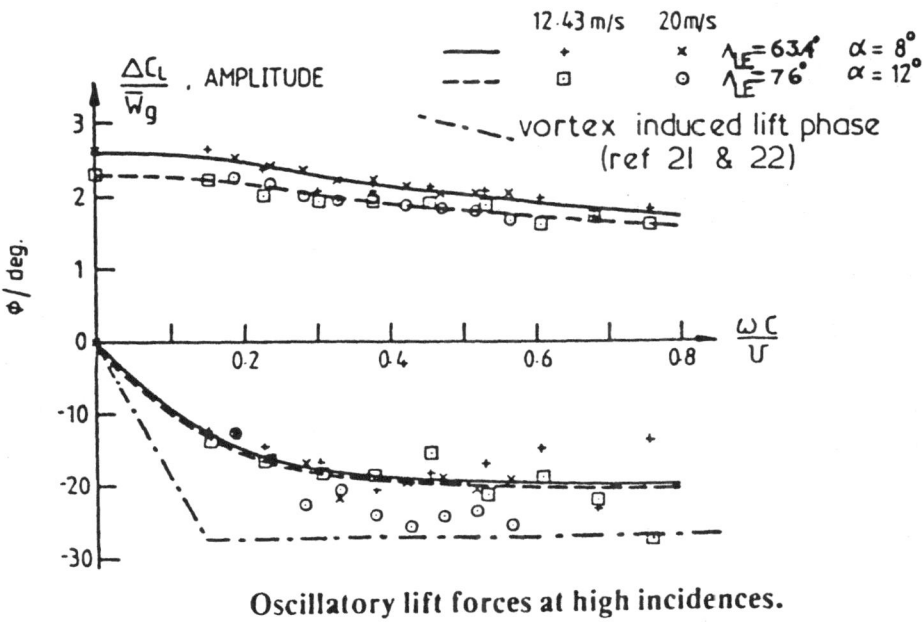

Oscillatory lift forces at high incidences.

Fig.27: Amplitude and phase angle of lift coefficient as the function of gust frequency, Patel [1980]

leading edge vortices in an accelerating flow. Lambourne and Bryer (1961) showed in their analysis that any pressure gradient in the freestream is magnified in the vortex core. If a delta wing is subjected to an accelerating flow, vortex breakdown should be delayed since a favorable pressure gradient would be imposed on the flow. However, Lambourne and Bryer observed that the breakdown location moved upstream when the flow in their wind tunnel was accelerating and vice versa during deceleration. This result was confirmed by Lee, Shih and Ho (1987). When the freestream returned to a steady speed, the bursting point returned to its usual position. To explain these observations, we recall that a favorable pressure gradient increases surface flux of vorticity in the boundary layer at the leading edge resulting in an increase in peak vorticity in the core. This increase in surface flux of vorticity produces a larger swirl angle causing the breakdown to migrate upstream. It appears that the usual adverse pressure gradient effect associated with vortex bursting disturbs the vortex by locally reducing the convection of vorticity along the core, thereby disrupting the vorticity balance.

When Lee, Shih and Ho (1987) imposed a *saw-toothed* free-stream velocity oscillation in the water tunnel, they did not observe any significant change in the breakdown location until the reduced oscillation period was less than $UT/C = 2$. The lift of the delta wing was measured to scale with the instantaneous acceleration with no phase lag as shown in Fig. 28, indicating that potential flow is important in this frequency range. These results again indicated that unsteady vortex effects have not come into play since the vortices have ample time to respond to the imposed disturbances.

4.3 Wing Rock

The phenomenon of wing rock have been observed during flights of aircraft with a delta wing platform marked by a sustained large amplitude oscillatory motion (Hwang and Pi 1979). A time history of the normal force and roll angle of an 80° delta wing tested in the wind tunnel is shown in Fig. 29 by Levin and Katz (1984). Whether a delta wing exhibited wing rock behavior or not depended on initial roll and angle of attack (Fig. 30). The roll amplitude and oscillation frequency were observed to depend on the angle of attack and freestream velocity as shown in Fig. 31. Ericsson (1984), suggested

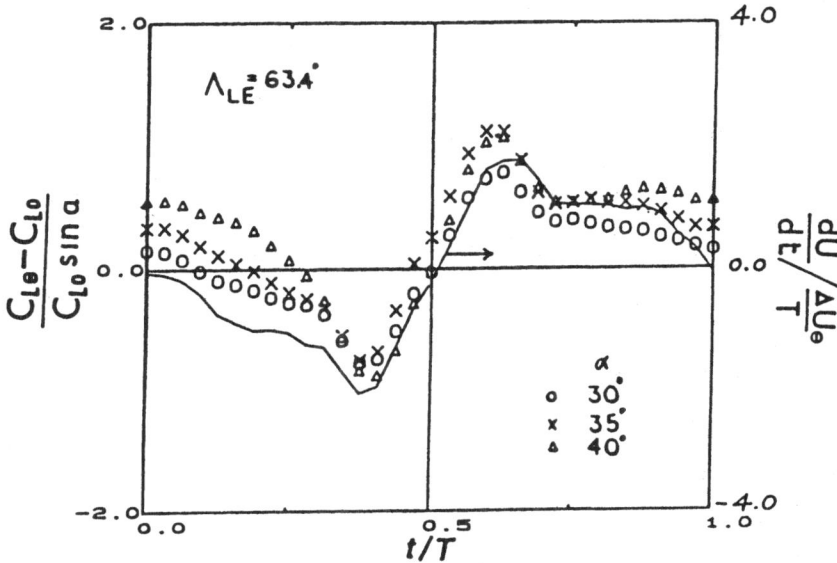

Fig.28: Unsteady lift in a time-varying flow, Lee, Shih and Ho [1987]

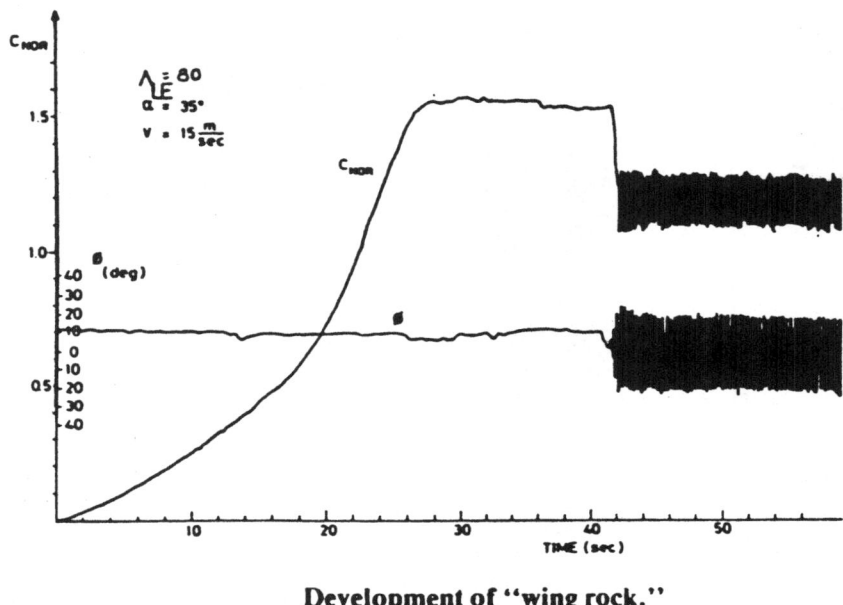

Development of "wing rock."

Fig.29: Time history of normal force and roll angle at the onset of wing rocking, Levin and Katz [1984]

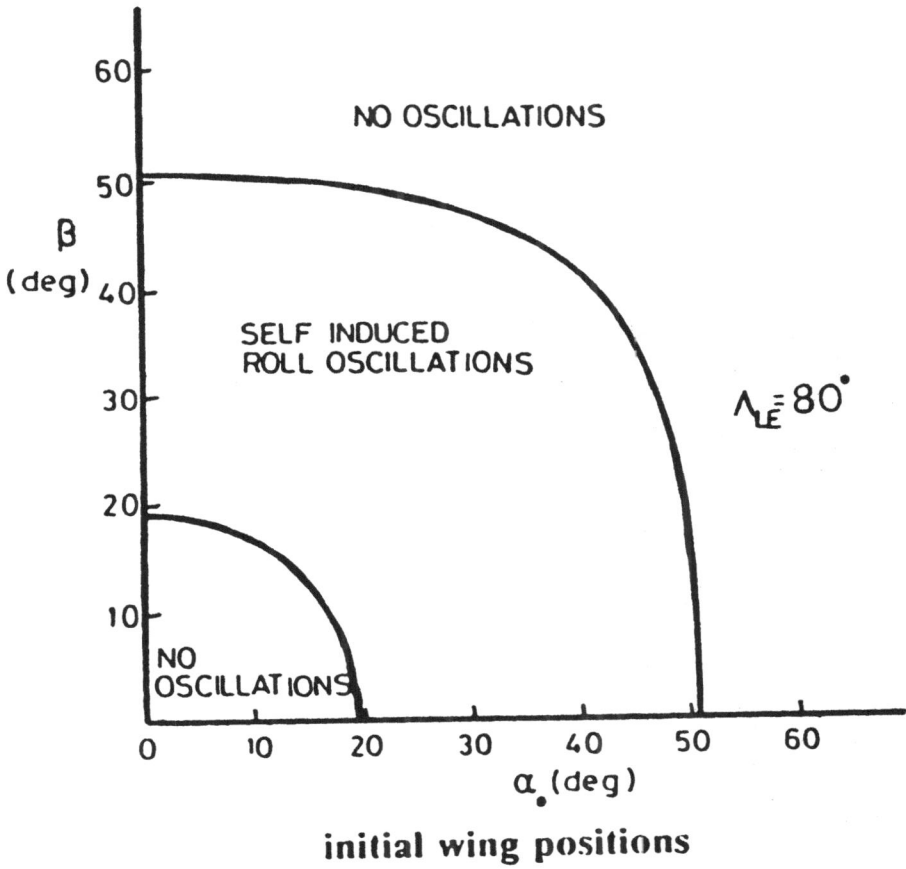

Fig.30: Wing rocking as a function of its initial condition, Levin and Katz [1984]

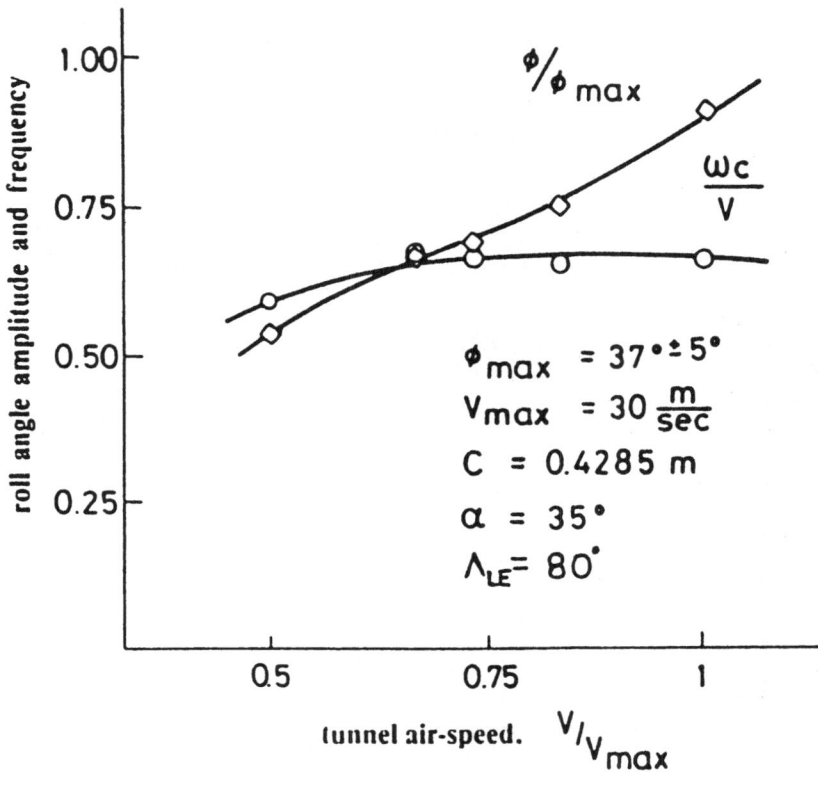

(A)

Fig.31: [a] Roll angle amplitude and frequency as a function of velocity, [b] Roll angle

amplitude and frequency as a function of angle of attack, Levin and Katz [1984]

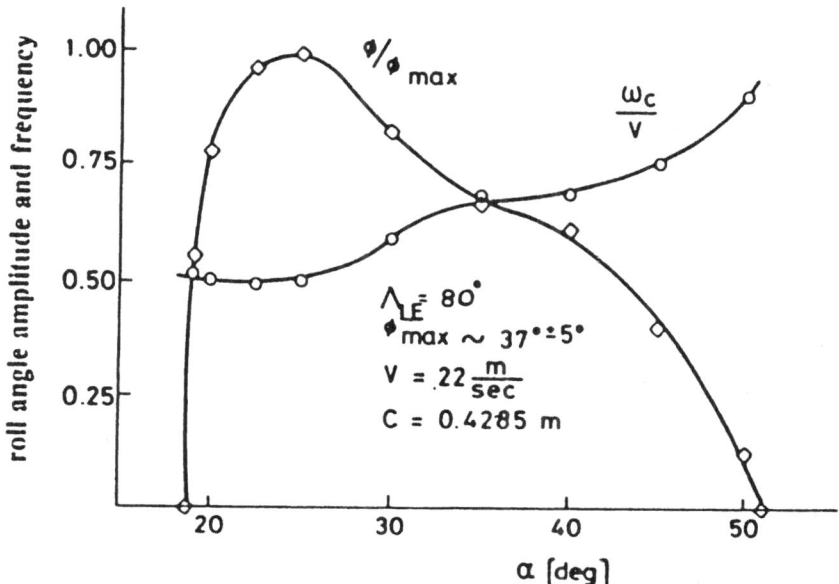

(B)

that asymmetric vortex shedding rather than vortex breakdown is the key mechanism leading to wing rock. He argued that vortex breakdown cannot produce wing rock because Levin and Katz (1984) observed wing roll before vortex breakdown and that the phenomenon is known to occur only on very slender wings. In addition, a loss of lift over the side of the wing with vortex breakdown produces roll which increases the effective angle of attack. Consequently, the breakdown location is advanced farther upstream resulting in no restoring moment. In the case of asymmetric vortex shedding, a limit cycle mechanism is produced when the wing rolls to one side resulting in an increase in the effective apex angle based on the regime chart in Fig. 32. This motion momentarily reduces the tendency for the vortex to shed asymmetrically. The leading edge vortex re-forms on the wing at a later time due to the vortex time lag effect discussed earlier and produces the necessary restoring rolling moment.

Another way of looking at the vortex shedding mechanism based on vorticity balance is as follows. When the side of the wing tilts downward due to roll, the vorticity generation is increased due to an increase in the local pressure gradient at the leading edge as the stagnation point shifts in the opposite direction. This leads to the formation of a stronger vortex which can resist convection downstream. The formation time is again determined by C/U which sets a limit of oscillation frequency of $\omega C/U = 1$ as indicated in Fig. 31. On the other side of the airfoil, the vortex is washed downstream due to an increase in convection as the leading edge vortex is exposed more to the freestream. This cycle is repeated when the wing rolls again in the opposite direction.

V. CONTROL OF LIFT

All the different schemes used to control the flow over delta wings reported here attempt to modify the leading edge vortex. Most methods can be classified under blowing, suction and mechanical flaps applied at strategic locations on the suction surface either steady or in an unsteady fashion. A discussion on potential applications of vortex flaps is given by Lamar and Campbell (1984). Another scheme which does not fall in the categories mentioned above is the effect of density and viscosity variations through heating. Although some methods produce better performance characteristics than others, no one

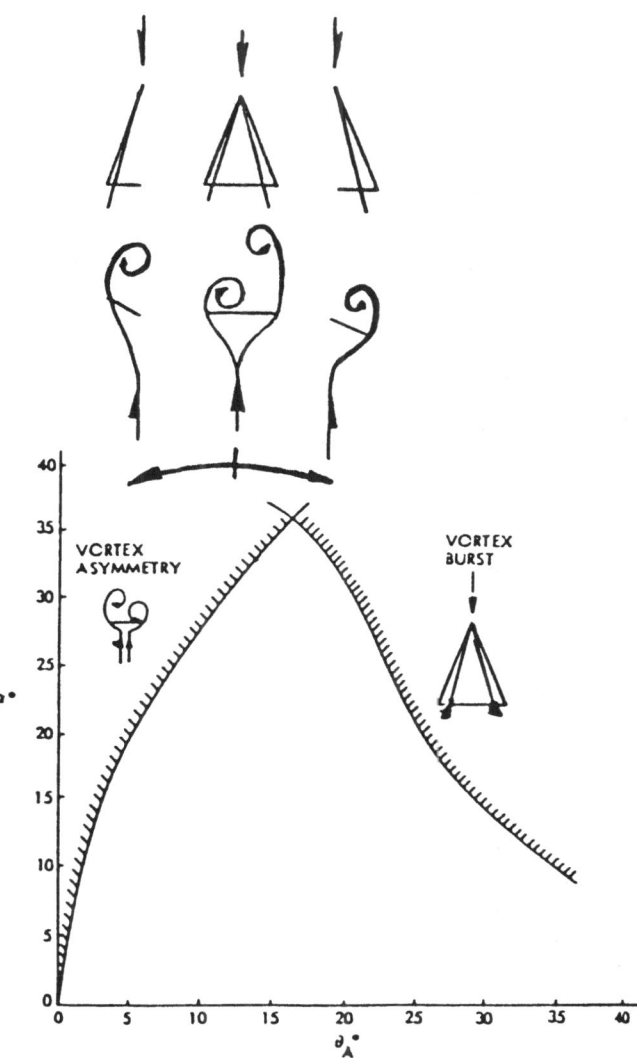

Boundaries for vortex asymmetry and vortex burst

Fig.32: Boundaries for vortex asymmetry and vortex burst. Ericsson [1984]

method is able to avoid some sort of penalty either because of increased drag or difficulty in implementation under flight conditions.

5.1 Blowing

Various attempts have been made to alter the separation vortex by blowing along the leading edge in different directions. Trebble (1966) was able to enhance the lift by blowing outward away from the wing along the edge. The effect of blowing produced a stronger vortex located further outboard. An increase in drag was also measured due to the reverse thrust generated by part of the high momentum fluid directed upstream. Bradley and Wray (1974) and Campbell (1976) achieved higher lift, a delay in stall and better drag polar as a result of spanwise blowing along the leading edge. Flow visualization pictures taken by Bradley and Wray (1974) of the vortex which exhibited a more coherent core and delay in breakdown is due to the increase in axial convection. When compared with Polhamus' theory, full vortex lift was achieved beyond the normal angle of attack for maximum lift (Fig. 33). Favorable results were also achieved by Wood and Roberts (1987) who directed the fluid tangentially upward past the round leading edge of a half delta wing. The pressure measurements shown in Fig. 34 suggest that leading edge separation was delayed for a small angle of attack and that the vortex was strengthened and localized at large angles. The reason for this is the increase in vorticity generation prior to separation due to changes in the local pressure gradient. Another method was attempted by Gad-el-Hak and Blackwelder (1987) who applied periodic blowing and suction at the leading edge. Although flow visualizations showed that the secondary vortices on the shear layer were more organized, no pressure data was available to deduce the effects on the aerodynamic forces acting on the wing.

5.2 Suction

Hummel (1967) investigated the effect of applying suction at the trailing edge and measured a decrease in pressure on the top surface while the bottom pressure profiles were not altered. A general increase in lift at high angle of attack was observed with no gain

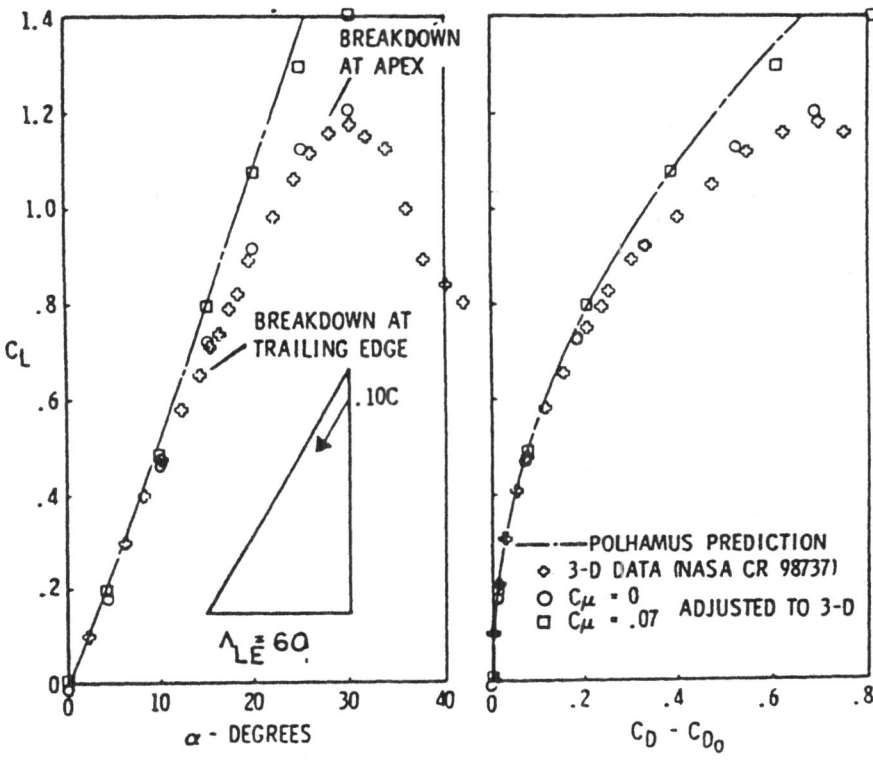

Fig.33: Measurements of lift coefficient with axial blowing, Bradley and Wray [1974]

416

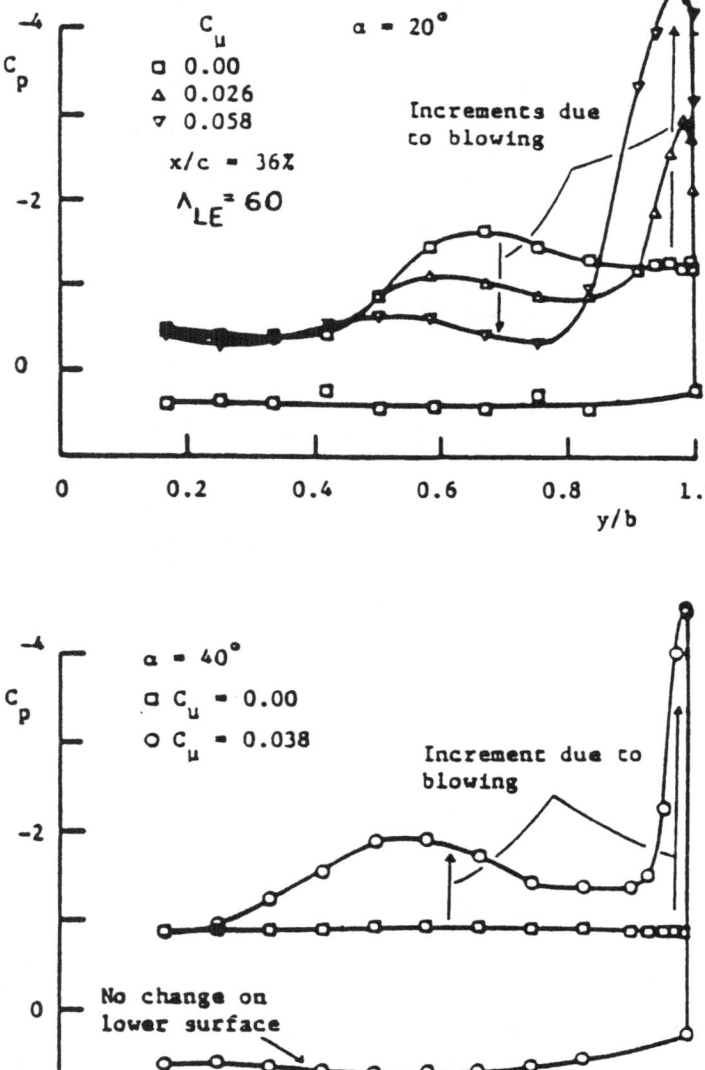

Fig.34: Pressure distribution of delta wing with tangential blowing, Wood and Roberts

[1987]

at small to moderate angles of attack (Fig. 35). This is reasonable since suction only affects the vortex which contributes only a small portion of the lift at small angles of attack. At larger angles, suction at the trailing edge reduces the local adverse pressure gradient and increases axial convection along the vortex. As a result, a delay in vortex breakdown can be expected resulting in higher lift.

5.3 Flaps

Both stationary and moving flaps of different shape and size have been placed at or near the leading edge in an attempt to modify the evolution of the separation vortex. An experiment with stationary flaps was performed by Wahls, Vess and Moskovitz (1986) who placed triangular shaped vertical fences at various locations close to the apex. Their flow visualizations showed the generation of streamwise vortices from the top of the fence. The new vortex eventually intertwined around the original vortex which was shed from the leading edge resulting in premature bursting. Rao and Buter (1983) also generated two pairs of streamwise vortices when they created an apex-flap through upward deflection of the apex at 25% chord. Lift was increased for small angles of attack due to the new leading edge vortices which were generated on the apex and normally would not exist at this angle. At moderate to high angles of attack, lift was lower than the basic wing. Although no visualizations were provided, this loss of lift was probably due to premature bursting of the main vortex as a result of interaction with the apex vortex, as has been observed by Wahls, Vess and Moskovitz (1986). In another attempt, Marchman (1981) investigated the effects of upward deflection of a leading edge flap. The measurements in Fig. 36 shows an increase of lift at low angles of attack but a loss of maximum lift. This is due to an effective increase in the swirl angle as a result of the deflected flaps. In addition to the lowered peak lift, an increase in drag was measured for all angles of attack reported. This type of flow control is probably more useful in conjunction with additional leading edge devices which can reduce the drag penalty. Of the ones tested by Rao and Johnson (1981) for this purpose, the combination of vortex plates and vertical fences produced the most drag reduction by creating a separation zone at the leading edge.

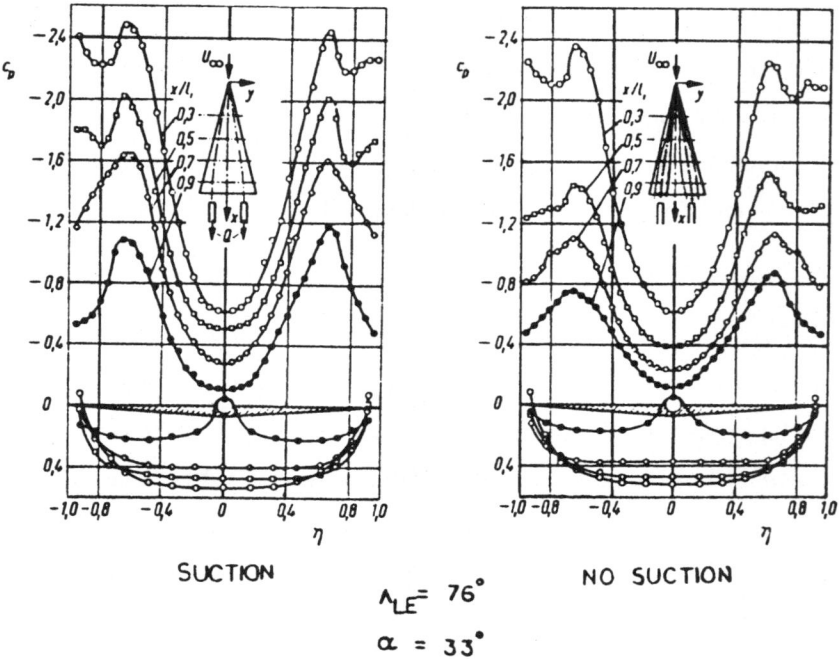

Fig.35: Surface presure distribution with trailing edge suction, Hummel [1967]

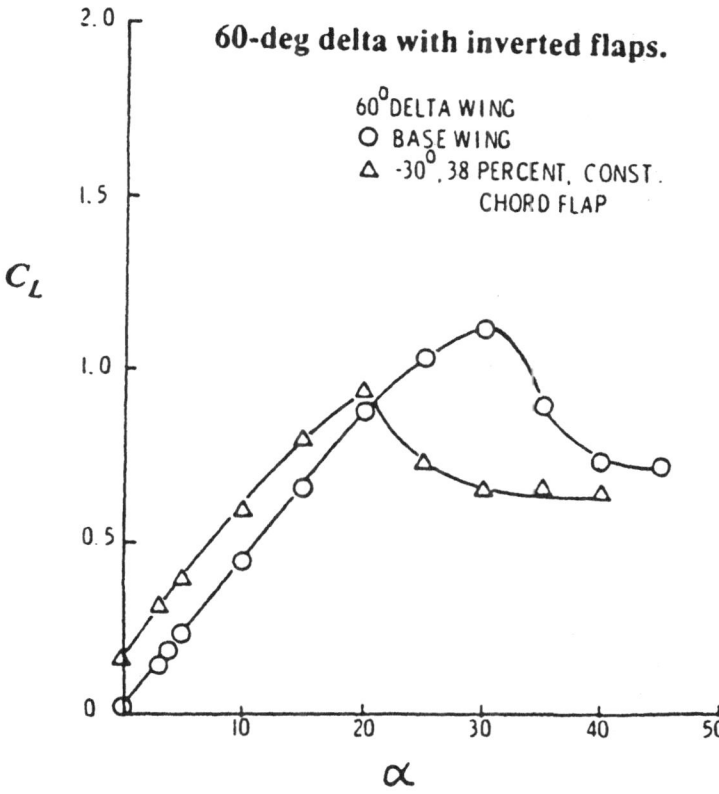

Fig.36: Lift coefficient of a wing with steady inverted flaps, Marchman [1981]

A concept borrowed from insect flight was the flapping delta wing experiment of Spedding, Maxworthy and Rignot (1987). Two triangular shaped extensions hinged along the leading edge were allowed to flap continuously at a frequency much faster than the response time of the original vortex. The idea was to generate a much stronger unsteady vortex through the flapping motion to enhance lift. Fig. 37 shows the increase in circulation over non-flapping delta wings as a function of the flapping frequency. Note that the reduced frequency was based on average radian frequency measured at the mean flap width located at $x/l = 0.5$.

5.4 Heating

Marchman (1975) looked at the effects of heating on delta wing performance. The surface was heated close to twice the free-stream temperature. Fig. 38 shows the drag polar for angles of attack up to 36°. These measurements showed that heating has virtually no affect on the lift and pitching moment suggesting that variations in density and viscosity does not play an important role in the generation of the leading edge vortex. However, an increase of up to 25% in drag was recorded at large angles of attack probably due to the increase viscosity on the boundary layer. This result confirmed the suggestion by Leibovich (1984) that vortex breakdown is governed by inviscid effects.

VI. CONCLUDING REMARKS

A review of experimental data for delta wings under both steady and unsteady conditions has been presented from a vortex dynamics point of view. Vorticity balance provides the framework for understanding the effects of different parameters on the wing lift. Stationary leading edge vortices result from the balance between vorticity surface flux and freestream convection. The surface flux of vorticity depends on the condition of the boundary layer on the leading edge prior to separation, while vorticity convection depends on the component of the freestream along the vortex. These vortices on the suction surface provide a major contribution to the lift of a delta wing. Hence, altering the leading edge vortices is a way to control lift.

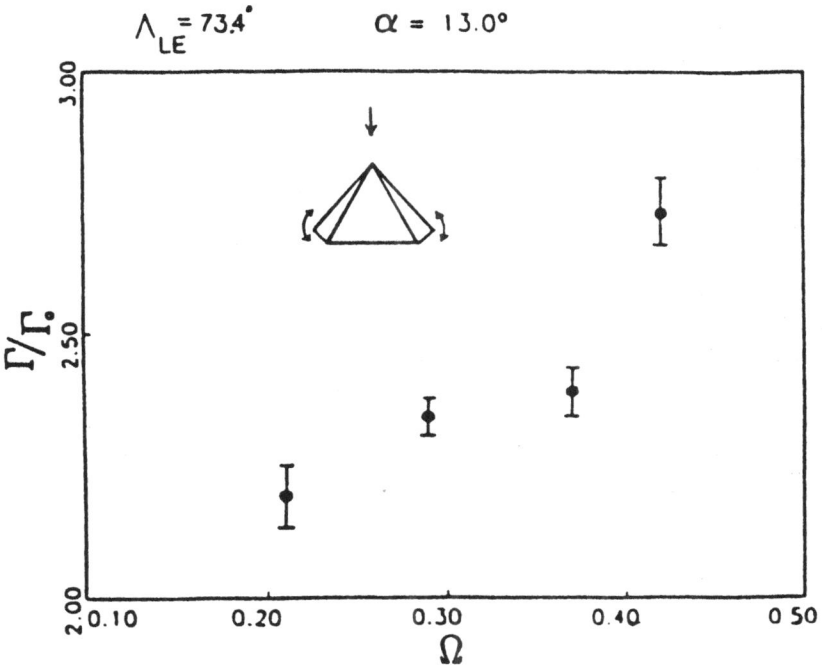

The change in normalised circulation with reduced
frequency for a flap in continuous sinusoidal
oscillation. The measurements were taken at the
maximum flap opening angle. $\beta_{max} = 70°$.

Fig.37   Changes in circulation by unsteady flap. Spedding, Maxworthy and Rignot [1987]

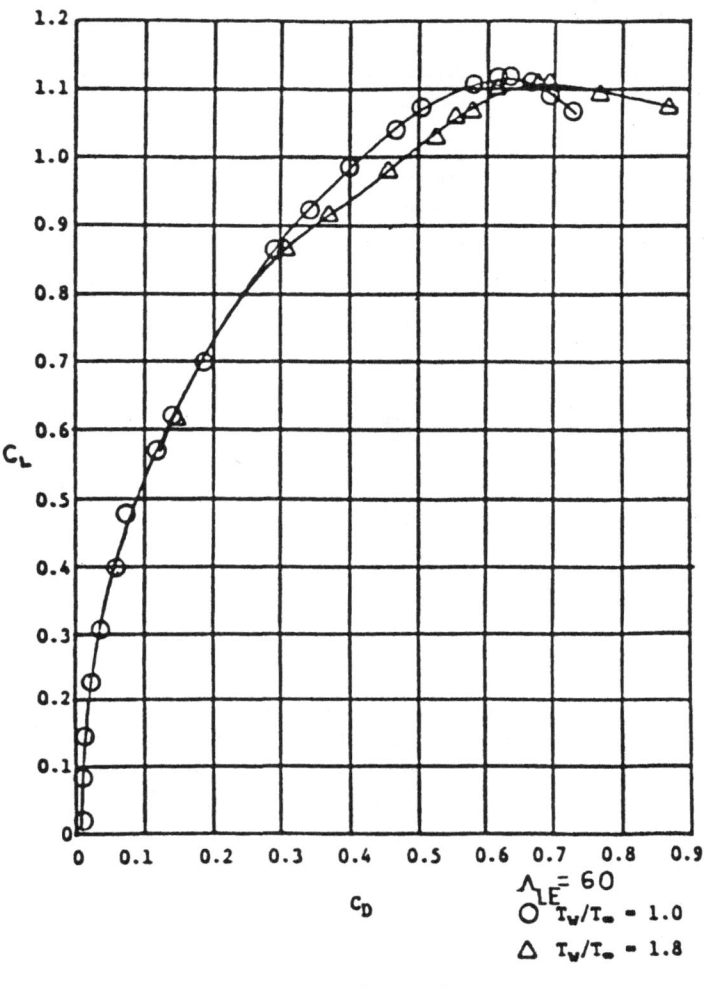

C_L vs C_D.

Fig 38: Lift and drag coefficients of delta wing with surfce heating, Marchman [1975]

ACKNOWLEDGEMENT

This work is supported by the Air Force Office of Scientific Research contract number (AFOSR F49620-85-C-0080). We are indebted to Dr. L.E. Ericsson and Dr. P. Freymuth for their valuable comments.

REFERENCES

Bartlett, G.E. and Vidal, R.J., "Experimental Investigation of Influence of Edge Shape on the Aerodynamic Characteristics of Low Aspect Ratio Wings at Low Speeds", J. Aero. Sci., Vol 22, 1955, pp 517-533. Bradley, R.G. and Wray, W.O., "A Conceptual Study of Leading-Edge-Vortex Enhancement by Blowing", J. Aircraft, Vol.11, No. 1, Jan. 1974, pp 33-38. Campbell, J.F., "Augmentation of Vortex Lift by Spanwise Blowing", J. Aircraft, Vol. 13, No. 9, Sept. 1976, pp 727-732.

Didden, N. and Ho, C.M., "Unsteady Separation in a Boundary Layer Produced by an Impinging Jet", J. of Fluid Mech., Vol. 160, 1985.

Earnshaw, P.B. and Lawford, J.A., "Low-Speed Wind Tunnel Experiments on a Series of Sharp-Edged Delta Wings", Aero. Res. Council, R & M No. 3424, 1964.

Elle, B.J., "An Investigation at Low Speed of the Flow near the Apex of Thin Delta Wings with Sharp Leading Edges", Aero. Res. Council, R & M No. 3176, 1961.

Erickson, G.E., "Water-Tunnel Studies of Leading-Edge Vortices", J. Aircraft, Vol. 19, No. 6, June 1982, pp. 442-448.

Ericsson, L.E., "The Fluid Mechanics of Slender Wing Rock", J. Aircraft, Vol. 21, No. 5, May 1984, pp. 322-328.

Ericsson, L.E. and Reding, J.P., "Approximate Nonlinear Slender Wing Aerodynamics", J. Aircraft, Vol. 14, No. 12, Dec. 1977, pp. 1197-1204.

Escudier, M.P. and Zehnder, N., "Vortex-Flow Regimes", J. Fluid Mech., Vol. 115, 1982, pp 105-121.

Fink, P.T. and Taylor, J., "Some Early Experiments on Vortex Separation", Aero. Res. Council, R & M No. 3489, 1967.

Freymuth, P., "Further Visualization of Combined Wing Tip and Starting Vortex Systems", AIAA J., Vol.25, No. 9, Sept. 1987, pp. 1153-1159.

Gad-el-Hak, M., "Unsteady Separation on Lifting Surfaces", Appl. Mech. Rev., Vol. 40, 1987, pp. 441-453.

Gad-el-Hak, M. and Blackwelder, R.F., "Control of the Discrete Vortices from a Delta Wing", AIAA J., Vol 25, 1987, pp. 1042-1049.

Gad-el-Hak, M. and Ho., C.M., "The Pitching Delta Wing", AIAA J., Vol. 23, No. 11, Nov. 1985, pp 1660-1665.

Harvey, L.K., "Some Measurements on a Yawed Slender Delta Wing with Leading-Edge Separation", Aero. Res. Council, R & M No. 3160, 1958.

Hummel, D., "Untersuchungen uber das Aufplatzen der Wirbel an schlanken Deltaflugeln", Z. Flugwiss., Vol. 13, No. 5, 1965, pp 158-168.

Hummel, D., "Zur Umstromung scharfkantiger schlanker Deltaflugel bei grossen Anstellwinkeln", Z. Flugwiss., Vol. 15, No. 10, 1967, pp 376-385.

Hwang, C. and Pi, W.S., "Some Observations on the Mechanism of Aircraft Wing Rock", J. Aircraft, Vol. 16, No. 6, Jun. 1979, pp 366-373.

Lamar, J.E., "Recent Studies of Subsonic Vortex Lift Including Parameters Affecting Stable Leading-Edge Vortex Flow", J. Aircraft, Vol. 14, No. 12, Dec. 1977, pp 1205-1211.

Lamar, J.E. and Campbell, J.F., "Vortex Flaps-Advanced Control Devices for Supercruise Fighters", Aerospace America, Jan. 1984.

Lambourne, N.C. and Bryer, D.W., "The Bursting of Leading-Edge Vortices-Some Observations and Discussion of the Phenomenon", Aero. Res. Council, R & M no. 3282, 1961.

Lambourne, N.C., Bryer, D.W. and Maybrey, J.F.M., "The Behaviour of the Leading-Edge Vortices over a Delta Wing Following a Sudden Change of Incidence", Aero. Res. Council, R & M no. 3645, 1969.

Lee, G.H., "Note on the Flow Around Delta Wings with Sharp Leading Edges", Aero. Res. Council, R & M No. 3070, 1955.

Lee, M., Shih, C. and Ho, C.M., "Response of a Delta Wing in Steady and Unsteady Flow", Proc. Forum on Unsteady Flow Separation, ASME 1987 Fluids Engineering Conference, FED, Vol. 52, pp 19-24.

Leibovich, S., "Vortex Stability and Breakdown: Survey and Extension", AIAA J., Vol. 22, No. 9, Sept. 1984, pp 1192-1206.

Levin, D. and Katz, J., "Dynamic Load Measurements with Delta Wings Undergoing Self-Induced Roll Oscillations", J. Aircraft, Vol. 21, No. 1, Jan. 1984, pp 30-36.

Maltby, R.L., Engler, P.B. and Keating, R.F.A., with addendum by Moss, G.F., "Some Exploratory Measurements by Leading Edge Vortex Positions on a Delta wing Oscillating in Heave", Aero. Res. Council. R & M No. 3410, 1963.

Marchman, J.F., "Effects of Heating on Leading Edge Vortices in Subsonic Flow", J. Aircraft, Vol. 12, No. 12, Feb. 1975, pp 121-123.

Marchman, J.F., "Aerodynamics of Inverted Leading-Edge Flaps on Delta Wings", J. Aircraft, Vol. 18, No. 12, Dec. 1981, pp 1051-1056.

McKernan, J.F. and Nelson, R.C., " An Investigation of the Breakdown of the Leading Edge Vortices on a Delta Wing at High Angles of Attack", AIAA paper no. 83-2114, 1983.

Payne, F.M., "The Structure of Leading Edge Vortex Flows Including Vortex Breakdown", Ph.D. Dissertation, Dept. Aerospace and Mechanical Eng., Univ. of Notre Dame, May 1987.

Payne, F.M., Ng, T.T., Nelson, R. C. and Schiff, L. B., " Visualization and Wake Surveys of Vortical Flow Over a Delta Wing", <u>AIAA J.</u>, Vol. 26, No. 1, Jan, 1988, pp. 137-143.

Parker, A.G., "Aerodynamic Characteristics of Slender Wings with Sharp Leading Edges-A Review", <u>J. Aircraft</u>, Vol. 13, No. 3, March 1976, pp 161-168.

Patel, M.H., "The Delta Wing in Oscillatory Gusts", <u>AIAA J.</u>, Vol. 18, No. 5, May 1980, pp 481-486.

Polhamus, E.C., "Predictions of Vortex-Lift Characteristics by a Leading-Edge-Suction Analogy", <u>J. Aircraft</u>, Vol. 8., No. 4, April 1971, pp 193-199.

Rao, D.M. and Buter, T.A., "Experimental and Computational Studies of a Delta Wing Apex-Flap", AIAA paper no. 83-1815.

Rao, D.M. and Johnson, T.D. Jr., "Investigation of Delta Wing Leading-Edge Devices", <u>J. Aircraft</u>, Vol. 18, No. 3, March 1981, pp 161-167.

Reynolds, W.C. and Carr, L.W., "Review of Unsteady, Driven, Separated Flows", AIAA paper no. 85-0527.

Rossow, V.J., "Lift Enhancement by an Externally Trapped Vortex", <u>J. Aircraft</u>, Vol. 15, No. 9, Sept. 1978, pp 618-625.

Sforza, P.M., Stasi, W., Pazienza, W. and Smorto, M., " Flow Measurements in Leading-Edge Devices", <u>AIAA J.</u>, Vol. 16, March 1978, pp 218-224.

Shih, C., Lee, M. and Ho, C.M., "Control of Separated Flow on a Symmetric Airfoil", <u>Proc. of IUTAM Conf.</u>, Banglore, India, Jan, 1987.

Spedding, G.R., Maxworthy, T. and Rignot, E., "Unsteady Vortex Flows Over Delta Wings", <u>Proc. 2nd. AFOSR Workshop on Unsteady and Separated Flows</u>, Colorado Springs, Colorado, July 1987.

Stanbrook, A, and Squire, L.C., "Possible Types of Flow at Swept Leading Edges", Aeronautical Quarterly, Vol. XV, 1964, pp 72-82.

Squire, L.C., "Flow Regimes over Delta Wings at Supersonic and Hypersonic Speeds", Aeronautical Quarterly, Vol. XXVII, 1976, pp 1-14.

Squire, L.C., Jones, J.G. and Stanbrook, A., "An Experimental Investigation of the Characteristics of some Plane and Cambered 65II° Delta Wings at Mach Numbers from 0.7 to 2.0", Aero. Res. Council, R & M No. 3305, 1961.

Trebble, W.J.G., "Exploratory Investigation of the Effects of Blowing from the Leading Edge of a Delta Wing", Aero. Res. Council, R & M No. 3518, 1966.

Wahls, R.A., Vess, R.J. and Moskovitz, C.A., "Experimental Investigation of Apex Fence Flaps on Delta Wings", J. Aircraft, Vol. 23, No. 10, Oct. 1986, pp 789-797.

Wedemeyer, E., "Vortex Breakdown", AGARD/VKI Lecture Series No.121, March 1982.

Wentz., W.H. Jr., "Effects of Leading Edge Camber on Low Speed Characteristics of Slender Delta Wings", NASA CR-2002, Oct. 1972.

Wentz, W.H. and Kohlman, D.L., "Vortex Breakdown on Slender Sharp-Edged Wings", J. Aircraft, Vol. 8, No. 3, March 1971, pp 156-161.

Wood, R.M. and Miller, D.S., "Fundamental Aerodynamic Characteristics of Delta Wings with Leading-Edge Vortex Flows", J. Aircraft, Vol. 22, No. 6, June 1985, pp 479-485.

Wood, N.J. and Roberts, L., "The Control of Vortical Lift on Delta Wings by Tangential Leading Edge Blowing", AIAA paper No.87-0158.

ACCOMPLISHED INSECT FLIERS

Marvin W. Luttges

Aerospace Engineering Sciences

University of Colorado, #429

Boulder, CO   80309

ABSTRACT

The flight characteristics of dragonflies, and to a lesser extent hawk moths, have been summarized. Wing kinematics, aerodynamic force generation and flow-wing interactions are presented. It is clear that these insects generate and use unsteady separated flow structures to support flight. Prominent vortex-wing interactions are routinely documented in conjunction with significant force generation. The wing geometry and kinematics dictate optimal unsteady flow generation as long as the wingbeat frequencies are maintained within prescribed ranges. The dragonfly appears able to readily switch between the use of unsteady flows and the use of more conventional steady state aerodynamics. The latter is used for gliding, a major element of dragonfly territoriality defense as seen in patrolling. The hawk moth appears to use similar unsteady flow strategies but doesn't exploit gliding. What we believe we have observed is (1) a mechanistic, self-correcting device for creating unsteady flows, (2) a set of devices for using these flows and (3) a set of principles for unsteady flow exploitation by other biological flight systems.

ACCOMPLISHED INSECT FLIERS

Interest in biological flight is shared by all who have watched birds soar lazily in the sky or have observed bees busy collecting nectar. The ease with which they move through the air is a marvel of biological engineering yet to be matched. But, it is a challenge that few engineers are able to completely ignore. The following discussion summarizes much of our recent work directed toward a better understanding of biological flight. In particular, we have elected to study the flight of dragonflies and, to a lesser extent, the flight of hawk moths. Both these insects are relatively accomplished fliers. And, both are attainable as fresh specimens in the local vicinity.

Why study the flight of insects, rather than that of birds or even flying mammals? Insects are more simple. They usually have rather simple wings. That is, wings that have a simple shape and that have few or no deployable devices. Birds, in contrast, can readily change wing shape and can alter the characteristics of the wing by selective deployment of feathers.   Birds can exert muscle control, independently, along the whole wing span. This muscle control can lead to selective deformations of the whole wing planform. It is notable that insects command a richness of flight modes or regimes in the absence of high levels of nervous system control.   They have simple nervous systems with few operational elements or neurons. Again, in contrast, birds have a relatively elaborate nervous system with much larger numbers of neurons. Finally, the insects we have chosen to study are relatively primitive. They appear to have been in existence, relatively unchanged, for approximately 250 million years. This again suggests a level of simplicity that is not shared by birds and it suggests a successful biological design that has endured despite uncountable challenges by competitive organisms over millions of years.

Insects are capable fliers by any standard. Most importantly, many insects possess the agility and maneuverability to completely avoid predation while in flight. Dragonflies can complete most of their major biological functions while in flight. They both feed and mate while flying. To complete these functions, they are able to exhibit hovering that includes rearward motions, they show rapid straight-away flight and they show highly maneuverable darting flight. In the latter case, it is clear that they are able to rapidly change from one maneuvering regime to another (Luttges et al., 1984). This agility permits episodes of rapid flight to be interspersed with those of hovering, twisting and turning in the air. The overall flight characteristics, then, are rather sophisticated yet are supported by modest flight hardware as well as modest flight controls as managed through a primitive ganglionic nervous system.

Despite these properties of insect flight, there are some drawbacks inherent to studies of these organisms. Being relatively small, the Reynolds numbers associated with insect flight mechanisms also are small. Flight episodes are performed in a viscous flow regime and in a regime characterized by transitional flows. The wing kinematics are usually a combination of large scale pitching and plunging motions sometimes made even more complex by the presence of sculling. Together these kinematics support the needed lift and thrust. They can change the vector of these forces as related to the center of gravity of the insect. The kinematics are somewhat difficult to interpret since the wings show little movement at the roots and maximum movement at the tips. Thus, even though the insects represent a special class of comparatively simple fliers that can be studied in some detail, they are still complex in terms of the fight mechanisms that they use and the controls that they are able to exercise over their flight. As will be seen later, there are advantages to being able to study insects of considerable size and of considerable flight sophistication.

A BIOLOGICAL FLIGHT PERSPECTIVE

Flight locomotion is not, of course, limited to insects. Various types of flight are exhibited by insects, reptiles, birds, mammals and fish. The agility and maneuverability varies. Typically, the flight episodes range from gliding to graceful flapping and from high speeds to hovering. Some flight regimes are reserved for special circumstances such as escape from immediate threat while others are the foundation for successful prey seeking and predation. Thus, biological flight has been used as a major locomotion adaptation of organisms in finding successful ways to move about and exploit or survive their environment. Regardless of the uses to which flight is put in the biological realm, it is important to recall that these locomotor adaptations have arisen in various species time after time. The hardware available to support flight has varied accordingly. Thus, it is not unreasonable to expect that biological systems through "trial and error" attempts over several hundred million years have tested many possible means for achieving useful flight.

Compared to engineered systems, biological flight systems have special kinds of design constraints. The adaptations must be generally successful in even the most early and primitive design stage. Otherwise, the organism doesn't survive long enough for additional iterations. This requirement is embedded within the concept of natural selection. It dictates that successful new locomotion designs such as those supporting flight be progressive and successive approximations of the final "optimized design." It is also clear that biological systems have limited power resources regardless of how this power is stored and used. This implies that the efforts or costs of flight be repaid in a fashion that either directly or indirectly benefits the user as a competitive or survival advantage. Finally, the design must be robust in terms of implementation. This simply means that flight design integration with biological needs must not be restrictive and that small implementation or wear flaws not be fatal. Thus, the flight observed in biological systems probably represents most of the means available for achieving flight within the biological domain. But, this is certain to be, for the reasons cited above, a somewhat small subset of all feasible types of flight.

Indeed, simple observations on biological flight systems indicate that each system has a different commitment to the use of flight. A flying fish, for example, uses flight for escape from predators. Insects use it for major migratory activities as well as for many other biological functions. Birds can use it for locomotion that carries them to new feeding areas or they can use it to achieve feeding via prey catching. Only those organisms that are able to use flight efficiently and effectively for their numerous life-sustaining activities do so in ubiquitous fashion. Few organisms are sufficiently accomplished flyers or are positioned in appropriate environmental niches to use flight as an extensive part of their total behavioral repertoire.

In the ensuing description of the flight of the dragonfly it should be noted that this insect uses flight to escape predation, to catch food, to patrol territory, to protect territory, to mate and, even, to lay eggs. Accordingly, much of the time the dragonfly is engaged in flight to serve one or another life- or species-sustaining purpose. Thus, the combination of flight regimes and behaviors must be consistent with other life systems of this insect, including such things as metabolism and thermal regulation needs. Such matches to design specifications are under constant pressure for optimization since each successive generation of a species competing for the same food supply evolves. Successive generations of potential predators evolve, as well. The success of the dragonfly in using the environmental niche in which it finds itself is attested to by the longevity of this species for over 200 million years.

Among all biological systems with the specialized characteristics that we wish to understand and model there is a special realization to be recognized. The characteristics of biological systems are often general "bootstrap" improvements constrained within existing system design. Because of the highly integrative nature of the way these systems are put together, each of the improvements in one system must be counterbalanced against the effect on all other systems. Clearly, a completely new approach to a problem is difficult within a single species. Such new designs are usually left to the development of new species or the development of new characteristics within an existing but different species. Thus, biological systems are characterized as systems in which a great deal of emphasis has been placed on constant but rather small design improvements --- systems in which a good deal of operational enhancement has been attained through years of trial and error testing. These years are coupled with species selection that favors the successful survival and reproduction of individuals having slightly better characteristics than other individuals of the same species existing at the same time.

DRAGONFLY FLIGHT

Most of the emphasis of this review is on dragonfly flight simply because this is the insect with which we have the greatest amount of experience. Comparisons will be made with other insects as appropriate. This is especially important since a considerable amount of research has been done with the Chalcid wasp as well as several other systems (Weis-Fogh, 1973).

Hardware. The dragonflies that we have studied range from 5 to 10 cm in body length. They typically weigh 250 to 450 mg and have wing spans of 6 to 9 cm. Total wing areas vary according to the chord lengths that range from 0.8 to 1.4 cm. The front wing attaches to the thorax and extends outward with a forward sweep angle of approximately 12-15 degrees. The aft wing attaches just behind the fore wing but extends outward with a rearward sweep angle of about 5-8 degrees. As seen in Fig. 1, the translucent wings are comprised of many tiny facets extending between opaque reinforcing members. The opaque lines are the tubular remnants of a complex trachea system and are reinforced with a very tough protein coating of chitin. These venations consist of a large (3% chord diameter) leading edge vein covered with comb tooth projections and lesser veins (< 2% chord) that are more like smooth tubing. As can be seen, the distribution of these venations is probably consistent with the aerodynamic loads normally encountered by these insects.

In side view, the opaque venations of the wings yield significant folds or corrugation in the wing surface. Valleys and ridges are formed between successive venations such that the surface of some wing specimens may show as much as a 5% chord thickness in these corrugated areas. The corrugation effects disappear near the trailing edge of the wings. However, even the tiny translucent facets show horizontal plane angles that differ from one to another.

The aspect ratio of the fore wings is approximately 9 whereas that of the wider aft wing is approximately 7. The widest portion or chord of the fore wing occurs at approximately half the distance along the span. The greatest chord dimension of the aft wing occurs near the wing root. In *Libella* specimens, the average chord of the fore wing is approximately 9.7 mm and the average chord of the aft wing is approximately 11.4 mm. Other dragonflies that have been tested have slightly smaller wing dimensions but the aft wing generally shows a 20% greater chord than the fore wing.

During the course of several years of testing we have used specimens with wings in various states of damage. The wings have been ripped and notched. In some instances dragonflies have been collected with significant shortening of one or another wing. Many specimens have escaped our netting attempts despite these somewhat damaged wings. In other words, we have been led to believe that many of the specific details of wing structure and shape may not be essential to acceptable, stable flight performance. This observation may attest to the robust character of either the flight control system, the aerodynamic system or to the combination of the two.

Dynamics. The wing kinematics of dragonflies consist of the combination of three different motions: pitching, plunging and sculling. The expression of these motions by the front and rear wings occur at slightly different times in the wingbeat

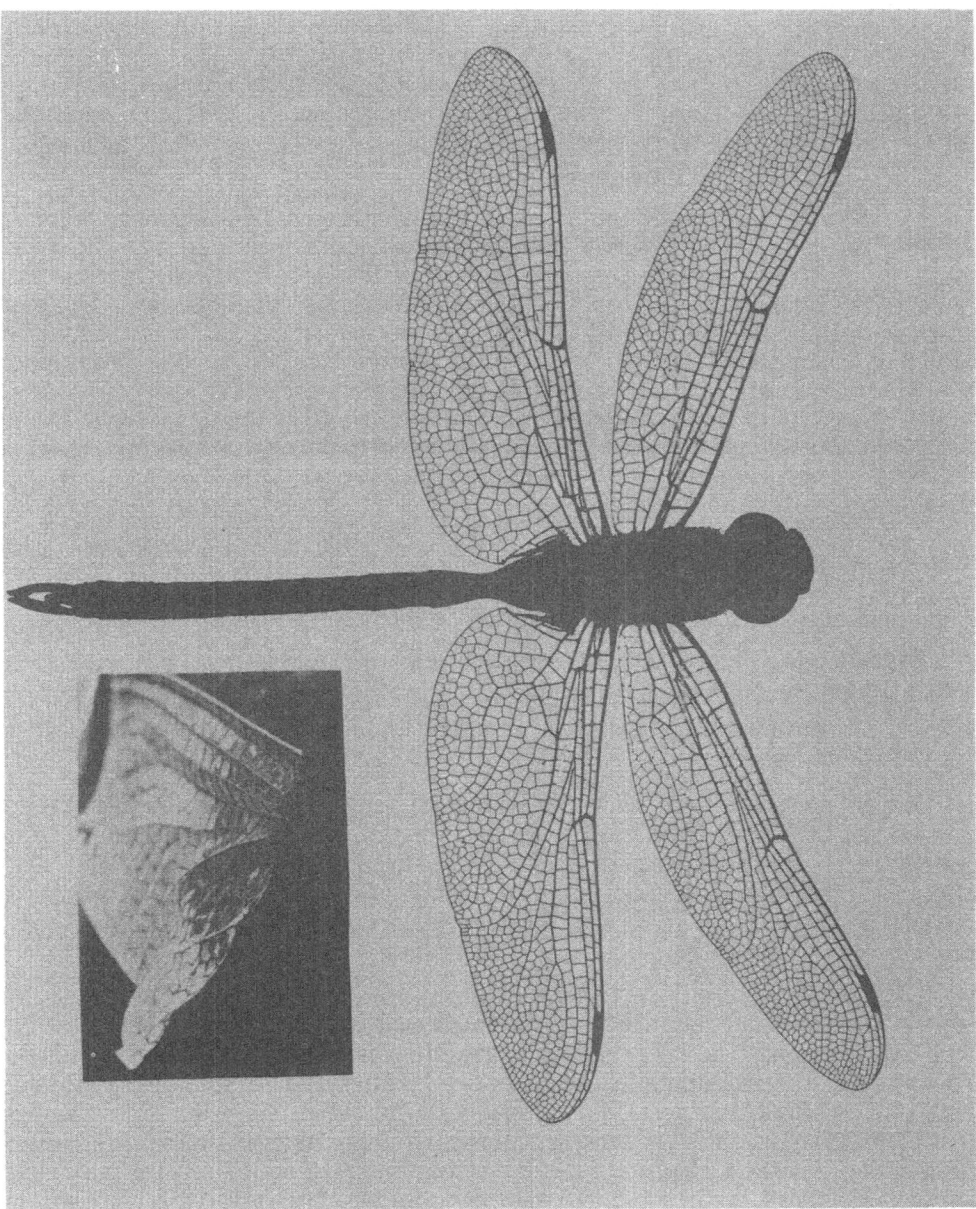

Figure 1. Shadowgraph of dragonfly, <u>Aeschan</u> palmata. Dark spanwise structures of the wings are venations clearly more numerous in the first quarter chord than the remaining chord locations. The small translucent facets are most obvious near the tip and trailing edge of the wings. Most corrugation or folding of the wings occurs at the chord locations having prominent venations(inset).

cycle. The aft wing can lead the motions of the fore wing by almost half a cycle or 180 degrees. This large phase difference has been photographed in hovering dragonflies using high shutter speeds and fast film (Fig. 2). High speed films of tethered dragonfly flight reveal significant phase angles between the fore and aft wings as the aft wing leads the fore wing through each wingbeat cycle. As will be shown later, the phase angles between the fore and aft wings appear to determine very different vectors for the aerodynamic forces generated.

The dominant wing motion in the dragonfly wingbeat is the *plunging* or flapping motion. It carries the wing tips of both the fore and aft wings from well above the horizontal plane of the body to well below it. The fore wing tip moves upward to a 60 degree angle with the horizontal and downward to a 30 degree angle. The rear wing tip traces a 40 degree angle above and another 40 degree angle below the body plane (Somps and Luttges, 1985). In times of escape flight, the amplitude of wing plunging increases. Clearly, the wing tips do not close with those of the wings on the other side of the body during either the upstroke or the downstroke of the wingbeat cycle. At wingbeat frequencies of 25-35 Hz., the total duration of a single wingbeat is approximately 28-40 msec. Wing tip upstroke velocities (~3.5 m/sec) exceeded downstroke velocities (~2.5 m/sec). These differential tip velocities occur whether the dragonflies were tested in a zero flow circumstance or a wind tunnel (V_∞=1-5 m/sec). In all of the tests of tethered dragonflies the observed plunging kinematics were essentially the same as those reported previously for unrestrained dragonflies filmed in natural habitats (Norberg, 1975).

The plunging motion is not a simple motion orthogonal to the plane of the body. Rather, this motion is combined with a promotion and remotion of the tips related to *sculling*. This motion carries both the fore and aft wing tips forward during downward plunging and rearward during upward plunging. On the downstroke the front wing tip moves 20 degrees forward of the root attachment to the body whereas the upstroke moves the front tip slightly more than 40 degrees rearward of the same attachment point. The resulting ellipsoidal paths have been plotted on a spherical surface in Fig. 3 (Saharon and Luttges, 1987). The combined plunging and sculling follow a slightly different path on the downstroke as compared to the upstroke. It is important to recognize that both plunging and sculling motions are anchored to a static point on the thorax.

The *pitching* motion consists of rapid pronation at the beginning of the downstroke and rapid supination at the beginning of the upstroke. The pitching changes the geometric angle of attack of the wings from close to 90 degrees during the upstroke to approximately 0 degrees on the downstroke. Although the front wing pitches upward about 15% more rapidly than the rear wing, the average 20,000 degrees/sec rate for both is quite high compared to those tested using airfoils (cf., Robinson and Luttges, 1984; Robinson and Luttges, 1983; Francis and Luttges, 1984). Since these rapid pitching motions are driven by mechanisms located at the wing roots, each of the motions yields a transient twist to the wing that propagates from the root to the wing tip. Thus, the pitching angle is shared by all spanwise components of the wings. Angles of attack occurring throughout the wingbeat are shown in Fig. 2.

Although these kinematics are complex when compared to conventional non-biological flight or lifting systems, they have been successfully modeled by a simple four bar kinematic device driven by a variable speed motor; with wing pitching

Figure 2. High speed photography of unrestrained dragonflies during hovering. These photos were chosen to show the large phase differences between fore and aft wings. Taken with a telephoto lens, these photographs were collected in instances of hovering when dragonflies were temporarily airborne next to favored reed perches. These and many other photos were used to corroborate the similarity of wing kinematics in unrestrained versus tethered dragonflies tested in the laboratory.

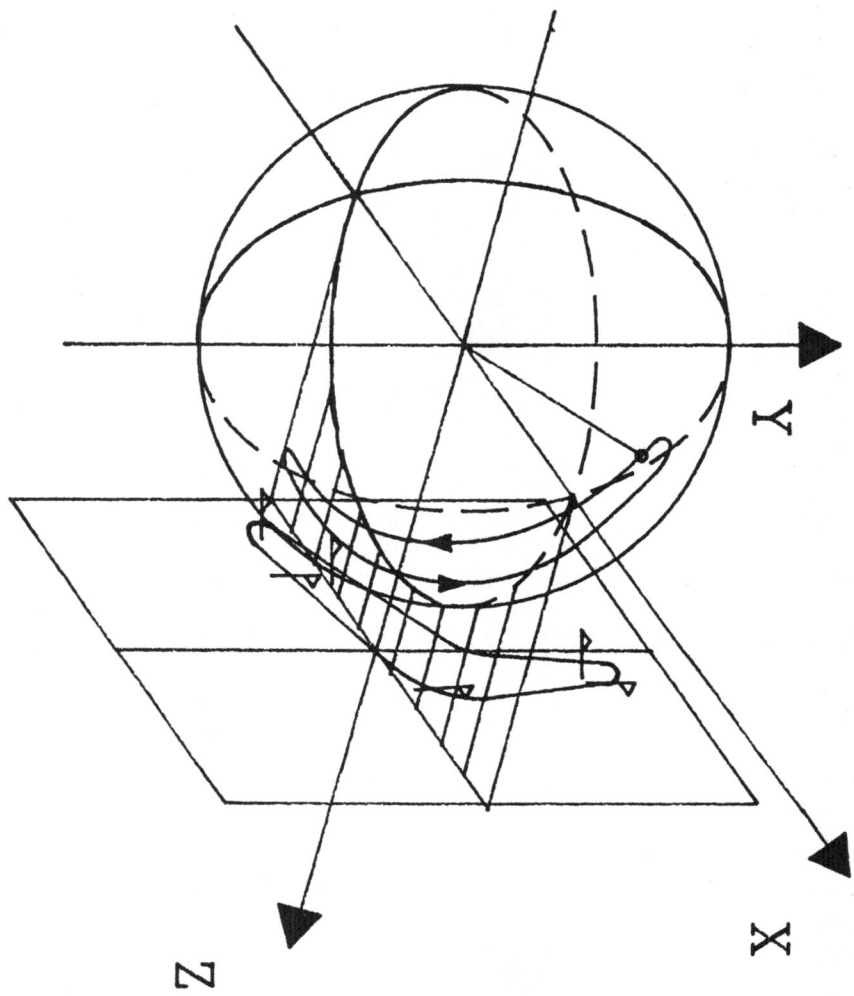

Figure 3. Dragonfly fore wing kinematics. Plunging and sculling are traced over a spherical surface with direction of the wing tip through the cycle indicated by arrows. The planar projection of the kinematics shows angles of wing pitch associated with portions of the beat cycle. A full description is provided elsewhere (Saharon and Luttges, 1987).

modeled separately using rotary solenoids (Saharon and Luttges, 1987). In some of the following comparisons of kinematics and aerodynamic forces, observations made on the model are used to enhance those made using actual dragonfly specimens.

Force Generation. In an attempt to characterize the force generation produced by tethered dragonflies, two different approaches have been used. A single dimension force balance has been used with dragonflies tested in a zero flow apparatus (Somps and Luttges, 1985). Then, a three-dimensional force balance was created (Fig. 4) to evaluate lift, thrust and lateral force generation by dragonflies tested in the wind tunnel (Reavis and Luttges, 1988). Throughout these tests, we were aware of the possibility that tethered dragonflies might exhibit flight different from that of unrestrained dragonflies.

The single dimension balance gave estimates of lift forces generated throughout the full wingbeat cycle. With peak forces of 5-7 grams being measured during some of the wingbeat cycles for specimens averaging 340 mg body weight, it was concluded that the dragonflies could produce transient lift values that were 15-20 times greater than body weights. These high lift values occur as the rear wing is approximately half way through the downstroke and the front wing is just pronating into the downstroke. Later in each wingbeat cycle, brief episodes of negative lift were recorded. Such negative lift was associated with both fore and aft wings moving through respective upstrokes. The large peaks and of aerodynamic lift valleys recorded at a frequency near 30 Hz. in these specimens were deemed inconsistent with the use of standard steady-state aerodynamics.

Three-dimensional force balance measurements were coupled with high speed films of dragonflies tested in the wind tunnel. Through extensive analysis it became clear that the dragonflies were employing several varieties of kinematics both within elicited flight episodes and between different episodes. As the vertical distances between fore and aft wing tips became large (consistent with a larger phase angle between wings), the force balance recorded a preponderance of lift force being generated. When the distances between wing tips remained small, comparatively large thrust forces were in evidence. These dragonfly specimens were *Aeschna palmetta* weighing an average of 600 gms. The maximum lift forces recorded were approximately 3.5 gm so these dragonflies were recorded to produce lift peaks only 5-6 times body weight during wind tunnel tests. It was apparent that simultaneously other aerodynamic forces were being generated, as well. The relation in Fig. 5, shows that across numerous tests the dragonflies produce reciprocal amounts of lift and thrust. With very high amounts of lift being generated, little thrust is recorded. And, with high thrust values little lift is recorded. As might have been expected, a dominance of lift is correlated with high phase angles between the fore and aft wings. Thrust dominance is associated with low fore and aft wing phase angles. As can be seen in Fig. 5, there were many instances in which comparable amounts of lift and thrust were produced.

Under the tethered test conditions, the dragonflies showed small lateral forces that were not strongly associated with either lift or thrust maxima.

One additional relationship was particularly striking. As these dragonflies increased wingbeat frequencies from 33 to 38 Hz. peak lift forces increased. It follows that with increased wingbeat frequencies the thrust forces show decreases relative to lift. Thus, each wingbeat appears to yield a unit value of increasing

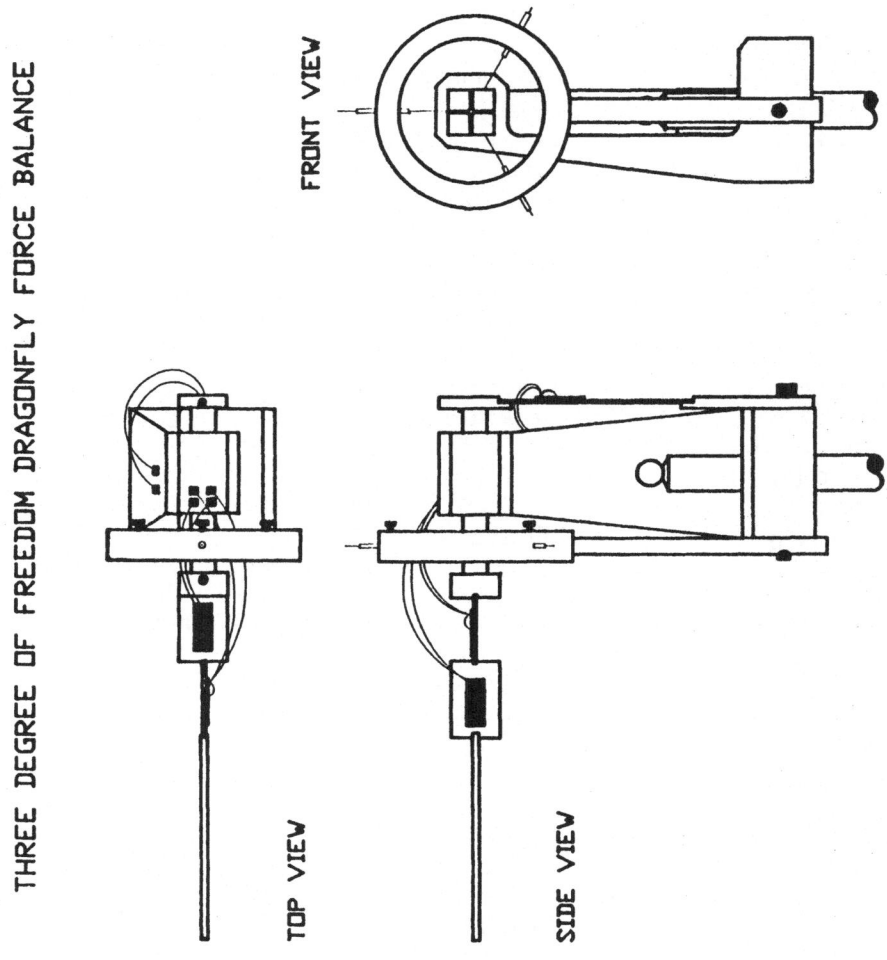

THREE DEGREE OF FREEDOM DRAGONFLY FORCE BALANCE

FRONT VIEW

TOP VIEW

SIDE VIEW

Figure 4. Diagram of dragonfly force balance used for measuring forces produced by tethered dragonflies during wind tunnel testing. Dragonflies previously glued (cyanoacrylate) to wood beams are attached to the hollow "sting" mount. Each strain gage is connected through a bridge to an amplifier. Data of simultaneous outputs from the three strain gages were recorded kymographically from the face of an oscilloscope that also displayed shutter openings from a high speed (500 frames/sec) 35 mm cameras. Force generation was, thus, recorded for the wing kinematics of all flight episodes.

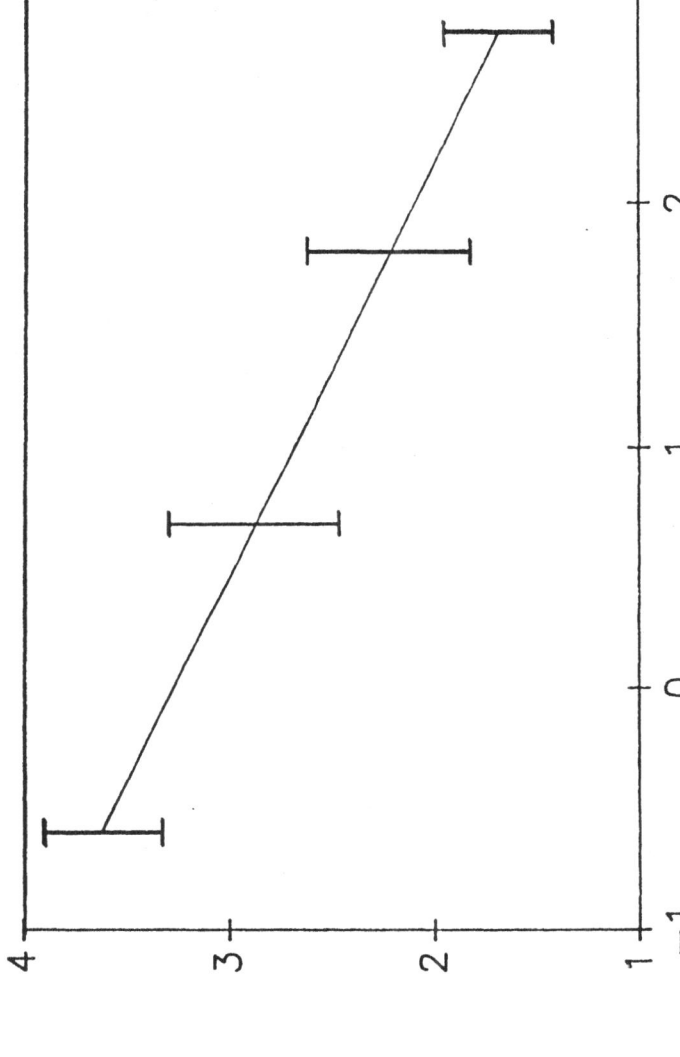

Figure 5. Cross plot of lift and thrust maxima recorded from one dragonfly specimen during wind tunnel testing. The reciprocity between lift and drag is readily seen despite the variance (bars) of the recorded forces.

aerodynamic force dependent on increasing wingbeat frequency. But, the use of this overall force as lift or thrust depends largely on fore and aft wing phase angles.

Wing-Flow Interactions. During the course of testing dragonflies in both the zero flow apparatus and the wind tunnel, flow visualization was completed. In some instances the flow visualization was done from a side view perspective and in others, from a rear view perspective. Thus, for the kinematic and force balance measures there exist a series of films that document the flow-wing interactions that support dragonfly flight.

The visualizations done in the zero flow apparatus required that smoke be delivered to the insects using a tube leading from the generator to the test apparatus. Thus, a dense smoke screen immersed the insect until flight episodes were elicited by vibration or direct stimulation of the leg (tarsal reflex). When a flight episode began, the induced flow carried the smoke over and around the dragonfly wings revealing wing-flow interactions and wake details (Somps and Luttges, 1985). In these early studies it became obvious that the dragonfly flight kinematics produced highly sculptured flows in which a number of discrete vortices could be identified. In Fig. 6, a typical flow visualization is provided from tests in the zero flow apparatus. As can be seen, the smoke reveals detailed vortex structures over the dragonfly wings. The result of multiple vortex interactions is evident in the wake where the residual flows of several successive wingbeat cycles are apparent. It is also apparent that the wing action of the dragonfly induces a significant and rather consistent overall flow condition in the zero flow apparatus. The rising smoke is attributed to the residual warmth of the flow produced during smoke generation.

To highlight the role of dragonfly wing kinematics in the production of specific flow field structures, two modifications were made to the test specimens. Perturbations of the flow field were simplified by clipping off all but a single fore wing. In addition, the dragonflies were implanted with electrodes that delivered electrical pulses directly to the thoracic flight muscular. This latter modification assured that the fore wing kinematics repeatedly exhibited the same wingbeat frequency as well as the same pitching, plunging and sculling characteristics. Fig. 7. shows the kind of flow structure that is elicited by this test protocol. The photograph resulted from multiple exposure, phase-locked (to the electrical stimulus) stroboscopic illumination of smoke laden flow. The formation of a vortex over the wing is very clear. Since multiple exposure photographs were used, each of the 6-7 exposures clearly resulted in the same form of vortex both in terms of the size and placement of this flow field structure. Thus, the wing kinematics are clearly capable of inducing specific vortex-wing interactions. Under these specific test conditions and phases of the wingbeat cycle, the major vortex is visualized as a chord-sized structure located above and slightly behind the forewing. Interestingly, the wake of the dragonfly exhibited much more turbulence than in the visualizations prepared for intact dragonfly specimens. The cohesive vortex structure in the presence of the fore wing rapidly transitioned to turbulence in the nearby wake.

In the absence of a freestream flow the dragonflies were difficult to stimulate into flight episodes. Also, as seen, the movement of smoke was almost wholly dependent on circulation induced by the wing motions. It also appeared that most of the flight episodes of the tethered insects were more like those of the escape

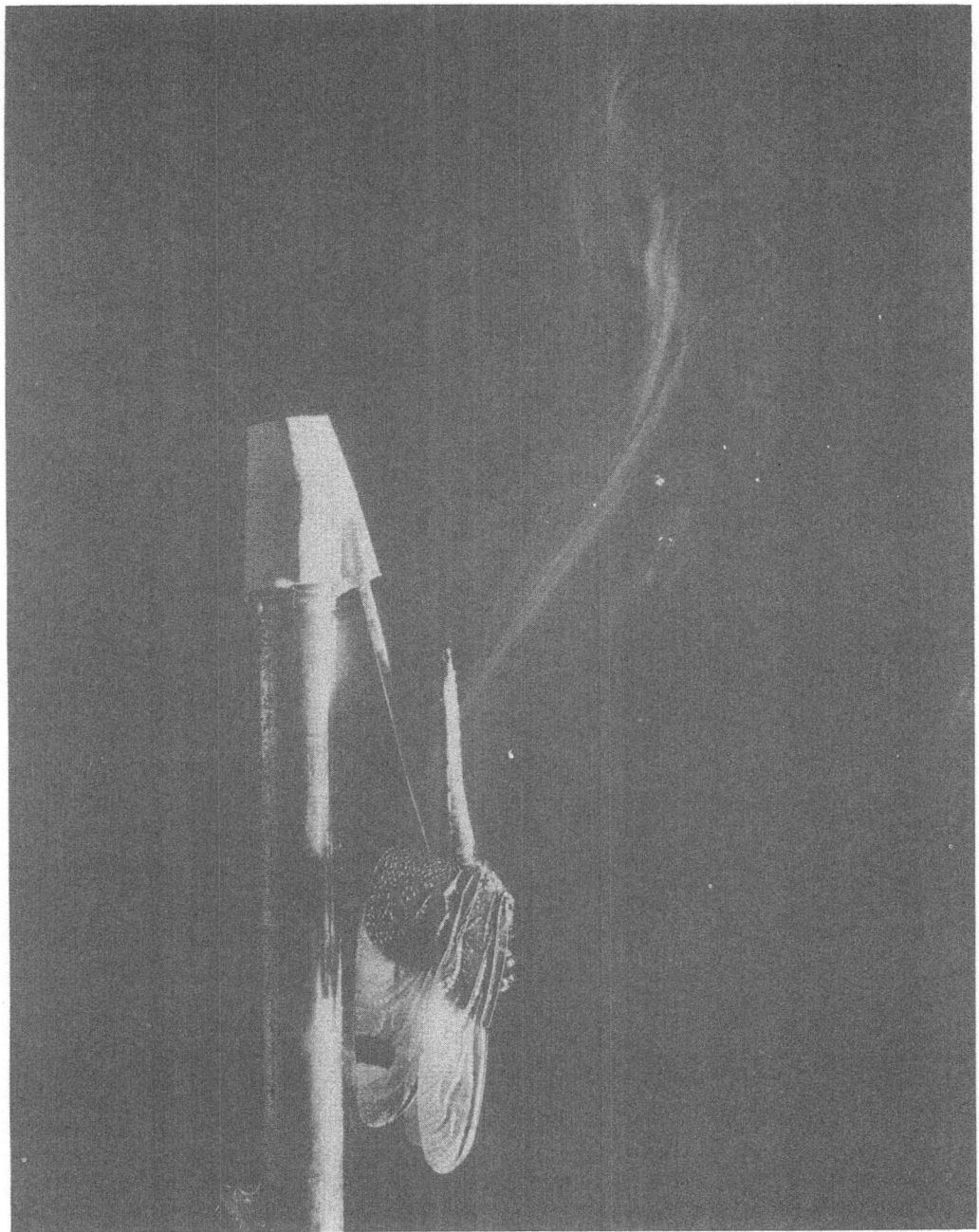

Figure 6. Photograph of dragonfly mounted in the zero flow apparatus with smoke delivered immediately ahead of the insect.

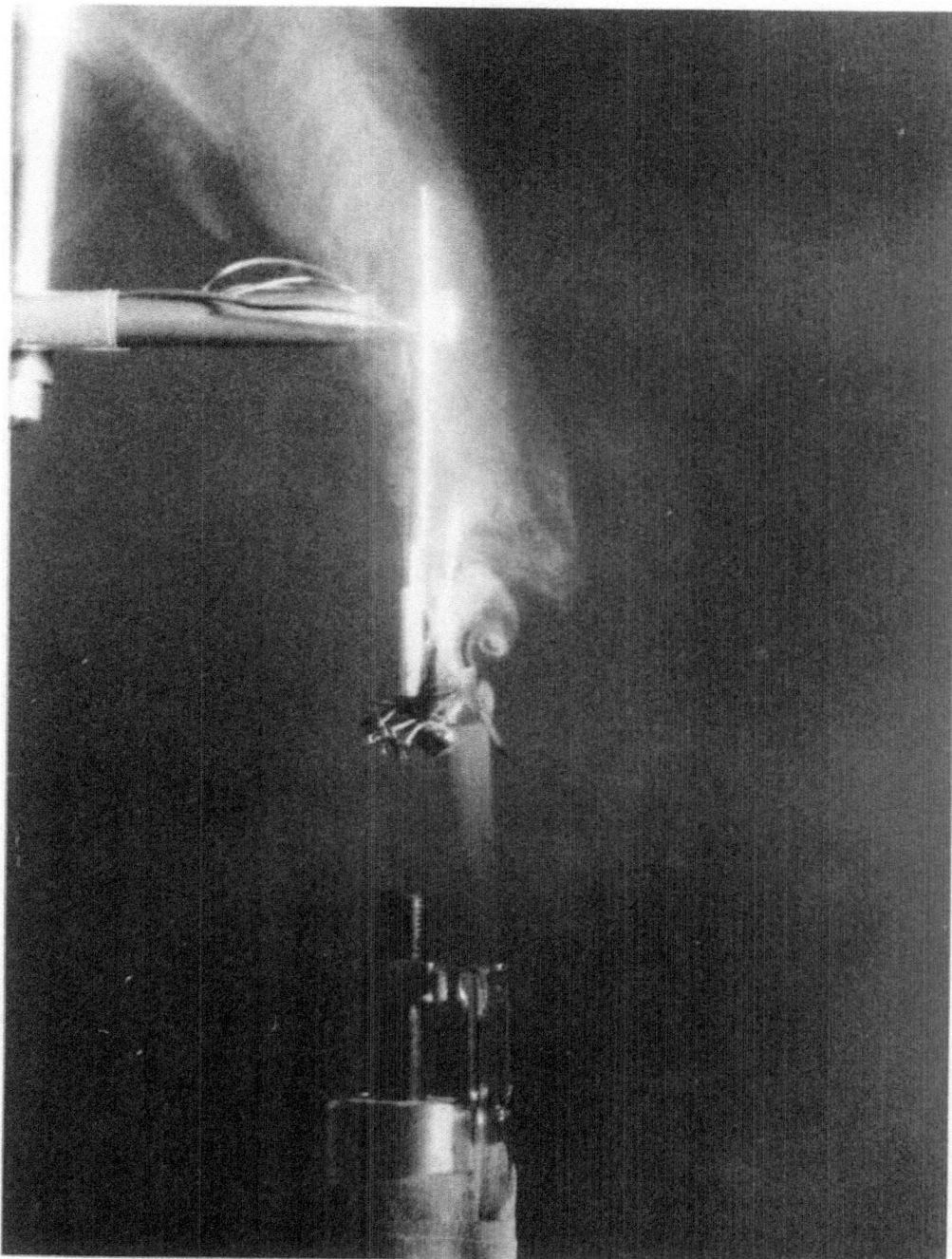

Figure 7. Multiexposure photograph of dragonfly mounted in zero flow apparatus with smoke delivery as in Fig. 6. Only a single fore wing remains and the insect musculature for flight is being driven electrically at 38 Hz. Stroboscopic flashes were phase-locked through a delay circuit to electrical stimuli.

flight mode than the hovering mode. Accordingly, the dragonflies were subsequently tested in a low speed, low turbulence wind tunnel. This tunnel had a 25 cm square test section constructed of clear cast acrylic. The dragonflies were tethered to a small mount (usually a force balance) near the centerline of the tunnel. A sliding vertical smoke wire was heated ohmically to pass a smoke sheet over the dragonfly wings at any desired span location. Fig. 8 shows stroboscopic single exposure visualizations for such test conditions.

The visualizations depict very complex but cohesive flow fields surrounding both the front and rear wings. The various separate flow structures combine into a wake consisting of amalgamated, repeating structures. The four plates reveal the wings at different stages of the full wingbeat cycle. The first plate shows the wings positioned such that the full complement of vortex structures is evident. Leading and trailing edge vortices are seen about both fore and aft wings. In these tests the wind tunnel was set with a freestream flow velocity of approximately 1.5 m/sec. The fore wing is nearing maximum upward plunging angles and the aft wing has begun the plunging downstroke. As will be recalled from the force balance data, this particular part of the wingbeat cycle occurs just prior to the appearance of the maximum lift peaks. This particular flow visualization is very similar to that collected from the mechanical model as shown in Fig. 9.

Returning to Fig. 8, the other three frames indicate the fate of the vortices formed in frame one. This fate was later corroborated with continuous high speed photography using a frame rate of 500/sec. In this abbreviated series of flow visualizations, the front wing is near the top of the wingbeat cycle and the aft wing is moving through the middle portion of the downstroke. The fore wing leading edge vortex appears at about midspan. It forms over the leading edge and then dwells there briefly before passing into the region of flow between fore and aft wings. At this time a prominent vortex is also shedding from the aft wing. Both vortices pass into the wake yielding a rearward moving jet flow. Careful analysis of these visualizations reveals that there are at least four prominent vortices produced during each wingbeat for the two wings: leading edge vortices produced by each wing as it pitches upwards and trailing edge vortices as each wing pitches downward. In addition, the wing tip vortex is clearly evident when it can be visualized passing into the smoke sheet. On the plunging downstroke, the tip vortex yields a large helical concavity traced out by the smoke. The interactions between all of these flow structures yields a rather complex but highly reproducible flow field.

Previous work on unsteady separated flows (Adler et al., 1983; Ashworth and Luttges, 1987; Ashworth et al., 1988; Shreck and Luttges, 1988) suggested that the three dimensional wing characteristics of the dragonfly might yield profound three dimensional effects in the unsteady flows produced by the dragonfly. To evaluate this possibility, simultaneous flow visualizations were prepared from both side and top view perspectives during wind tunnel testing of the dragonflies. Fig. 10 shows the visualizations for different points during the wingbeat cycles for paired photographs having different perspectives. The inboard flows are remarkably two dimensional throughout the wingbeat cycle. The wing tip flows show typical helical circulation through those portions of the cycle successfully visualized. Since the upper and lower extremes of the wingbeat cycle carry the wing tips away from the sheet of flow positioned to intercept the tips, these flow effects are not clearly

Figure 8. Single exposure flow visualizations for a tethered dragonfly in the 25 X 25 cm wind tunnel. Flow velocity approximately 1.5 m/sec. A smoke wire upstream was positioned such that a sheet of smoke lines intercepted the left fore and aft wings approximately midspan.

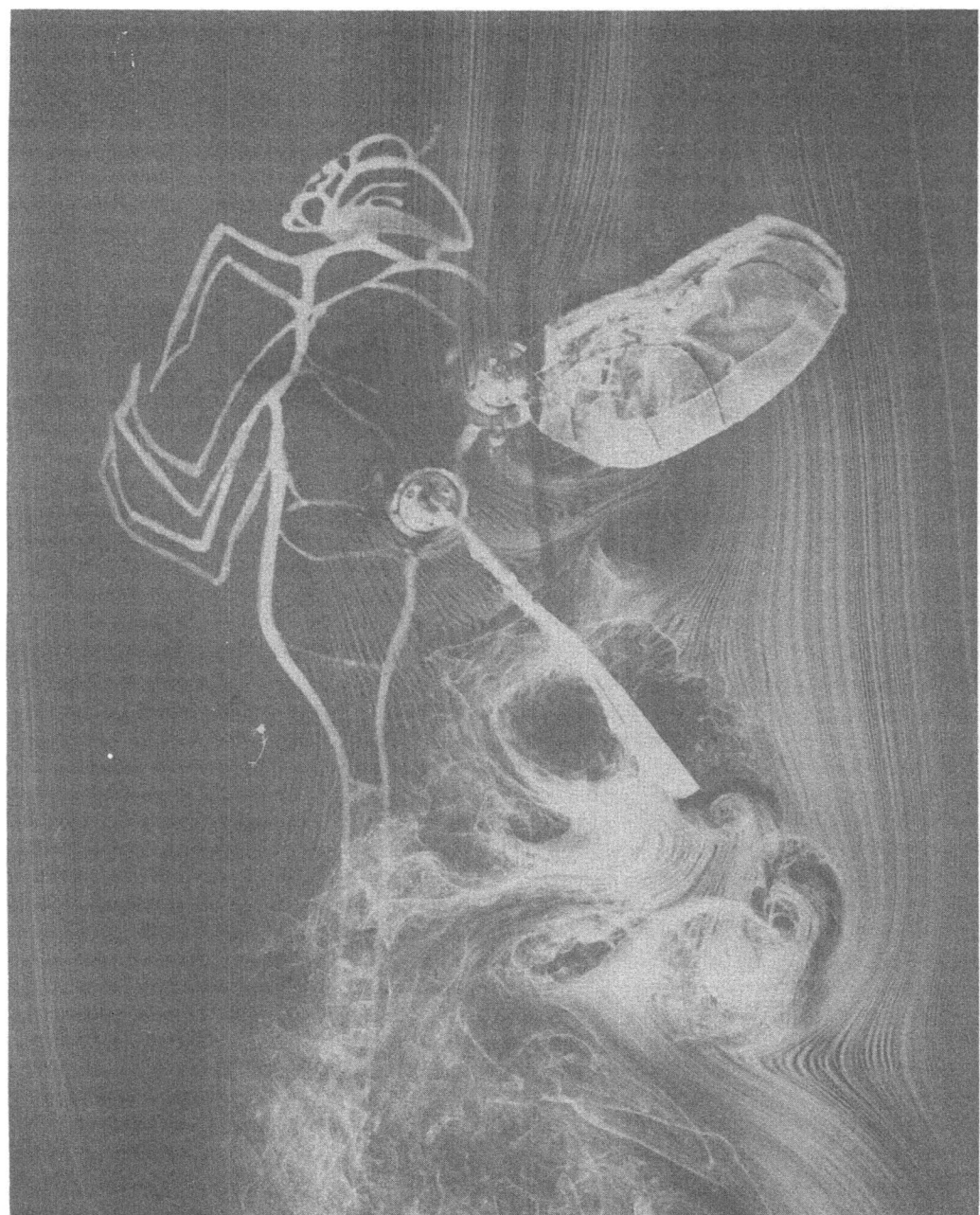

Figure 9. Flow visualization for a mechanical model that mimics dragonfly wing kinematics. The tunnel velocity (about 2 m/sec) was adjusted to yield plunging kinematics similar to those of the dragonfly. Smoke lines were produced by an upstream smoke wire.

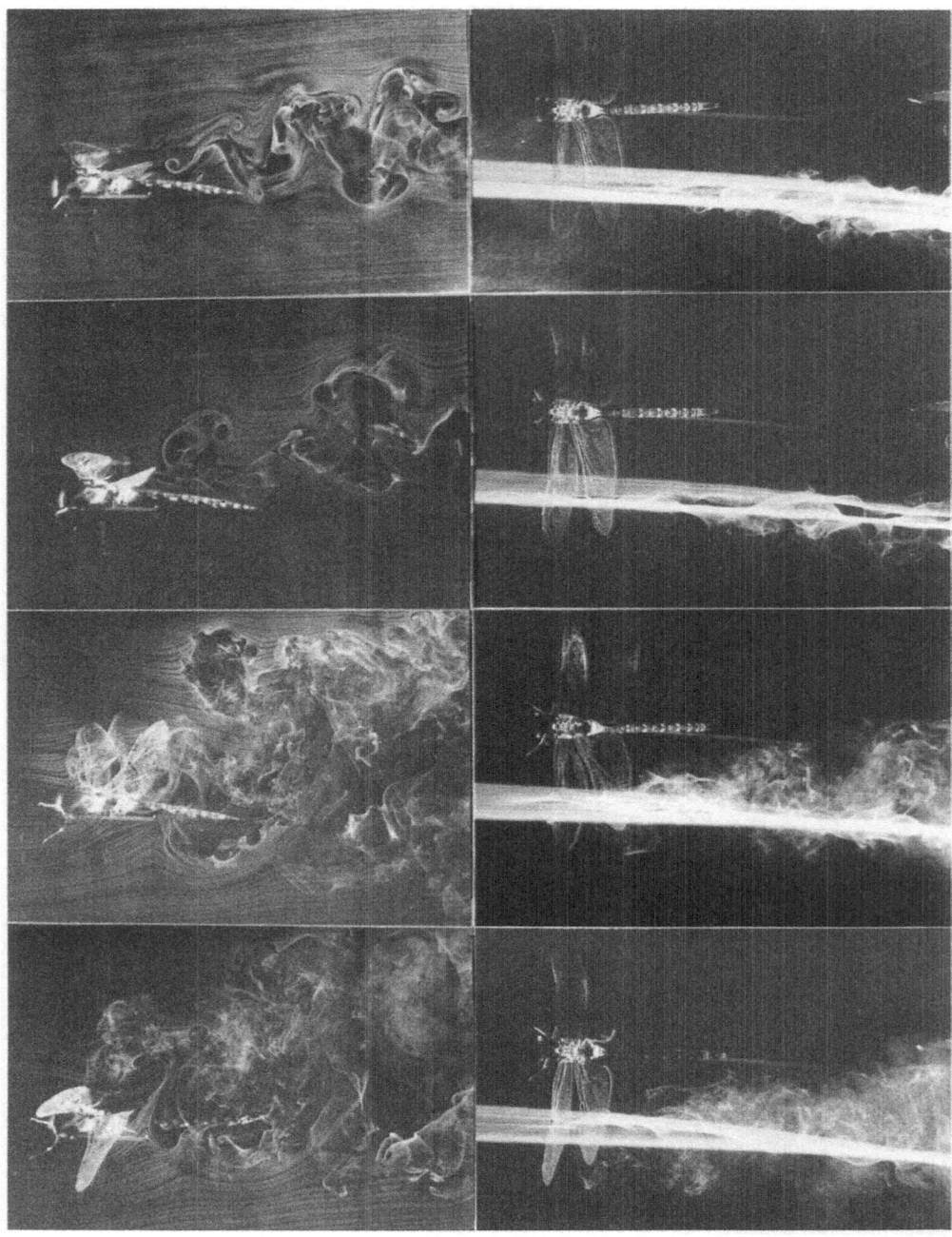

Figure 10. Simultaneous flow visualizations from side and top views of a tethered dragonfly in wind tunnel tests. Flow velocities were appoximately 1.5 m/sec.

visualized. In those cases where a vortex exists near the wing surface, the maximum inboard deflections of the smoke lines are less than one chord length. Smoke lines that moved outward beyond the tip to become part of the helical wing tip flow, show displacements that are approximately a half chord in length. In the wake flow well downstream of the aft wing the inboard-outboard dispersion continues to grow. Thus, all top view perspectives show the cohesive smoke sheet moving from left to right in the visualizations with clear indications of smoke having been drawn inboard (upward) and outboard (downward) during specific portions of the wingbeat cycle. At the right of these visualizations the evidence for wake distortions of the smoke is very clear with turbulence appearing everywhere.

All of the foregoing visualizations were obtained using single frame photographic exposures that permit high resolution. To obtain a better appreciation of the dynamic interactions between the wing and the fluid, Fig. 11 shows a series of frames taken from a 500 frame per sec movie. The same major flow field structures are evident. When the sheet of visualization smoke is passed slightly outboard of the midspan location, the wing tip vortex created by the fore wing appears as the sculptured cusp above the dragonfly. As the fore wing continues into the downstroke the smoke sheet intercepts more of the inboard span. The visualization shows the presence of the leading edge vortex at this point in the wingbeat cycle. At the beginning of the upstroke the forewing pitches upward revealing both the inboard leading edge vortex and the outboard wing tip flow. The aft wing is more difficult to visualize in this series. The most prominent feature in the flow elicited by the aft wing is actually an interaction with the flow perturbation of the fore wing. It is most readily seen as the fore wing approaches the top of the cycle and the rear wing is in the middle of the downstroke. It is easy to discern the flow being delivered from the front wing to a lower, downstream position over the rear wing. The large vortex structure resides between the fore and aft wings momentarily prior to shedding into the wake. This large vortex structure has grown while dwelling over the rear wing and continues to grow as it accelerates into the wake. Other flow structures are less clearly defined in these visualizations.

Another series of flow visualizations is shown in Fig. 12. In this series the smoke is set to intercept the wing further inboard, near the root of the dragonfly wing. Most apparent is the absence of the sculptured flow created by the wing tip vortex. Beginning in the first frame of the series (marked in 2 msec increments), the aft wings are moving downward mid-way through the downstroke and the fore wings are just beginning the downstroke. A vortex can be seen just behind the fore wings. Following this vortex, a series of shear flow structures appear between the vortex and a massive downwash that pushes over the rear wing and the tail of the dragonfly. This sequence is repeated in frames 4 msec and 44 msec with the vortices prominent and in frames 26 msec and 66 msec with the downwash most evident. Small variations in the appearance of the flow field structures can be attributed to the smoke sheet having moved a little more inboard in the later compared to the earlier visualizations of the series.

Apparent in all of the visualizations was the ability of the dragonflies to change wing kinematics from one wingbeat cycle to the next. Some of these behaviors were undoubtedly associated with the restraint used in tethered specimens tested in the tunnel. Because of the kinematic variability and the diminutive size of the

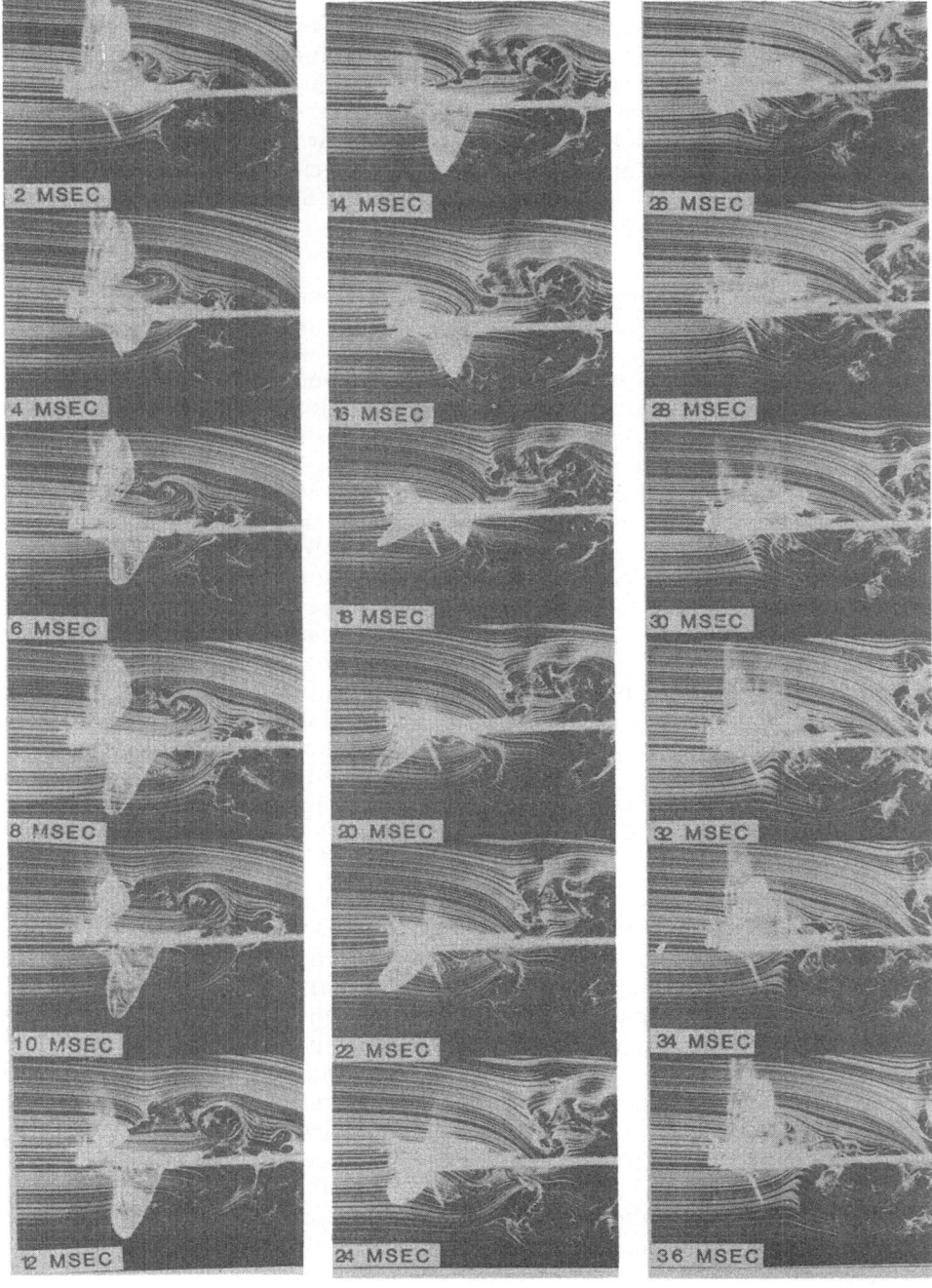

Figure 11. Continuous flight kinematics and flow visualizations obtained during a dragonfly flight episode at 500 frames/sec. Stroboscopic lighting (10 μ sec) was used. Smoke intercepted the wing at about midspan.

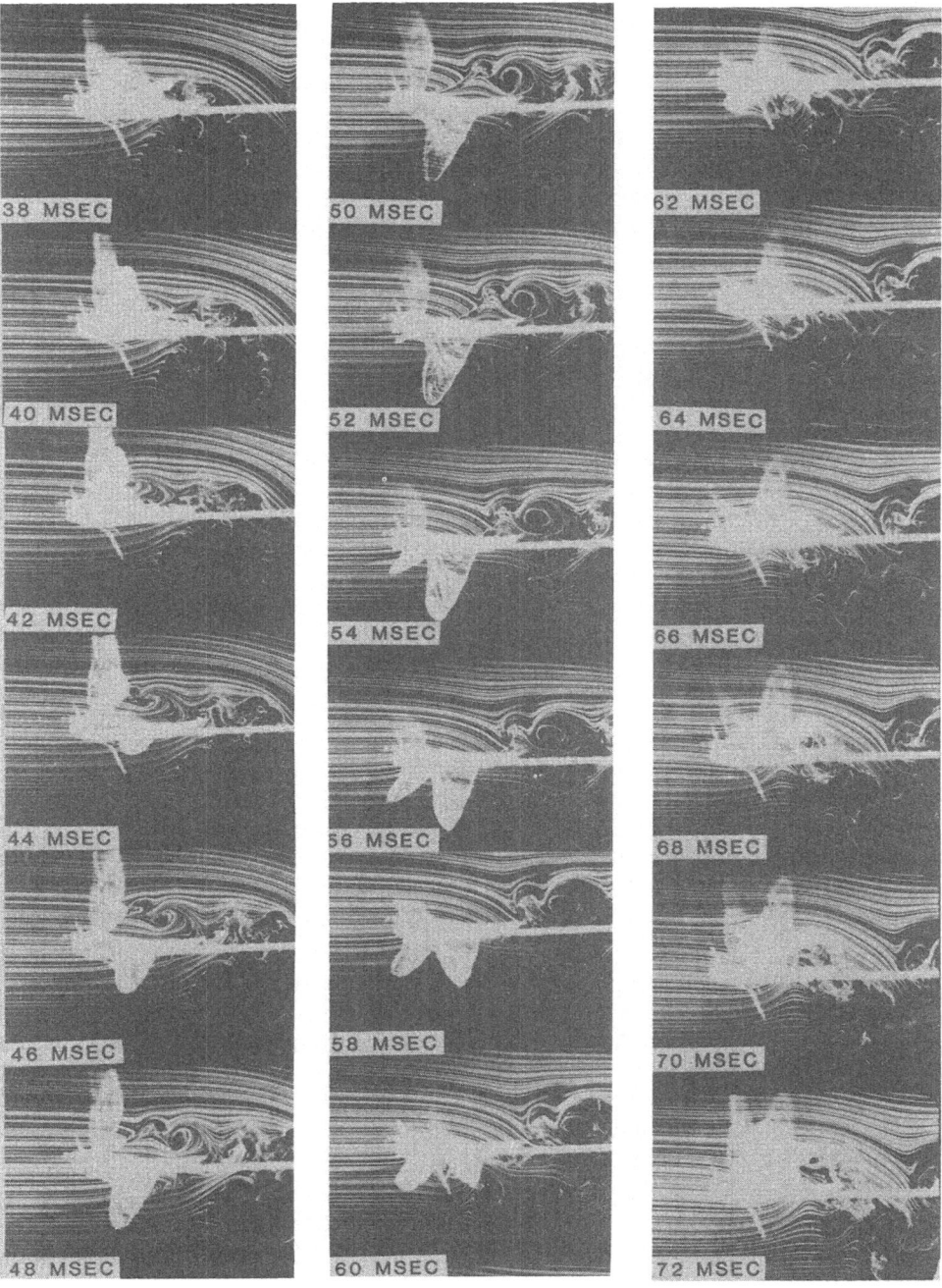

Figure 12. Same as in Fig. 11, except the smoke intercepts the wing further toward the wing root.

dragonflies the testing episodes are not as clear as they might otherwise be. The recognition of this problem had led us to design and build a mechanical simulation of the dragonfly wings and wingbeat characteristics. To better visualize the flow-wing interactions some visualizations from the mechanical model are presented in Fig. 13. The model has been described in detail elsewhere (Saharon and Luttges, 1987; Saharon and Luttges, 1988). It incorporates a scaled set of fore and aft wings that are moulded into corrugated shapes matching those of the dragonfly wings. In addition, it has incorporated all of the kinematic characteristics described earlier. The dynamics of the model can be changed to scale wingbeat frequency to the freestream velocity of the wind tunnel, to scale pitching rates to the wing flapping or plunging rates and to show varying phase angles between the fore and aft wings.

Comparisons of the visualizations from the mechanical model to those of the dragonfly specimens reveal excellent agreement in the form and quality of the elicited flow fields. The leading edge vortices of the forewing are evident and the sculptured cusps of the helical wing tip vortices are equally prominent. The flow field structures are carried into the wake and are accompanied by the strong downwash over the aft wing. The test conditions show an enhanced constructive interference between these flow field structures. Notably, a different set of test conditions can give destructive interference effects (Saharon and Luttges, 1988). The appearance of the strong downwash flow occurs at that time in the wingbeat cycle that had previously been shown to be associated with the maximum lift peaks (Reavis and Luttges, 1988).

HAWK MOTH FLIGHT

The hawk moth is a less well known flyer than the dragonfly. In many parts of the country it can be seen mornings and evenings hovering about flowers much like a small hummingbird. Distinguished by a dull brown background color with orange abdominal rings, this insect hovers then rapidly accelerates over to another flower where it again hovers while collecting nectar. We have not observed this insect engaged in any significant amounts of gliding. Since the insect is relatively large we elected to contrast some of the flight characteristics of this insect with those described above for the dragonflies. The work to date, is much less extensive than that accomplished with dragonflies.

Hardware. Like the dragonfly, the hawk moth possesses tandem wings. But, the fore wing is much larger than the aft wing. Also, the two wings are latched together with miniature hooks such that the two wings act as a single deformable surface. The maximum chord of this single effective wing is approximately 2 cm and is adjacent to the wing root. The aspect ratio is no more than 3-4. The whole insect weighs approximately 250-300 mg. Wing surface is relatively flat with small scale-like surface roughness. Overall wing contour supports a significant amount of camber. A photograph of the major wing structures is provided in Fig. 14. The thick body of the insect is obvious as are the triangular wing tips.

Kinematics. The wingbeat of the hawk moth has been visualized in the wind tunnel using techniques developed for the dragonfly. The tests were conducted with the insects mounted to the force balance with a slight angle of the thorax relative to the oncoming flow. The dragonfly, of course, flies with the body parallel to the horizon. As seen in Fig. 14, the wing pitches upward on the upstroke and downward on the downstroke. The rest of the wing forms a surface that appears to

10t/T   **Downstroke**   **Upstroke**

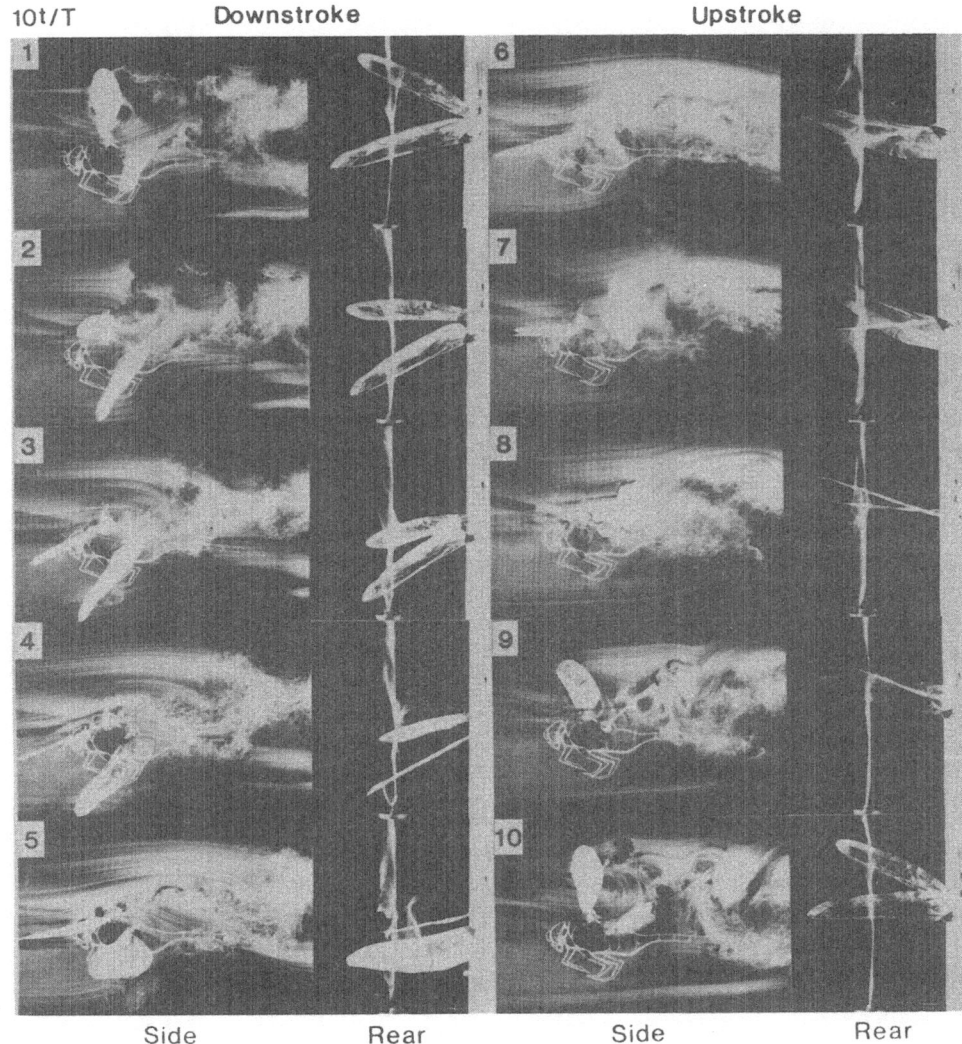

Side   Rear   Side   Rear

Figure 13. **Characteristic flow perturbations produced by the mechanical model of dragonfly flight kinematics.**

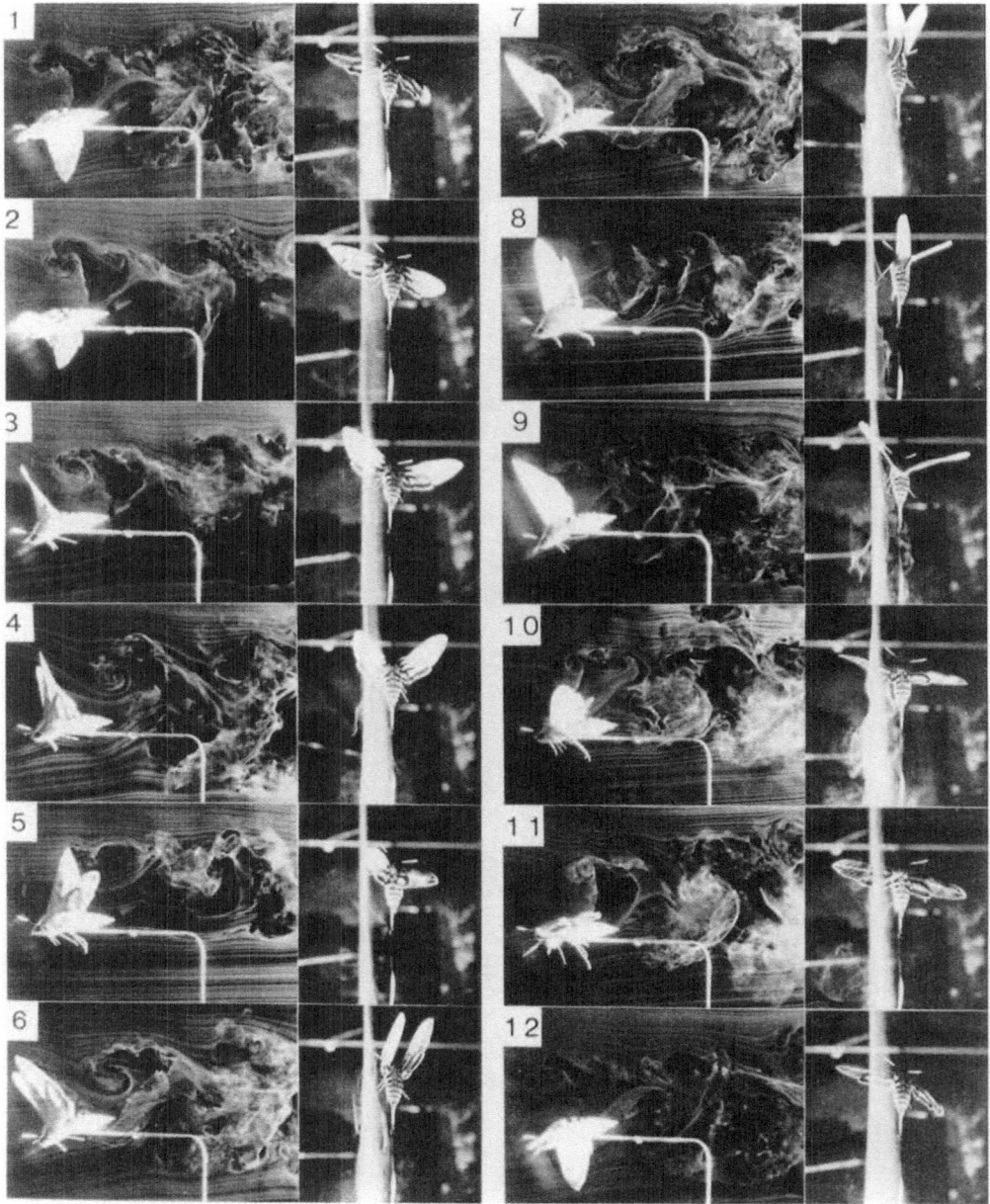

Figure 14. Single exposures of tethered hawk moth flight and flow perturbation characteristics during wind tunnel tests. Plates on the left are side views obtained simultaneously with the top views shown to the right. Each column has been arranged to depict most phases of the flight kinematics.

mimic the motions of the dragonfly wings except that more wing camber is in evidence. In the photo series of Fig. 14, smoke is positioned to intercept the wing at about mid-span on the middle of the plunging downstroke and upstroke. In side perspectives the smoke lines move from left to right and in top-rear perspectives the smoke sheet appears as a line moving from the top to the bottom of each plate. Both perspectives are exposed simultaneously using the same stroboscopic lighting as seen through two different cameras. In this series the specimen was exhibiting a wingbeat frequency of 45 Hz in the wind tunnel having a 1.6 m/sec freestream velocity. A whole wingbeat series is shown but these pictures were prepared from single exposures obtained across different actual wingbeats.

The wing kinematics show many of the characteristics described for the dragonfly. From the plane of the body the plunging consists of the wings moving upwards to approximately 75 degrees and then downward from the body to approximately 60 degrees. The upward moving wing shows remotion and the downward plunging wing shows promotion. As stated above the pitching angles occur at the reversal of the direction of the plunging kinematics. Based upon the observed changes in geometric angles of attack, it is estimated that pitching upward rates approach 65,000 degrees/sec. Pitch rates of the downstroke are approximately half those of the upstroke. Supination leads to a twisting of the wing while pronation does not. Most interestingly, through all of the visualizations analyzed there was no evidence for the "clap" of the two wings above the body of the moth (cf., Weis-Fogh, 1973). Yet the wing pronation combined with the downward plunging and forward sculling appeared similar to kinematics previously described in the Chalcid wasp as a "fling."

Force Generation. The moths were tested in the wind tunnel using the force balance previously used for dragonflies. The results are not as well documented as those for the dragonfly since we have tested limited numbers of specimens, to date. Peak lift forces occurred as the wing pronated into the downstroke and were sustained as the wing passed through the middle of the downstroke. Thereafter, the lift decreased until the bottom of the plunging downstroke was attained. As the wing supinated upward a small, short-lived peak of lift occurred. Otherwise, the decreasing amounts of lift were in evidence until the wings approached the top of the upstroke. The lift history suggests that the underlying fluid dynamic mechanisms may be related to unsteady separated flows. The related thrust measurements are equally periodic with maxima that occur and then decline just prior to the top of the upstroke. In the tests conducted where the specimens are mounted with bodies parallel to the oncoming freestream, the values of the measured thrust peaks were routinely less than 30% the values of the lift peaks. The magnitudes of the lift peaks were approximately five times the weight of the specimens being tested. In the present testing circumstances it was not possible to compare changes in lift with substantial changes in thrust since lift always dominated. Finally, the moths appeared to exhibit very significant lateral forces whereas the dragonflies rarely did so. The episodes of lateral force peaks occurred at periods during the wingbeat cycle when the wing just begins to supinate. In many instances these peaks can be quite large, approaching those observed for the lift. During any elicited flight episode (often lasting for 30 sec or more) the moths showed a variety of kinematics and associated forces. The attempts to fly asymmetrically to escape the tether were a constant feature of all of the observed flight episodes.

Wing-Flow Interactions. The wingbeat of the moth produces most of the same flow disturbances recorded for the dragonfly. The prominent sculpturing of the smoke sheet by the wing tip vortex is evident. The leading edge vortex over the body of the moths is quite distinctive. But, the downwash of flow so prominent between fore and aft wings of the dragonfly has no apparent counterpart in the flow visualizations collected for the moths. The three-dimensionality of the flow perturbations appears to be larger in the moth than in the dragonfly but it has not been possible to rule out the role of "turning" escape maneuvers in the moth as a major contributor to this observation. In some instances the flow over the left wing of the insect has been observed to cross over the rear of the insect to pass into the wake on the right side of the insect. These observations could not be matched by a symmetric (unmarked by smoke) flow since the crossover is clearly unperturbed by flow moving in an opposite direction. Again, these observations are taken as evidence for different escape maneuvers being practiced by moths compared to dragonflies.

Overall, the moths appear to produce flow structures that are just as reproducible as those observed in dragonflies. Diffuse flows occur only in the wake of the moth at distances of a body length or more downstream. Cohesive flow structures are in evidence everywhere despite the fact that these structures emanate from tip and inboard flows and that these flow structures are quite complex.

CONCLUSIONS

Observations on both dragonflies and hawk moths show that they both generate aerodynamic force generation histories and flow field structures consistent with the use of unsteady separated flows. Both insects appear to derive aerodynamic forces from the production of bound vortices and the use of the circulation of these vortices prior to the time that they pass into the wake. The flow structure of the vortices is largely two-dimensional while in the presence of the wing (or wings) that produce them. Neither insect appears to depend upon the use of the Weis-Fogh clap and fling mechanism (Maxworthy, 1981) to achieve the aerodynamic forces that they generate. Nevertheless, both insects show the periodic, transient production of lift peaks that are many-fold the weight of the insect. The wing kinematics of the two insects appear to have many features in common including high pitch rates, similar plunging rates and similar sculling angles.

The hawk moth and the dragonfly differ in the combinations of aerodynamic forces that they generate. The dragonfly generates most force in the lift and thrust domains while the hawk moth generates lift and lateral forces. The tethering of both insects in the wind tunnel, of course, prevents them from changing orientation relative to the oncoming freestream. The influence of this limitation remains a matter of speculation. The wings used to achieve the observed flow field structures and force measurements differ in that the dragonfly has relatively high aspect ratio fore and aft wings while the moth has lower aspect ratio wings and fore-aft wings that are latched together. It may be significant that we have not observed moths to show any significant amount of gliding either in the wind tunnel tests or in natural observations. Also, the moths tend to show large changes in body orientation when seen hovering near sources of nectar. During rapid escape flight, the moths are more often seen with bodies oriented parallel to the horizon.

A clue to the importance of the kinematics in using unsteady flows is that neither insect shows large variations in wingbeat frequencies. They show frequencies that are rather narrowly circumscribed. In the dragonfly we have attempted to use electrical stimulation to elicit slower and faster wingbeat frequencies, to no avail. The specimens simply showed small amplitude plunging motions when stimulation forced wingbeats clearly at rates outside the normal wingbeat rates. Simple models tested in our laboratories (Kliss et al., in press) show that such wingbeat frequencies may be related to the use of vorticity. Slower wing beat frequencies don't yield cohesive vortex structures while higher frequencies may yield structures that are small and mutually disruptive one to another. Controlled stroke lengths and frequencies yield cohesive structures that dwell for rather long periods in the presence of the surfaces that produced them. In the insects evaluated here, the wing stroke distance is controlled by wing length and plunging angles. A reduced plunging angle can simply move optimal flow vortex generation outboard whereas larger plunging angles move the vortex generation site inboard. The anchored wing root allows the flow vorticity to "find" the optimal span location for cohesive vortex formation. Sculling kinematics have much the same effect. In contrast, the pitching occurs almost instantaneously along the whole wing span. The result is that vorticity thereafter accumulates in the area of cohesive flow movement. Thus, the vorticity actually supports and enhances the flow structures initiated by combined pitching and plunging.

In general, we believe that insects operating in the viscous flow regime must use this regime effectively. Basically the insects are faced with large vorticity production and with few ways to shed this vorticity. Vortices not only aid in the large scale shedding of vorticity but provide a source of aerodynamic force generation that is readily exploited. The dragonfly appears to be able to use such unsteady separation structures at some times, not others. The moth, in contrast, appears to use these mechanisms exclusively. Then, the moths seek places to rest for extended periods. As might be expected from the previous discussions, other insects are likely to use unsteady flow structures in yet other ways.

Interestingly, even giant Manta Rays of the ocean may use similar mechanisms to move efficiently through their viscous environment. It seems certain that other examples of use can be found, as well.

What has been presented here is a rather terse summary of the work of many colleagues and myself over almost a half decade. The insects described are, perhaps, the most well understood practitioners using unsteady separated flows. Many control features of such use are "built in", by the hardware and kinematic matches. Other control features are yet to be deciphered. The challenge is to bring what the insects teach us to the service of our engineering flight efforts.

ACKNOWLEDGEMENTS

This work was supported in part by grant F49620-84-C-0065 from the Air Force Office of Scientific Research, Dr. Hank Helin, Program Manager and by grant N00014-85-K-0053 from the Office of Naval Research, Dr. Frank Hempel, Program Manager.

The collaboration of many valued student colleagues is in evidence and much appreciated. The special efforts of Suzanne Walts were essential to this work.

Bibliography

Adler, J.N., Robinson, M.C., Luttges, N.W. and Kennedy, D.A.(1983) Visualizing unsteady separated flows. *Third International Symposium on Flow Visualization, Proceedings* Vol III, 806-811.

Ashworth, J. and Luttges, M.W. (1986) Comparisons in three-dimensionality in the unsteady flows elicited by straight and swept wings. *AIAA Paper* No. 86-2280-CP.

Ashworth, J., Mouch, T. and Luttges, M.W. (1988) Application of forced unsteady aerodynamics to a forward swept wing X-29 model. *AIAA Paper* No. 88-0563.

Francis, M. and Luttges, M.W. (Eds.) (1984) *Unsteady Separated Flows*. University of Colorado, Boulder.

Kliss, M., Somps, C. and Luttges, M.W. (in press) Stable vortex structures: A flat plate model of dragonfly hovering. *J. Theor. Biol.*

Luttges, M.W., Somps, C., Kliss, M. and Robinson, M.(1984) Unsteady separated flows: generation and use by insects. In *Unsteady Separated Flows*(Francis, M. and Luttges, M. Eds.) University of Colorado, Boulder, 1984. 127-136.

Maxworthy, T. (1981) The fluid dynamics of insect flight. *Ann. Rev. Fluid Mech.*, 13, 329-350.

Norberg, R.A. (1975) Hovering flight of the dragonfly Aeschna Juncea L.: Kinematics and aerodynamics. In *Swimming and Flying in Nature* (Wu, Y., Brokaw, C. and Brennen, C., Eds.) Plenum Press, New York.

Reavis, M.A. and Luttges, M.W. (1988) Aerodynamic forces produced by a dragonfly. *AIAA Paper* No. 88-0330

Robinson, M. and Luttges, M.W. (1983) Unsteady flow separation and attachment induced by pitching airfoils. *AIAA Paper No. 83-0131.*

Robinson, M.C. and Luttges, M.W. (1984) Forced and common vorticity about oscillating airfoils. In *Unsteady Separated Flows* (Francis, M. and Luttges, M. Eds.) University of Colorado, Boulder, 117-126.

Weis-Fogh, T. (1973) Quick estimates of flight fitness in hovering animals, including novel mechanisms for lift production. *J. Exp. Biol.*, 59, 169-230.

Saharon, D. and Luttges, M.W. (1988) Visualization of unsteady separated flow produced by mechanically driven dragonfly wing kinematics model. *AIAA Paper* No. 88-0569.

Saharon, D. and Luttges, M.W. (1987) Three-dimensional flow produced by a pitching-plunging model dragonfly wing. *AIAA Paper* No. 87-0121.

Schreck, S.J. and Luttges, M.W. (1988) Unsteady separated flow structure: Extended K range and oscillations through zero pitch angle. *AIAA Paper* No. 88-0325.

Somps, C. and Luttges, M.W. (1985) Dragonfly flight: Novel uses of unsteady separated flows. *Science*, 228, 1326-9.

THE AEROACOUSTICS OF TRAILING EDGES

William K. Blake
Jonathan L. Gershfeld
David Taylor Research Center
Bethesda, MD 20084-5000

CONTENTS

Page

458

FIGURES

ABSTRACT

Trailing edge sound is an aeroacoustic phenomenon
which contributes to the noise from lifting surface
flaps, rotating machines and certain turbulent nozzle
flows. Part I of this paper is a survey which describes the physics of aerodynamic sound generation by
trailing edge flows; it is particularly relevant to
lifting surfaces. The survey will describe the
importance of the geometry of the surface, its upstream
boundary layer, and its trailing edge wake in determining the nature of the aeroacoustic sources. Techniques
for measurement will be discussed and recent results
will be brought forth which will further elucidate the
relevant features of flow structure. Part 2 of the
paper will deal with the trailing edge flow sources in
more depth. A distinction is to be made between
continuous-spectrum surface pressures that are locally
generated by the immediate separated flow and narrower
band pressures which are developed by superimposed
orderly structures developing in the wake. The paper
will also examine this distinction as well as the
relevance of existing analytical models of trailing
edge sound to each mechanism, and it will compare those
theories to measurements of the sound generated by the
separated trailing edge flow. General agreement is
found between measurement and theory for both tonal and
random aerodynamic sound depending on the frequency and
the geometry of the trailing edge.

1.0 INTRODUCTION

Aerodynamically generated sound from turbulent flow that is contiguous to a boundary is quadruple in nature as long as both the bounding surface and the flow field are spatially homogeneous in the plane of the surface and as long as the surface contains no inhomogeneities in impedance. Under such a restriction, the surface is a simple reflector of the flow sources. As is well known, such quadruple sources are relatively inefficient radiators of sound. When the surface is terminated as a half plane so that the flow passes from above the surface to a free fluid, sound is radiated by acoustically more efficient dipoles that are formed at the edge. Thus, the edge becomes a scatterer with a more efficient sound radiation than would be provided by the simple flow quadruples. The range of flow-acoustic problems which are concerned with the interaction of the flow with a trailing edge in this manner are encompassed in a subject area that has become known as trailing edge noise.

Figure 1 illustrates the various physical arrangements which generate trailing edge sound. The first three are wall jets in which the plane is either flat (in the "half-plane" sense) or includes a flap which protrudes into or away from the flow direction. The fourth is the airfoil geometry. The reader may readily confirm the relative inefficiency of a subsonic free jet in relation to the wall jet by blowing across the edge of a sheet of paper or straight edge and noting the more strongly audible broadband sound that is generated in the exercise of edge blowing. This is just a trivial example of the blown flap problem of airframes; which problem initiated most of the early work in the area of trailing edge sound. Figure 2 shows a more scientific example in the form of measurements of sound from a converging wall jet compared with that emitted by the jet alone; this example was taken from Scharton et al (1973). It is to be emphasized that, to first order, the sound generation mechanism is acoustic in nature and the sound level is directly related to the relative impedances of the surface and the fluid, among other factors. The actual radiation is, however, also related to the structure of the turbulence in the immediate vicinity of the edge; this is controlled by viscous effects which prevail as the flow interacts with the edge. Accordingly one-sided flows such as wall jets and two-sided flows such as airfoils may generate

significantly different sound levels at a given Mach number
because of differing flow structures in these classes of flows.
This paper will concentrate on the sounds emitted from lifting
surfaces by the trailing edge mechanism; we shall focus on the
relevance of flow structure to the nature of the radiated sound.

We shall show how characteristics of the edge dipole differ
depending on whether the source is controlled by grazing fine
scale turbulent flow or whether it is controlled by large scale
wake structures.

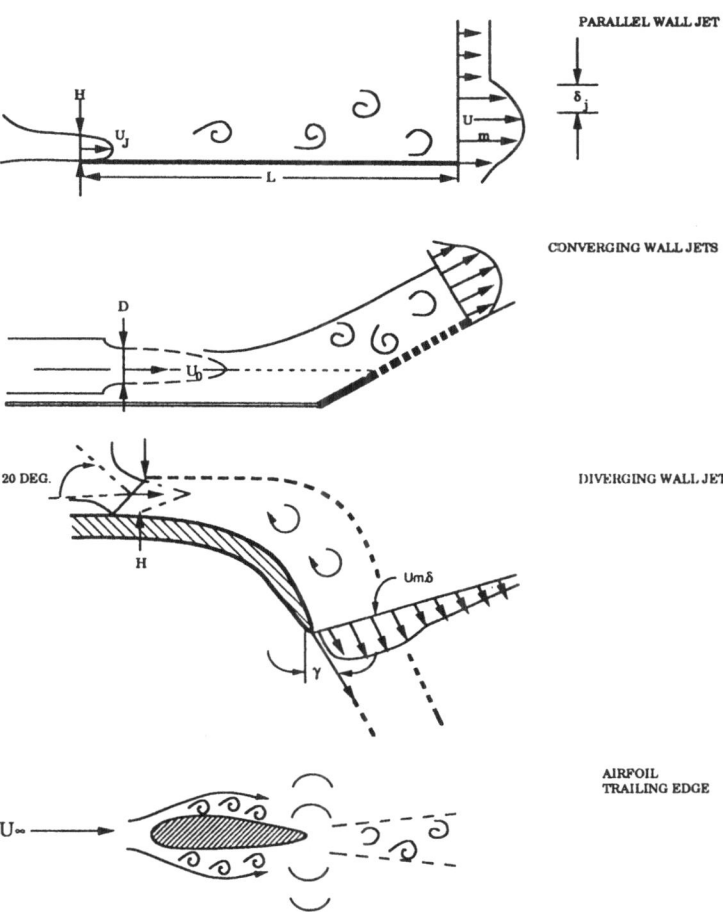

Figure 1. Sound From Effectively Rigid Turbulent-Flow Airfoils.

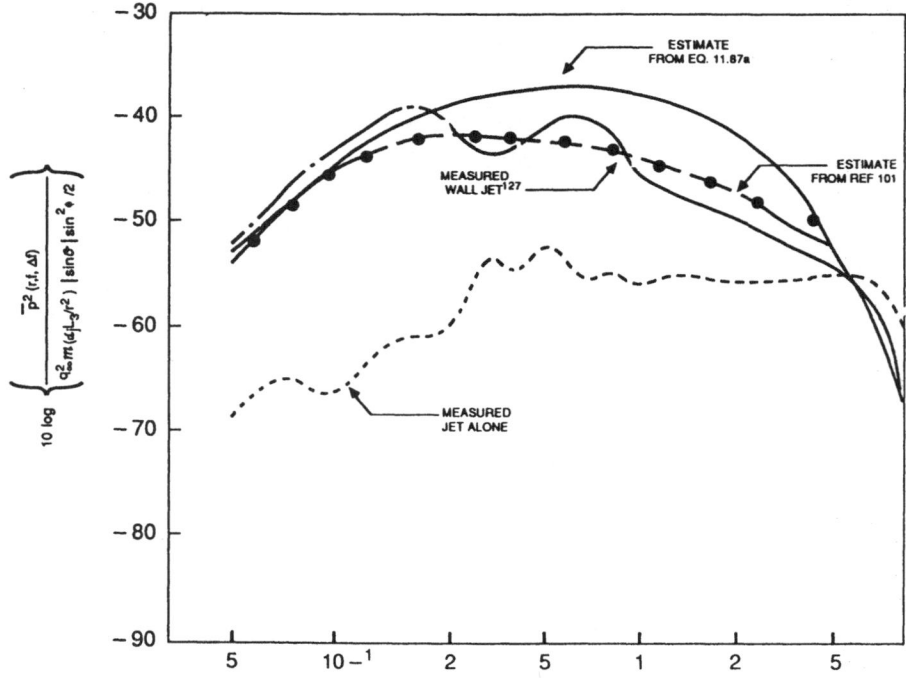

Figure 2. Measured and Predicted Noise from Wall Jet of Sharton.

PART I. THEORY AND BEHAVIOR OF TRAILING EDGE SOUND
2.0 HISTORY AND PHYSICAL ISSUES OF LIFTING SURFACES

The nature of sound radiated from lifting surfaces is well-connected with the character of the flow; laminar or turbulent flow, sharp or blunt trailing edges, and angle of attack are all aspects which control the sound. Figure 3 shows in an elementary way what these interrelationships are; h is the geometric thickness of the trailing edge, y_f is the wake thickness, and U_∞ is the free stream velocity. The reader may wish to refer to this chart as a reference in reading various parts of this chapter. Sound from trailing edge flow from airfoils has classically taken

REYNOLDS NUMBER	ANGLE OF ATTACK	TRAILING EDGE FORM	SPECTRAL CHARACTER
LAMINAR FLOW $\Re_c \lesssim 2 \times 10^5$	SMALL $\alpha < O[10°]$	BLUNT h	TONAL, $f \sim U$ $fh/U = CONST$
		SHARP	TONAL, $f\,\delta/U = CONST$ $f \sim U^{3/2}$, $\delta \sim U^{-1/2}$
	LARGE $\alpha > O[10°]$	BLUNT OR SHARP	CONTINUOUS SPECTRUM
TURBULENT FLOW $\Re_c > 2 \times 10^5$	SMALL $\alpha < O[10°]$	SHARP	CONTINUOUS SPECTRUM
		BLUNT, $y_f \rangle 3000\,\upsilon/U$	$fy_f/U = CONST$ $f \sim U$ $y_f \approx 0.8h$
	LARGE $\alpha > O[10°]$	BLUNT OR SHARP	CONTINUOUS SPECTRUM

Figure 3. Diagram Showing Relationships Between Reynolds Number, Angle of Attack, and Trailing Edge Bluntness that Govern Frequency Spectral Characteristics.

two forms. In the first, a tonal sound is generated when periodic wake vorticity results from Helmholtz instability in the trailing wake. This sound is normally associated either with laminar-flow airfoils or blunt-edged lifting surfaces on which flow separation is clearly defined. This wake-induced flow source is closely akin to the aeolian tone of the circular cylinder in a cross-flow. The second source is continuous-spectrum and it is due to the convection of upstream turbulent flow past the edge as previously demonstrated, in the simple case of blowing across the paper edge. Considerations of this source are generally reserved for higher Reynolds numbers on lifting

surfaces with sharp trailing edges. As noted above, the sound
from wall jets and blown flaps is also commonly regarded as due
to this source.

The existence of a given form of trailing edge sound is
strongly dependent on the overall nature of the flow that exists
on the lifting surfaces. Figure 4 shows diagrams of the various
patterns which could exist and which are due to the multiple
forms of boundary layer and vortex flows that occur on lifting
surfaces. These flows depend on the cross section geometry of
the surface and its lifting (pressure) distribution, Reynolds
number, and the existence of a tip vortex. At low Reynolds
number a leading edge separation may occur and this may serve as
a turbulent boundary layer trip. Depending on both loading and
Reynolds number, a three-dimensional separation zone can exist.
This zone will be less extensive as long as the static pressure
gradient along the chord is small to moderate as determined by
both angles of attack and geometry of the section and Reynolds
number. For sharp-edged surfaces which have light trailing edge
loading, separation will not be likely. If the edge is blunted,
localized turbulent flow separation may be expected. On three-
dimensional surfaces with lifting tips, a tip vortex can be
formed the influence of which is imposed on the flow to some
distance inboard of the tip.

Focusing for a moment on the lightly loaded lifting surface,
we note that the flow is dependent on the geometry of the trailing edge. Figure 5 illustrates the two extremes of a fully
separated flow at a blunt edge and a completely attached flow at
an unloaded sharp edge. The third illustration is an
intermediate case in which the wake is unsymmetrically shed from
a nonsymmetrically curved edge. Further discussions of these
flows will appear later in this paper as appropriate. In the
case of the blunt edge, vortex shedding is periodic (or nearly so
at lower Reynolds numbers), and it generates a nearly tonal
pressure disturbance. Figure 6 from Brooks and Hodgeson (1981)
gives examples of the spectral characteristics of both surface
pressures and radiated sound which may be generated by a lifting
surface on which the trailing edge has been blunted. In this
case it appears that a disturbance over a narrower band of frequencies has been superimposed on a pressure field which occupies
a much broader frequency band, i.e. is a more continuous-spectrum. This latter pressure component is generated by the

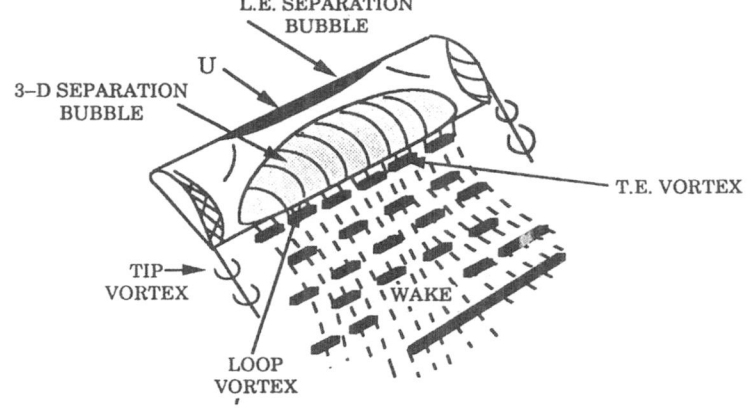

**Overall Pattern of Vortex Flows
(Winklemann & Barlow (1980))**

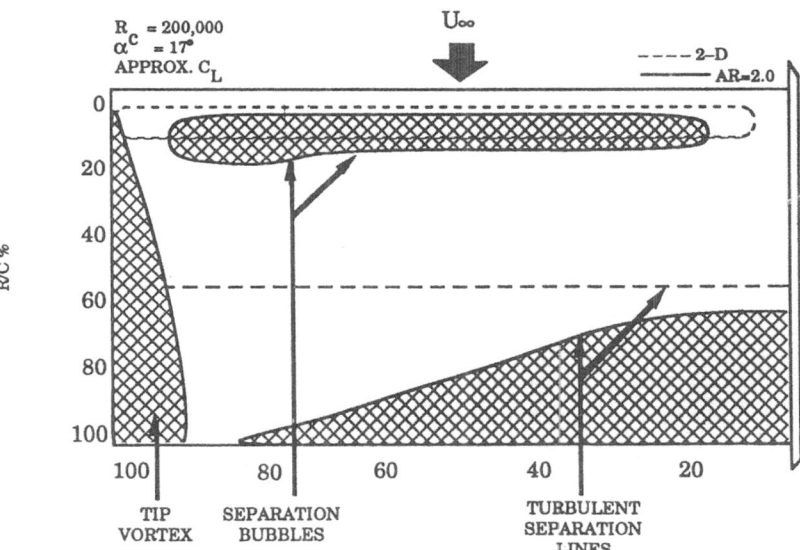

2–D & 3–D Flow Regimes – (Bastedo & Mueller (1986))

Figure 4. Low – Reynolds Number Flow Regimes on Lifting Surfaces.

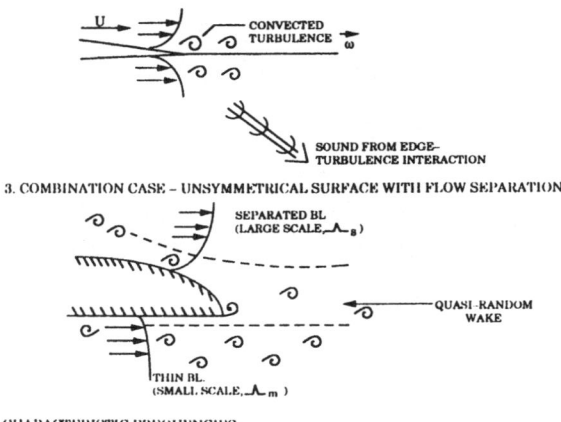

1. VORTEX SHEDDING DUE TO WAKE
   INSTABILITIES:

   PRESSURE
   FLUCTUATIONS

   VORTICITY, ω

   SOUND FROM PERIODIC
   FORMATION OF VORTICITY

2. AEROACOUSTIC SCATTERING

   U

   CONVECTED
   TURBULENCE ω

   SOUND FROM EDGE–
   TURBULENCE INTERACTION

3. COMBINATION CASE – UNSYMMETRICAL SURFACE WITH FLOW SEPARATION

   SEPARATED BL
   (LARGE SCALE, Λ_s)

   QUASI-RANDOM
   WAKE

   THIN BL.
   (SMALL SCALE, Λ_m)

   CHARACTERISTIC FREQUENCIES:
   UPPER SIDE U/Λ_s
   LOWER SIDE $U/\Lambda_m > U/\Lambda_s$

Figure 5. Near-Wake Structures and Sound at Trailing Edges.

attached boundary layer. A significant feature that is illustrated here is the relatively more important "source strength" of
orderly wake-induced pressures compared with the "source
strength" of the turbulent boundary layer pressure. This is
shown by comparing the surface and radiated pressures and noting
the relatively greater prominence of the narrowband "bump" in the
spectrum of the radiated sound than in the surface pressures.
The frequency spectral character of so-called "vortex-shedding"
pressures is dependent on the orderliness of the wake vorticity
that is formed at the trailing edge. For blunt edges with relatively thin upstream boundary layers, the trailing edge wake is a
prominent vortex street and the pressures are concentrated at a
frequency

$$f_s = \mathcal{S} U_\infty / y_f \qquad\qquad (1)$$

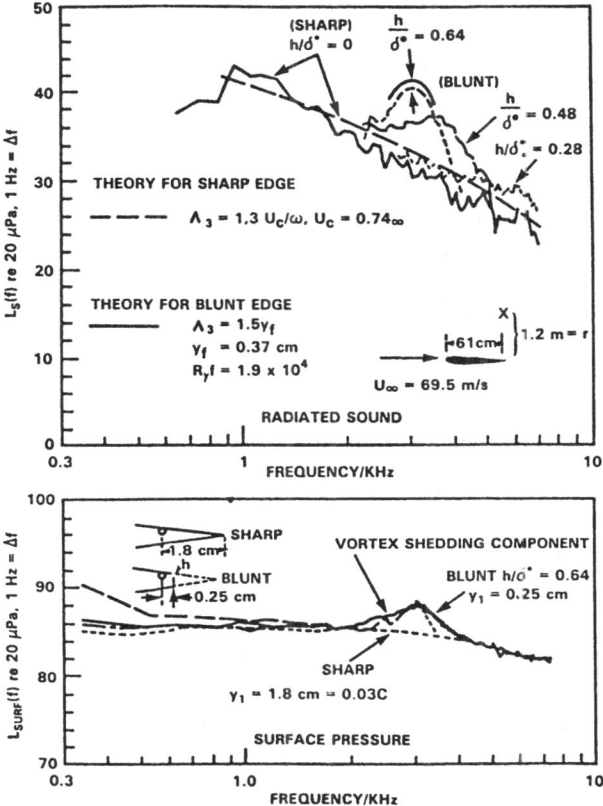

Figure 6. Radiated Sound and Surface Pressure for Sharp and Blunt Trailing Edges of a NACA 0012 Airfoil with Leading Edge Tripping, Data Taken from Brooks and Hodgeson (1981).

where y_1 is the wake thickness, U_∞ is the free stream velocity and $\mathcal{S}$ is a number (≈ 0.2), called the Strouhal number. The factor y_1 shown diagrammatically in Figure 7, is a property of the near wake and is the lateral distance between the wake shear layers as described by the maxima in the intensities of the fluctuating velocities. This factor will be further discussed in later sections of this paper. As Reynolds number decreases, as the upstream boundary layer thickness, δ, is thickened, or as the thickness of the trailing edge, h, is decreased, the magnitude of "vortex shedding" pressure is decreased and its relative

470

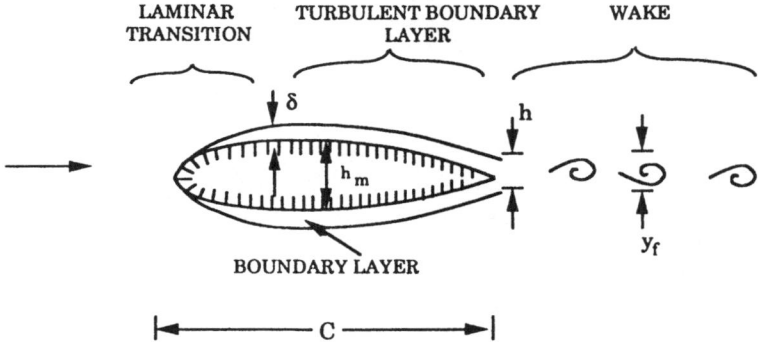

Figure 7. Characteristic Dimensions of a Lifting Surface and Typical Regimes of Flow.

frequency bandwidth or quality factor increases. The behavior
with both Reynolds number and trailing edge geometry is
summarized in Figure 8.

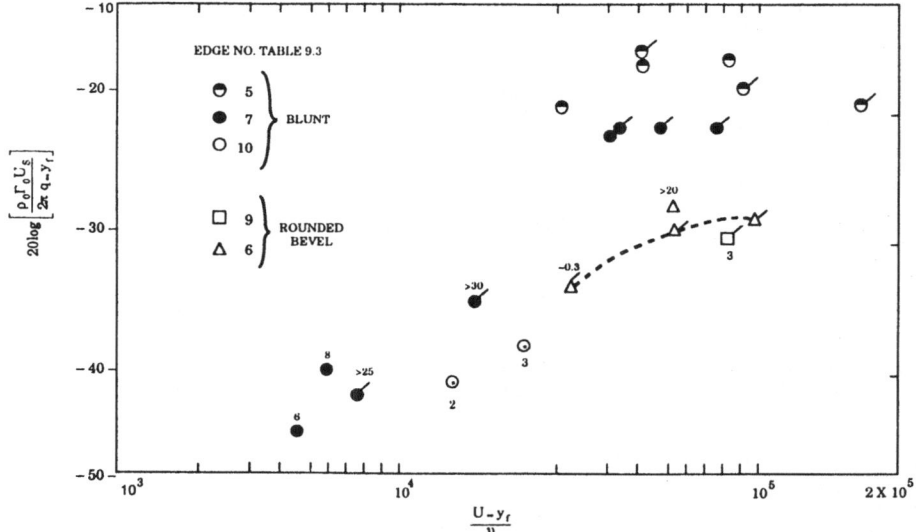

Figure 8. Mean-Square Vortex Strengths for Various Training Edges with Vortex Shedding. Unflagged
Points Obtained with Wake Intensity Measurements. Flagged Points Obtained with Surface
Pressure Measurements. Numbers Denote Quality Factors $f_s/\Delta f_s = Q$ for $Q < 30$.

Comprehensive review surveys of this subject with regard to
theoretical developments have been given by Brooks and Schlinker
(1983) and by Blake (1986) so only a few remarks, retrospective
of the important milestones in the development of the subject,
will be given here. Although first considered by Powell (1959),
the sound generation by turbulent trailing edge flow did not
really get widespread attention until Ffowcs Williams and Hall
(1970). The first quantitative relationships that set the format

for prediction and elaborated on the aeroacoustic scattering mechanism that governs this sound were published by Chase (1972, 1975). Amiet (1976) and Howe (1978) subsequently extended the theory to collectively consider fluid compressibility, finite chord and the importance of a trailing edge Kutta condition in establishing the acoustic source strength. Kutta conditions, in this regard, have been applied in the analyses in the classical way. The interaction of the flow disturbances with a sharp edge would create an analytical singularity in pressure that is removed by launching a second disturbance in the wake. The analytical modelling of edge flows in this manner has been confirmed in the case of a wall jet by Yu and Tam (1978) who observed that when upstream disturbances interact with the edge, the wake flow responds by generating another disturbance. More recently, modelling of Atassi (1984) used the Kutta condition to examine the frequencies of aeroacoustic tones emitted from laminar flow airfoils. The downstream vorticity developed from a wake instability launched at the trailing edge induces a velocity component across the edge. A Kutta condition applied to an acoustic disturbance that is generated at the edge results in a coupling of aerodynamic and aeroacoustic disturbances which control the allowed frequencies.

Perhaps, the first experimental evaluations of turbulent trailing edge noise were published by Grosche (1970) who considered the sound generated by a blown flap. The principal interest in this subject was in connection with edge blowing as this sound mechanism was thought to be important in VTOL aircraft. The first comprehensive experimental studies of sound from trailing edge flow of lifting surfaces were published by Schlinker (1977), but Yu and Tam (1978), Yu and Joshi (1979) and Brooks and Hodgeson (1981) collectively conducted measurement programs that tested and validated important features of the experiments. These features provide for sound radiated as upstream eddies are convected past the sharp trailing edge. In the analytical modeling, a tangency condition (Kutta condition) at the edge forces a second vortex to be shed in the wake. Sound is generated by edge dipoles which are established in the process. The strengths of the edge dipoles depend on the relative convection velocities of the incident and shed vorticity. Closed-form relationships between the surface and farfield pressures disclosed by the theory were confirmed by the experiments.

The essential Mach number dependence, i.e. M_∞^6 for compact air-foils and M_∞^5 for chords exceeding the acoustic wave length, and directivity patterns were confirmed by the measurements.

Sounds from vortex shedding that result from the Helmholtz wake instability have been given extensive attention from many perspectives for many years. Most recently consideration of vortex shedding when the upstream flow is turbulent and the edge is blunt has been considered by Blake (1985, 1986). This work shows that the surface pressures increase to a maximum as measurement location approaches the edge from upstream. Pressures on opposite sides of the airfoil are 180° out of phase as are those generated in the mechanism of aeroacoustic scattering.

The real distinction between these two forms of trailing edge noise thus seems to be in the chord-wise dependence of the surface pressures that are connected with the aeroacoustic sources. In the aeroacoustic scattering mechanism, the differential in pressure on opposite sides of the surface seems to vanish monotonically as the edge is approached, while in the instability mode, this pressure differential increases close to the edge only to be alleviated by some viscous effect, the nature of which has still not been given any consideration.

The new work that will be presented in this paper extends previous work by considering geometries of the trailing edge which promote a coexistent mix of flows. In this case the situation is similar to that depicted in Figure 6, but now caused by shape rather than by low Reynolds number. No periodic disturbances are formed, but two pressure fields with distinctly different frequency spectral forms, spatial variations, and correlation properties are associated with co-existing flows of different character. These surfaces are not clearly catalogued as their trailing edge flows are necessarily affected by both local geometry and upstream flow structure. It can be said, however, that on these shapes both forms of aeroacoustic source mechanism may coexist and they may both be continuous-spectrum. In the cases examined, the trailing edges are unsymmetric so that the natures of the flows on opposite sides is examined and their modes of interaction are considered. The paper will describe the surface pressure fields in context of both aerodynamic and aeroacoustic features. It will then examine the radiated sound and compare it, in a dimensionless fashion, to some previously collected measurements, and it will compare the measured sound with theoretical prediction. Measurements of cross-spectral densities between surface pressure and sound pressure will be used to verify the governing surface pressure-to-sound pressure Green function and to establish the effective aeroacoustic spanwise correlation length of the edge dipoles.

3.0 THEORY AND ANALYSIS OF TRAILING EDGE FLOWS

The purpose of this section is not to present a comprehensive and rigorous review of theory. Rather, we will examine a few results and their physical meanings particularly as they relate to various experimental data.

3.1 THE FUNDAMENTAL HALF-PLANE PROBLEM

The high-frequency continuous-spectrum sound emitted from the interaction of a trailing edge with the surface boundary layer turbulence is fundamentally a scattering phenomenon in which a high-wave-number disturbance incident on the edge generates a low-wave-number, acoustically propagating, pressure. In this section we shall review the conclusions of the mathematical fundamentals so that in later sections analytically derived prediction formulas may be set down with limited mathematical discussion. As the sophistication of analytical modeling of flow-edge interactions evolved following the work of Ffowcs Williams and Hall (1970) to include physically realistic acoustic and aerodynamic interactions, so did the mathematical complexity.

We first assume that the turbulent flow is convected past the edge (leading of trailing) and that the only interaction of importance is acoustic. That is to say, the disturbances created by the flow-edge interaction in no way feedback on the turbulence, no new aerodynamic disturbances are created. The surface is a half-plane, as shown in Figure 9 with the flow occupying the region $y_2 > 0$, in Figure 10. The mean flow of turbulence is parallel to the y_1, y_3 plane, but its mean direction makes an angle α with the trailing edge. The vertical extent of the turbulent zone is confined to the region $0 < y_2 < \delta$. Figure 11 depicts possible orientations of the vorticity vector $\vec{\omega}$; the component $(\vec{\omega} \times \vec{u})_2$ will be shown below to be particularly important in sound generation.

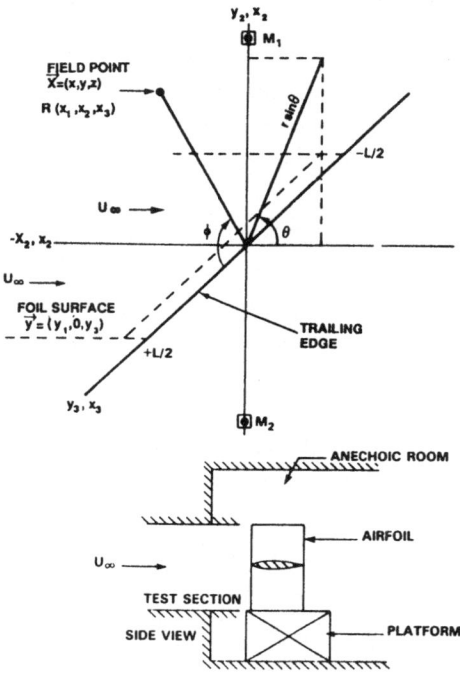

Figure 9. Coordinate System for Trailing Edge Radiated Noise and Schematic of the Wind Tunnel Arrangement of Gershfeld and Blake (1988)

A starting point for the analysis is the reduced wave equation, for isentropic flow, Howe (1975):

$$\nabla^2 B + k_o^2 B = -\nabla \cdot (\vec{\omega} \times \vec{u}), \qquad (2)$$

where $B = p_o/\rho + u^2/2$. The Helmholtz equation for B is

$$B(x,\omega) = \iiint [\nabla \cdot (\vec{\omega} \times \vec{u})] \, G(\vec{x},\vec{y},\vec{\omega}) dV$$

$$+ \iint_s \left[G(\vec{x},\vec{y},\omega) \frac{\partial B}{\partial n} - B(\vec{y},\vec{\omega}) \frac{\partial G(\vec{x},\vec{y},\omega)}{\partial n} \right] dS(\vec{y}) \qquad (3)$$

where the surface integral extends over the body. The Green function is the standard one for the body in an unbounded acoustic medium; it satisfies both the acoustic wave equation and the boundary conditions for either $\partial G/\partial n$ or G on the surface, depending on whether p or $\partial p/\partial n$, respectively, is specified in the problem. An alternative form for equation (3) for the case in which the motion of the surface is specified is:

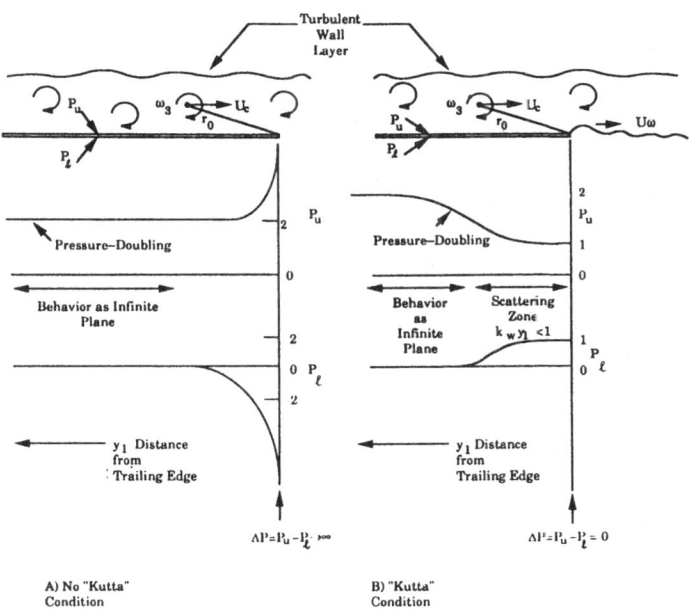

Figure 10. The Geometry of a Wall Jet Which Is Incident on the Trailing Edge of a Semi-Infinite Half-Plane (Blake (1986)).

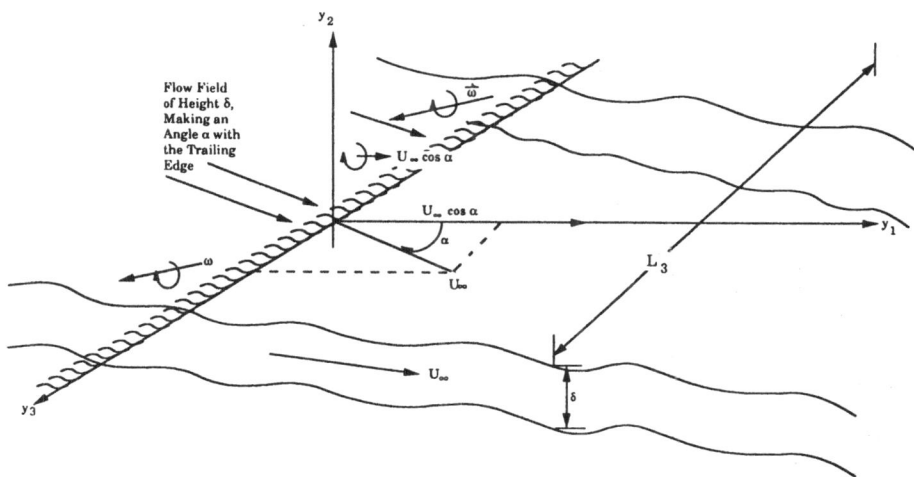

Figure 11. Behaviors of Surface Pressures Induced by Vortex-Edge Interactions and Associated with Alternative Pressure Boundary Conditions at Edges. The case shown is a one-sided wall-flow passing an edge such as a blow flap or wall jet. Pressures p_u and p_1, are referred to a value of 2 on the wetted side far from the edge.

$$B(\vec{x},\omega) = - \iiint (\vec{\omega} \times \vec{u}) \cdot \nabla_y G(\vec{x},\vec{y},\omega) \, dV(\vec{y}) - \iint_s \frac{\partial u_n}{\partial t} G(\vec{x},\vec{y},\omega) \, dS(\vec{y}) \quad (4)$$

It can be shown that (Blake (1986)) ∇_y represents the del operation with respect to $\vec{y}$ and u_n is the motion of the surface normal to its plane. Since the motion of the surface $u_n(\vec{y},\omega)$ is specified, the appropriate boundary condition for $G(\vec{x},\vec{y},\omega)$ on S is

$$\frac{\partial G}{\partial n}(\vec{x},\vec{y},\omega)|_y \text{ on } S = 0 \qquad (5)$$

$B(\vec{x},\omega)$ is then given by the sum of a volume integral over the source region and contribution due to the motion of the surface.

For the present, we shall assume that the surface is a rigid half-plane. As a practical matter, the analysis will apply to lifting surfaces of large chord and without fluid-induced vibration. The Green function for this case that satisfies equation (5) has been worked out by Ffowcs Williams and Hall (1970) for the special case of sources near the edge of a half-plane as illustrated in Figures 9 and 10. The Green function is composed of two parts; a function that describes the incident wave from the sources and a scattered field. It is the scattered-field component that governs the interesting behavior from the perspective of flow-induced sound. For the case such that the sources at y are near the edge and the flow point at x is far from the edge, the function (see MacDonald (1915) Crighton and Leppington (1970), Howe, (1975), and Bowman et al. (1969)) is

$$G(\vec{x},\vec{y},\omega) = \frac{e^{ik_0 r}}{4\pi r} \left[1 + \frac{2e^{-i\pi/4}}{\sqrt{\pi}} \sqrt{2k_0 r_0 \sin \phi_0} \sin(\theta_0/2) \sqrt{\sin \phi} \sin(\theta/2) \right] \qquad (6)$$

for sources near the trailing edge $[k_0 r_0 \ll 1, (|\vec{y}| = r_0)$ and $k_0 r \gg 1, (|\vec{x}| = r)]$. . The coordinate system is shown in Figure 9. The second term in the brackets describes the scattered field of the edge.

The sources in the flow are assumed to be convected along the half-plane, as shown in Figure 9 or 10. The mean flow velocity vector U makes an angle α with the edge in the y_1, y_3 plane. The edge may be either a leading or a trailing edge since at this point we assume no hydrodynamic interaction between the flow and edge. This simply means that no additional flow disturbances are created as the flow passed the edge of the half plane, and this may be regarded either as the case of a lack of "wetting" of the lifting surface by the flow or as a case in which a Kutta condition is not applied to the edge. Since the mean flow vector U lies in the 1,3 plane, the source vector $\vec{\omega} \times \vec{u}$ may be simplified somewhat to retain only terms that are linear in the disturbance quality; i.e., we rewrite $\vec{u} = \vec{U} + \vec{u}$, where the ensemble average of the velocity fluctuation vanishes, i.e. $\langle \vec{u} \rangle = 0$. All the fluctuations are thus embodied in $\vec{\omega}$, thus,

$$\vec{\omega} \times u = (\omega_2 \overline{U}_3) i + (\overline{U}_1 \omega_3 - \overline{U}_3 \omega_1) j + (-\omega_2 \overline{U}_1) k = \vec{\omega} \times \overline{U} \qquad (7)$$

where $U = (U_1, 0, U_3)$ and U_1 is the component of velocity directed across the edge. For a general vorticity patch that encounters the edge, has three possible components so the quantities $\hat{i}, \hat{j}, \hat{k}$, represent unit vectors in the y_1, y_2, y_3 directions, respectively. In (7) we have also assumed that mean vorticity $\nabla \times \overline{U}$ is zero.

Note that $G(\vec{x}, \vec{y}, \omega)$ is independent of y_3; $r_0 \sin \phi_0$ is the radial distance of the sources from the edge in a cylindrical coordinate system and is not a function of y_3. Therefore any contribution from $(\vec{\omega} \times \vec{u})_3$ is zero.

The component $(\vec{\omega} \times \vec{u})_1 \simeq \omega_2 \overline{U}_3$ represents a vortex filament whose axis is normal to the plane of the edge and interacts with it at a spanwise velocity $U_3 = U_\infty \sin \alpha$, as shown in Figure 10. This component can be assumed to vanish (Blake (1986)) for a spanwise symmetric flow in which there is no net variation in velocity or vorticity along the span of the edge. In this section the mean velocity $U = U_\infty$ is presumed to be independent of y_2.

Thus in this simple analytical model of a two-dimensional surface, which is a representation of most trailing edge flows, the only component of $\vec{\omega} \times \vec{u}$ contributing to dipole sound radiation is that directed normal to the plane of the lifting surface, i.e., $(\vec{\omega} \times \vec{U})_2 = \omega_1 \times U_3 - \omega_3 \times U_2$. The streamwise vorticity term is $\omega_1 = \partial u_3 / \partial y_2 - \partial u_2 / \partial y_3$ but if this has zero mean and if $U_3 = 0$, then the term $\omega_1 \times U_3$ must be second order.

The spanwise vorticity

$$\omega_3 = \partial u_2 / \partial y_1 - \partial u_1 / \partial y_2$$

cannot be considered to vanish on either symmetry or kinematic grounds; the product $\omega_3 \times U_1$ is first order due to the dominance of U_1, thus it provides the first-order dipole source. Equation (4) for a rigid surface reduces to

$$P_a(x, \omega) = -\rho_0 \iiint (\omega_3 \times U_\infty \cos \alpha) \frac{\partial G}{\partial y_2} (\vec{x}, \vec{y}, \omega) \, dV(\vec{y}) \tag{8}$$

where, α is the angle of yaw that the flow makes to the edge, see Figure 10. The frequency of encounter of a given wave number component k_1 is $\omega = (U_\infty \cos \alpha) k_1$.

The gradient of the Green function is:

$$\frac{\partial G}{\partial y_2} = \frac{e^{ik_0 r}}{4\pi r} \frac{e^{-i\pi/4}}{2\sqrt{\pi}} \left[\frac{2k_0}{r_0 \sin \phi_0} \right]^{1/2} \cos \theta_0 / 2 \sqrt{\sin \theta} \, \sin \theta/2 \tag{9}$$

Dimensional analysis of this equation yields a parametric relationship for the spectrum of sound pressure $\Phi_{PR}(\vec{r}, \omega)$ which is of the form

$$\Phi_{Pr}(\vec{r}, \omega) \approx \rho_0^2 U_\infty^4 \left(\frac{U_\infty}{c_0} \right) \left(\frac{\delta}{r} \right)^2 \frac{\Phi_{uu}(\omega)}{U_\infty^2} \cos^3 \alpha \left(|\sin \phi| \sin^2 \theta/2 \right) \tag{10}$$

where δ is a length scale of the turbulence field. Equation (10) shows that the sound pressure level in proportional bands (i.e., $\Delta \omega \alpha \omega$) is in direct proportion to the turbulence spectrum level $\Phi_{uu}(\omega)$, the square of the length scale of the turbulence region, and U_∞^5. It also shows that the sound pressure may be reduced by yawing or sweeping the edge by an angle α to the inflow. At yaw angles of 45°, the sound spectrum level is reduced to nearly one third (-5 dB) with no yaw. Equation (10) actually applies to either leading or trailing edge noise at high frequencies, although the magnitude of sound depends on the details of the flow and the aerodynamic mechanism of flow-edge interactions including the extent to which a Kutta condition applies or does not apply to the edge (as we shall discuss). A particularly notable feature of equation (10) is the fifth-power dependence of the sound pressure on velocity. Note that dipole sound from compact bodies increases as the sixth power of the velocity. This relationship applies to airfoils in the limit $k_0 C \ll 1$, as verified by measurements of Hersh and Meecham (1973) and Siddon (1973). The effect of the extended rigid surface is to reduce by one order of magnitude the dependence of the sound pressure on the Mach number because the surface baffles the fluid cancellations on one side of the dipole.

3.2   THE KUTTA CONDITION

Equation (6) can also be interpreted as the potential flow pressure at (r_0, ϕ_0, θ_0) due to a source at (r, ϕ, θ). On the rigid plane surface in the plane $y_1 < 0$ (i.e., $r_0 \sin \phi_0 > 0$ and $\theta_0 \pm \pi$), the normal velocity of the fluid, expressed by $\partial G/\partial y_2$ is zero. This velocity in the wake ($\theta_0 = 0$) becomes singular, however, as $y_1^{-1/2}$ [or as $(r_0 \sin \phi_0)^{-1/2}$] as the edge is approached, i.e., as $y_1 \to 0$ This singularity in $\partial G/\partial y_2$ as $y_1 \to 0$ implies that the pressure differential between the upper and lower surface also becomes singular. When a Kutta condition is applied in the analysis, this singularity is generally removed by adding functions that exactly cancel the growing singularity as $y_1 \to 0$. A Kutta condition may be justified since it provides for the shedding of vorticity downstream of the edge into the wake. The magnitude of this vorticity is determined by the requirement that the pressure differential induced across the surface by the wake is just sufficient to exactly cancel the singular pressure differential induced by the approaching upstream vortex. The flow thus "wets" the edge and responds to its presence. The shed vorticity in the wake is convected at a velocity U_w. The distinction between boundary conditions is shown schematically in Figure 11 (Blake (1986)). When no Kutta condition is applied in analysis, a singular pressure is allowed at the edge and the pressures on opposite sides are out of phase. When the Kutta condition is applied, the singularity is removed. A "complete" Kutta condition removes the differential pressure completely. The measured pressures show that the differential in surface pressure between the upper and lower surfaces increases as $y_1^{-1/2}$ in two physical circumstances: at the leading edge of a lifting surface responding to upstream flow inhomogeneities (so-called leading edge noise) and at blunt trailing edges downstream of which a vortex street is formed in the wake. These results suggest that no

Kutta condition should be applied in pertinent analyses. Such apparent singularities are not formed on the surface of wall jets for which the upstream boundary flow is turbulent and at sharp trailing edges of airfoils at high Reynolds number, indicating that a Kutta condition is appropriate in these cases. Thus, the application of Kutta conditions must be done with care, and one mathematical form may not be universally valid for all types of flows. Howe (1976) has examined the implications of a mathematical trailing edge Kutta condition for the two-dimensional problem of an infinitely long vortex filament with axis parallel to the edge y_3 and approaching the trailing edge from the surface side ($y_1 < 0$) by moving in the y_1 direction at a velocity U_c. The geometry illustrated in Figure 6 is conceptually that of physical flows. When a trailing edge singularity is permitted in the analysis, i.e., for no Kutta condition, the radiated pressure is given by

$$p_a(x,t) = \frac{\rho_0 \Gamma_3 U_c \sin \theta/2}{2\pi \sqrt{r}} \left[\frac{\cos \theta_0/2}{r_0^{1/2}} \right]_{t - r/c_0} \tag{11}$$

where the bracket denotes the location of the vortex is that the earlier time $t - r/c_0$ and where Γ_3 is the circulation of the vortex. This form closely resembles the three-dimensional result that would result from substituting equation (9) into equation (8), but it has the $1/\sqrt{r}$ geometrical spreading loss that is characteristic of two-dimensional acoustics problems. When a complete Kutta condition is applied, i.e., when the pressure differential at $y_1 = 0$ is taken to be zero, the sound pressure emitted is reduced from the case of no-Kutta condition by an amount ($1 - U_w/U_c$) because of the required shed vorticity. Thus the sound pressure is given by

$$p_a(x,t) = \frac{\rho_0 \Gamma_3 U_c \sin \theta/2}{2\pi \sqrt{r}} \left(1 - \frac{U_w}{U_c} \right) \left[\frac{\cos \theta_0/2}{r_0^{1/2}} \right]_{t - r/c_0} \tag{12}$$

Therefore, if the convection velocity U_w is equal to U_c, the radiated sound is identically zero. Observations by Yu and Tam (1978) of the vortex structure in a wall jet disclose that, in response to upstream eddies, wake eddies are shed at a velocity $U_w \approx 0.6 U_c$. More recent results of correlation analyses by Gershfeld, Blake and Kinsley (1988) discussed later in connection with Figure 11, disclose diminishing pressures as the edge is

approached. These results suggest that a Kutta condition should
apply to such flows involving upstream boundary layer turbulence
and that the sound pressure radiated could be significantly less
than the value given by theories based on classical acoustic
diffraction such as that used above due to the generation of
additional cancelling sources.

In summary, these results, like the results of measurements
to be described later, suggest that edge boundary conditions
permitting a $1//y_1$ dependence of the differential surface pressure as $y_1 \rightarrow 0$ apply to leading edge noise and to vortex shedding
noise. The singularity must be removed by the application of a
Kutta condition for those cases involving wall jets, blown flaps,
and the upstream boundary layer turbulence convected past the
edge. Essentially, this condition amounts to the requirement
that the flow leave the edge tangentially with respect to both
the mean and the instantaneous velocities in the immediate
vicinity of the edge. For either leading or trailing edge flows,
the essential dependence of the sound on the flow parameters is
still that given by equation (10) but with differing coefficients
of proportionality.

3.3 ACOUSTIC TONES FROM VORTEX SHEDDING BY RIGID SURFACES

When trailing edge flow consists of a discrete vortex shedding at frequency $f_s = \omega_s/2\pi$ the vorticity field can be modelled
as a sinusoidally-varying continuous sheet lying in the
$x_1 > 0$, $y_2 = 0$ plane, or $\theta = 0$ plane (see Blake (1986).

Physically, the vortex street is neither purely periodic nor
correlated along the entire length of the edge. The modelling
is, however, facilitated by assuming periodicity, and two-
dimensional flows local to the edge in the sense that the wake
width is small compared with the effective (or correlation)
lengths of the vortices along the span.

Thus

$$\omega_3(y_1, y_2, t) = \gamma_0(y_2) e^{ik_c(y_1 - U_c t)} dy_1 \tag{13}$$

where γ_0 is the elemental vortex circulation. The delta function
is the kronecker delta function defined as $\int \delta(y_2) dy_2 = 1$. γ_0 is
currently known experimentally for the cases given in Figure 8
when note is taken that

$$\gamma_0 = \frac{\Gamma_0 \omega_s}{2U_c} \tag{14}$$

where γ_0 is the circulation of the vortices. The mean square radiated pressure at the vortex shedding frequency is thus

$$\overline{p^2}_R(r,\omega_s) = \frac{1}{16\pi^2}\left(\frac{k_o}{k_w}\right)|\sin\phi|\sin^2\left(\tfrac{\theta}{2}\right)\rho_0^2\gamma_0^2\left(\frac{2\Lambda_3 L_3}{r^2}\right)U_c^2 \tag{15}$$

where $\qquad\qquad\qquad \omega_s y_f/U_\infty \cong 1 \tag{16}$

and where $\qquad\qquad\qquad \gamma_0^2 = \Gamma_0^2 k_w^2/4 \tag{17}$

To be precise, the velocity scale U_ω in these relationships, is one local to the potential flow at the trailing edge. This velocity scale is, say, $U_\omega=U_0\sqrt{1-C_p}$ where C_p is the coefficient of static pressure in the separation zone of the edge and U_0 is the free stream velocity to the section.

The flow is assumed to be correlated along the span for distances which may be represented as a correlation length where

$$2\Lambda_3 = \int_{-\infty}^{\infty}\frac{<\omega_3(y_3+r_3,t)\,\omega_3(y_3,t)>dr_3}{\omega_3^2} \tag{18}$$

and $2\Lambda_3$ is presumed to be larger than the physical width of the actual wake. The bracketed term is the normalized spatial correlation function of the wake disturbances along the flow span. Mathematically it is represented as a measure vorticity correlation, it may be physically determined by measurements of velocity fluctuations in the wake. Thus the wake is modelled as an infinitesimally thin sheet, whence the delta function $\delta(y_2)$ which is physically thin enough that $y_1 << \ell_t$ and that $y_f/C << 1$.

The mean square surface pressure at a point $y_1 < 0$, $p_s^2(y_1)$ is given by (Blake 1986).

$$\overline{p_s^2}(y_1) \approx \frac{\pi}{16}\left(\frac{U_c}{U_\infty}\right)^2\frac{y_f}{|y_1|}\cdot\rho_0^2 U_\infty^2\left(\Gamma_0/2\pi y_f\right)^2 \tag{19}$$

which shows the singular behavior as $y_1\to 0$ as would be expected if a Kutta condition is not applied.

The relationships (15) and (19) assume that the wake extends from the edge at $y_1 = 0$; their ratio gives an equation that does not include the vortex strength. Thus

$$\frac{\overline{p_r^2(\vec{r})}}{\overline{p_s^2(y_1)}} \approx \frac{1}{2\pi^2} \frac{U_\infty}{c_0} \frac{2L_3}{y_f} \frac{|y_1 - y_s|}{r^2} \Lambda_3 |\sin\phi| \sin^2\frac{\theta}{2} \qquad (20)$$

relates the surface pressures to sound pressures for trailing edge vortex shedding. In this equation y_1 is replaced by $y_1 - y_s$, because, as we shall discuss below on non-symmetrical edges, the appropriate coordinate for describing p_s in equation(19) must be measured from a stagnation point on the edge y_s. For symmetrical blunt edges $y_s = 0$ on which the vortex shedding is also symmetrical. Equation (20) also applies for the "randomized" vortex shedding as occurs on the bevelled edges of Gershfeld et al (1988) and of Brooks and Hodgeson (1981), Figure 6, in which a tone is not generated. In such cases the ratio of mean square pressures may be replaced by ratios of auto spectra.

3.4 CONTINUOUS SPECTRUM SOUND: ACOUSTIC SCATTERING THEORY

The theory of sound pressure from turbulent flow past trailing edges that resulted in equation (10) applies to practical edge flows. In fact, equation (10) suggests qualitative dependencies on flow-acoustic parameters that hold empirically for a great deal of measured sound pressure levels for a variety of airfoils and wall jets. The possibility of quantitative prediction of aerodynamic sound from fundamental theory is quite another matter, because equations such as (10) do not incorporate sufficient description of the turbulent field relating to the vorticity distribution near the edge. The most useful theoretical formulations for prediction purposes are those which yield the spectrum of radiated sound pressure in terms of the spectrum of aerodynamic surface pressure and its spanwise integral scale. Their physical implications have been fully reviewed by Blake (1986) and that review shall not be repeated here.

A reasonably complete analysis of the three-dimensional acoustic problem of trailing-edge noise from sharp edges has been given by Howe (1978). The approach solves the homogeneous wave equation (2), in which the sources are controlled by the $\omega_3 \times \overline{U}$ term, for subsonic flow. The turbulent sources are postulated to be in a region above and adjacent to the sharp-edged plate and in the wake downstream of the edge. The solution of the problem includes both the far-field and surface pressures generated from

the edge-flow interaction Figures 9 and 10 show the appropriate
geometry.

Howe (1975, 1978) applies a Kutta condition at the trailing
edge that relieves the singular pressure by shedding vorticity in
precisely the same manner as described in Section 3.2. Postu-
lating that for flow on one side of the surface the source region
above and downstream of the edge constitutes a region of non-
vanishing $\nabla(\vec{\omega} \times \vec{U})$, Howe obtains an expression for the far field
sound pressure:

$$P_a(r,\theta,\phi,t) = \frac{-i\rho_0 \sin(\theta/2)\sqrt{\sin\phi}}{r2^{1/2}} \int_{-\infty}^{\infty} \frac{dk_1}{k_1} \int_0^Z dZ\left(1-\frac{U_w}{U_c}\right) \left(\frac{U_c}{U_\infty}\right)^{1/2}$$

$$\times (\vec{\mu} \cdot (\vec{\omega} \times U_c)) \, e^{-k_1 z} e^{i(k_0 r - \omega t)}$$

$$(21)$$

$$\vec{\mu} = (k_1, -i|k_1|, k_0\cos\phi); \quad \vec{\omega}(\vec{k}_{13},Z,\omega) = \frac{1}{(2\pi)^3} \int_{-\infty}^{\infty} e^{-i(\vec{k}_{13}\cdot\vec{y}_{13} - \omega t)} \omega(y,t)d^2\vec{y}_{13}dt$$

where $\vec{k}_{13}$ is a wave vector in the plane of the surface, (k_1,k_3),
$\vec{y} = (y_1, z, y_3)$, and $U_c - U_w$ arises from the application of the Kutta
condition as noted above. If the wake vorticity is convected at
the same velocity as the incident vorticity, $\vec{\omega}$, then $U_c = U_w$ and
equation (21) shows that there will be no sound at all.
Typically, $U_c > U_w$ so one expects sound to be generated. If the
Kutta condition had not been applied, then $1 - U_w/U_c$ is replaced
by unity, see also equations (11) and (12).

Assumptions made in the derivation of equation (21) are
summarized as follows:

(1) The eddy field is frozen during the time that it translates past the edge;

(2) The eddy convection velocity $U_c(y_2)$ is equal to the
local mean velocity in the boundary layer.

(3) There is no correlation between eddies that translate
at different values of $U_c(y_2)$.

(4) The wake vorticity created in response to the eddy-edge
interaction by the imposition of the Kutta condition is concentrated in a thin sheet $\delta(y_2)$ and is convected in frozen fashion
at velocity U_w; the plate has zero thickness.

(5) The source term $\vec{\omega} \times \vec{u}$ is such that $\vec{\omega} \times \vec{u} \approx \omega_3 \times \vec{U}$.

From the foregoing results we can find a useful relationship
between the frequency spectra of radiated sound and of the
surface pressure on planes that are terminated by rigid knife
edges. This relationship will make it unnecessary to devise

prediction schemes in terms, of absolute values of vorticity measurements.

The frequency spectral density of the radiated sound pressure at a point in the far field can now be written in terms of the surface pressures. The spectrum of the radiated sound,

$$\Phi_{P_{rad}}(r,\omega) = \frac{1}{8\pi^2} \sin^2\frac{\theta}{2} |\sin\phi| \; \mathcal{M}_c \frac{L_3 \; 2\Lambda_3}{r^2} \; \mathcal{P}(\omega)$$

(24)

where $\mathcal{P}(\omega)$ represents the integrated influence of the entire vortical source region and $\mathcal{M}_c$ represents the average convection Mach number of turbulence past the edge. The function $\mathcal{P}(\omega)$ is dependent on the strength of the upstream vorticity and its geometric relationship to the edge and relative convection velocities of upstream velocity and wake vorticity. $\mathcal{P}(\omega)$ is dimensionally in the form of a pressure spectrum at a point on the edge; i.e. it scales parametrically as

$$\mathcal{P}(\omega) = (\rho_0 U_\infty^2)^2 \phi(\omega)$$

(25)

where $\phi(\omega)$ has dimensions of time and is defined for ω extending from $-\infty$ to ∞. This leads to a nondimensionalization of the sound pressure which is of the form

$$\frac{\Phi_{PR}(\vec{r},\omega)}{q^2 \mathcal{M}_c \sin^2\frac{\theta}{2} |\sin\phi| \frac{L_3 y_f}{r^2} F(\frac{\omega y_f}{U_\infty})}$$

(26)

where y_f is a measure of the turbulence length in the near wake scale. This may be the wake thickness y_f or the boundary layer thickness near the edge, δ, but one may also use the boundary layer thickness or some indirect variable like the air foil thickness, h_m. This latter scale may be selected as an expediency in cases where direct measurements of the flow are not available. In geometrically and dynamically similar flows it may be stated that

$$y_f \sim \delta \sim h_m.$$

(27)

Figure 12, from Blake (1986) shows the examples radiated sound spectra from a series of sharp trailing edges of NACA airfoils.

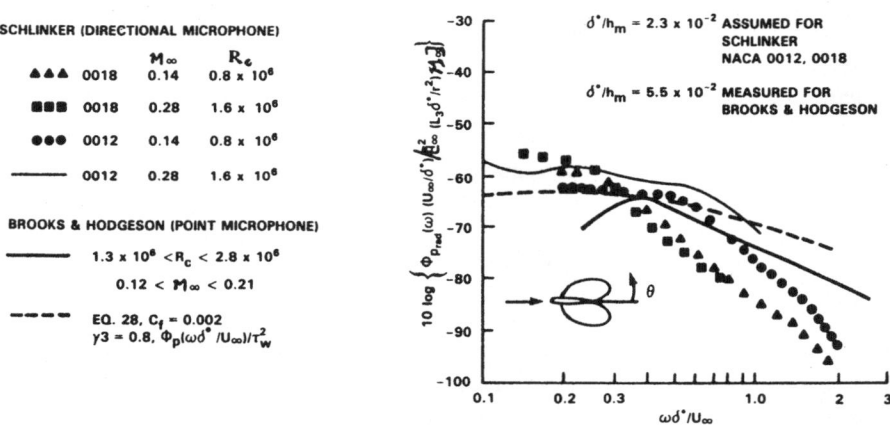

Figure 12. Spectra of Trailing-Edge Noise from NACA Airfoils in Uniform Flow, see Blake (1986).

Also shown is an estimation of the sound that would be generated by the idealized flow of homogeneous boundary layers past a rigid half plane with uncorrelated flows on both sides. That sound is given by

$$\frac{\Phi_{PR}(\vec{r},\omega)\,U_\infty/\delta^*}{q_\infty^2\,\mathcal{M}_\infty\,(L_3\,\delta^*/r^2)\sin^2\theta/2|\sin\phi|} = \frac{C_f^2}{2\pi^2\gamma_3}\left(\frac{U_c}{U_p}\right)^2\left[\frac{\Phi_{PP}(\omega\delta^*/U_\infty)(\omega\delta^*/U_\infty)^{-1}}{\tau_w^2}\right] \quad (28)$$

where C_f is the wall shear coefficient, $\gamma_3 = 0.8$, is determined by the spanwise cross spectrum of the wall pressures upstream of the edge, and $\Phi_{PP}(\omega\delta^*/U_\infty)/\tau_w^2$ is the nondimensionalized spectrum of wall pressure at a point. δ^* is the displacement thickness of the boundary layer at the trailing edge. The theoretical spectrum (28) agrees better with the measurement for $\omega\delta^*/U_\infty < 1$ than for higher frequencies. The reasons for this have not been resolved, but the actual flow may have less of a high frequency contribution because of the typical adverse pressure gradient on the after section of the airfoils. This gradient could have the effect of reducing high frequency pressure fluctuations as observed in boundary layer studies (Blake 1986).

4.0 REVIEW OF TECHNIQUES FOR MEASURING RADIATED SOUND

One of the most limiting factors in making measurements of aerodynamic sound in experimental facilities is contamination from the facility background noise. This is particularly so when dealing with continuous spectrum sound. Such sound is generally weakly radiated and has very little spectral character making it difficult to distinguish it from facility background noise. The background noise may also have a similar dependence on speed as that of the test source.

This problem is particularly important in the study of turbulent trailing edge noise from airfoils. Various authors have attempted to overcome problems of background noise; these will be briefly surveyed below. This is not intended to be an exhaustive review, but some example references will be given. These methods generally fall into three broad categories: measurements with
- single conventional microphones in the far field
- directive microphones in the far field (beamforming)
- cross correlation of microphones in the far field (multiplicative processing).

Figure 13 shows illustrations of the various alternative forms of instrumentation which have been used to measure far field radiated sound from airfoils and the relationship of sound to surface pressure.

Measurements with single microphones in the far field are generally limited by facility background noise. Most of the earliest measurements were made in this, the most straight-forward, manner. Examples are those of Yu and Joshi (1979) and of Clark and Ribner (1969). Aravamudan and Harris (1979) and Hersh and Hayden (1971), used this method for rotating blades for which the ratio of source velocity to facility velocity is made as large as possible thus providing a large ratio of signal-to-noise. This method is also used with the many measurements on wall jets or in cases when the aerodynamic source is generated within a jet. This last method provides a reasonably modest facility that benefits from the fact that subsonic dipoles are more efficient radiators than subsonic jets. Space does not

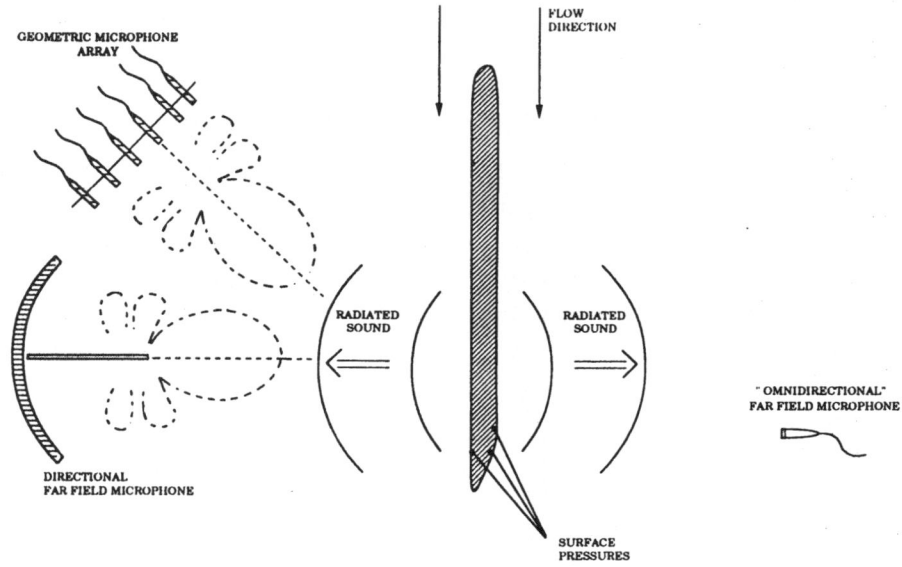

FLOW
DIRECTION

GEOMETRIC MICROPHONE
ARRAY

RADIATED
SOUND

RADIATED
SOUND

"OMNIDIRECTIONAL"
FAR FIELD MICROPHONE

DIRECTIONAL
FAR FIELD MICROPHONE

SURFACE
PRESSURES

Figure 13. Diagram of Types of Sensors Used in Measuring Surface and Far Field Sound Pressures.
Dashed Lines Illustrate the Type of Directivity Patterns to be Expected of Directive Sensors.

permit the full survey of such work which, really, has been
surveyed elsewhere (Blake (1986)). The most significant drawback
of this method is the unavoidable constraint on the size of the
test body.

Most work on larger sources, e.g., lifting surfaces has had
to rely on benefits of acoustic signal processing. The directional microphone relies on the geometric focussing of a backing
surface which acts as an acoustic mirror. Lucas and Muir (1982)
have recently presented the general theory of such devices although to the authors' knowledge Grosche, Stiewitt, and Bender
(1976) were the first users for aeroacoustics measurements.
Schinkler (1977) and Kendall (1978) have used it to measure trailing edge noise and flap noise, respectively. The directional
microphone becomes most effective at frequencies high enough so
that the wavelength of the sound is smaller than the diameter of
the reflector. The reflector may be a sector of a spherical
surface, but other geometries may be used. At higher frequencies, diffraction may limit performance by adding additional
receiving lobes. The lower limit frequency for practical gain is

that for which the acoustic wave length is less than roughly 1/5 the diameter of the reflector.

Geometric arrays of microphones have been used by Soderman and Noble (1975) Billingsley and Kinns (1976) and Brooks, Marcolini, and Pope (1987a) for aeroacoustic studies. Brooks, Marcolini, Pope (1987b) have used this method to examine noise from helicopter rotors. Though the microphone array is electronically more complex than the directional microphone, it is potentially more versatile. The lower limiting frequency for directivity is

$$f_{\ell} = c_0/2L_A \qquad (29)$$

where L_A is the smallest dimension of the array and c_0 is the speed of sound. Above a frequency

$$f_u = c_0/2L_S \qquad (30)$$

where L_s is the spacing of the microphones, aliasing lobes begin to give ambiguity in discriminating between sources. Thus, from f_l to f_u the microphone array will discriminate unambiguously between the sources. The ability of the array to discriminate between the trailing edge sound and isotropic background noise increases as the number of receivers, L_A/L_S. Performance may of course, be degraded if the facility background is not diffuse. These rules only apply when the distance from the source to the center of the array exceeds, roughly $4L_A^2/\lambda_0$ where $\lambda_0 = c_0/f$. Improved performance is possible if the elements of the array are electronically "shaded". The reader is referred to texts on underwater acoustics for more details.

The third manner of improving the signal-to-noise ratio in aeroacoustic measurements, and that used for Part II of this chapter, is by cross correlating two or more far-field microphones. In cases for which the background noise is diffuse, the cross-correlation of sound pressures at two points that are separated a distance of at least 2 wavelengths apart will be fairly small (say <0.1). Any considerable directionality in the ambient sound field could result in a larger correlation coefficient. However, if directionality exists in the background sound field it will be discernable by a phase shift

$$\chi_s = \frac{2\pi fr_m}{c_{TR}} \tag{31}$$

where $c_{TR} = c_o/\cos\theta_{TR}$ is the trace wave speed of a wave arriving at angle θ_{TR} to a line connecting the microphones. Brooks and Hodgeson (1981), Brooks and Marcolini (1986), and Gershfeld and Blake (1988), have used this method for measurements of lifting surface sound. This technique is best used when advantage is taken of the features of the sound field of interest. For example, in the application of trailing edge sound the directivity is known to be normal to the surface. Microphones placed in the far field on opposite sides as illustrated in Figure 9 will give a cross spectrum phase α_s that is constant, $\alpha_s = 180°$, at all frequencies for which the measurement is dominated by airfoil sound. Letting p_n be the facility noise, and P_a the airfoil sound, then the cross correlation of the signal from microphones 1 and 2 in Figure 9 is

$$\langle p_{s_1} p_{s_2}\rangle = \langle p_{a_1} p_{a_2}\rangle + \langle p_{n_1} p_{n_2}\rangle = \langle p_{a_1} p_{a_2}\rangle \tag{32}$$

when the noise field is perfectly diffuse, $\langle p_{n1}p_{n2}\rangle = 0$ when $fr_m/c_o \gg 1$, where r_m is the distance between the microphones. In practical experiments, we are interested in cases for which $\overline{p_n^2} \geq \overline{p_a^2}$ i.e., negative signal-to-noise ratio environments. Figure 14 shows an example from Gershfeld et al (1988) in which the sound levels at a single microphone exceed these of trailing edge source by roughly 5 dB. By the cross correlation technique the signal-to-noise ratio of the measurement (S/N) was enhanced by the amount shown.

The cross correlation technique may be substantially supported when the far field microphone signal is correlated with a microphone on the surface. In this case, the phase between the two sensors will be approximately

$$\alpha_{ss} = 2\pi fr_{ss}/c_o + \alpha_o + 2\pi fy_{ss}/U_c \tag{33}$$

where α_o is some reference, r_{ss} is the distance from the acoustic source center to the far field; y_{ss} is the distance on the surface between the measurement of surface pressure and the acoustic source center where aerodynamic disturbances of convection

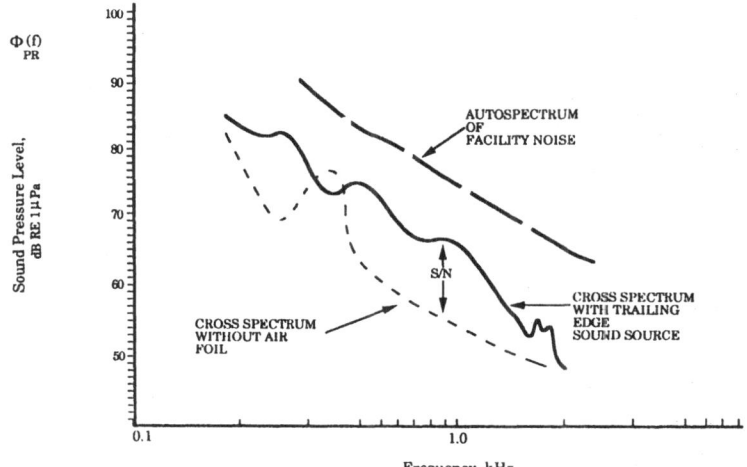

Figure 14. Cross Spectrum of Sound Field with and without Airfoil Using Microphones M1 and M2 on Opposite Sides of Trailing Edge. Measurement Taken at U_∞ = 150 ft/sec and Shows Coherent Signal to Noise Ratio for Thin-Edged Foil Shown inFigure 14.

velocity U_c are converted to sound, i.e. near y_1=0. As long as y_{ss} is small (in this case the surface measurement is very close to the edge) the variation of α_{ss} with frequency will be a sensitive indicator of how much the far field signal is dominated by the sources. This method of measurement may also be conducted in an environment for which

$$\overline{p_n^2} \geq \overline{p_a^2}$$

The first test of significance in measurements of this type thus could be the α_s = 180° phase shift test of the radiated sound field. Then the second test could be that of measurable α_{ss} between surface and far field transducers. For those frequencies for which α_s is clear, but α_{ss} is unclear the measurement may be doubtful, yet not necessarily invalid. This is because the measurement of α_{ss} will also be subjected to the influences of non-radiating components of the turbulent surface pressures which will appear as noise in the measurement. One advantage of the cross-correlation method is that it does not suffer the same geometric limitations that occur in the directive array method.

In the methods using the directional array, the measurement may be influenced by the refractive effects of the flow if the array is in a still air region. If r_0 is the source-to-receiver distance, the shift in the apparent source location will be downstream a distance Δx where $\Delta x = r_a U_\omega / c_0$

where U_ω is the velocity of the flow, - Grosche, Stiewitt and Binder (1976) have confirmed this behavior.

The measurements of Gershfeld and Blake (1988) that will be expanded on in Part II of this paper were obtained with the cross-correlation method. The size of the airfoil (dictated by requirements of the highest possible Reynolds number) coupled with the size of the facility, forced the frequencies of interest into a region for which an array length-to-source range was not an optimum relationship. Both conditions on α_s and on α_{ss} were used in interpretation of the data.

PART II   NEW WORK ON DETAILS OF FLOW-EDGE INTERACTION

In this part we expand on work recently published by Gershfeld et
al. (1988). This work carries the knowledge of trailing edge
aeroacoustics into a realm of non ideal flows which are unsymmetric on the two sides of the airfoil and which are formed at
trailing edges which are neither squared-off blunt nor sharp.

5.0 MEASUREMENT PROCEDURES FOR THE NEW WORK

New work has been done to more clearly elucidate the
relationship between vortex structure and the quality of sound
produced by trailing edge flows. This work, recently presented
at the National Fluid Dynamics Conference (Gershfeld et al
(1988)), examines the noise and flow structure of trailing edges
which are unsymmetrically leveled. Figure 15 shows sketches of
the two trailing edges which were installed on a standard airfoil
form.

Figure 15. Aft Half of Airfoil Sections of Two-Dimensional Foils with a Blunt Trailing Edge, and Thin
Trailing Edge. The Front Half of Both Airfoils Are Identical.

These edges were configured to provide varying mixes of orderly
wake structure and locally-generated turbulence due to separated
and nearly-separated flow. The blunter edge generated the more
prominent vortex flow. The two trailing edges that were the
subjects of the new work described in this paper were attached to
foils which were two-dimensional, and were made of solid
mahogany. Each foil had 60 μ in. r.m.s. roughness applied in a
0.5 inch strip at 0.05 chord from the leading edge to trip the
boundary layer. The upper side of both foils shown in Figure 15
generated a separated flow field in the trailing edge region.
The blunter-edged foil generated a nearly-periodic vortex street
that gave rise to nearly-tonal surface pressures in the trailing
edge region and associated vortex shedding.

The fluctuating flow field quantities were measured in the
Anechoic Flow Facility (AFF) of the David Taylor Research Center.
The wind tunnel test section (8 ft. X 8 ft. cross section) has a
low turbulence level that exits into an anechoic room (20 ft)
which facilitates aeroacoustic measurements and a maximum velocity of roughly 200 f/s. The wind tunnel is described in a report
by De Metz et al. (1973).

The trailing edge wall pressure spectra were measured with
small 0.125 in. diameter strain-gage type Kulite LQ-125 transducers similar to those described by Brooks and Hodgeson (1981).
Small parallel chordwise grooves 0.25 in. wide were routed at
various spacings in the trailing edge region for transducer
placement either parallel to the edge (transverse separations =
r_3) or chordwise (streamwise = r_1). The transducers were flush-
mounted by fairing over the grooves with plasticine clay so that
only the sensor faces remained exposed. The pressure transducer
signals were amplified with Ectron series 750 Differential Ampli-
fiers. They were post-amplified with Ithaco Model 451M125 ampli-
fiers.

Aerodynamic sound measurements were made using a cross-
spectral technique involving two microphones at symmetric field
points to the foil so that only the coherent sound of the foil
was measured as was discussed above. A pair of Bruel and Kjaer
(B&K) 0.5 in. diameter phase matched condenser-type omni-
directional microphones, (cartridge type 4165) were used for this
purpose. A B&K 2633 preamplifier and Ithaco AC amplifier were
used to condition the signal. Microphones were placed in the
anechoic room, outside of the jet shear layer on opposite sides
of the foil aligned normal to the trailing edge of the foil.
Figure 9 shows the coordinate geometry of the microphones
placement relative to the edge of the foil. A sketch is also
included to show the foil orientation in the anechoic room. The
microphones were aligned slightly downstream of the actual edge,
to account for the shear layer refraction affect on sound waves
passing from the jet (moving fluid) to the stagnant region of the
microphones.

Multiplicative signal processing as described in Section 4.0
was performed with a Hewlett Packard 3562A Dual Channel Dynamic
Signal Analyzer. Ensemble averages ranging from 256 to 4,000
were taken depending on the physical coherency expected in a
particular measurement.

6.0 SURFACE PRESSURES ON TRAILING EDGES

6.1 AUTOSPECTRA OF PRESSURES AT A POINT

On trailing edges which generate a weak vortex structure together with a super imposed random field, the spectrum of surface pressures may appear as shown in the lower half Figure 6. The discernable bump that is centered on a frequency $\omega y_f / U_\infty \approx 1$ is due to the ordered vortex structure in the wake caused by instabilities while the flat portion of the spectrum is due to the locally-generated boundary layer structure. On the edges shown in Figure 15, it appears that the influence of the vortex street structure on the surface pressure is even less apparent. This is illustrated in Figure 16 by the examples of auto spectra on the two sides of the surface.

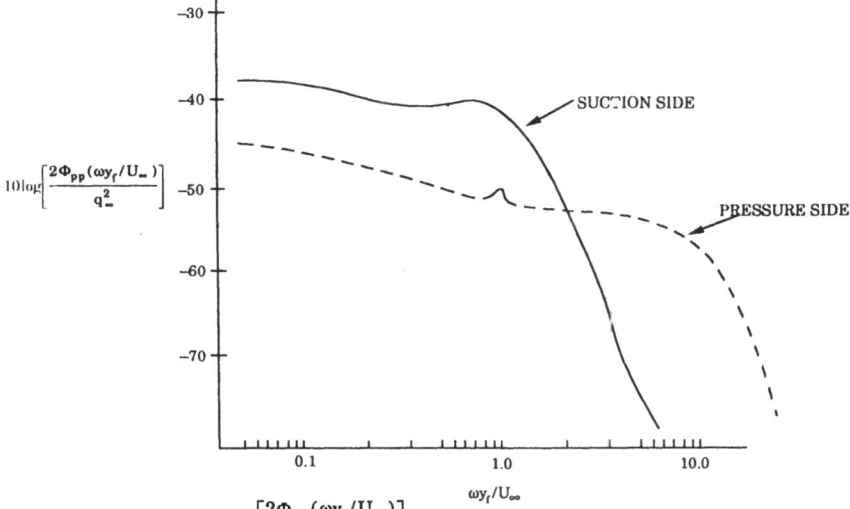

Figure 16. Autospectra of $\left[\dfrac{2\Phi_{pp}(\omega y_f / U_\infty)}{q_\infty^2}\right]$ for Pressure and Suction Sides of Blunter Edge Shown in Figure 15.

Further discussion of these data will be given later in this section. When such a mix of ordered- and locally-generated random structure occurs, the surface pressure fields which are generated must be considered separately.

In the discussion which follows we will be dealing separately with pressures which are in the vicinity of $\omega y_f / U_\infty \approx 1$ that may be generated by the vortex-street structure and the pressures at other frequencies. It is useful to point out first some general structural features pertinent to the wake-edge structure. Cross-

correlations between the fluctuating wall pressure and turbulent
velocities were previously measured by Blake (1975, 1984) in the
separated zone of a similar edge. These measurements, some of
which are summarized in Figure 17, disclosed that the velocity
fluctuations in the free shear layer appear to be the dominant
sources of turbulent wall pressures at the trailing edge. The
measurements were conducted in the wake of a flat strut with a
chord-to-thickness ratio of roughly 20. The upper diagrams show
the velocity profiles and static pressure distribution in the
vicinity of the edge. The lower 3 diagrams relate to detailed
correlation measurements. The separation occurs between locations D and F as indicated by the flattening of the static pressure distribution. The fluctuating velocities show a pair of
peaks along locii y_u and y_l with a lateral displacement y_u-y_l
which first decreases, then increases as the wake spreads.

The region for which y_u-y_l decreases is that for which the
vortex structures form; Bearman (1967) has shown that for
squared-off edges the magnitudes of the fluctuating velocities
which are determined by the developing vortex structures increase
as y_u-y_l decreases with distance downstream of the edge. This is
the "formation zone" and the maximum in y_u-y_l defines the wake
length scale y_l. The relevance of wake structures to the surface
pressures is disclosed by pressure-velocity correlations, <pu>
with the fluctuating velocities measured along y_u and y_l. As
shown in Figure 17, the magnitude of <pu> increases as the
velocity probe is moved downstream and the sign of the
correlation alternates between positive and negative with further
placements downstream in the wake. These correlations are the
maximum magnitudes of the moving axis correlation as determined
by the convection time delay $\bar{\tau} = r_1/U_c$ where r_1 is the streamwise
distance between the pressure and velocity probes and U_c is the
convection velocity of the fluctuating disturbances downstream of
the edge. As this character of the correlation is also manifest-
ed by the periodic vortex shedding of squared off edges, it was
interpreted (see Blake (1975, 1984)) as locating the average
position of vortices in a wake structure that resembles a von
Karman vortex street. It turns out that this more-ordered vortex
structure, or weak von Karman vortex street, is responsible for
surface pressures localized in frequency near $\omega y_f/U_\infty$~1. Simultaneously, the more local random turbulent field generates

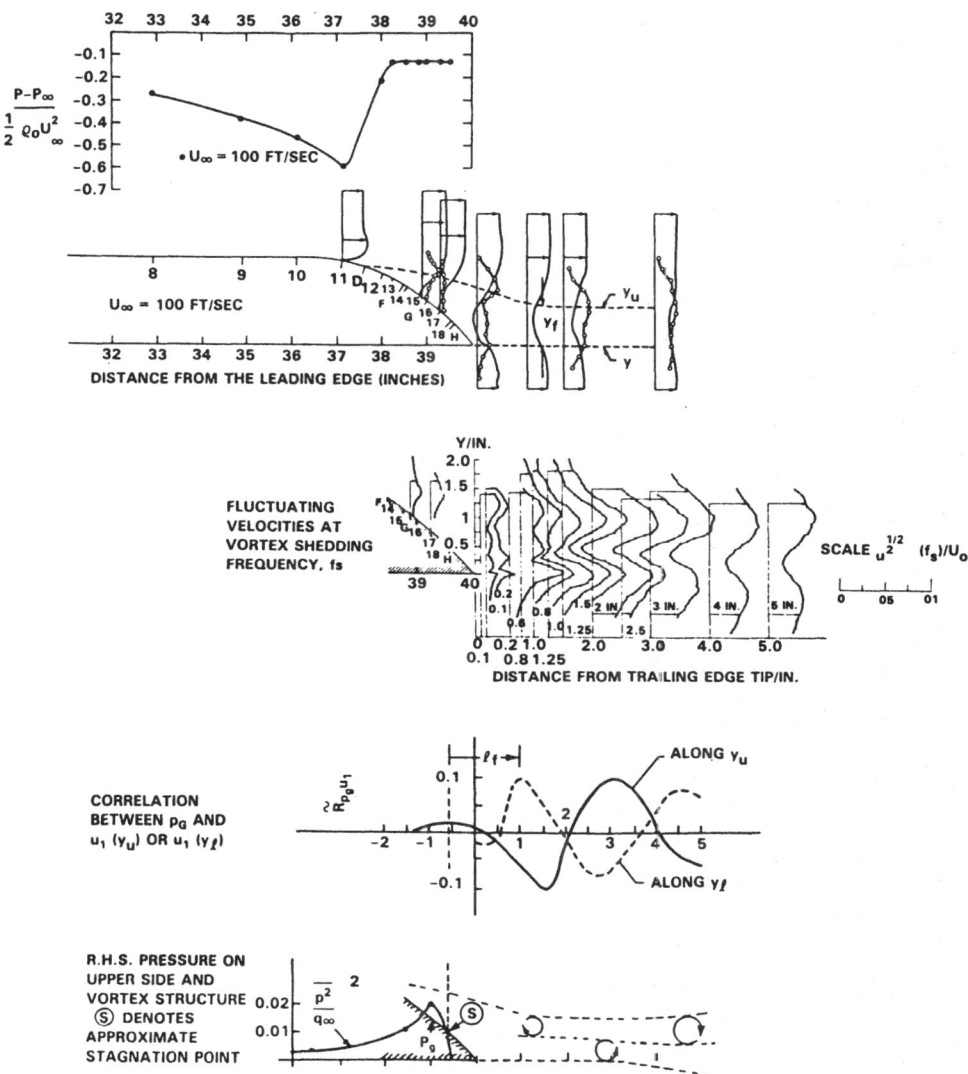

Figure 17. Flow structure and pressure distributions on a beveled trailing edge

continuous-spectrum pressures at other frequencies. Also, shown
in Figure 17, is an indication of the chordwise profile of fluctuating pressure at $\omega y_f/U_\infty \approx 1$ along the suction side of the edge.
The magnitude drops off roughly as prescribed by equation (20)
for the periodic vortex street.   The point(s) denotes the
coordinate y_s in that equation.

Figure 18 shows some additional measurements of similar
correlations downstream of another, thinner, trailing edge with a
smaller zone of separated flow.

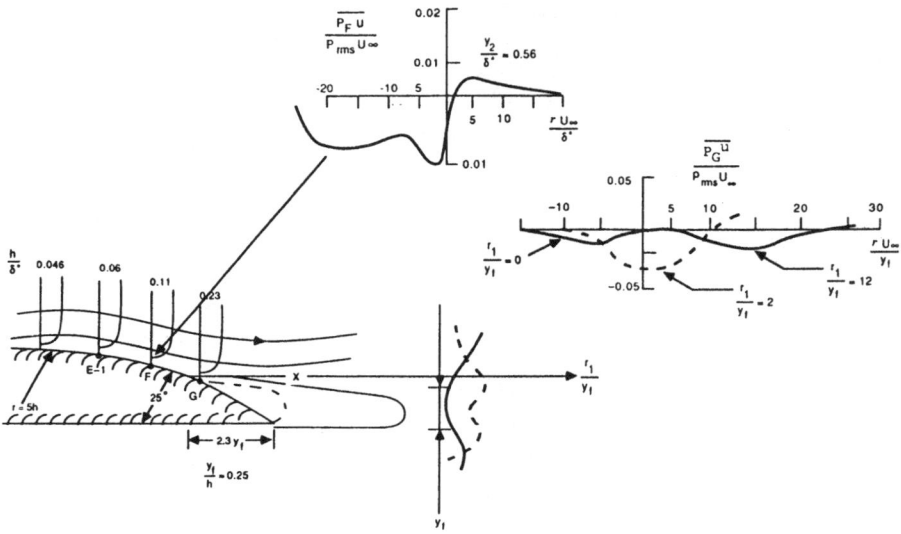

Figure 18. The Development of the Large-Scale Wake Structure Downstream of a Thin Nonsymmetric
Wake, as Indicated by the Behavior of Pressure Velocity Correlations.

Just upstream, of separation, the pressure velocity correlation
is typical of those obtained with boundary layers, and it is to
be noted that the correlation is principally negative and continues over long time delays. With the velocity probe downstream
of the edge, the correlations are largely negative. These negative dips of the correlation translate to larger values of time
delay as separation distance from the edge, r_1, increases. This
largely negative behavior is also typical of the turbulent
boundary layer as noted above when the height of the probe is of
order $\delta*$ or greater, which location is in the Coles' (1956) wake-
like flow region of the layer. Note also the increase of $<-pu>$
as r_1 increases to $r_1=2y_1$ followed by a decrease as r_1 increases

beyond this point. This behavior further illustrates the evolu-
tionary nature of the pressure-producing velocity fluctuations in
the wake.

In the case of this edge, however, no oscillations in the
correlations are apparent as the velocity probe is moved downstream in the wake, thus no vortex street-like structure is
disclosed by the correlation. The measured frequency spectrum of
surface pressures on this edge are more broadband and locally
generated by the separated flow.

The more recently-conducted measurements on the edges
sketched in Figure 15 involve both types of flow, and their
relevance to both sound and surface pressure was examined. We
consider first pressures on the blunter of the two edges. Correlations of wake velocity and surface pressure, such as those
discussed above, were conducted and disclosed an apparent vortex-
street-like structure and an associated value of y_f. Figure 19
shows the spectra at various points within and outside of the
separated flow zone on the suction side.
The spectra are all non-dimensionalized on the common variables
U_∞ and y_f. On this side, the vortex street pressures were not

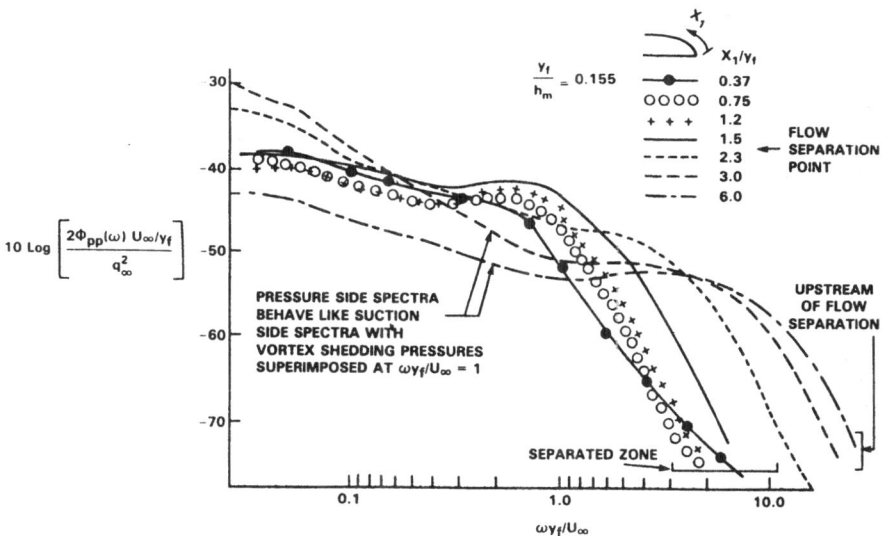

Figure 19. Chordwise variation of surface pressure spectra on suction side of blunt-edged airfoil at
$\Re_{yf} = 5.1 \times 10^4$.

evident as previously illustrated in Figure 16 because of dominance from the locally-generated separation. However, the
spectral magnitudes of these continuous-spectrum pressures at the
vortex shedding frequency varied with distance upstream of the
separation point as $p^2 \sim 1/x$, for $x_1 > 1.5y_f$ (see Figure 20), while
the gradient of static pressure in this region was positive (or
adverse). On the pressure side, on which the flow remained fully
attached, the spectrum in Figure 16 shows a clear bump at
$\omega y_f/U_\infty = 1$ much as that shown in Figure 6, but the continuous part
of the spectrum was significantly lower than that on the suction
side. Thus, on this side, we have a superposition of pressures
near $\omega y_f/U_\infty = 1$ imposed by the wake structure and continuous
spectrum pressure locally generated by the attached boundary
layer.

The random pressures produced on the suction side of the
trailing edge appear to be generated by a similar source mechanism as that which controls the generation of boundary layer
pressures by the outer boundary layer region. Upstream of the
separation region for chordwise positions away from the edge, for
$x_1/y_f = 3$ and 6, the surface pressure spectra take on a dimensionless form.

$$\frac{2\Phi_{pp}(\omega)\, U_\infty/y_f}{q_\infty^2} \quad \text{vs.} \quad \frac{\omega y_f}{U_\infty}$$

on either the suction side or the pressure side, the magnitude of
which approaches 10^{-5} at low dimensionless frequencies. In the
separation zone on the suction side for positions $x_1 < 2.3y_f$ the
pressure spectral densities behave in a similar fashion as
reported by Blake (1984), but they approach

$$\frac{2\Phi_{pp}(\omega)U_\infty/y_f}{q_\infty^2} \cong 10^{-4} \quad \text{for} \quad \frac{\omega y_f}{U_\infty} < 1$$

which is a substantially larger level than pressures beneath a
homogeneous boundary layer, or elsewhere on the airfoil as noted
above. At higher frequencies in the separation zone, the pressures
decrease. Actually, the pressure fluctuations in the separation
zone are quite typical of those occuring beneath adverse pressure
gradient turbulent boundary layers.

We can now summarize the behavior of pressures on this edge
near $\omega y_f/U_\infty = 1$ which are generated by the large-scale near-wake

vortex structure. Figure 20 shows the relevant features of the
wake geometry including the length scale y_f disclosed from the
contraction of measured loci y_u and y_l.

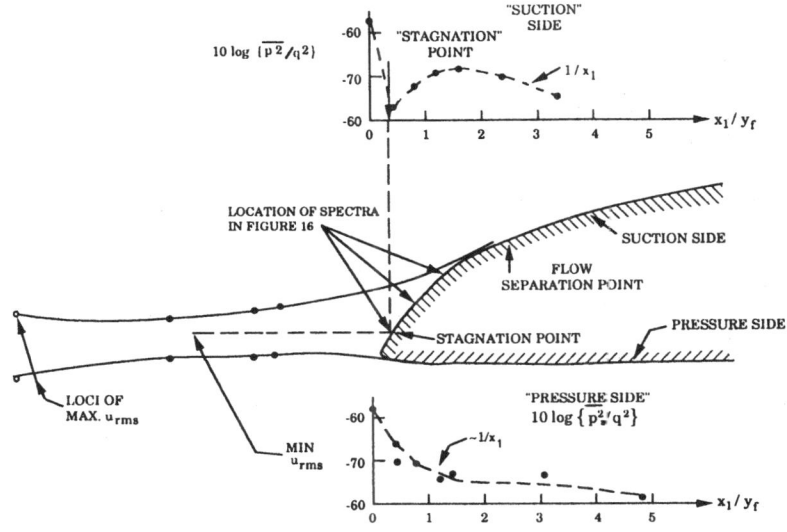

Figure 20. Surface Pressure and Wake Loci for Blunt-Edged Foil with Vortex Shedding.

The values of pressures at this frequency are also shown for the
pressure and suction sides. The minimum value occurs at the
stagnation point x_s; on either side of this point these pressures
increase and then decrease as predicted by the theory (equation -
19) for the vortex-generated pressures. For the thicker edge,
Reynolds numbers based on y_f and U_∞ ranged from 1.6×10^4 to 6.8×10^4. This range is to be noted in Figure 8 as in a region of increasing Γ_0 with Reynolds number. The locations of the spectra
shown in Figure 16 are also illustrated.

Surface pressure spectra for the thin-edged foil at various
chordwise locations are given in Figure 21. The thin-edged foil
on the suction side has similar limiting values of spectral
levels near the edge and away from the edge, as described above,
although the variation through and upstream of separation is
smoother than for the blunter-edged foil. Because of the thinner
wake behind this trailing edge, the values of $y_f U_\infty/\nu$ ranged from
0.9×10^4 to 3.6×10^4. In Figure 8 this can be seen to be a
range in which the relatively lower values of Γ_0 could be

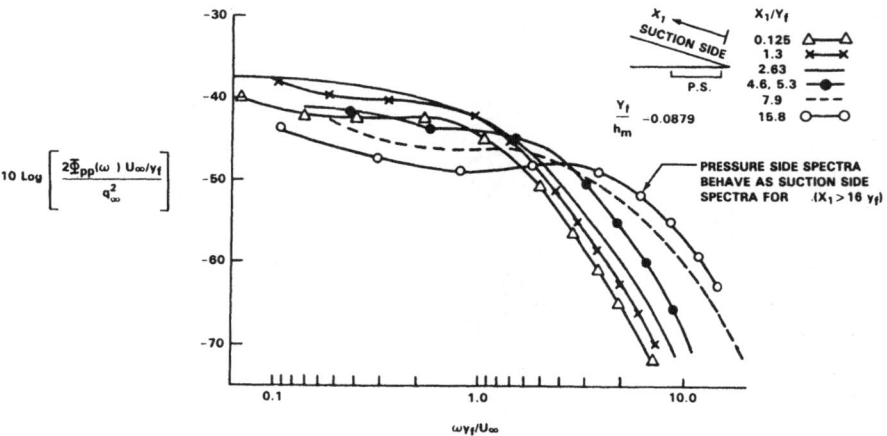

Figure 21. Surface Pressure Spectra at Various Chordwise Locations on Suction Side of Thin Trailing Edge Foil.

expected if quasi-periodic vortex shedding occurred. No discernable vortex-street-like behavior could be observed for this edge, through correlation measurements, however. Comparison of Figures 18 and 21 shows that the blunter edge did generate a more clear "hump" on the suction side pressure fluctuations.

6.2 STATISTICAL MEASURES OF SURFACE PRESSURE

As shown by the theories for trailing edge sound, equations (15) and (24), the autospectrum of radiated sound is directly proportional to the frequency-dependent spanwise integral scale $\Lambda_3(\omega)$. This is defined by equation (18) in terms of a vorticity distribution, but the spanwise correlation length may in principle be determined in a physical aerodynamics sense by measurements of velocity fluctuations or of surface pressure fluctuations. The question of the relevance of a given aerodynamic measurement to the aeroacoustic sources will be examined in detail in Section 7. For the present, it is to be kept in mind that these aerodynamic measurements are often made to provide an input to the prediction of sound from a first-principles point of view.

Spanwise scales of the eddy pressure field that convect past the trailing edge are characterized by the integral scales, $\Lambda_3(\omega)$, for the suction and pressure sides of the blunt and thin-edged foils and ranges of values are given for the edges of Figure 15 in Figures 22 and 23 respectively. They were obtained at each frequency by graphically integrating over spanwise transducer separation distance, r_3, using a curve fit through the locus of values of the square root of the coherence function, $\sqrt{\gamma_3^2}$. For different separation distances, r_3, $\sqrt{\gamma_3^2}$ is a normalized amplitude of a cross-spectral density function and is defined as:

$$\sqrt{\gamma_3^2(r_3,\omega,x_1)} = \frac{\left|\Phi_{pp}(r_3,\omega,x_1)\right|}{\left|\Phi_{pp}(x_1,x_3,\omega)\right|^{1/2}\left|\Phi_{pp}(x_1+r_3,\omega)\right|^{1/2}} \qquad (34)$$

where the auto spectral functions in the denominator are defined in equation (1). The cross spectrum in the numerator is formally defined as:

$$\left|\Phi_{pp}(r_3,\omega,x_1)\right| = \frac{1}{2\pi}\int_{-\infty}^{+\infty} <p(x_1,t)\,p(x_1+r_3,\,t+\tau)>e^{i\omega t}\,d\tau \qquad (35)$$

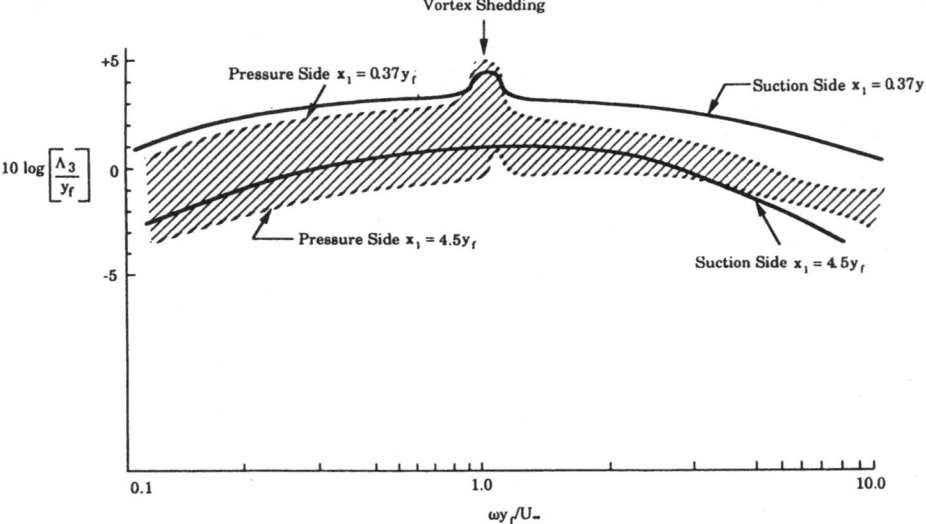

Figure 22. Range of Lateral Integral Scales, Λ_3, for Suction Side (Solid Lines), and Pressure Side (Dashed Lines) of Blunt-Edged Foil at $\Re_{yf} = 5.1 \times 10^4$.

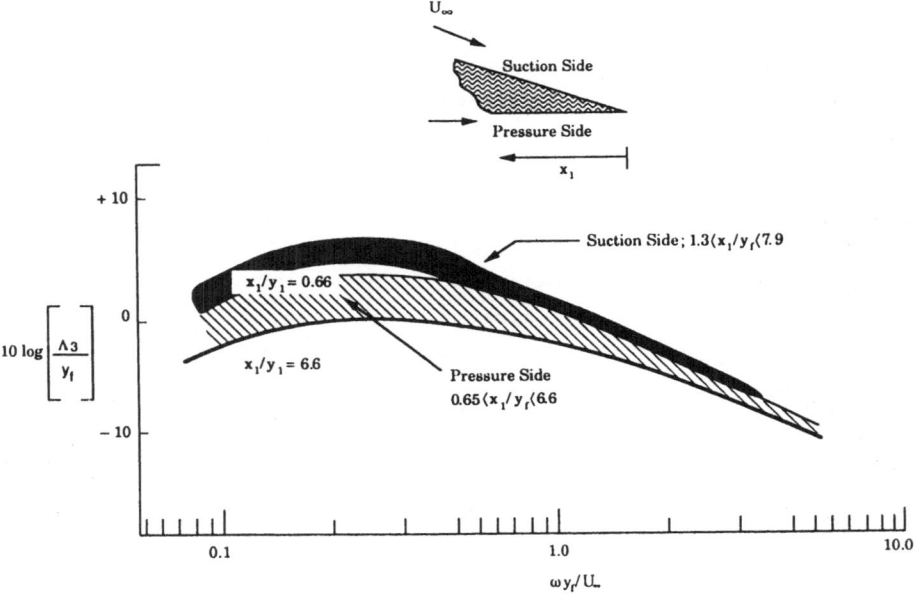

Figure 23. Range of Spanwise (Lateral) Integral Scales, Λ_3, for Suction Side and Pressure Side of Thin-Edged Foil, $\Re_{yf} = 2.9 \times 10^4$.

As shown in the notation of equation (34) $\sqrt{\gamma_3^2}$ is in general a function of the chordwise location of measurement, x_1, because of spatial non-homogeneity of the developing adverse pressure gradient at the edge. By definition at each frequency, for $r_3=0$, $\sqrt{\gamma_3^2(\omega, r_3=0)} = 1$. To assist in the graphical integration for the lateral integral scale, $\Lambda_3(\omega)$ was defined at each streamwise location, x_1, by an exponential function running through the measured values at various r_3 with constant $(\omega y_f/U_\infty)$ as a parameter. Thus

$$\Lambda_3(x_1, \omega y_f/U_\infty) = \int_0^\infty e^{-k(\omega y_f/U_\infty)r_3}\, dr_3$$

$$\Lambda_3^{-1} = k(\omega y_f/U_\infty) \tag{36}$$

This procedure was one of convenience and no other significance placed in this particular functional form. As a practical matter, the scatter of data was pronounced enough that other less-convenient functional forms could have been used with equal ability to determine Λ_3.

Qualitatively, the integral scales of the eddy pressure field are largest for frequencies near the vortex shedding frequency, occurring at edge $\omega y_f/U_\infty = 1$, while at other frequencies, the scales for the random pressures increase for distances, x_1, closer to the edge in accordance with the gradual thickening of the boundary Λ_3 layer. Figure 22 for the blunter edge shows this increased near $\omega y_f/U_\infty = 1$. The values of Λ_3, for both edges fall off nearly reciprocally with frequency for $\omega y_f/U_\infty = 1$. On the pressure side, Λ_3, is only slightly smaller than on the suction side considering the radically different natures of the autospectra on the two sides: The magnitude of normalized pressure cross spectrum for positions on opposite sides at $x_1=3.9y_f$ upstream of the trailing edge of the thin-edged foil are given by:

$$\frac{\Phi_{u\ell}(\omega, x_1)}{\sqrt{\Phi_{uu}(\omega, x_1)\Phi_{\ell\ell}(\omega, x_1)}}$$

where the subscript u = upper, ℓ = lower side. On this foil the formation of orderly wake structure is suppressed so that the surface of foil serves to baffle the flow sources upstream of the edge on opposite sides of the foil. Thus, only the contribution to the pressures due to the aeroacoustic near field of the trail-

ing edge flow-noise can account for the pressure coherence on opposite sides. Since on the thinner of the edges in Figure 15 the
formation of a vortex-street-like wake does not occur, we expect
that this edge flow is more typical of the classical boundary
layer-edge interaction.

Figure 24 shows that the theory

$$\frac{\left|\Phi_{u}\ell(\omega,x_{1},)\right|}{\sqrt{\Phi_{uu}(\omega,x_{1},)\,\Phi_{\ell\ell}(\omega,x_{1},)}} = \frac{\left|\cos\left(\left|k_{c}y_{1}\right|+\pi/4\right)\right|}{\sqrt{\pi k_{c} y_{1}}} \qquad (37)$$

where $k_{c} = \omega/U_{c}$ for upstream pressure fields of equal magnitude
agrees well (dashed line) with the experimental data (solid line)
for the <u>thin</u>-edged foil a distance $x_1 = 3.9\ y_1$ upstream of the
edge.

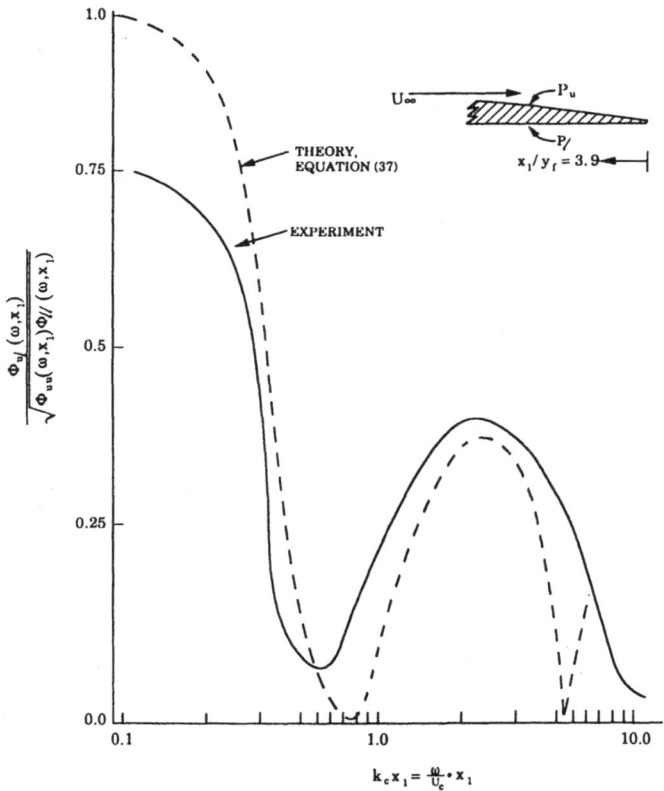

Figure 24. Opposing Side Trailing Edge Pressure Coherence Measurement (Solid Lines) and Theory
(Dashed Lines) for Thin-Edged Foil at $x_1/y_f = 3.9$, $\Re_{yf} = 2.9 \times 10^4$.

Brooks and Hodgeson previously experimentally verified this contribution due to the upstream eddy pressure field plus the near field of the scattered pressure field for non-separating trailing edge flows on a NACA 0012 foil.

Cross correlation measurements of pressures on opposite sides for the blunt-edged foil revealed qualitatively similar behavior as the thin-edged foil for all frequencies except those near the vortex shedding frequency. Near this frequency $\omega y_f / U_\infty = 1$, the pressure arising from wake instabilities are highly correlated on opposite sides by the aerodynamic induction field of the wake. These pressures yield a normalized cross spectrum that is constant and equal to 1 with phase π for all values of x_1 for which the wake-induced pressures dominate the boundary layer pressures.

Returning to the aeroacoustic scattering mechanism, we consider now the experimental evidence that the Kutta condition applies for the pressure on the thinner edge of Figure 15. According to the theory that incorporates a trailing edge Kutta conditions, the magnitude of the total (aerodynamic plus scattered) pressure field near the edge should decrease as $k_c x_1 = \omega x_1 / U_c \longrightarrow 0$. This reduction is due to a coherent cancellation of the incident pressure field with its own scattered pressure field. Alternatively, away from the edge, for $k_c x_1$ large, the pressures are doubled due to the specular reflection by the surface. This behavior is illustrated in Figure 11; see section 3.2. As the eddies approach the edge, the continuity of pressure on opposite sides relieve the surface pressure so that the pressure on the "flow" side is reduced by a factor of two. The "scattered" pressure components on both sides are in phase as $k_c x_1 \longrightarrow 0$ giving a vanishing induced pressure differential. This behavior is qualitatively observed for the flow on the suction side of thin-edged foil in Figure 25. The theoretical prediction (dashed line) and measurements (solid line) of the pressure autospectrum for a two-sided flow is normalized by its value at $k_c x_1$, = 0.14. The asymptotic decrease of the pressure field at the edge as well as the fact that the measured phase of the cross spectrum approaches zero suggests that a Kutta condition is applicable for non-singing separated trailing edge flows. The calculated relative spectrum levels approach 4(6dB) from unity (0dB) at distances that are far from the edge. Near $k_c x_1 = 0.2$ the theoretical value is slightly higher than its asymptote due to

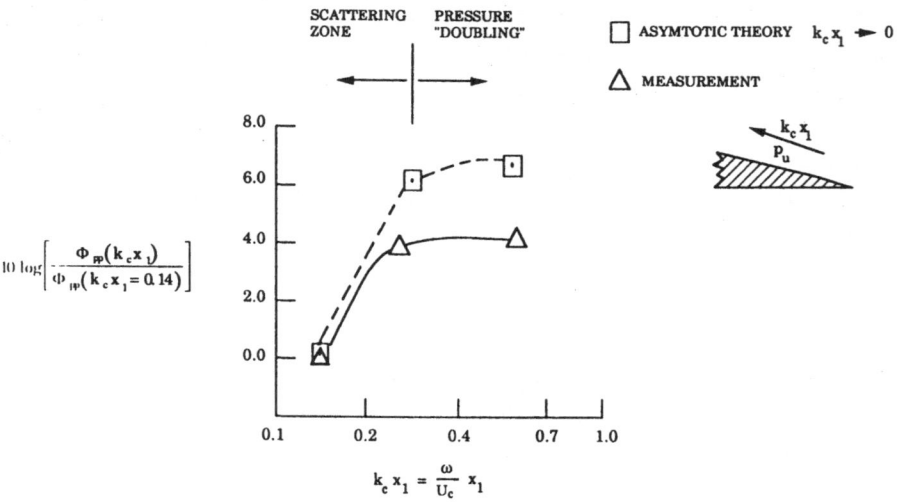

Figure 25. Reduction in Net Wall Pressure Spectrum Very Near the Edge for the Suction Side of Thin-Edged Foil.

the local effects of diffraction. The convection velocity U_c used this discussion was determined from chordwise spatial correlations the wall pressure. This behavior of the continuous-spectrum aeroacoustic scattering mechanism contrasts strongly with that of the vortex wake-like mechanism of the blunter edge. As Figure 20 shows, the pressure field near $\omega y_f / U_\infty = 1$ on this edge does not satisfy a Kutta condition of the potential flow type, rather the pressure increases as $x_1^{-1/2}$ as $x_1 \rightarrow 0$ becoming limited by some viscous flow mechanism that has yet to be formally described.

7.0 RADIATED SOUND FROM TRAILING EDGES WITH SEPARATED FLOW

Tonal noise due to vortex shedding for blunt-edge foils and random noise due to scattering of the upstream evanescent, non-propagating boundary layer pressures at the trailing edges of both thin-edged and thick-edged foils may be predicted for simple cases using the existing theories reviewed in Section 3. These theories have been supported by measurement on simple airfoils with sharp trailing edges. The aerodynamic sound is estimable in terms of the surface pressure spectra very near the edge and spanwise integral scales such as those that were reported in

Section 6. This section will make similar comparisons, but it will also test some of the other features of the analyses as they relate to the flow source strengths.

7.1 SOUND DUE TO VORTEX SHEDDING FROM WAKE INSTABILITY

We first focus on the sound radiated in a narrow frequency range near the frequency of vortex shedding, $\omega y_1/U_\infty = 1$. Equation (20) may be used to obtain the sound spectrum, $\Phi_{PR}(\vec{r},\omega)$, at the field point $\vec{r}$, from the vortex shedding contribution. The aerodynamic nearly-tonal pressure field on the edge of the foil is measured at a chordwise distance x_1 upstream of the stagnation point, x_s in the region of $\overline{p^2} \approx |x_1 - x_s|^{-1}$. Figure 6 shows the results of one such prediction with this equation. In the new work, Gershfeld and Blake (1988), a comparison of the predicted vortex shedding noise using surface pressures measured at alternative positions lying in the zone $0.75 y_1 < (x_1 - x_S) < 4.5\ y_f$ on the pressure side of the foil trailing edge, illustrated in Figure 20, are compared to the aerodynamic sound in Figure 26. The acoustic source strength is again characterized by the auto-spectrum of surface pressure given in Figure 20, at frequencies near $\omega y_f/U_\infty = 1$ and together with the spanwise integral scales, Λ_3 given in Figure 22 for frequencies near $\omega y_f/U_\infty = 1$. The theory predicts the sound levels to within 2 dB for surface pressures measured closest to the edge, i.e. for $x_1 = 0.25$ in. This is comparable in accuracy to the previous prediction (Blake-(1984) of vortex shedding noise for Brooks and Hodgeson's (1981) NACA 0012 foil with truncated edge as shown here in Figure 6. If the point of measuring surface pressure is taken further away form the apex, the theory under-predicts the radiated noise. Even though the surface pressure intensities, $\Phi_{pp}(\omega, x_1 - x_s)$ fall off inversely with $(x_1 - x_s)$, as expected from theory (Blake (1984)) and Equation (19), (Figure 20), the measured integral scales Λ_3 are not constant (Figure 22) for distances, $x_1 - x_s$, away from the edge. This variation in Λ_3 may be due to increasing dominance over the vortex-shedding pressures by the greater intensity turbulent pressures at upstream distance from the edges. Thus, measured pressures on this edge obtained upstream of the apex are contaminated by flow sources which do not contribute to the vortex-shedding noise.

7.2 RANDOM AERODYNAMIC SOUND

We now consider the remainder of the spectrum of radiated sound that is dominated by the mechanism of aeroacoustic scattering the trailing edge. Howe's (1978) theory in Section 3.4, may be used to relate the acoustic radiation to the surface pressure spectrum $\Phi_{PS}(\omega, x_1)$. In a simplified form, the scattered random sound pressure, $\Phi_{PR}(\omega)$, is given in dimensionless form for $k_c = \dfrac{\omega}{U_c} \ll k_0 = \dfrac{\omega}{c_0}$

$$\frac{\Phi_{PR}(r,\omega)\, U_\infty / y_f}{2\pi q_\infty^2 M_c} = \frac{\Phi_{pp}(\omega, x_1)\, U_\infty / y_f}{2\pi q_\infty^2} \cdot \frac{2\Lambda_3(\omega y_f / U_\infty)}{y_f} \cdot \frac{L_3 y_f}{r^2} \cdot \frac{\sin^2 \theta/2 |\sin \phi|}{\alpha(k_c x_1)(2\pi)^2} \quad (38)$$

where
$$\alpha(k_c x_1) = \left[1 + \sin(y_2)\operatorname{erf}\left(\sqrt{ik_c x_1}\right)\right]^2$$

erf = error function

The sign $(y_2) = +$ for surface pressures measured on the same side of the edge plane as the flow side and sign $(y_2) = -$ for surface pressure measured on the opposite side of the flow side. For two-sided flows for which surface pressure statistics are on both sides, sign $(y_2) = +$ for surface and radiated pressures on the same side and sign $(y_2) = -$ for surface and radiated pressures on the opposite sides; the radiated pressure spectrum is then the sum of two terms each representing a contribution from one side. Equation 38 is consistent with equation 24 and was used to derive equation 28.

Figure 6 shows the measured and predicted radiated sound for the NACA 0012 airfoil experiment of Brooks and Hodgeson (1981). The measured spectra of aerodynamic sound are also compared in dimensionless form for a broad range of Reynolds numbers for the beveled-edge foils of Gershfeld and Blake (1988). In Figures 26 and 27 these comparisons show that the agreement between predicted and measured levels is dependent on the location selected for measuring the surface pressure. The surface pressures which were measured on both the suction and pressure sides of the foils were used to make the predictions in Figures 26 and 27. Over the range of frequency for which comparison is possible, the measured aerodynamic sound for both foils appears to best correspond to the theory when calculations are based on surface pressures measured in the separation zone of the foil on the suction side. This indicates that the sound is dominated by flow sources in the separated flow region of the suction side,

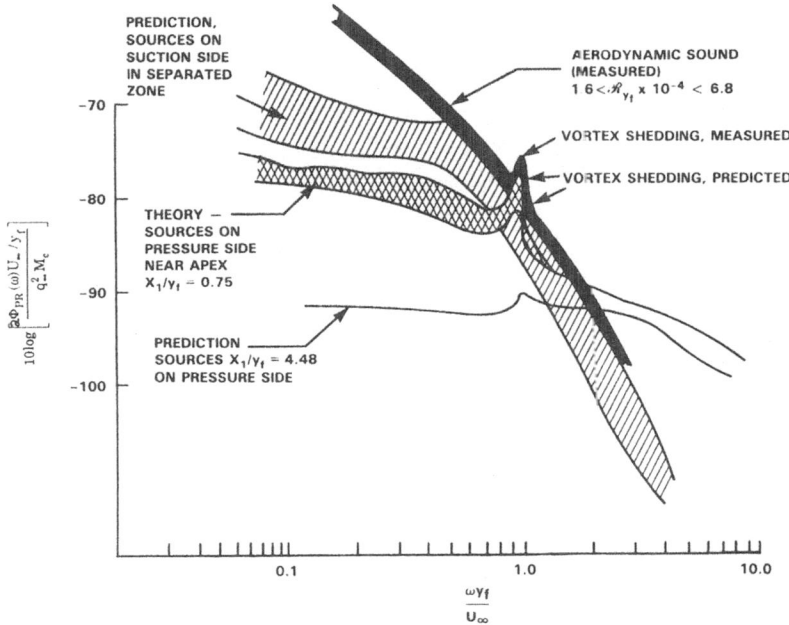

Figure 26. Prediction and Measured Dimensionless Aerodynamic Sound for Blunt-Edged Foil.

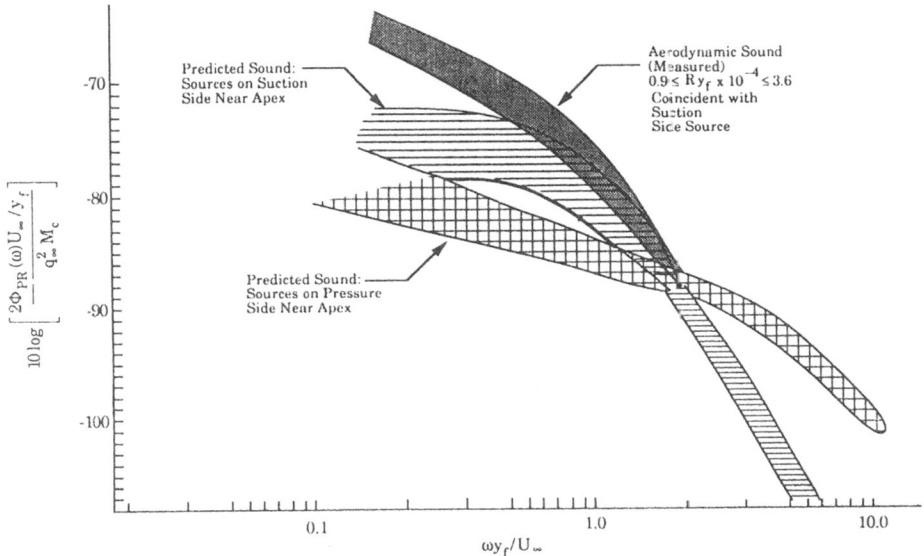

Figure 27. Predicted and Measured Radiated Noise for Thin-Edged Foil.

but the mechanism is still that of aeroacoustic scattering. The
mechanism of generating continuous spectrum sound by the aeroacoustic scattering mechanism is thus similar to that occurring
in nonseparating edge flows, as studied by Brooks and Hodgeson
(1981), but in the case of edges with separated flow the sources
appear to be dominated by flow on one side and determined by the
larger scales of the separated zone. Additional evidence is provided in Figure 28 for the apparent dominance of the radiated
sound by the separated flow on the upper side of the edge.

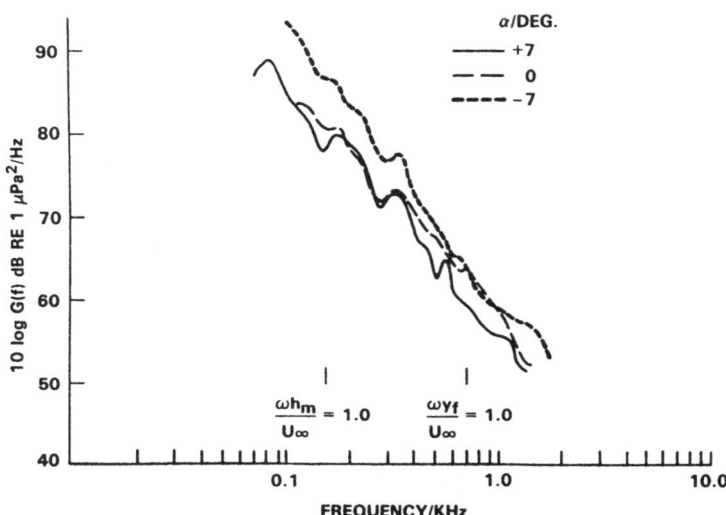

Figure 28. Measured radiated noise as a function of angle of attack for the thin-edged foil.

The spectrum of sound shifts to higher frequencies as the angle
of attack is decreased from +7 to -7 degrees. Values of $\omega y_f / U_\infty = 1.0$
for the $\alpha = 0^\circ$ case and of $\omega h_m / U_\infty = 1.0$ for all three cases are shown
for cross referencing with Figure 12. The boundary layer on the
suction side of the foil thins as angle of attack decreases; in
turn this should reduce the near wake scale y_f. This shifts the
characteristic frequencies to higher values as predicted in the
non-dimensionalization which gives $\omega \alpha U_\infty / y_f$. Finally, the
collapse of the dimensionless measured radiated sound (the shaded
region) shows in an effective way that, the overall sound intensity obeys the theoretical U_∞^5 dependence. The differences
between the measured and the predicted spectra, particularly at
$\omega y_f / U_\infty > 1$ possibly relate to questions of the relevance of the

lateral integral scale, Λ_3, that was measured aerodynamically.
This will be discussed in the next section.

7.3 SPANWISE INTEGRAL LENGTHS AND TRANSFER FUNCTIONS DETERMINED FROM THE ACOUSTIC FIELD

Experimental techniques have been devised for indirectly confirming essential features of the theoretical formulations of trailing edge sound. The theory can, in principal, be tested by direct
substitution of empirically-determined values of parameters into
equations (20) and (38); however in a gross sense, experimental uncertainties in all the parameters involved make careful scrutiny
impossible. It is possible to isolate separately the transfer function between surface pressure and far-field pressure that can be
compared with theory because it relates directly to the half-
plane green function. Furthermore, it turns out that the study
of the appropriately nondimensionalized cross correlation functions between the surface pressures at the trailing edge and the
far field acoustic pressure can divulge an independent measure of
the spanwise integral scale which may be compared to the direct
aerodynamic measurements of Λ_3 shown in Figures 22 and 23. The
measurements that will be described below thus may be used to
further test the theory as well as to obtain independent measure
of spanwise correlation as it relates to sound production.

Correlation functions between sensors in the source zone and
in the acoustic far field have been used to identify propagation,
correlation lengths, and transfer functions. We will review some
of these measurements following a brief discussion of theory.

To fix ideas, we will first review the two-dimensional
acoustic dipoles which are acoustically compact in the flow
direction. In that problem, which is applicable to the aeolian
tones of anulor circular, the far field sound pressure at the
field point $\vec{r} = (\vec{r}, \omega)$, $p_r(\vec{r})$, is related to the flow-induced
force per unit span, $f_s(x)$, where x is a co-ordinate along the
span. This force per unit span can be uniquely described by a
surface pressure distribution $p_s(x)$ as long as it is a deterministic in the flow (chordwise direction). Thus $f_s(x)$ and $P_r(\vec{r})$
are connected by a relationship of the form

$$p_r(\vec{r}) = \int_{-L_3/2}^{L_3/2} G(\vec{r}, x) f_s(x) \; dx$$

where $G(\vec{r},x)$ and $p_s(x)$ are connected through the geometry of the surface. In this relationship, a periodic time function has been suppressed. The transfer function $G(\vec{r},x)$ is well-known in the case of the cylinder vortex shedding and it is determined by the appropriate Green function for the rigid cylinder (see, for example Blake (1986)), but its exact form is not yet relevant to this discussion. The autospectrum of the radiated sound is given by

$$<p_r^2(\vec{r})> \approx \int_{-L_3/2}^{L_3/2} \int_{-L_3/2}^{L_3/2} G(\vec{r},x_1)\ G(\vec{r},x_2)< f_s(x_1)f_s(x_2)> dx_1 dx_2$$

If the pressure field on the surface of the cylinder is spatially stationary and homogeneous along the span, then the cross spectrum of the force per unit span $<f_s(x_1)f_s(x_2)>$ is spatially stationary and homogenous in (x_1,x_2). We can write

$$<f_s(x_1)\,f_s(x_2)> = \overline{f_s^2}\,R_{ff}(x_1 - x_2) = \overline{f_s^2}\,R_{pp}(r_x)$$

where $\overline{f_s^2}$ is the mean square of f_s. For a long-enough cylinder, the spatial correlation function $R_{ff}(r_x)$ has the property $R_{ff}(0)=1$ and $R_{ff}(L/2)=0$. This allows us to define the spanwise integral scale using the unbounded form

$$2\Lambda_3 = \int_{-\infty}^{\infty} R_{ff}(r_x)dr_x$$

We then have

$$<\overline{p_r^2}(r)> = \int_{-L_3/2}^{L_3/2} |G(\vec{r},x)|^2 \overline{f_s^2}\,2\Lambda_3\,dx \qquad (39)$$

$$<\overline{p_r^2}(r)> = |\overline{G(\vec{r},x)}|^2 \overline{p_s^2}\,2\Lambda_3\,L_3$$

or

$$= |G(\vec{r})|^2 \overline{f_s^2}\,2\Lambda_3 L_3 \qquad (40)$$

since $G(\vec{r},x)$ is not a function of spanwise location for a cylinder that is long enough to neglect end effects. The vinculum over the transfer function represents a spatially-averaged value over the span of the source region.

In similar fashion we consider the cross spectrum between the force per unit length, and the radiated sound; thus,

$$<p_r(\vec{r})\,f_s(x)> = \int_{-L_3/2}^{L_3/2} G(\vec{r},x)< f_s^2(x_1)\,f_s^2(x)> dx_1$$

$$= G(\vec{r},x)\,\overline{f_s^2}\,2\Lambda_3 \qquad (42)$$

This cross spectrum has a real and imaginary part because of the complex nature of $G(\vec{r},x) = G(\vec{r})$.

Ratios of $\langle p_r(\vec{r})\, f_s(x) \rangle$ to various combinations of $p_r^2(\vec{r})$ and f_s^2 may now be used to find $G(\vec{r},x) = G(\vec{r})$ and $2\Lambda_3$. Thus we have

$$\frac{\langle p_r(\vec{r})\, f_s(x) \rangle^2}{\overline{p_r^2(\vec{r})}\;\overline{f_s^2}} = \frac{2\Lambda_3}{L_3} = \frac{\ell_c}{L_3}$$

or in the notations of this paper.

$$\Phi^2_{\substack{pp\\rs}}(\vec{r},x)\big/\big[\Phi_{RR}(\vec{r})\,\Phi_{ss}(x)\big] = \ell_c/L_3 \tag{43}$$

where ℓ_c is simply the "correlation length" of the forcing function, $2\Lambda_3$. The other ratio of interest is

$$\frac{\langle p_r(\vec{r})f_s(x) \rangle^2}{\overline{p_r^2(\vec{r})}} = \frac{G(\vec{r})\;\overline{f_s^2}\,\ell_c}{|G(\vec{r})|^2\overline{p^2}\,\ell_c\,L_3} = \frac{1}{|G(\vec{r})|\,L_3}$$

or

$$\frac{\Phi_{\substack{pp\\rs}}(\vec{r},x)}{\Phi_{PR}(\vec{r},\omega)} = \frac{1}{|G(\vec{r})|L_3} \tag{44}$$

Equation (43) is just a restatement of similar relationships that were published by Clark and Ribner (1969) and Sidden (1973). The use of the force-to-sound pressure correlation gives the effective areoacoustic spanwise correlation length. This may not necessarily be identical to the integral scale given by equation (39) that would be determined from measurements of the aerodynamic forces. It is to be noted that the correlation function $\langle p_r(\vec{r})f_s(x) \rangle$ includes the phase $(\omega r_1/c_0)$ and the magnitude of the transfer function, $G(\vec{r})$.

In a similar fashion, we may consider the fully three-dimensional acoustic problem which may involve a two-dimensional Green function as discussed in connection with equation (3). The governing equation may thus be of the form

$$p_r(\vec{r}) = \int_s G_T(\vec{r},\vec{y})p_s(\vec{y})d^2\vec{y} \tag{45}$$

where $p_s(\vec{y})$ represents the surface pressure at a location $\vec{y}$ on the surface and $\vec{r}$ is a three-dimensional field vector. This may be seen in connection with equation (3) as follows. Let the surface be rigid, thus affording $\partial B/\partial n = 0$; consider the first order disturbance giving $B=p$; consider the wake contribution that determines the volume integration over quadruples to be dominated by the surface integration of the $B(\vec{y},\omega)$ on the surface. The dominant contribution to the radiation is thus

$$B(\vec{x},\omega) = p_r(\vec{x},\omega) = -\iint_s p_s(\vec{y},\omega) \cdot \frac{\partial G(\vec{x},\vec{y},\omega)}{\partial n} \; d^2(\vec{y}) \qquad (46)$$

Thus the transfer function defined above is related to the half-plane green function by

$$G_T(\vec{r},\vec{y}) = \frac{-\partial\, G(\vec{x},\vec{y},\omega)}{\partial n} \qquad (47)$$

As above, the $e^{i\omega t}$ time dependence has been suppressed. The transfer function $G_1(\vec{r},\vec{y})$ may be independent of spanwise location y_3 for the spanwise-homogenous aeroacoustic problem, but it may not be independent of chordwise location y_1. This is because the surface is not acoustically compact in the streamwise direction and because of the chordwise development of the flow in the vicinity of the trailing edge. Accordingly the formulation of the one-dimensional aeroacoustic dipole cannot be paralleled in a simple yet rigorous manner. However, if we ascribe to the problem an effective chordwise region, say δc which controls the integral of equation (45), and an aeroacoustic correlation area A_c, ot the surface pressure cross spectrum, where in the notation of Section 6.2,

$$\Phi_{PP}(\omega)\, A_c = \iint_{-\infty}^{\infty} \Phi_{PP}(r_1,r_3)\; dr_1 dr_3 \qquad (48)$$

we can develop notional expressions for the two-dimensional
surface that parallel those of the simpler case. Thus, we have
the autospectrum of the far-field sound pressure

$$\Phi_{PR}(\vec{r},\omega) \cong \overline{|G_T(\vec{r},\vec{y}_e)|^2}^{\Delta C} \overline{\Phi_{PP}(\vec{y}_e,\omega)}^{\Delta C} \Delta C L_3 \, A_c \tag{49}$$

where $\overline{|G_T(\vec{r},\vec{y}_e)|^2}^{\Delta C}$ is the magnitude of the
transfer function squared evaluated at a location $\vec{y}_e = (y_{e1}, \; y_{e3})$
near the trailing edge, say within the chordwise region $-\delta C < y_{e1} < 0$
and $-L_3/2 < y_{e3} < L_3/2$. $\overline{\Phi_{pp}(\vec{y}_e,\omega)}^{\Delta C}$ is the average autospectrum of wall
pressure in this region. The cross spectrum between surface
pressure and far field sound gives

$$\Phi_{PP_{rs}}(\vec{r},\vec{y}_e') = < p_r(\vec{r}) \; p_s(\vec{y}_e) >$$

$$\cong \overline{G_T(\vec{r},\vec{y}_e)}^{\Delta C} \overline{\Phi_{PP}(\vec{y}_e,\omega)}^{\Delta C} A_c \tag{50}$$

We can now form the appropriate ratios; thus

$$\frac{|\Phi_{PP_{rs}}(\vec{r},\vec{y}_e)|^2}{\Phi_{RR}(\vec{r},\omega)\Phi_{PP}(\vec{y}_e,\omega)} =$$

$$\frac{\overline{|G_T(\vec{r},\vec{y}_e)|^2}^{\Delta C}}{\overline{|G_T^2(\vec{r},\vec{y}_e)|}^{\Delta C}} \cdot \frac{A_c}{\Delta C \, L_3} \cong \frac{A_c}{\Delta C \cdot L_3} \tag{51}$$

and

$$\frac{\Phi_{PP_{rs}}(\vec{r},\vec{y}_e)}{\Phi_{PR}(\vec{r},\omega)} \cong \frac{\overline{G_T(\vec{r},\vec{y}_e)}^{\Delta C}}{\overline{|G_T^2(\vec{r},\vec{y}_e)|}^{\Delta C} \cdot \Delta C \cdot L_3}$$

$$= \frac{1}{\sqrt{\overline{|G_T(\vec{r},\vec{y}_e)|}^{\Delta C}} \cdot \Delta C \cdot L_3} \tag{52}$$

Equations (50) and (51) are the analogs of equations (43) and
(44) for the one-dimensional dipole. The approximations that are
implicit in the case of the two-dimensional surface are less
serious if variations in the transfer function (or Green function
for the half-plane) with chordwise dimensions are small over
δC. Theory shows these variations to be small within a distance
of the edge of $x_1 < U_c/\omega$.

Yu and Joshi (1979) have made cross correlations between
surface and far field pressure which confirm many aspects of the
theory. The correlations were made in the time domain so that
they do not disclose the frequency-dependent transfer function.
What those measurements show are clear dependencies that are
brought out by equations (11) and (12) and equation (33). The
measurements in Figure 29 were obtained with a series of microphones installed in the surface of a NACA airfoil.

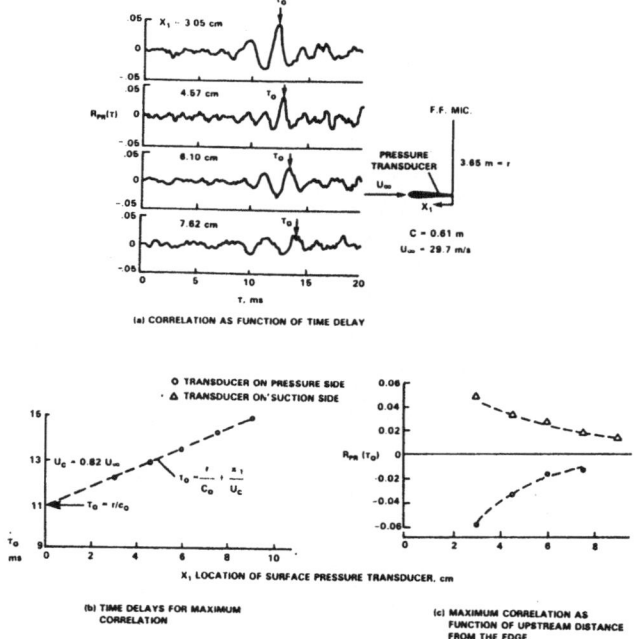

Figure 29. Cross correlations between Farfield sound and surface pressure fluctuations on a NACA
63-012 airfoil at 0° angle of attack. Pressure transducers alternately on suction side and
pressure side. (Ref Yu and Joshi (1979)).

Cross correlations between the surface pressures at various x_1, with the radiated pressure disclose a correlation peak at a time delay τ_0 which increases as x_1, moves away from the edge. The surface pressure - to-far field pressure time delay consists of the convection time for a pressure-producing eddy to reach the edge plus a propagation time from the edge to the farfield. Thus $\tau_0 = x_1/U_c + r/c_0$ where U_c is the convection velocity of the pressure-producing eddy. Figure 29b shows this behavior with $U_c = 0.82\ U_\omega$; equation (33) expresses the frequency domain phase shift that is the analog of the increasing time delay. Note that the correlation is defined as

$$R_{PR}(\vec{x},\tau) = \frac{< p_s(\vec{x}_1, t+\tau)\, p_r(\vec{r}, t) >}{(\overline{p_s^2}\ \overline{p_r^2})^{1/2}} \qquad (53)$$

The value of this correlation at $\tau = \tau_0$ decreases monatomically as x_1 increases away from the edge. This is because, as illustrated by equations (11) and (12) the acoustic influence of the edge falls off with the distance x_1 that an eddy is from the edge. Figure 29(c) shows this influence very nicely; it also shows the 180° phase shift between opposite sides of the foil. In the context of the Gershfeld et al (1988) measurements, Yu and Joshi's surface pressure measurements were made over a range of $2.3 < x_1/y_1 < 7$. The aerodynamic correlation lengths in Figures 22 and 23 were found to decrease systematically as x_1 increased in this range.

To relate these developments to the problem of trailing edge noise with flow separation we note that the theory for either the mechanism of viscous wake instability or the mechanism of trailing edge scattering, the source is deterministic in the chordwise direction and random in the spanwise direction. This is because the interaction between the flow and the edge that produces sound creates trailing edge dipoles which are randomly distributed along the edge span, but acoustically concentrated near the edge by the flow-edge interaction. A significant difference does occur in the relative magnitudes in radiating pressures and non-radiating pressures which may be measured on these classes of edges. In the case of the viscous wake instability, the surface pressures provide a much more distinct indicator of the source strength than in the case of trailing edge scattering. In the latter scattering problem, the scattered pressures that directly relate to the sources are weaker and really are superimposed on the random pseudo-sound pressures. They are less distinct and thus more aptly "masked" by the non-radiating pseudo-sound pressures especially from the immediate location of the edge.

A further complication arises when flow separation occurs upstream of the apex of the edge, as it does in the cases considered by Gershfeld and Blake (1988) and illustrated in Figure 15. In this flow, the sources due to both viscous wake instability and aeroacoustic scattering are superimposed, thus surface pressure measurements made at the edge include both flow types. As noted above, the sound is generated by an integration over the source zone yet pressure measurements are dominated by local flow effects which may not be related to the acoustic sources. It is to be noted that the case of separated flow on an unsymmetrically beveled edge has been treated analytically (Howe (1988)) as an extension of the classical problem, and it confirms this view that the flow-generated lift dipoles are connected with the sources within a distance U_c/ω of the edge.

Thus, for interpretation of current measurements, the trailing edge source may be regarded as a line dipole along the trailing edge and we may apply equations (43) and (44) (noting that $P_s(\vec{y}_e)$ near the edges replaces the f_s of that analysis) with

$$|G_T(\vec{r})| = \frac{\sqrt{\dfrac{U_s}{c_0}}\,\sin(\theta/2)\sqrt{\sin\phi}}{\sqrt{2\pi}\,r\left(y_f/(x_1-x_s)\right)} \tag{54}$$

in the case of "tonal" sources caused by Helmholtz instabilities on the near wake and

$$|G_T(\vec{r})| = \frac{2\sin(\theta/2)\,\sqrt{\sin\phi}}{\sqrt{2}\,r}\,\mathcal{M}_c^{1/2} \tag{55}$$

in the case of sources induced by aeroacoustic scattering. Equation (54) relates directly to equation (20); equation (55) from equation (38) in the limit $k_x x_1 \to 0$. In both cases the chordwise dependence of the sources and the precise form of the Green function are linked to the geometry of the body in this paper, equation (6) for the half-plane. Accordingly, equations

(51) and (44) are fully equivalent as are equations (50) and (43) when note is taken of the fact that $A_c = \Delta C \ell c$ for such chordwise-deterministic sources.

Figures 30 and 31 show the magnitude and phase of cross spectral density

$$\frac{\Phi_{P_R P_s}(\vec{r}, \vec{y}_c)}{\sqrt{\Phi_{P_r}(\vec{r}, \omega) \ \Phi_{PP}(\vec{y}_c, \omega)}}$$

for the blunter of the edges shown in Figure 15. This function shows clearly the enhanced value at frequencies near $\omega y_f/U_\infty = 1$, i.e., in the vortex-shedding zone. The phase also shows acoustic propagation $k_0 r = \omega r/c_0$ that must occur for the phase of an edge-to-field point transfer function.

A comparison of the acoustically deduced integral scales determined from this cross spectrum using equation (43) with integral scales determined wholly from the surface pressure (Figure 22) are given in Figure 32 for the blunt-edged foil of Figure 15. The surface pressures were measured near the apex on the pressure side. Close agreement between the measurements is shown for the blunt-edged foil for the frequency determined by Hemholtz wake instability vortex shedding and the more randomly-generated aeroacoustic integral scales at all other frequencies.

522

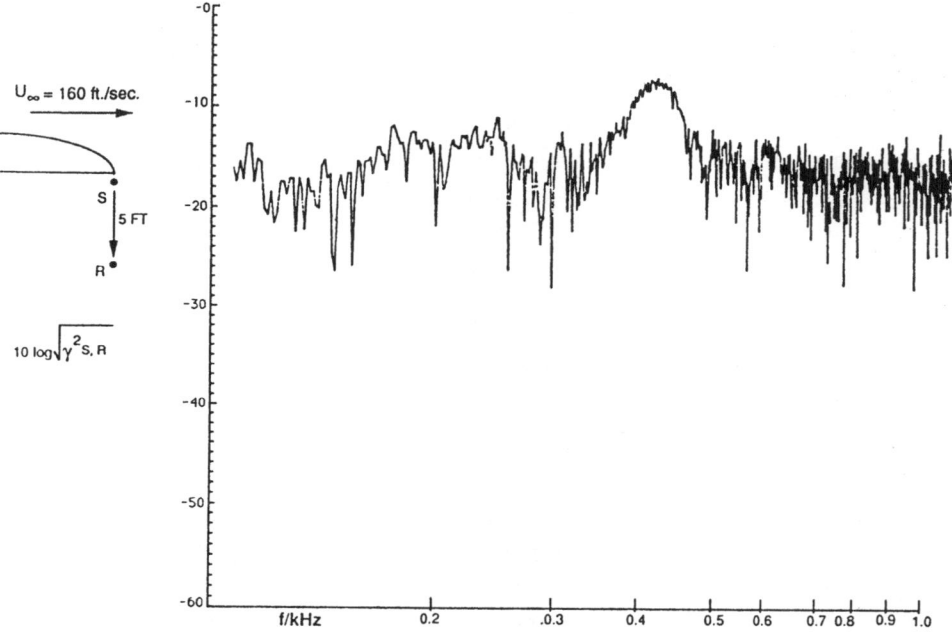

Figure 30. Coherence In, Db, of the Pressures Near the Apex of the Blunt-Edged Foil, on the Pressure Side, S, and the Acoustic Field Point, R = 5 ft. at U_∞ = 160 ft/sec.

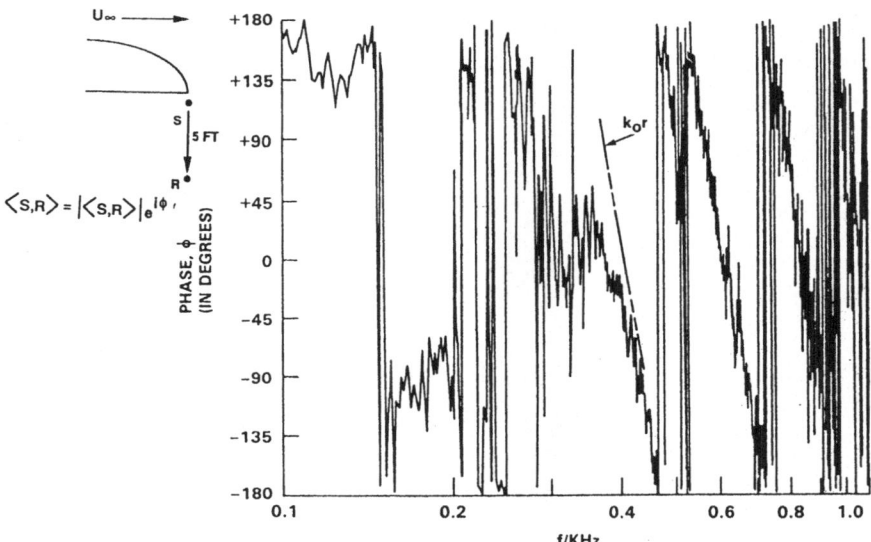

Figure 31. Phase From Cross Spectrum of Boundary Layer Pressures Near the Apex on Pressure Side of Blunt Trailing-Edged Foil, (Point S) and the Acoustic Field (Point R) 5 ft Away.

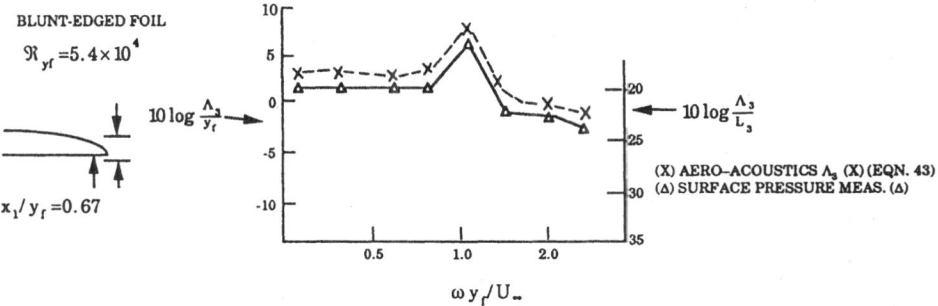

Figure 32. Comparison of Spanwise Integral Scales From Acoustic Cross Spectral Measurements (Eq. 12, Dashed Line), and Surface Pressure Measurements (Solid Line, Δ).

Figure 33 shows a comparison of the theoretical transfer function (equation (55)) with that deduced from measurement for the thin-edged foil. The measurements agree at the lower frequencies but start to diverge at the higher frequencies.

Figure 34 shows the measured and theoretical transfer functions both vortex shedding (equation (54)) and trailing edge noise (equation (55)) for the blunt-edged foil. The estimation of the transfer function for the vortex shedding noise has a range of uncertainty to the uncertainty of selecting the value of $x_1 - x_s$. Two measurements are included to show the effect of analysis bandwidth on the measurement. The dotted line represent data taken over to 10KHz range given in analysis band $\Delta f = 12.5$ Hz. the solid line data was taken over a 1 kHz range with a narrower analysis band, $f = 1.25$ Hz.

Clearly, agreement between measurement and theory for both types of sources is much more apparent in the case of the stronger sources associated with the blunt trailing edge. In the case of the thin edge, however, the measurement and theory diverge as frequency increases. The reason for this divergence is unknown, but it may be related to weaker radiated sound signal and greater statistical uncertainty in the cross spectral density measurements.

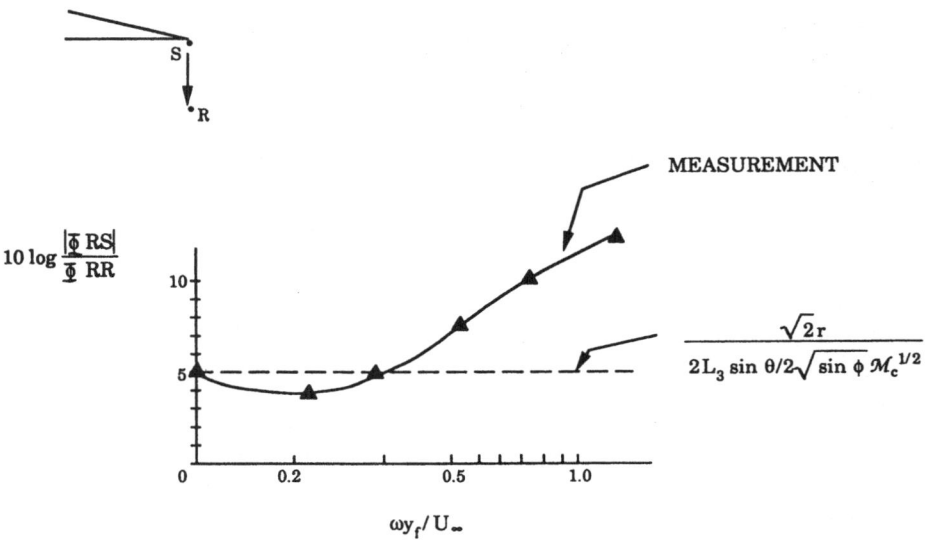

Figure 33. Greens Function Determination from Source(s) and Radiated Noise (R) Measurements for Thin-Edged Foil at $\mathfrak{R}_{yf} = 2.9 \times 10^4$.

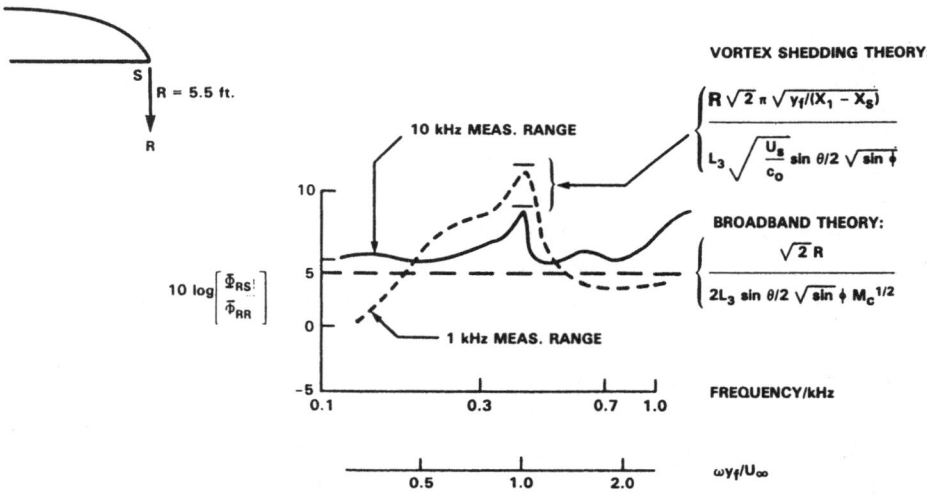

Figure 34. Comparison of Measured and Calculated Green's Functions for Vortex Shedding and Trailing Edge Noise for Blunt-Edge Foil, $\mathfrak{R}_{yf} = 5.4 \times 10^4$. Separate measurements with different bandwidths disclose uncertainties of the experimental results.

8.0 TIP VORTEX SOUND

We conclude this chapter with a brief discussion of the sound generated by tip flows. The initial work in this area was done by Kendall (1978) who quantified the noise from the sides of lifting flaps. This work preceded an analysis by Howe (1980). Sound from the sides of flaps was found dominate that from the trailing edge source when the angle of attack (or incidence angle) is of order 10 degrees and larger. Kendall reports no data for smaller incidence angles.

Work on the tip flows from lifting surfaces has been very sparse in comparison with that done on trailing edge noise. George, Najjar, and Kim (1980) proposed a semi-empirical theoretical model that was based on a survey of various studies of tip flows. They postulated that the induction of the tip vortex caused a cross flow to the tip and the occurrence of an edge-type source. The source strength was then proportional to the component of cross flow velocity projected normal to the edge. The recent experimental work of Brooks and Marcolini (1986) is the first to attempt the quantification of tip vortex sound of airfoils. They evaluate some features of the flow structure at the tips of lifting 3-dimensional airfoils and relate these features to the apparent sound that is produced. The measurements of the sound was attempted indirectly largely by inference by comparing the sound from the 3-dimensional foil to that from a 2-dimensional one of the same cross section. The component due to the tip flow was determined by difference. Figure 35 illustrates some of the flow features and gives an example: the spectrum shows levels from the two and three dimensional foils as well as the deduced component from the tip vortex.

The importance of tip vortex flow noise at high frequencies that is shown in Figure 35 agrees qualitatively with the analytical result of George et al (1980). The experimental result disagrees with the theory in disclosing the magnitudes of the actual levels and characteristic frequencies that would be predicted. The Brooks and Marolini (1986) results and methods point out the difficulties to be expected in experimentally identifying the relative importance of competing sources that occur simultaneously. They also show that significant levels could be inferred only when the tip loading was significant requiring incidence angles of order 10 degrees.

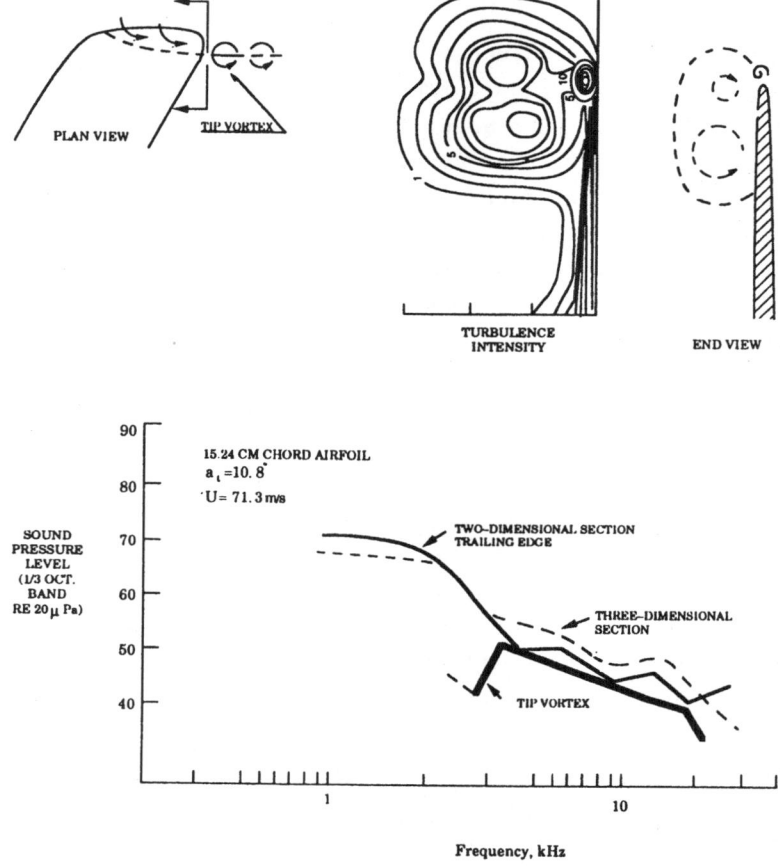

Figure 35. Tip Vortex Sound (Ref. Brooks & Marcolini (1986)).

9.0 CONCLUSIONS

The fundamental understanding of the aeroacoustic behavior of simple trailing edges seems well in hand. In the simplest case of cusp trailing edges with no upstream flow separation, the principal of aeroacoustic scattering and the importance of a Kutta condition are experimentally validated physical features of the process of sound generation. The unsteady aerodynamics of both acoustically small and large chord lifting surfaces of large aspect ratio has been formally worked out. The formulations lead to pragmatic relationships for quantifying both the magnitude and directivity of aeroacoustic trailing edge sound. It is, in principle, possible to infer relationships between gross boundary

layer feature and the sound radiated from effectively two
dimensional, separating, sharp edged foils.

There are at least three unresolved issues of both a scientific and a practical nature that remain, however. The first
question relates to the engineering significance of the theory.
On the practical side, the observed sound from families of real
lifting surfaces, as illustrated in Figure 15, does not conform
very neatly to a simple formula for nondimensionalization that is
suggested by the theory. This lack of collapse is probably
related to unknown (or unaccounted for) variabilities in the
natures of the boundary layers on individual sections. A general
statement for these lifting surface examples is that the measured
sound occurs at lower frequencies than the idealized theory
accounts for. Thus it appears that the engineering utility of
the theory for making predictions is currently limited by the
limited experiences with actual lifting sections.

A second question is raised by the three dimensionality of
lifting surfaces flow in real cases. The importance of tip
flows has only been touched on by the currently available
research. It is not possible to make any a priori prediction of
the sound from tip flows. This relates to rotors, such as those
in helicopter application, where lifting is carried to the tips.

Finally, in a scientific sense, the Kutta condition really
applies to the potential, or outer flow of the surface. This
condition is a boundedness condition that simply removes a pressure singularity at the edge. Realistically, it begs the question of how the singularity is relieved by the physical flow. The
"relief" occurs on a small spatial scale at the generation of the
near wake that is interior to the potential flow region. This is
a particularly provoking aspect of the blunter trailing edges
that have been discussed in Part II of this paper. Essentially,
the entire dynamical process of intermittent separation and
reattachment that is observed experimentally is a physical satis-
faction of the Kutta condition of the outer flow. On such trailing edges, there appears two super-imposed types of aeroacoustic
flow sources which raise some specific questions. How does one
physically account for the simultaneous existence of two pressure
fields on a given edge with these fields each indicative of
differing boundary conditions (i.e. Kutta and no-Kutta)? How do
these flow types couple with and influence each other? How would
a theoretical analyst have anticipated this dual nature? Answers

to all these questions can lead to better understandings of the
relationships among leading, geometry, and viscous edge flow as
they relate to aerodynamic sound production.

Acknowledgement

A review paper of this type would not be possible without a
continuous collaboration between the authors and their colleagues
in the field. Accordingly, the authors wish to show gratitude at
this time for many discussions with T. Brooks, M.S. Howe,
D. Chase, D. Crighton and R. Schlinker who have made valuable
contributions to the field. We are also grateful to
Drs. L. Maga, C. Knisley, T. Huang, and P. Purtell for their
participation in the experimental phases of the work done at DTRC
and to T. Thornhill, M. Evans, for important assistance in
editing the manuscript.

10. References

Amiet, R. "Noise Due To Turbulent Flow Past a Trailing Edge", J.
Sound Vib. 47, 387-393 (1976).

Aravamudan, K.S. and Harris, W.L. "Low-Frequency Broadband Noise
Generated by a Model Rotor", J. Acoust. Soc. Am. 66,
522-533 (1979).

Atassi, H.M. "Feedback in Separated Flows Over Symmetric Airfoils", A1AA 9th Aeroacoustics Conference, Williamsburg,
VA (1984).

Bastedo, W.G. Jr. and Mueller, T.J. "Spanwise Variation of Laminar Separation Bubbles on Wings at Low Reynold Numbers", J.
Aircraft 23, 687-694 (1986).

Beaman, P.W. "Investigation of the Flow Behind a Two-Dimensional
Model With a Blunt Trailing Edge and Fitted With Splitter
Plates", J. Fluid Mech. 21, 241-255 (1965/1967).

Billingsley, J. and Kinns, R. " The Acoustic Telescope", J. Sound
Vib. 48, 485-510 (1976).

Blake, W.K. "A Statistical Description of Pressure and Velocity
Fields at the Trailing Edges of a Flat Strut", NSRDC Report 4241
(Dec. 1975).

Blake, W.K. "Trailing Edge Flow and Aerodynamic Sound", Part I
and Part 2, DTNSRDC Report - 83/113 (Dec. 1984).

Blake, W.K. "Structure of Trailing Edge Flow Related To Sound
Generation." DTRC Report 83/113 (1985).

Blake, W.K. "Mechanics of Flow-Induced Sound and Vibration",
Vol S. Academic Press (1986).

Bowman, J.J., Senior, T.B.A., and Uslenghi, P.L.E.
"Electromagnetic and Acoustic Scattering by Simple Shapes",
North-Holland, Amsterdam (1969).

Brooks, T.F. and T.H. Hodgeson, "Trailing Edge Noise Prediction
Using Measured Surface Pressures", J. Sound Vib., 78,
pp. 69-117 (1981).

Brooks, T.F. and Marcolini, M.A. "Airfoil Tip Vortex Formation
Noise", AIAA Journal 24, 246-252 (1986).

Brooks, T.F., Marcolini, M.A. and Pope, D.S. "Main Rotor Broadband Noise Study in the DNW", AHS Specialists Meeting on Aerodynamics and Aeroacoustics, Arlington, TX (1987b).

Brooks, T.F. and Schlinker, R.H. "Progress in Rotor Broadband
Noise Research", Vertica 7, 287-307 1983).

Brooks, T.F., Marcolini, M.A. and Pope, D.S. "A Directional Array
Approach for the Measurement of Rotor Noise Source Distributions
With Controlled Spatial Resolution", J. Sound Vib. 112, 192-197
(1987).

Chase D.M. "Sound Radiated by Turbulent flow off a Rigid Half-
Plane as Obtained From a Wave Vector Spectrum of Hydrodynamic
Pressure", J. Acous. Soc. Am. 52, 1011-1023 (1972).

Chase D.M. "Noise Radiated From An Edge In Turbulent Flow", A1AA J. 13, 1041-1047 (1975).

Clark, P.J.F. and Ribner, H.S. "Direct Correlation of Fluctuating Lift With Radiated Sound for An Airfoil in Turbulent Flow", J. Acoust. Soc. Am. 46, 802-805 (1969).

Coles, D., "The Law of the Wake in the Turbulent Boundary Layer J. Fluid Mech. 1, 191-226 (1956).

Crighton, D.G. and Leppington, F.G. "Scattering of Aerodynamic Noise By a Semi-Infinite Compliant Plate", J. Fluid Mech. 43, 721-736 (1970).

DeMetz, et al, "An Experimental Study of the Intermittent Properties of the Boundary Layer Pressure Field During Transition on a Flat Plate", NSRDC Report, 4140 (Nov. 1973).

Ffowcs Williams, J.E. and Hall, L.H. "Aerodynamics Sound Generation by Turbulent Flow in the Vicinity of a Scattering Half-Plane", F. Fluid Mech 40, 657-670 (1970).

George, A.R., Najjar, F.E., and Kim, Y.N. "Noise Due To Tip Vortex Formation on Lifting Rotors", A1AA 6th Aeroacoustics Conference, Hartford, Conn. (1980).

Gershfeld, J.L., Blake, W.K., and Knisely, C.W. "Trailing Edge Flows and Aerodynamic Sound", First Fluid Dynamics Conference, Cincinnati, OH. (1988).

Grosche, F.R. "On The Generation of Sound Resulting from the Passage of a Turbulent Air Jet Over a Flat Plate of Finite Dimensions", Royal Aircraft Establishment Libr. Trans. No. 1460 (1970).

Grosche, F.R., Stiewitt, H., Bender, B. "On Aero-Acoustic Measurements in Wind Tunnels by Means of a Highly Directional Microphone System", 3rd A1AA Aero-Acoustics Conference, Palo Alto, CA, (1976).

Hersh, A.S. and Hayden, R.E. "Aerodynamic Sound Radiation With and Without Leading Edge Serrations", NASA-CR-114370 (1971)

Hersh, A.S. and Meecham, W.C. "Sound Directivity Radiated from Small Airfoils", J. Acoust. Soc. Am. 53, 602-606 (1973).

Howe, M.S. "Contributions to the Theory of Aerodynamic Sound With Application To Excess Jet Noise and the Theory of the Flute", J. Fluid Mech. 71, 625-0677 (1975).

Howe, M.S. "The Influence of Vortex Shedding on the Generation of Sound By Convected Turbulence", J. Fluid Mech. 76, 711-740 (1976).

Howe, M.S. "A Review of Trailing Edge Noise", J. Sound Vib. 61, pp. 437-465 (1978).

Howe, M.S. "On the Generation of Side-Edge Flap Noise", BDT Beranek and Newman Report for NASA Contract NASI-16142 (1980).

Howe, M.S. "The Influence of Surface Rounding on Trailing Edge Noise", BBN report 6175 (1988).

Kendall, J.M. "Measurements of Noise Produced by Flow Past Lifting Surfaces" AlAA Paper 78-239 (1978).

Lucus, B.G. and Muir, T.G. "The Field of A Focusing Source" J. Acoust. Soc. Am. 72, 1289-1296 (1982).

Mac Donald, H.M. "A Class of Diffraction Problems", Proc Lan. Math Soc. 14, 410-427 (1915).

Powell, A. "On the Aerodynamic Noise of a Rigid Flat Plate Moving at Zero Incidence", of Acoust. Soc. Am. 31, 1649-1653, (1959).

Scharton, T.D., Pinkel, B. and Wilby, J.F. "A Study of Trailing Edge Blowing As a Mean of Reducing Noise Generated by the Interaction of Flow With a Surface." NASA CR-132270 (1973).

Schlinker, R.H. "Airfoil Trailing Edge Noise Measurements with a Direction Microphone", A1AA 9th Aeroacoustic Conference, Paper 77-1269 (1977).

Siddon, T.E. "Surface Dipole Strength by Cross Correlation Method", J. Acous. Soc. Am., 53, pp. 619-633 (1973).

Soderman, P.T. and Noble, S.N. " Directional Microphone Array for Acoustic Studies of Wind Tunnel Models", A1AA Journal 12, 168-173 (1975).

Winklemann, A.E. and Barlow, J.B. "A Flow Field Model for a Rectangular Platform Wing", A1AA Journal 18, 1006-1008 (1980).

Yu, J.C. and Tam, C.K.W. "An Experimental Investigation of Trailing Edge Noise Mechanism", A1AA J. 16, 1048-1052 (1978).

Yu, J.C. and Joshi, M.C. "On Sound Radiation From the Trailing Edge of an Isolated Airfoil in the Uniform Flow", A1AA 5th Aeroacoustics Conference, Paper 79-0603, Seattle, Wash (1979).

Lecture Notes in Engineering

Edited by C.A. Brebbia and S.A. Orszag

Lecture Notes in Engineering

Lecture Notes in Engineering

Edited by C.A. Brebbia and S.A. Orszag